海洋文化翻译思政教材

海洋文化翻译教程

A Coursebook on Marine Culture Translation

主　编　郭艳玲　李明秋
副主编　于　冰　林雅琴　张艺玲
　　　　张　恒　关晓云　王　可

中国矿业大学出版社
·徐州·

内 容 提 要

本书按照教育部《高等学校课程思政建设指导纲要》文件的要求,以"立德树人"为宗旨,深入加强思想政治教育、传播海洋文化知识、提高翻译能力为根本任务,同时,以助推中国海洋文化走向世界为目标,以培养具有国际视野与家国情怀的海洋翻译人才为宗旨撰写而成。每个单元分别设置了思想政治教育的内容,从不同的角度关心海洋、认识海洋、经略海洋,建设我们的"海洋强国"。

本书可作为高等院校翻译专业、英语专业研究生及本科生的教材使用,还可作为翻译研究爱好者学习的参考用书。

图书在版编目(CIP)数据

海洋文化翻译教程：汉英对照 / 郭艳玲,李明秋主编. — 徐州：中国矿业大学出版社,2024.4

ISBN 978 - 7 - 5646 - 6011 - 6

Ⅰ.①海… Ⅱ.①郭…②李… Ⅲ.①海洋－文化－翻译－教材－汉、英 Ⅳ.①P7

中国国家版本馆 CIP 数据核字(2023)第 210318 号

书 名	海洋文化翻译教程
主 编	郭艳玲 李明秋
责任编辑	何晓明 何 戈
出版发行	中国矿业大学出版社有限责任公司
	(江苏省徐州市解放南路 邮编 221008)
营销热线	(0516)83885370 83884103
出版服务	(0516)83995789 83884920
网 址	http://www.cumtp.com E-mail:cumtpvip@cumtp.com
印 刷	苏州市古得堡数码印刷有限公司
开 本	787 mm×1092 mm 1/16 印张 31.5 字数 1048 千字
版次印次	2024 年 4 月第 1 版 2024 年 4 月第 1 次印刷
定 价	68.00 元

(图书出现印装质量问题,本社负责调换)

前　言

语言是世界文化的载体，翻译架起沟通世界的桥梁。翻译不仅可以推动不同语言文化交流的革新与发展，还在思想启蒙、知识普及、对外交流、经贸往来、科技发展、文化传播乃至社会演进中扮演着重要的角色。翻译人才应具备良好的政治品格、精湛的翻译技能、跨文化的传播能力、各个行业与专业领域知识、广阔的国际视野等能力素质。所以，培养高素质的翻译人才更不能缺少一部适应时代背景的思政型专业性翻译教材。

本教程按照教育部《高等学校课程思政建设指导纲要》文件的要求，以"立德树人"为宗旨，以深入加强思想政治教育、传播海洋文化知识、提高翻译能力为根本任务，同时，以助推中国海洋文化走向世界为目标，以培养具有国际视野与家国情怀的海洋翻译人才为宗旨撰写而成。每个单元分别设置了思想政治教育内容，从不同的角度关心海洋、认识海洋、经略海洋，建设我们的"海洋强国"。

本教程以"立德树人"为宗旨，全方位贯穿思想政治教育、传播海洋文化知识，以提高翻译专业、英语专业研究生、本科生以及翻译研究爱好者等学习者的海洋知识翻译能力为根本目的，使学习者比较系统地掌握各种不同的翻译策略、方法与技巧，扩展海洋文化方面的知识，深入地了解海洋领域的翻译技能。培养具有家国情怀、国际视野的翻译人才，提升海洋专业领域的文化翻译能力。

全书设置了包括 The Ocean and Human Civilization 海洋与人类文明、History of the Ocean 海洋历史、Cultural Exchanges along the Maritime Road 海路文化交流、Marine Folklife 海洋民俗生活、Ocean Ambience 海洋风情、Marine Tourism 海洋旅游、Marine Economic Activities 海洋经济活动、Ocean Art 海洋艺术、Fishery Culture 渔业文化、Marine Mythology 海洋神话、Harbour Culture 海港文化、Sea Fiction 海洋小说、Marine Science 海洋科学、Ocean Exploration 海洋探索、Marine Cultural Industry 海洋文化产业等共 15 个单元。

在编写过程中，我们一直坚持思政性、知识性、海洋性和翻译性相结合的原则，思政内容结合各个单元主题，书写翻译的家国情怀；翻译策略、方法与技巧的选择重在为学生们提供翻译的各种方法；翻译练习为学生们提供了多个不同的练习模式,提高学生们的翻译实践能力;翻译家论翻译为学生们介绍了翻译大家的翻译观点与翻译名言;翻译简析分别对课文中的重点句子该如何翻译进行解析;参考译文为学生们进行课文翻译提供了参考;文化背景知识丰富了学生们对单元主题的了解。本教程的内容比较丰富,每个单元的设计以提高学生的翻译能力为宗旨,教学过程中教师可根据课文内容的教学重点进行安排与设计。

希望本教程能够为翻译专业、英语专业研究生、本科生及翻译研究爱好者等学习者带来学习翻译的"资源大餐",引领学习者心有祖国、具有国际视野与家国情怀的思想政治理念,为我国的跨学科专业性复合型翻译人才培养提供更为前沿的翻译专业书籍,为外语教学同仁提供更为广阔的海洋翻译领域思路与借鉴,成为翻译学习者的好帮手、好资源、智慧库。

本教程由大连海洋大学郭艳玲、李明秋总体设计与统稿,与于冰、林雅琴、张艺玲、张恒、关晓云和王可等六位老师共同编写完成。全书共计104.8万字,其中郭艳玲撰写15.3万字、李明秋撰写15.3万字、于冰撰写13.6万字、林雅琴撰写13.6万字、张艺玲撰写13.6万字、张恒撰写13.6万字、关晓云撰写13.6万字、王可撰写6.2万字。为打造这部海洋专业性精品书籍,全体撰写教师付出了巨大的努力与心血,旨在帮助学习者能够从更多的翻译角度去了解和掌握海洋相关的翻译知识,希望我们的努力能够为推动课程思政、翻译教学改革,探索翻译人才培养模式提供新思路和新举措。

本教程是2022年辽宁省教科文卫体系统郭艳玲职工创新工作室、大连海洋大学涉海外语翻译团队(C202113)、辽宁省社会科学规划基金项目"语用学视角下辽宁外宣翻译的话语体系构建研究"(L21BYY009)、辽宁省教育厅科学研究经费项目(面上项目)"生态翻译学视域下的生态定位与译者行为研究"(LJKR0264)的研究成果。

当然,由于水平有限,在编纂和撰写方面进行了新的尝试,不当甚至纰漏、谬误之处在所难免,恳请业界专家、学者和读者朋友们不吝指正。另外,书中内容引用了翻译相关的书籍、网络等媒体资源与专家学者的学术观点,在这里深表谢意!

郭艳玲　李明秋
二〇二三年八月于大连海洋大学

目 录

Unit One　The Ocean and Human Civilization 海洋与人类文明 ………… 1
　Ideological and Political Education：推动海洋命运共同体 建设"海洋强国" ……… 1
　Text A　Social Well-being and How the Ocean Contributes to It ……… 1
　Ideological and Political Education：提高海洋开发能力 实现民族伟大复兴 ……… 21
　Text B　The Human Relationship with the Ocean ……… 21
　Cultural Background Knowledge：The Ocean and Human Civilization ……… 31

Unit Two　History of the Ocean 海洋历史 ……… 32
　Ideological and Political Education：维护海洋和平安宁 构建海洋命运共同体 ……… 32
　Text A　History of the Ocean ……… 32
　Ideological and Political Education：推进海洋研究探索 建设"海洋强国" ……… 48
　Text B　A History of the Study of Marine Biology ……… 48
　Cultural Background Knowledge：History of the Ocean ……… 58

Unit Three　Cultural Exchanges along the Maritime Road 海路文化交流 ……… 60
　Ideological and Political Education：开创新型海洋文明 建设海上丝绸之路 ……… 60
　Text A　The Maritime Silk Road ……… 60
　Ideological and Political Education：鉴航海进步发展史 展大国风采 ……… 77
　Text B　Maritime Rules of the Road ……… 77
　Cultural Background Knowledge：Cultural Exchanges along the Maritime Road ……… 88

Unit Four　Marine Folklife 海洋民俗生活 ……… 89
　Ideological and Political Education：海洋民俗傍海而生 因海而兴 ……… 89
　Text A　Pirate Cultural Relics ……… 89
　Ideological and Political Education：关注海洋民俗 保护海洋民俗文化 ……… 106
　Text B　Goan Houses：Memory of Home ……… 106
　Cultural Background Knowledge：Marine Folklife ……… 116

Unit Five　Ocean Ambience 海洋风情 ……… 118
　Ideological and Political Education：依自然风情 展经略海洋 ……… 118
　Text A　Native Hawaiian Culture ……… 118

Ideological and Political Education：推动海洋发展 依海富国 ·············· 137
　　　Text B　Oceania's Indigenous Peoples ·· 137
　　　Cultural Background Knowledge：Ocean Ambience ····························· 148

Unit Six　Marine Tourism 海洋旅游 ·· 150
　　　Ideological and Political Education：探索海洋蓝色治理 拓展海洋旅游新业态 ······ 150
　　　Text A　The Challenges to Sustainability in Island Tourism ··············· 150
　　　Ideological and Political Education：讲好海洋文化故事 打造旅游品牌效应 ······ 168
　　　Text B　Sustainable Travel：Eight Best Islands for Ecotourism ········· 168
　　　Cultural Background Knowledge：Marine Tourism ······························ 179

Unit Seven　Marine Economic Activities 海洋经济活动 ··························· 180
　　　Ideological and Political Education：发展蓝色经济 助力海洋强国建设 ······ 180
　　　Text A　Ocean-based Industries in Developing Countries ················· 180
　　　Ideological and Political Education：打造海洋的"绿水青山" ················· 198
　　　Text B　Key Pressures on Ocean and Ecosystem Services ················ 198
　　　Cultural Background Knowledge：Marine Economic Activities ············ 209

Unit Eight　Ocean Art 海洋艺术 ··· 210
　　　Ideological and Political Education：乘新时代长风 开海洋艺术新局 ······· 210
　　　Text A　Ocean Artworks ··· 210
　　　Ideological and Political Education：用海洋艺术讲好中国故事 ················ 230
　　　Text B　Four Underwater Art Museums ·· 230
　　　Cultural Background Knowledge：Ocean Art ·· 240

Unit Nine　Fishery Culture 渔业文化 ··· 242
　　　Ideological and Political Education：传承渔业文化 建设现代化渔业强国 ······ 242
　　　Text A　Eliminating Government Support to Illegal, Unreported and
　　　　　　　Unregulated Fishing ··· 242
　　　Ideological and Political Education：转型升级 助力渔业绿色发展 ············ 260
　　　Text B　Pacific Monuments to Fishing ··· 260
　　　Cultural Background Knowledge：Fishery Culture ································ 271

Unit Ten　Marine Mythology 海洋神话 ·· 272
　　　Ideological and Political Education：追溯神话传承密码 构筑共有精神家园 ···· 272
　　　Text A　Women in the Marine Mythology of Ancient Mediterranean ···· 272
　　　Ideological and Political Education：从神话到现实 书写文化认同 ············ 292
　　　Text B　Why Noah's Ark will Never Be Found ·································· 292
　　　Cultural Background Knowledge：Marine Mythology ···························· 303

Unit Eleven Harbour Culture 海港文化 ... 304
Ideological and Political Education：从"中国制造"到"中国智造" ... 304
Text A   Reconstruction and Reuse of Port Heritage ... 304
Ideological and Political Education：传承中华优秀文化 增强文化自信 ... 321
Text B   The Port of Rotterdam ... 321
Cultural Background Knowledge：Harbour Culture ... 331

Unit Twelve Sea Fiction 海洋小说 ... 332
Ideological and Political Education：建设海洋生态文明 完成五大重点任务 ... 332
Text A   Mysteries on the Island ... 332
Ideological and Political Education：保护海洋生态环境 建设美丽海湾 ... 348
Text B   Taking a Ferry ... 348
Cultural Background Knowledge：Sea Fiction ... 357

Unit Thirteen Marine Science 海洋科学 ... 359
Ideological and Political Education：促进海洋产业发展 强海兴国 ... 359
Text A   Marine Science and Marine Robotics ... 359
Ideological and Political Education：奋斗者号下潜 展中国海洋科技成就 ... 378
Text B   Human-Robot Spaceflight Exploration ... 378
Cultural Background Knowledge：Marine Science ... 389

Unit Fourteen Ocean Exploration 海洋探索 ... 391
Ideological and Political Education：提高深海探测能力 助力海洋强国建设 ... 391
Text A   International Ocean Exploration ... 391
Ideological and Political Education：探海格局形成 展中国海洋探索成就 ... 412
Text B   The Ocean Unexplored and Unprotected ... 412
Cultural Background Knowledge：Ocean Exploration ... 424

Unit Fifteen Marine Cultural Industry 海洋文化产业 ... 425
Ideological and Political Education：整治发展无序状态 推动产业高质量发展 ... 425
Text A   Maritime Cultural Landscape ... 425
Ideological and Political Education：加快发展海洋文化产业 维护国家主权 ... 441
Text B   Maritime Cultural Heritage Linked to Women ... 441
Cultural Background Knowledge：Marine Cultural Industry ... 453

参考答案 ... 454

参考文献 ... 495

Unit One
The Ocean and Human Civilization

海洋与人类文明

Text A

Social Well-Being and How the Ocean Contributes to It

By Edward H. Allison, John Kurien and Yoshitaka Ota

Ideological and Political Education：推动海洋命运共同体 建设"海洋强国"

党的二十大报告提出："从现在起，中国共产党的中心任务就是团结带领全国各族人民全面建成社会主义现代化强国、实现第二个百年奋斗目标，以中国式现代化全面推进中华民族伟大复兴。"同时强调到2035年要建成教育强国、科技强国、人才强国、文化强国、体育强国、健康中国，国家文化软实力显著增强。二十大报告中有多处关于"强国"的表述，其中，关于"海洋强国"，报告提出要"发展海洋经济，保护海洋生态环境，加快建设海洋强国"。

十年前，党的十八大报告提出了"建设海洋强国"的战略目标。党的十九大报告对"建设海洋强国"做出了重大的战略部署。十年来，在海洋科技创新、培育壮大海洋新兴产业、拓展海洋经济发展新空间、参与全球海洋治理等方面，我国均取得了突破性进展。在海洋科技创新领域，我们持续深耕建设海洋强国的关键领域，着力推动海洋科技向创新引领型转变。深水、绿色、安全等海洋高技术领域自主创新不断取得新突破，多个领域跻身世界前列。未来五年，我们将聚焦优化国家海洋科研力量布局，加快实现高水平海洋科技自立自强。在培育壮大海洋新兴产业方面，十年来，通过海洋经济创新发展示范，中央财政先后投入近90亿元支持海洋新兴产业发展，引导创新要素向优势区域集聚，提升了海洋科技创新能力，成果应用产业化、资本化速度明显加快，2012—2020年海洋新兴产业增加值年均增速达14%。二十大报告提出"坚持把发展经济的着力点放在实体经济上"。国务院批复的《"十四五"海洋经济发展规划》明确把海洋经济发展的着力点放在实体经济上，打造竞争有力的现代海洋产业体系，特别是要推动海洋新兴产业蓬勃发展。在拓展海洋经济发展新空间方面，中国提倡秉持和平、主权、普惠、共治原则，把深海、极地、外空、互联网等领域打造成各方合作的新疆域。近年来，深海资源调查勘探取得积极进展，极地认知和保护利用能力不断增强。在深度参与全球海洋治理方面，海洋命运共同体是人类命运共同体理念的丰富和发展，是人类命运共同体理念在海洋领域的具体实践。作为发展中的海洋大国，我们必须全面推动海洋命运共同体理念走深走实，深度参与并支持全球海洋治理。

1. Economic activity in the ocean is growing rapidly. If the **upsurge** in economic activity is to lead to an upsurge in human **well-being**, then its emergent and potential future impacts must be understood. In order to build that understanding, we must first unpack the concept of well-being and identify the ways the ocean contributes to well-being in all its **dimensions**. It is these "human relationships with the ocean" that we seek to **characterize**.

2. Ocean values and their contributions to human well-being at multiple levels are **outlined**, with the material, relational and subjective dimensions outlined. Although we include an outline of ocean contributions to material well-being, the focus of this study is on the ocean's contribution to relational and subjective dimensions of well-being. Other Blue Papers focus mostly on the material dimensions of well-being when they connect to human values. However, human well-being is only achieved if attention is paid to all three dimensions. Different individuals and cultures ascribe different levels of **priority** to these values, but no society **discounts** them entirely.

3. Subjective well-being has entered national economic policy as measures of economic and social performance, also popularly known as the "economics of happiness". The material, relational and subjective dimensions of well-being are, however, interconnected or "co-constitutive", and, like all such classifications, the **boundaries** between **categories** are **porous**. For example, seafood provides for material needs for nutrients, protein, energy, income and profit, particularly in maritime South and Southeast Asia, coastal West Africa and the Pacific islands. But it also contributes to relational and subjective well-being through association with religious observance (e.g. fish at Easter in Catholic societies in South America), status (e.g. seafood banquets in Chinese culture) and feelings of connection to place (e.g. the importance of wild salmon to both Native/First Nation and settler coastal populations in the U.S. and Canadian Pacific Northwest). Note also that most well-being classifications are made at a single scale (e.g. the individual, household or nation-state) while here we consider multiple scales. This classification is therefore **schematic** and each dimension and scale is illustrated by a small number of examples only, due to considerations of space.

4. The aim here is to establish a new **conceptual** framework that links ocean services or benefits to human well-being in ways that account for the plurality of human values. Note that ocean services or benefits described here extend beyond ocean ecosystem services. An ecosystem services approach values only some of what the ocean contributes to human well-being. Many of the ocean's contributions are not directly related to its ecology but instead related to the ocean as a space both material (having area, volume and **fluidity**) and non-material (as a place of consciousness and imagination).

5. The ocean has and will continue to contribute to human well-being in many ways. While many of these contributions, across all three dimensions, rely on maintaining ocean ecosystem health, it is too simple to say that all of them do. Ensuring ocean health is, by itself, not enough to safeguard and improve human well-being. It is also important to continue to maintain and build the kinds of social and cultural connections to the ocean that have improved human knowledge, understanding, cooperation, security, meaning and happiness in the past. If the majority of those who would benefit from an ecologically healthy ocean are excluded from it, this will not lead to improved human well-being for all. Thus, maintaining ocean health and maintaining inclusive ocean access should be the dual aims of governing the future ocean.

6. Although we have separated out the dimensions of well-being in order to explain and explore them, it is important to **reinforce** that they are interrelated or "co-constitutive" (i.e. each dimension builds on the others). As Sarah White notes, "Rather than dividing 'subjective' from 'objective', subjective, material and relational dimensions of well-being are revealed as co-constitutive. Well-being is emergent, the outcome of accommodation and interaction that happens in and over time through the dynamic interplay of personal, societal and environmental structures and processes, interacting at a range of scales, in ways that are both reinforcing and in tension".

7. Such reinforcing feedbacks between dimensions of well-being can be found in the ways Pacific Islanders think about their relationship with the ocean. As Epeli Hau'ofa wrote, the ocean provides material, relational and subjective "goods" in inseparable and historically constructed ways: "'Oceania' **denotes** a sea of islands with their inhabitants. The world of our ancestors was a large sea full of places to explore, to make their homes in, to breed generations of seafarers like themselves. People in this environment were at home in the sea."

8. Similarly, a sense of being part of a community and a sense of place contribute to the "social embeddedness" of coastal communities engaged in small-scale fishing. Coastal and sea-dwelling communities have strong social ties and distinct cultures from which they derive well-being. These identities and concepts of embeddedness **straddle** both the relational and subjective dimensions of well-being—the sense of belonging to a group, such as fishers or a coastal community, of being able to depend on your group during emergencies, times of loss and crisis due to the strength of social relations and networks, but also subjective feelings of pride and self-worth in one's occupation, community or **ethnicity**.

9. We will outline some case-study examples of institutions—both contemporary and

traditional—that illustrate how the different dimensions of well-being and the multiple spatial scales at which they **accrue** and are **intertwined** in the institutions that have evolved to govern human-ocean interactions in practice. Above all, we consider each set of relationships separately.

10. The ocean contributes to provisioning human needs and regulating the earth's environment to ensure human flourishing and biodiversity conservation. These contributions are linked to the ocean as both a place and as an ecological system. While the ability to extract minerals or transport goods is largely independent of ocean ecosystem health, these activities certainly impact upon it. The challenge for the future ocean economy is to ensure that governance of provisioning and regulatory goods and services—such as food production and climate **mitigation** measures—do not threaten ocean health. Future ocean governance also has to ensure that human relational and subjective well-being are supported rather than undermined by the "blue acceleration". Most modern, state-based and global ocean governance institutions have formed with the goal of managing access to and use of ocean resources and ocean space for material use and, to a degree, to regulate relationships between private enterprises (e.g. property rights and trade and commercial laws) and between states (e.g. regional seas agreements, freedom of navigation agreements). They seldom consider subjective well-being.

11. Ocean governance institutions are also mostly designed to regulate commercial activities at sea. However, people do more than extract resources, trade goods or migrate across the ocean, they also interact with the marine environment and the marine species and ecosystems in a multitude of ways that may be rooted in material elements of the ocean, such as seafood, beaches, waves and reefs, but which are enjoyed, both consumptively and non-consumptively, for non-material purposes. These interactions—whether we experience them directly as beachcombers, rock-poolers, snorkelers, divers or recreational fishers, or vicariously through aquarium visits or viewing television series such as David Attenborough's *The Blue Planet*—create sets of relationships with ocean nature that respond to a range of human material, relational, spiritual and emotional needs. As people rise out of poverty globally, such interactions may engage an increasing number of us.

12. Prior to COVID-19, a **burgeoning** area of tourism research was devoted to understanding ways to cater to the preferences of China's growing number of newly middle-class beach tourists, both domestically and internationally. It is encouraging that over 80 percent of surveyed beach tourists in Qingdao, China, would be willing to pay a tourist tax in order to maintain beach and water quality at their destination. The

global rise of beach and marine tourism, instead of being seen solely as a threat, might be considered an opportunity to bring the well-being benefits of the ocean to a growing proportion of the global population, and to engage ever more people in the cause of sustaining the global ocean.

13. The ways the ocean contributes to relational well-being are more concrete and better understood at smaller scales: the social **cohesion** of traditional fishing communities and how this contributes to economic, social and cultural life is well-studied, understood and increasingly legally **mandated** in the form of devolved management and community resource rights. At higher spatial scales, the relationships are more abstract but nevertheless important: for example, the need to share the oceanic realm has fostered certain moral norms that have spread onto land, such as the principles of neutrality, **truce** and rendering assistance to others in need. These were all **codified** at sea before they became part of the broader moral and legal framework for inter-state governance, and in some cases (e.g. rendering assistance to others in need) they remain more strongly upheld in oceanic than terrestrial contexts. This became very evident when fishers in the Indian state of Kerala took the lead to rescue thousands of inland folk in the 2018 floods because they felt it was the "right thing to do".

14. Of the three main dimensions of well-being, the elements of "subjective" well-being are the most difficult to **ascribe** monetary value to and therefore to incorporate into traditional sectoral economic planning, though some of them have been considered in social accounts and happiness and well-being indices. We know that these are some of the concepts and emotions that give life meaning, purpose and value beyond the meeting of basic **physiological** and economic needs and beyond the sociopolitical necessities of cooperating with others. For these reasons, they are worthy of policy consideration.

15. Because they are difficult to value—and even to **articulate**—the subjective elements of well-being may be dismissed as unimportant. Yet people have used symbols of belief or identity as a pretext to fight wars or have gone to war driven by socially constructed moral concepts such as honour. Political and legal regimes are built around symbols such as flags. As this study is being carried out, statues that symbolize economic and social progress to some and colonial oppression and enslavement to others are being fought over as the U.S.-initiated Black Lives Matter movement **ignites** a worldwide reckoning on racism and colonial history. The symbolic value of the ocean and its organisms to coastal societies—and the extent to which people from these societies are willing to defend them—should therefore not be dismissed lightly, since it provides opportunity for both conservation and development.

16. Several policy implications arise from an understanding of the symbolic value of marine organisms. The first is that dominant global sensibilities and relationships to animals may be regarded as an imposition of cultural values if forced upon all people. There are lessons for wider global ocean governance from attempts to implement universal bans on the harvest of whales and other marine mammals, with nation-states and Indigenous Peoples who pursue traditional whaling activity resisting these bans in various ways and maintaining their cultural relationship to whales as food as well as cultural keystone species invested with complex symbolic meaning. The principle of free prior informed consent (PIC) is relevant here. PIC is a negotiated or treaty-based procedural right for Indigenous Peoples in relation to development or natural resource exploitation proposals and their effect on Indigenous lands, culture and traditions. PIC relates to the public trust doctrine—the main legal concept for governments' **fiduciary** obligation to protect and sustainably manage natural places held in common by the public citizenry. It is especially relevant as legal support for citizen participation in official decisions made about the marine space when government trustee obligations are breached.

17. Subjective well-being is also driven by anxieties, with psychologists identifying six existential ones: identity, happiness, isolation, meaning in life, freedom and death. All these anxieties can be either confronted or relieved (or both) in our relationships with the ocean, and with nature more generally, whether that relationship is professional, residential, consumptive or recreational. We observe that groups whose lives are closely entwined with marine resource use (fisherfolk, mariners, Indigenous Peoples, marine tourism and recreation professionals) have complex, multidimensional relationships with the ocean which are often deeply spiritual and strongly inform social and cultural identities.

18. While separating out the different dimensions of well-being enables them to be identified in any policy context, it is also useful to consider how they relate to and reinforce each other in a sectoral context. Consider, for example, the values embodied in small-scale fisheries.

19. Human health also combines all three dimensions of well-being. The physical and mental aspects of people's health is affected positively by a clean ocean, which can be enjoyed by seafood consumption, trips to the seaside, swimming or pleasure cruises on the ocean. It is negatively affected by a polluted ocean (mercury and microplastics in seafood chains, oil spills, coastal industries, etc.). The relational aspect of health has to do with a sense of community, social cohesion, and so on, for example following the disintegration of coastal communities due to loss of fish stocks, as happened with the

collapse of the Canadian cod in Newfoundland in the 1990s. The subjective aspect of health has to do with the emotional state of being, in this case with the kind of feelings towards the ocean that are evoked by relationships to the environment and to marine species. By examining issues from both a sectoral perspective (in this case, health) and a well-being perspective, the ramifications of different policy choices can be examined, and **synergies** and trade-offs between dimensions of well-being can be identified.

20. Finally, it is important to reinforce that cultures, along with their symbols, spirituality, aesthetics and ethics, are not static. Even the ways emotions are elicited and expressed—how we show anger, fear, hope or love—change over time. Governing the ocean to maintain well-being is not, therefore, about preserving the status quo or returning to the past. It is about finding ways to maintain a diverse and inclusive set of relationships with the ocean and among ocean nations and peoples. It is these relationships that have generated—and will continue to generate—curiosity, awe, wonder, spirituality and aesthetic appreciation, as well as food, energy and wealth. Supporting these "ocean contributions to people" means allowing people to (re) discover and interact with the ocean in ways that build on their own histories and their existing maritime relationships. Such relationships may be highlighted and promoted under existing slogans and campaigns such as those **extolling** "ocean pride", "ocean optimism" and "ocean literacy", though they may need extending to become more inclusive. The aim of such campaigns should be to reconnect people with the ocean, and raise awareness of its importance to our history, our present and our future.

(2,497 words)

https://oceanpanel.org/blue-papers

Extracted from *The Human Relationship with Our Ocean Planet*, 13-28, 2020.

New Words

1. upsurge [ˈʌpsɜːdʒ]

    n. a sudden or abrupt strong increase 剧增；高涨

    At the time of the voice of human cloning on the upsurge, scientific groups launched the movement of anti-human cloning.

    过去克隆人的呼声高涨时，科学界曾掀起了一场"反克隆人运动"。

2. well-being [ˈwel biːɪŋ]

    n. a contented state of being happy, healthy and prosperous 幸福；幸福感

    Striking a balance between life and work is critical to improving people's overall well-being.

    工作与生活之间的平衡对于提升人们的整体幸福感而言至关重要。

3. emergent [ɪˈmɜːdʒənt]

 a. occurring unexpectedly and requiring urgent action 突发性的

 It is an emergent property that results from the interactions among the people in the group.

 它是一种突发性的属性,是群体中人们相互作用的结果。

4. potential [pəˈtenʃ(ə)l]

 a. existing in possibility 潜在的;可能的

 Potential customers are softened up with free gifts before the sales talk.

 谈生意之前,先送给潜在的客户一些赠品,以联络感情。

5. dimension [daɪˈmenʃ(ə)n]

 n. the magnitude of something in a particular direction 维度;方面

 Loutherbourg, like other contemporary painters, wanted to add the dimension of time to his paintings.

 像其他当代画家一样,卢德堡想要在他的绘画中加入时间这一维度。

6. characterize [ˈkærəktəraɪz]

 v. describe or portray the character or the qualities or peculiarities of 刻画……的性格(或特征)

 The real and proper use of the word "romantic" is simply to characterize an improbable or unaccustomed degree of beauty, sublimity, or virtue.

 其实浪漫这个词真正的、正确的用途是刻画一种罕见的或不寻常的美丽、崇高或美德的程度。

7. outline [ˈaʊtlaɪn]

 v. describe roughly or briefly or give the main points or summary of 概述;描画……的轮廓

 Senator Bill Bradley outlined his own tax cut, giving families $350 in tax credits per child.

 参议员比尔·布拉德利概述了自己的减税计划,给家庭中每一个孩子$350的税款补助。

8. priority [praɪˈɒrəti]

 n. status established in order of importance or urgency 优先事项;优先;优先权

 Education has been given a high-priority rating by the new administration.

 新一届政府将教育放在了高度优先的地位。

9. discount [ˈdɪskaʊnt]

 v. bar from attention or consideration 不重视;忽视

 I think that you don't really have self possession if you choose to live in a society because you cannot just discount the people around you.

 我认为如果你生活在社会中,就不可能完全拥有自主权,因为你不能忽视其他人的存在。

10. boundary [ˈbaʊndri]

 n. the line or plane indicating the limit or extent of something 边界;界限

 Compared with the long-standing friendship between the two countries, their boundary dispute is only an issue of a temporary and limited nature.

两国之间的边界争论，比起两国的悠久友谊来说，是个暂时性和局部性的问题。

11. category [ˈkætəɡəri]

   n. a collection of things sharing a common attribute 种类；范畴

   Music shops should arrange their recordings in simple alphabetical order, rather than by category.

   音像店应按照简单的字母顺序摆放其唱片，而不是按种类。

12. porous [ˈpɔːrəs]

   a. allowing passage in and out 能渗透的；有孔的

   Swimming is helpful for bones that are porous and weak.

   游泳有助于改善骨质疏松和骨骼脆弱。

13. schematic [skiːˈmætɪk]

   a. represented in simplified or symbolic form 概要的；略图的

   We reproduced a same working schematic diagram.

   我们复制了一份同样的施工简图。

14. conceptual [kənˈseptʃuəl]

   a. being or characterized by concepts or their formation 概念的

   By the combination of radicals or semantic elements, single compound words can express complex conceptual relations, often of an abstract universal character.

   通过部首或语义元素的组合，单个复合词可以表达复杂的概念关系，通常具有抽象的普遍性。

15. fluidity [fluˈɪdəti]

   n. the property of flowing easily 流动性；易变性

   Because of the natural fluidity and immediacy of marketing, we are constantly building new areas of knowledge.

   因为市场固有的流动性和即时性，所以我们经常构建新知识领域。

16. reinforce [riːɪnˈfɔːs]

   v. strengthen and support with rewards 加强；强化

   The delegation hopes to reinforce the idea that human rights are not purely internal matters.

   代表团希望强化人权不完全是国家内部事务的观念。

17. denote [dɪˈnəʊt]

   v. be a sign or indication of 标志；表示；意指

   Solidarity, however, does not necessarily denote acceptance or equality, let alone full citizenship.

   团结，并不一定意味着认可或者平等，更别提充分享有公民权。

18. straddle [ˈstrædl]

   v. range or extend over; occupy a certain area 横跨；跨越（不同时期、群体或领域）

   The tower's two bases straddle Edgar Street and then spiral up, joining in the middle before splitting off into two atria.

塔楼的两个基座跨越埃德加街,然后螺旋上升,在分成两个大厅前在中间合二为一。

19. ethnicity [eθˈnɪsəti]

    n. an ethnic quality or affiliation resulting from racial or cultural ties 种族渊源;种族特点

    The government uses a classification system that includes both race and ethnicity.

    政府采用一种既包括人种又包括种族特点的类别体系。

20. accrue [əˈkruː]

    v. grow by addition (利益、好处等)产生,形成;(钱不断地)累积,增加

    While they may use a credit card for convenience, affluent people never let interest charges accrue.

    虽然有钱人可能为了方便而使用信用卡,但是他们从来不让利息费累积。

21. intertwine [ɪntəˈtwaɪn]

    v. spin or twist together so as to form a cord 缠绕;纠缠;交织

    Three major narratives intertwine within Foucault's text, *Madness and Civilization*.

    福考特的文章《疯狂与文明》中交织着三种主要的叙述。

22. mitigation [mɪtɪˈgeɪʃn]

    n. the action of lessening in severity or intensity 减轻;缓和

    In mitigation, the defence lawyer said his client was seriously depressed at the time of the assault.

    为了减轻罪行,辩护律师说他的当事人在袭击人的时候精神极度压抑。

23. vicariously [vɪˈkeərɪəsli]

    ad. indirectly, as, by, or through a substitute 代理地;间接感受到地

    People often live vicariously through the adventures of their more socially active peers.

    人们通常对社交中更为活跃的同伴的冒险经历感同身受。

24. burgeoning [ˈbɜːdʒənɪŋ]

    a. growing and flourishing 迅速发展的;繁荣的

    Pushed by a burgeoning middle class, cellphone usage is surging across Africa.

    受中产阶级迅速发展的推动,非洲各地的手机使用率正急剧增长。

25. cohesion [kəʊˈhiːʒ(ə)n]

    n. the state of cohering or sticking together 凝聚力;团结

    By 1990, it was clear that the cohesion of the armed forces was rapidly breaking down.

    到1990年时,武装部队凝聚力明显地迅速瓦解。

26. mandate [ˈmændeɪt]

    v. assign authority to 授权;(领土)(由国际联盟)授权托管

    He'd been mandated by the West African Economic Community to go in and to enforce a ceasefire.

    他受西非经济共同体授权去介入并执行停火协定。

27. truce [truːs]

    n. a state of peace agreed to between opponents so they can discuss peace terms 停战协

定;休战;停战期

That peace often looks like a brief truce before the next plunge into war.

那种和平看起来常常像是投入新一轮战争前的短暂停战期。

28. codify ['kəʊdɪfaɪ]

v. organize into a code or system, such as a body of law 编纂;将……编成法典

Legislation to codify network neutrality failed to pass, and carriers backed off their plans for a tiered Internet.

编纂网络中立性的法案未获通过,运营商撤回其分层互联网的计划。

29. terrestrial [təˈrestrɪəl]

a. of or relating to or inhabiting the land as opposed to the sea or air 地球上的;陆地上的

The polar bear is the largest terrestrial carnivore, being more than twice as big as the Siberian tiger.

北极熊是陆地上最大的肉食性动物,比西伯利亚虎都要大两倍多。

30. ascribe [əˈskraɪb]

v. attribute or credit to 把……归因于;认为……具有

The world always seems to ascribe financial success to superior intelligence.

世人似乎总是把金融方面的成功归因于胜人一筹的智力。

31. physiological [fɪzɪəˈlɒdʒɪk(ə)l]

a. of or relating to the biological study of physiology 生理的;生理学的

Yawning did nothing special to their state of physiological activity.

打哈欠对他们的生理活动状态没有什么特殊的影响。

32. articulate [ɑːˈtɪkjuleɪt]

v. express or state clearly 明确表达;清楚说明

The president has been accused of failing to articulate an overall vision in foreign affairs.

总统被指责没能清楚地说明对外交事务的总体设想。

33. ignite [ɪgˈnaɪt]

v. arouse or excite feelings and passions 引发;激起

In these inflationary days, such expenses easily ignite public anger.

在这个通货膨胀的时代,这样的开支很容易引发公愤。

34. fiduciary [fɪˈdjuːʃəri]

a. relating to or of the nature of a legal trust 信托的;信用的

They have a case against their directors for breach of fiduciary duty.

他们对董事们违反信托义务提起诉讼。

35. ramification [ræmɪfɪˈkeɪʃn]

n. an arrangement of branching parts 衍生物;分支

Virtual self is a ramification of real self in virtual environment.

网络中的虚拟自我是现实自我在网络虚拟环境中的衍生物。

36. synergy [ˈsɪnədʒi]

n. the working together of two things to produce an effect greater than the sum of their individual effects（两个或多个组织共同协作后产生的）协同增效作用

Of course, there's quite obviously a lot of synergy between the two companies.

当然，这两家公司之间显然有许多的协同增效作用。

37. extol [ɪkˈstəʊl]

v. praise, glorify, or honor 颂扬；赞美；赞颂

Doctors often extol the virtues of eating less fat.

医生常常宣扬少吃脂肪的好处。

Phrases and Expressions

focus on 集中于
ascribe......to 把……归于；归因于；认为……是
contribute to 有助于；捐献；促成
account for 对……做出解释
rely on 依靠；依赖
benefit from 得益于
a multitude of 大批的，众多的
exclude......from 把……排除在外
be engaged in 参与；从事于
derive......from 从……得到；从……获取
interact with 与……相互作用
prior to 在……之前；居先
cater to 迎合；为……服务

Terminology

material well-being 物质幸福感
subjective well-being 主观幸福感
relational well-being 关系幸福感
social embeddedness 社会嵌入性
dynamic interplay 动态相互作用
biodiversity conservation 生物多样性保护
blue acceleration 人类对海洋资源的使用增加
COVID-19 新型冠状病毒肺炎（coronavirus disease 2019）
principle of free prior informed consent（PIC）自由事先知情同意原则

Proper Names

Blue Papers 蓝皮书（用于官方文件时，一般指英国议会的出版物）
Easter 复活节
First Nation 第一民族（指加拿大的原住民印第安人）

Pacific Islanders 太平洋岛民；太平洋群岛的人
David Attenborough's *The Blue Planet* 大卫·爱登堡的《蓝色星球》
U.S.-initiated Black Lives Matter movement 美国发起的"黑人的命也是命"运动
Indigenous Peoples 土著民族

一、翻译策略/方法/技巧：增译法

增译法作为一种翻译技巧，由来已久。在西方翻译史中，早期对增译法的使用见于《圣经》的翻译活动。中国最早关于增译法的记载见于佛经翻译，如后秦僧人鸠摩罗什的翻译"对于原本，有增有损，求达求雅"。不同时期的翻译学者赋予了"在翻译中增加有其意而无其形的词汇"这一翻译技巧不同的称名，如增词（法）、增补（法）、增添、增删、削增、增减、加词等。通过梳理相关文献发现，称名"增译法"的认可度更高。在翻译的过程中，根据上下文的需要，增加的不仅是词和短语，还涉及小句、句群、篇章等，因此称名"增译法"涵盖的范围比其他的称名如"增词（法）"更大。同时，"增译"与其他的翻译技巧"X译"（如省译、分译、合译等）构成同级范畴。从术语体系与术语结构上看，"增译"的称名方式更客观，更具科学性。当代翻译学者郭富强对增译法给出了系统的定义："增译法就是增加一些原文中有其意而无其形的词汇、短语、从句或整句，这样可以提高译文在语法、结构和表达的完整性，同时更加契合汉语的表达习惯，使得目的语与源语在内容、形式和精神等方面都能达到对等。"

增译法的使用原则是：内隐于外显，增词不增意。在翻译过程中，根据语义或者修辞的需要，将原文隐含的信息增补出来。在充分尊重原文含义的前提下，通过适当增添原文中无其形但有其意的词，能使译文表达更为准确清晰。但是切忌过度翻译，增译出一些不必要的词语，对原文画蛇添足，使译文显得冗长拖沓，甚至词不达意。英汉两种语言分属两大语系，中西方文化以及思维方式截然不同，英汉表达差异显著。具体来说，汉语重"意合"，语法呈隐性，句式较灵活；而英语重"形合"，语法呈显性，结构严密，层次分明。英汉互译过程中，通过增译动词、名词、逻辑关联词、语气助词、形容词、解释性文字等，能有效解决由于英汉两种语言的巨大差异所带来的翻译问题，使译文清晰、准确、流畅。

以下是一些常见的增译法应用例子：

1. 增译动词

MPAs help improve conditions of marine environments such as <u>enhancing</u> biological parameters, survival rates of juvenile fish, species diversity, fish biomass, density and species richness.

海洋保护区有助于改善海洋环境条件，如优化生物系统参数，提高幼鱼的存活率，<u>提升</u>物种多样性，<u>增加</u>鱼类生物量、种群密度和物种丰富度。

2. 增译名词

The government has long been concerned with the local unemployment.

政府一直关心当地<u>失业</u>问题。

3. 增译逻辑关联词

Those coastal areas that do not have marine reserves are often associated with the collapse of fisheries and damaged marine ecosystems which do not attract more tourists.

在未实施海洋保护区的沿海地区,不仅渔业产业面临崩溃,而且由于海洋生态系统受损,旅游业也无法吸引更多的游客。

4. 增译解释性文字

Students on campus are electronically linked to each other, to professors and to their classwork 24/7 in an ever-flowing river of information and communication.

通过源源不断的信息流,学生之间、师生之间以及学生和课堂作业之间建立了电子化的联系,一天 24 小时,一周 7 天,从不间断。

(摘自:罗新璋,陈应年.翻译论集[M].北京:商务印书馆,2009:3;郭富强.英汉翻译理论与实践[M].北京:机械工业出版社,2004:89.)

二、译例与练习

Translation

1. If the upsurge in economic activity is to lead to an upsurge in human well-being, then its emergent and potential future impacts must be understood. (Para.1)

如果经济活动的激增将导致人类幸福感的激增,那么必须理解其正在出现的和潜在的未来影响。

2. The material, relational and subjective dimensions of well-being are, however, interconnected or "co-constitutive", and, like all such classifications, the boundaries between categories are porous. (Para.3)

然而,幸福感的物质维度、关系维度和主观维度是相互关联的或"相互建构的",而且就像所有这种分类一样,类别之间的边界是模糊的。

【英汉两种语言的搭配不尽相同,"porous boundaries"在汉语中不可表达为"多孔的边界",而应为"模糊的边界"。类似的例子,"porous medium"译为汉语时应为"疏松介质","porous structure"在汉语中应为"多孔结构","porous borders"在汉语中应为"漏洞百出的边境线"。翻译时应根据英汉两种语言各自常用的固定词组和常见搭配选择词汇,否则容易出现误译。】

3. It is also important to continue to maintain and build the kinds of social and cultural connections to the ocean that have improved human knowledge, understanding, cooperation, security, meaning and happiness in the past. (Para.5)

过去,人类与海洋的种种社会和文化联系增进了人类的知识、理解、合作、安全、意义和幸福。继续保持并建立这些联系也很重要。

【此句包含一个由"that"引导的限制性定语从句,先行词为"the kinds of social and cultural connections to the ocean",句子较长,翻译成汉语时要根据意群适当地断句,所以在关系代词"that"前将句子断开,并增添主语,同时根据汉语主要信息放在句尾的习惯,调整语序,使译文更符合逻辑。】

4. These identities and concepts of embeddedness straddle both the relational and subjective dimensions of well-being—the sense of belonging to a group, such as fishers or a coastal community, of being able to depend on your group during emergencies, times of loss and crisis due to the strength of social relations and networks, but also subjective feelings of pride and self-worth in one's occupation, community or ethnicity. (Para. 8)

这些嵌入性身份和概念横跨幸福感的关系维度和主观维度。它们是渔民或沿海社区对群体的归属感，是社会关系和网络力量带来紧急情况、损失和危机时，对群体的依赖感，但也是对自己的职业、社区或种族的自豪感和自我价值的主观感受。

5. Future ocean governance also has to ensure that human relational and subjective well-being are supported rather than undermined by the "blue acceleration". (Para. 10)

未来的海洋治理还必须确保人类关系幸福感和主观幸福感得到"蓝色加速"的支持，而不是破坏。

【21世纪初，人类对世界海洋的争夺急剧增加，并且没有放缓的迹象。科学家称这一急剧上升的现象为"蓝色加速"。"蓝色加速"实际上是一场海洋资源与空间的争夺赛，它为全球可持续发展带来风险与机遇。】

6. These interactions—whether we experience them directly as beachcombers, rockpoolers, snorkelers, divers or recreational fishers, or vicariously through aquarium visits or viewing television series such as David Attenborough's *The Blue Planet*—create sets of relationships with ocean nature that respond to a range of human material, relational, spiritual and emotional needs. (Para. 11)

我们无论是通过海滩拾荒、岩石泳池游泳、浮潜、潜水或休闲捕鱼直接体验，还是通过参观水族馆或观看大卫·爱登堡的《蓝色星球》等电视系列片来间接体验，这些互动都创造了一系列人类与海洋自然的关系，以回应人类一系列物质、关系、精神和情感需求。

7. The ways the ocean contributes to relational well-being are more concrete and better understood at smaller scales: the social cohesion of traditional fishing communities and how this contributes to economic, social and cultural life is well-studied, understood and increasingly legally mandated in the form of devolved management and community resource rights. (Para. 13)

在更小的范围内，海洋促进关系幸福感的方式更具体，也更好理解：传统渔业社区的社会凝聚力及其对经济、社会和文化生活的贡献得到充分的研究和理解，并越来越多地以下放管理和社区资源权利的形式得到法律授权。

8. Of the three main dimensions of well-being, the elements of "subjective" well-being are the most difficult to ascribe monetary value to and therefore to incorporate into traditional sectoral economic planning, though some of them have been considered in social accounts and happiness and well-being indices. (Para. 14)

在幸福感的三个主要维度中，"主观"幸福感的要素是最难用货币价值来衡量的，因此也最难纳入传统的部门性经济规划之中，尽管其中一些要素已经在社会核算、幸福和幸福指数方面得到考量。

9. There are lessons for wider global ocean governance from attempts to implement universal bans on the harvest of whales and other marine mammals, with nation-states and Indigenous Peoples who pursue traditional whaling activity resisting these bans in various ways and maintaining their cultural relationship to whales as food as well as cultural keystone species invested with complex symbolic meaning. (Para. 16)

全面禁止捕鲸和其他海洋哺乳动物的尝试，为更广泛的全球海洋治理提供了经验教训。从事传统捕鲸活动的民族、国家和土著人民将鲸作为食物，并将其视为具有复杂象征意义的文化基石物种，他们以各种方式抵制这些禁令，并保持他们与鲸的文化关系。

10. The physical and mental aspects of people's health is affected positively by a clean ocean, which can be enjoyed by seafood consumption, trips to the seaside, swimming or pleasure cruises on the ocean. (Para. 19)

人们的身心健康受到清洁海洋的积极影响，人们可以通过食用海鲜、去海边旅行、在海洋上游泳或巡游来享受清洁海洋带来的乐趣。

Exercises

1. Fill in the blanks with the proper given words, and then translate the sentences into Chinese.

accommodation   dynamic   extract   regulatory   incorporate   initiate

1) How can the hormone adrenaline that does not act directly on the brain have a _____ effect on brain function?

2) People who are talked to have equally positive experiences as those who _____ a conversation.

3) The local government will immediately provide temporary _____ for up to three thousand homeless people.

4) Barcelona has become one of the most _____ and prosperous cities in the Mediterranean in recent years.

5) Technological change is construed as the outcome of negotiations among interested parties who seek to _____ their own interests into the design and configuration of the machinery.

6) There was a machine in the kitchen which could _____ the juice of two hundred oranges in half an hour if a little button was pressed two hundred times by a butler's thumb.

2. Translate the following sentences into Chinese.

1) Existing maritime communications, transport and tourism industries have expanded rapidly, as have the territorial claims and information needs around the ocean.

2) The ocean has played a key role in sustaining the flow of knowledge and social and cultural exchange among societies and countries.

3) Sharing these resources can help maintain regional political and economic stability, which contributes to the well-being of populations around regional seas and ocean rims.

4) Ocean peoples have long shared many social norms, many of which are now codified in modern law. These maritime codes of conduct were designed to improve safety and well-being at sea and facilitate travel and commerce.

5) A series of connections, formed over centuries of trade among seafaring peoples of the Americas, Europe and Africa, led to loose coalitions of fugitives from state violence and exploitative work.

6) In many contemporary cultures, beaches and the seaside have strong associations with family holidays and childhood memories, with romance and with togetherness in old age.

7) Knowing how to row a boat, fix an engine, cross the surf or spear a fish are among the many practical skills that imbue a sense of pride in maritime occupations.

8) Fishers often emphasize their need for independence and choose to stay in the fisheries, even when more lucrative work is available, because they value the independence and freedom of working for themselves, or with friends and family, and are unwilling to submit to working indoors, being bound to a timetable and reporting to a boss.

9) The ocean and coasts have inspired the visual and creative arts, and humans have felt the need for aesthetic expression since at least the time the first cave paintings were created.

10) A sense of well-being that comes from feeling at one with the ocean may also be achieved by those who are not of the ocean.

3. Translate the following sentences into English.

1) 海洋是你工作的地方，它也可能是你感觉最自由、最能掌控自己命运、最有能力、最受他人重视的地方。

2) 太平洋岛屿的鲨鱼被赋予了精神力量，被认为是祖先的守护者或守护神，可以保护他们免受海洋中不可预知力量的影响。

3) 在过去的50年时间里，新的海洋产业迅速发展，如海水养殖、深海碳氢化合物和矿物钻探、海水淡化和海上风力发电场。

4) 风能传统上用于海上航行，但现在越来越多地通过海上风电场为陆地上的人类活动发电。

5) 人们把海洋看作当代社会未充分利用的原材料来源。

6) 长期以来，航运和渔业为沿海和岛屿经济体创造财富和就业、维持人民生计做出了重要贡献。

7) 在5 000到3 000年前，这些社会开始相互联系，当时商人们学会了利用季风进行跨洋贸易，而不是沿着海岸进行贸易。

8) 被封控的人群在允许的情况下涌向海滩,因而人们担忧北半球夏季会出现第二波新冠肺炎感染。

9) 随着人们开始考虑如何重建经济和重启社会生活,未来几个月将有机会强调"蓝色空间"对人们的重要性,并确保人们能够利用"蓝色空间"获得幸福感。

10) 沿海人口的增长速度大约是全国增长速度的 2 倍,那里的人口密度是世界平均水平的 2 倍。

4. Choose the best paragraph translation. And then answer why you choose the first translation or the second one.

海洋、航行在海洋中的船只以及栖息在海洋空间中的鱼类和哺乳动物,都是艺术灵感的无限源泉。我们看到、听到、闻到海洋时,与海洋互动时,会感受到某些情感。描绘海洋的艺术作品会让人产生相似的情感。艺术可以只是一个地点、时间或事件的记录,但其目的通常是引起某种情感反应。观赏海洋艺术可以丰富我们的生活,帮助我们看到我们的生命是多么珍贵。海洋艺术在欧洲历史上占有重要地位,在全球其他地区也同样如此。

译文一:

The sea, the ships that sail in it, and the fish and mammals that inhabit its space are all infinite sources of artistic inspiration. We experience certain emotions when we see, hear, smell, and interact with the ocean. Artworks depicting the ocean evoke similar emotions. Art can be just a record of a place, time or event, but the purpose is usually to elicit some kind of emotional response. Watching marine art can enrich our lives and help us see how precious our lives are. Maritime art plays an important role in European history, as it does in other parts of the globe.

译文二:

The sea, the ships that navigate it and the fish and mammals that inhabit the marine space are a limitless source of artistic inspiration. Artistic representations of the ocean can create similar emotions that we experience from seeing, hearing, smelling and interacting with the ocean. Art can simply be a record of a place, time or event, but its purpose is usually to create an emotional response. Contemplating marine art can enrich us, help us see how precious all our lives are. Marine art has a prominent place in European history but equally so in other parts of the globe.

5. Translate the following passage into Chinese.

Subjective well-being is also driven by anxieties, with psychologists identifying six existential ones: identity, happiness, isolation, meaning in life, freedom and death. All these anxieties can be either confronted or relieved (or both) in our relationships with the ocean—and with nature more generally—whether that relationship is professional, residential, consumptive or recreational. We observe that groups whose lives are closely entwined with marine resource use (fisherfolk, mariners, Indigenous Peoples, marine tourism and recreation professionals) have complex, multidimensional relationships with the ocean which are often deeply spiritual and strongly inform social and cultural identities.

6. Translate the following passage into English.

中国政府在2016年的"十三五"规划中提到了"蓝色经济",但中国当代海洋经济政策可以追溯到20世纪70年代末改革开放刚开始之时。21世纪初,中国蓝色经济加速发展。自1996年批准《联合国海洋法公约》以来,中国建立了众多专属经济区,呼吁"实施海洋开发",并发布了各种海洋经济发展五年规划,最终目标是成为"海洋强国",拥有军事防御能力、强大的海洋经济和先进的海洋科学技术。2019年,国务院总理李克强将中国蓝色经济的国家愿景概括为"大力发展蓝色经济,保护海洋环境,建设海洋强国"。

三、翻译家论翻译

王佐良(1916—1995)博古通今,学贯中西,多才多艺,著述等身,成就卓著,在国内外学术界享有盛誉。他不仅是一位著名教育家、杰出语言学家、出色作家和诗人、英国文学研究界权威、比较文学研究开创者、优秀编辑工作者,还是一位卓越的翻译家和翻译理论工作者。他曾任中国译协理事和北京市译协副会长。王佐良对翻译和翻译研究有着特殊的情结,究其原因,似可归结为两点:一是个人喜欢,二是研究需要。他曾说:"我是喜欢翻译的。有时候,当我写完了一篇所谓的'研究'论文,我总是感到:与其论述一个外国作家,不如把他的作品翻译一点过来,也许对读者更有用。"王佐良一生译作以数量可观的经典诗歌和散文为主,也有小说(短篇)和戏剧;既有英译汉,也有汉译英。他的译文篇篇都是精品,语言新鲜隽永,耐人寻味。他边翻译边思考,提出了自己独到的翻译主张,为我国翻译研究做出了重要贡献。

王佐良在翻译理论方面很有建树,他的翻译主张汇成了文体翻译观、文化翻译观、译诗观、理论与实践统一观、新时期翻译观等"五位一体"的王佐良翻译思想体系。这里着重学习他的文化翻译观。

王佐良认为,翻译与文化密不可分,翻译可以促进文化繁荣,文化繁荣往往会带来翻译高潮。他认为,"翻译,特别是文学翻译对于任何民族文学、任何民族文化都有莫大好处。不仅是打开了若干朝外的门窗,还能给民族文学以新的生命力。如果去掉翻译,每个民族的文化都将大为贫乏,整个世界也将失去光泽,宛如脱了锦袍,只剩下单调的内衣"。

一个人如果不了解语言当中的社会文化,也就无法真正掌握语言。具体到翻译,王佐良认为,译者的第一个困难是对原文的了解。虽然人类有很多共同的东西,无论怎样难的原文,总有了解的可能,这也使翻译成为可能,但原文中也总含有若干外国人不易了解的东西,这又使深入了解外国文化十分必要。不仅如此,他还指出,作为译者,仅仅了解外国文化仍然不够,还必须深入了解自己民族的文化,因为翻译不仅仅是双语交流,它更是一种跨文化交流,翻译的目的是突破语言障碍,实现并促进文化交流。而这种"文化交流"在本质上是双向互动的。一言以蔽之,"翻译者必须是一个真正意义上的文化人"。

王佐良认为,译者不仅仅是一个真正意义上的文化人,还要不断地把两种文化加以比较,这是因为翻译是一个动态的过程,翻译中译者面对的最大、最直接的困难就是两种文化的不同,他处理的是个别的词,他面对的则是两大片文化。这样,译者在寻找与原文相当的"对等词"的过程中,就要做一番比较,因为真正的对等应该是在各自文化里的含义、作用、范围、情感色彩、影响等等都相当,否则,就会望文生义,跌入陷阱。此外,王佐良还强调指出,译者所做的文化比较远比一般人细致、深入,因为这里还有"译文要适合社会场合"的问题,即译者必须在注意语言与社会文化关系的基础上,根据原文的要求,运用各种不同的语类、文体知识,不断寻找适合社会场合的"对等说法"。

王佐良既是翻译实践家也是翻译理论家,为我国的翻译事业做出了巨大贡献。在翻译实践方面,他的译作不论在语意、语体,还是在文体、审美特征上,都与原文在形式、内容和风格上实现了高度的统一,堪称形神兼备的佳译,很好践行了其"一切照原作,雅俗如之,深浅如之,口气如之,文体如之"的翻译主张。他的译作措辞得体,简练精确。行文流畅,脉络清晰,篇篇皆精品,对翻译界产生深远影响,诚如有人指出:"王佐良的译文好似一座高高的山峰,令后来者难以翻越。后来者无论怎么努力也无法逃出王译的影子,也就只好绕道而行,结果到头来,译出的东西几乎成了王译的解释或延伸,原文的简约和精辟不见了踪影。"

(摘自:黎昌抱,王佐良.翻译风格研究[M].北京:光明日报出版社,2009:1-9.)

Text B

The Human Relationship with the Ocean
By Edward H. Allison, John Kurien and Yoshitaka Ota

Ideological and Political Education：提高海洋开发能力 实现民族伟大复兴

习近平总书记高度重视海洋强国建设,围绕海洋事业多次发表重要讲话,他强调"建设海洋强国是实现中华民族伟大复兴的重大战略任务"。下面是习近平总书记的部分相关重要论述,一起来学习。

建设海洋强国是实现中华民族伟大复兴的重大战略任务

1. 建设海洋强国是中国特色社会主义事业的重要组成部分。党的十八大作出了建设海洋强国的重大部署。实施这一重大部署,对推动经济持续健康发展,对维护国家主权、安全、发展利益,对实现全面建成小康社会目标、进而实现中华民族伟大复兴都具有重大而深远的意义。——2013年7月30日,十八届中央政治局第八次集体学习时讲话

2. 海洋事业关系民族生存发展状态,关系国家兴衰安危。——2013年8月28日至31日,习近平在辽宁考察时的讲话

3. 我国是一个海洋大国,海域面积十分辽阔。一定要向海洋进军,加快建设海洋强国。——2018年4月12日,习近平在海南考察时的讲话

4. 建设海洋强国是实现中华民族伟大复兴的重大战略任务。要推动海洋科技实现高水平自立自强,加强原创性、引领性科技攻关,把装备制造牢牢抓在自己手里,努力用我们自己的装备开发油气资源,提高能源自给率,保障国家能源安全。——2022年4月10日至13日,习近平在海南考察时的讲话

不断提高海洋开发能力,使海洋经济成为新的增长点

5. 要顺应建设海洋强国的需要,加快培育海洋工程制造业这一战略性新兴产业,不断提高海洋开发能力,使海洋经济成为新的增长点。——2013年8月28日至31日,习近平在辽宁考察时的讲话

6. 海洋经济发展前途无量。建设海洋强国,必须进一步关心海洋、认识海洋、经略海洋,加快海洋科技创新步伐。——2018年6月12日至14日,习近平在山东考察时的讲话

7. 要高度重视海洋生态文明建设,加强海洋环境污染防治,保护海洋生物多样性,实现海洋资源有序开发利用,为子孙后代留下一片碧海蓝天。——2019年10月15日,习近平致2019中国海洋经济博览会的贺信

> **构建蓝色经济伙伴关系**
>
> 8. 海洋对于人类社会生存和发展具有重要意义。海洋孕育了生命、联通了世界、促进了发展。我们人类居住的这个蓝色星球,不是被海洋分割成了各个孤岛,而是被海洋连结成了命运共同体,各国人民安危与共。海洋的和平安宁关乎世界各国安危和利益,需要共同维护,倍加珍惜。——2019年4月23日,习近平在青岛集体会见应邀出席中国人民解放军海军成立70周年多国海军活动的外方代表团团长时的讲话

1. The human relationship with the ocean is diverse and complex. It is built on values that are often non-monetary, and which contribute to non-material dimensions of well-being. These values are essential to broader human **flourishing**. They include contributions to cultural and social and legal identity; a sense of place; occupational pride and self-respect; spirituality; mental and bodily health; and human security. The plurality of these values and interests matters to individuals and societies and could be more strongly represented in high-level ocean policy discussions.

2. People across the world have diverse economic, socio-legal, **institutional**, social and cultural relationships with the ocean—both its **littoral** zones and the open sea spaces through which people have traditionally navigated, migrated, fished, traded, played and sought **solace**, spiritual **enlightenment**, adventure, material enrichment, social identity, cultural expression, artistic inspiration or good health. These relationships are reflected in formal and informal institutions (polices, laws, social norms) that regulate many of these activities, including those that regulate access to resources. These institutions represent a series of prior claims and rights to the use and enjoyment of the ocean by coastal and maritime societies.

3. By taking account of the range of ways coastal and maritime societies use, enjoy and govern coastal seas and ocean basins, we are better placed to design a sustainable ocean economy that is fair and equitable and that reflects "the future we want". Policymakers should consider the full range of human relationships with the ocean. The economic investment strategies and governance actions envisaged in contemporary ocean policy and planning can transform those relationships and will thus change the nature and distributions of the values that humanity derives from its interactions with the oceanic realm.

4. How can humanity's diverse relationships with the ocean be supported to flourish in the future, so that the ocean can make sustainable contributions to human well-being? This is the **overarching** policy question to which this study responds. Policy research has made significant

advances in assessing the ocean's ability to generate economic goods and services. The complementary **perspectives** presented here aim to draw attention to the wider role that the ocean has played—and will continue to play—in sustaining and reproducing other human values such as social and cultural identity, individual and collective well-being, sense of place and belonging, and human emotions such as curiosity, spirituality, awe and a sense of adventure.

5. From a brief survey of the past and current range of human relationships with the ocean and how they contribute to human well-being, and by examining the economic and policy implications of these relationships, we will argue that a sustainable ocean economy can contribute not only to the sustainable and equitable growth of economic goods and services but also to human well-being and flourishing more generally. Thus, the ocean can play a **catalytic** role in the next phase of human development, enhancing human capabilities and freedoms, and thereby contribute to meeting the UN Sustainable Development Goals (SDGs).

6. It is not our intent here to **document** every way that people and the ocean interact, for good or ill. Other studies in this series examine in detail how we might sustain and grow marine food production; how climate change has impacted the ocean and how humanity may respond; how we might better deal with human rights violations and other criminal activities and inequities at sea; how pollution threatens the ocean and how we might control it better; what opportunities exist to improve the financing and governance of the ocean economy, and so on. These issues and solution pathways all impact the plurality of people-ocean relationships and may undermine some and enhance others, in part depending on how they affect existing ocean-related economic inequalities. Our point here is that the relational and subjective elements of people-ocean relationships have not yet been fully articulated in policy **arenas** and are therefore not yet fully considered in plans to respond to these ocean threats or to seize ocean economic and conservation opportunities.

7. Drawing on brief **overviews** of representative social and legal institutions that have developed in different maritime societies, we identify how different societies have governed oceanic spaces and volumes and how these governance mechanisms reflect the diversity of "ocean values" held by different peoples. We use these overviews of the diversity of human relationships with the ocean, the examples of historically and culturally grounded sea **tenure** arrangements, and contemporary policy debates around the "blue economy", "blue justice" and "blue **degrowth**", to identify a series of opportunities for action to build a sustainable ocean economy and a future human relationship with the ocean that reflects the breadth and plurality of world views and values of current and future ocean citizens, and that acknowledges the diversity of social

identities of the people for whom the ocean matters.

8. At the time of this study, the world was **reeling** from the impacts of the COVID-19 pandemic, which, by 16 August 2020, had infected around 21.3 million people and resulted in 761,779 deaths. We felt it necessary to consider how relationships between people and the ocean may be affected by the public health measures taken to slow the spread of the virus and the economic and social consequences of both the disease itself and measures taken to contain it. Accordingly, we briefly consider what is known about impacts on the maritime economy and on human-ocean relationships.

9. It also cannot be overlooked that humanity is embarking on an ocean governance transformation at a time when action on climate change is critical. The ocean offers many opportunities to reduce greenhouse gas emissions and increase carbon capture and storage. Ocean-related climate change impacts are likely to **exacerbate** existing inequalities within coastal communities, with vulnerable populations being those living in low-lying areas of the tropics, on small oceanic islands and in the Arctic, as well as those whose livelihoods are tied to fisheries affected by global environmental change. Most sectors of the ocean economy will be negatively impacted by climate change, and tele-connected climate and economic processes mean that oceanic changes also have impacts inland. Investments in building adaptive capacity in ways that respond to different peoples' values will be required, and these should be kept in mind when considering how the human relationship with the ocean is understood, assessed and governed.

(1,051 words)

https://oceanpanel.org/blue-papers

Extracted from *The Human Relationship with Our Ocean Planet*, 2-4, 2020.

New Words

1. flourish ['flʌrɪʃ]
    v. grow stronger or gain in wealth 繁荣；昌盛
    They have created an environment in which productivity should flourish.
    他们创造了一种可以大大提高生产力的环境。
2. plurality [plʊə'ræləti]
    n. the state of being plural or a large indefinite number 多数；复数
    Franklin had won with a plurality in electoral votes of 449 to 82.
    富兰克林在选举团投票中以449票对82票的多数票获胜。

3. institutional [ˌɪnstɪˈtjuːʃənl]

   a. relating to or constituting or involving an institution 机构的；制度的

   Outside the protected environment of institutional care, he could not survive.

   离开福利机构照顾下的这个受保护的环境，他无法生存。

4. littoral [ˈlɪtərəl]

   a. of or relating to a coastal or shore region 沿海的；海滨的

   Indeed before the age of tourism not many littoral folk learned to swim because the risk of drowning was too high.

   的确，在大规模旅游业时代到来之前，海边居民中学会游泳的人并不多，因为在海中溺水身亡的危险性太高了。

5. solace [ˈsɒləs]

   n. comfort in disappointment or misery 安慰；慰藉

   Optimists will also find some solace in the IMF's comparative analysis.

   乐观主义者还将从国际货币基金组织的比对分析中找到些许慰藉。

6. enlightenment [ɪnˈlaɪt(ə)nmənt]

   n. education that results in understanding and the spread of knowledge 启迪；教导；启蒙运动

   The newspapers provided little enlightenment about the cause of the accident.

   报纸对事故原因并未解释清楚。

7. envisage [ɪnˈvɪzɪdʒ]

   v. form a mental image of something that is not present or that is not the case 设想；展望；正视

   The future beckons and you are beginning to envisage its attraction after the darkness you have experienced.

   未来在召唤，在你们经历了黑暗之后，你们正在开始正视它的吸引力。

8. overarching [ˌəʊvərˈɑːtʃɪŋ]

   a. central or dominant 首要的；支配一切的；包罗万象的

   When the overarching objective is poverty reduction, if you miss the poor, you miss the point.

   当减贫是首要目标时，如果遗漏了穷人，也就错失了要点。

9. perspective [pəˈspektɪv]

   n. a way of regarding situations or topics etc. （观察问题的）视角；观点

   Getting input from others not only offers a fresh perspective and thought process, it often also includes riskier choices.

   听取他人的意见，不仅可以提供一个全新的视角和思考过程，而且通常还包括一些更冒险的选择。

10. awe [ɔː]

    n. an overwhelming feeling of wonder or admiration 敬畏；惊叹

    Brought before the evidence of his crimes, he was awe-struck and could say nothing in his defence.

罪证面前,他张口结舌,无言以对。

11. catalytic [ˌkætəˈlɪtɪk]

a. relating to or causing or involving catalysis 接触反应的;起催化作用的

Lower production also means less demand for palladium and platinum used in the production of catalytic converters.

减少生产,意味着对钯和铂的需求将减少,这两种金属主要用于生产汽车的催化转换器。

12. phase [feɪz]

n. any distinct time period in a sequence of events(发展或变化的)阶段;时期

The wedding marked the beginning of a new phase in Emma's life.

婚礼标志着埃玛生活新阶段的开始。

13. document [ˈdɒkjumənt]

v. record in detail 记录;记载(详情)

The relationship between getting more sleep and making better food choices is well-documented.

更充足的睡眠和选择更好的食物之间的关系有据可查。

14. arena [əˈriːnə]

n. a particular environment or walk of life 竞技场;斗争场所;活动舞台

Our country needs more scientists who are willing to step out in the public arena and offer their opinions on important matters.

我们的国家需要更多的科学家,他们愿意走到公众舞台上,提供他们对重要事情的看法。

15. overview [ˈəʊvəvjuː]

n. a general summary of a subject 概述;综述

Professor Henderson, could you give us a brief overview of what you do, where you work and your main area of research?

亨德森教授,您能给我们简单介绍一下你的工作、工作地点和主要研究领域吗?

16. tenure [ˈtenjə(r)]

n. the right to hold property(土地的)居住权;保有权

Studies have shown that farmers in developing countries who have achieved certain levels of education, wealth, and security of land tenure are more likely to adopt such technologies.

研究表明,在发展中国家,拥有一定教育水平、财富和土地所有权保障的农民更有可能采用这类技术。

17. degrowth [deˈɡrəʊθ]

n. regressive growth 退行生长

The second international conference on degrowth economics met recently in Barcelona.

日前,第二次国际逆生长经济会议在巴塞罗那举行。

18. reel [riːl]

v. to feel very shocked, upset, or confused 震惊;困惑

The mixture of sights, smells and sounds around her made her senses reel.

四周的物象、气味和声音纷至沓来,使她晕头转向。

19. pandemic [pæn'demɪk]

n. an epidemic that is geographically widespread; occurring throughout a region or even throughout the world 大流行病

From climate change to the ongoing pandemic and beyond, the issues facing today's world are increasingly complex and dynamic.

从气候变化到持续存在的大流行病以及其他问题,当今世界面对的问题越来越复杂多变。

20. exacerbate [ɪɡ'zæsəbeɪt]

v. to make a problem worse 使恶化;使加剧

The Food and Drug Administration has warned that some heart medications may exacerbate it.

食品药品监督管理局提醒,某些心脏药物会加重这种疾病。

Phrases and Expressions

take account of 考虑到;顾及
draw attention to 吸引对……的注意力
for good or ill 无论好坏
in part 部分地;在某种程度上
draw on 利用;吸收
result in 导致;结果是
embark on 从事;着手;登上船
keep in mind 记住

Terminology

ocean basin 海洋盆地(海盆、洋盆)
blue growth 蓝色增长(2012年9月13日欧委会发布了官方文件《蓝色增长——海洋和海洋可持续增长的机会》)
blue degrowth 蓝色去增长(去增长意味着经济收缩而不是增长,致力于使用更少的能源和资源,并将福祉置于利润之上。这一理论的基本设想是,通过推行去增长政策,经济可以变得更可持续,并以此来帮助自身、帮助公民与全地球)

Proper Names

the UN Sustainable Development Goals(SDGs) 联合国可持续发展目标

一、翻译简析

1. People across the world have diverse economic, socio-legal, institutional, social and cultural

relationships with the ocean—both its littoral zones and the open sea spaces through which people have traditionally navigated, migrated, fished, traded, played and sought solace, spiritual enlightenment, adventure, material enrichment, social identity, cultural expression, artistic inspiration or good health. (Para. 2)

世界各地人民与海洋有着不同的经济、社会法律、体制、社会和文化关系，无论是沿海地区还是公海空间，人们历来都是通过海洋进行航行、迁徙、捕鱼、贸易、玩耍，也是通过海洋寻求心灵慰藉、精神启迪、冒险奇遇、物质丰富、社会认同、文化表达、艺术灵感或身心健康。

【这个句子并列结构很多，并列层次关系较复杂，为了标清不同层次的语义关系，可以添加标示性词语，还可以使用不同的标点符号。原文的定语从句中第一层并列是"navigated, migrated, fished, traded, played and sought"这几个动词的并列，第二层是"solace, spiritual enlightenment, adventure, material enrichment, social identity, cultural expression, artistic inspiration or good health"这几个名词或名词短语的并列，它们都是"sought"的宾语。翻译时为了让读者清楚这些层次关系，添加了动词"进行"并将并列的动词都转换成名词，其后使用顿号，另一个动词"寻求"之前增加"也是通过海洋"，它的宾语之后也都使用顿号，这样处理，句子结构一清二楚。】

2. These relationships are reflected in formal and informal institutions (polices, laws, social norms) that regulate many of these activities, including those that regulate access to resources. (Para. 2)

这些关系反映在规范其中许多活动的正式和非正式制度（政策、法律、社会规范）中，包括规范资源获取途径的制度。

【英汉两种语言中都有一词多义和一词多类的现象，选择词义时，首先应根据上下文确定词义，另外还要考虑词语之间的搭配，注意词的褒贬、语气的轻重。本句中的"institutions"有"机构""团体""制度""习俗""建立"等含义，但根据上下文和逻辑关系，这里只能选择"制度"这个含义。】

3. At the time of this study, the world was reeling from the impacts of the COVID-19 pandemic, which, by 16 August 2020, had infected around 21.3 million people and resulted in 761,779 deaths. (Para. 8)

进行本研究时，世界正遭受新型冠状病毒肺炎大流行的重创，截至2020年8月16日，已有约2 130万人感染，761 779人死亡。

【本句中"which"引导的定语从句较长，如果译成前置定语，不符合汉语的表达习惯，所以此处把定语从句译成并列的分句，置于先行词"the COVID-19 pandemic"的后面。】

4. Accordingly, we briefly consider what is known about impacts on the maritime economy and on human-ocean relationships. (Para. 8)

因此，关于疫情对海洋经济和人与海洋关系的影响，我们简要地考虑一些已知情况。

【英语和汉语有不同的语言习惯，英语的逻辑顺序一般是先主要后次要，即先表态后叙事、先果后因等，把信息重心放在句首，而汉语一般按照事物的自然逻辑发展顺序，先叙事后表态、先因后果、先条件后结论，将信息焦点放在句末。所以翻译本句时，把句中的次要信息介词短语"about……"调整到句首,把主要信息句子主干"we briefly consider what is known"调整到句尾,符合汉语逻辑习惯。】

5. Ocean-related climate change impacts are likely to exacerbate existing inequalities within coastal communities, with vulnerable populations being those living in low-lying areas of the tropics, on small oceanic islands and in the Arctic, as well as those whose livelihoods are tied to fisheries affected by global environmental change. (Para. 9)

与海洋相关的气候变化影响可能会加剧沿海社区内部现有的不平等现象,易受影响人群包括生活在热带低洼地区、海洋小岛和北极的人群,还有些易受影响人群,他们的生计与受全球环境变化影响的渔业息息相关。

【本句中包含一个由"whose"引导的限制性定语从句,从句较长,如果翻译成前置定语会使整个句子的译文不通顺、难理解,所以翻译时将定语从句才处理成并列的分句,放在先行词"those"的后面,同时把"as well as those"具体翻译成"还有些易受影响人群"。】

二、参考译文

人类与海洋的关系

1. 人类与海洋的关系多样而复杂。它建立的基础通常是非货币价值观,有助于非物质层面幸福感的产生。这些价值观对促进人类繁荣至关重要,包括促进文化、社会和法律认同,地方感、职业自豪感,自尊、灵性、身心健康以及人类安全。这些价值观和利益的多样性对个人和社会都很重要,可以在高级别海洋政策讨论中得到更有力的体现。

2. 世界各地人民与海洋有着不同的经济、社会法律、体制、社会和文化关系,无论是沿海地区还是公海空间,人们历来都是通过海洋进行航行、迁徙、捕鱼、贸易、玩耍,也是通过海洋寻求心灵慰藉、精神启迪、冒险奇遇、物质丰富、社会认同、文化表达、艺术灵感或身心健康。这些关系反映在规范其中许多活动的正式和非正式制度(政策、法律、社会规范)中,包括规范资源获取途径的制度。这些制度表明沿海和海洋社会对享用海洋资源拥有一系列优先权。

3. 通过考虑沿海和海洋社会使用、享受和管理近岸海域和海洋盆地的各种方式,我们更有能力设计可持续的海洋经济,它公平公正,能够反映"我们想要的未来"。政策制定者应该全面考虑人类与海洋的关系。当代海洋政策和规划中设想的经济投资战略和治理行动可以改变这些关系,从而改变这些价值观的性质和分布。人类从与海洋领域的相互作用中获得了这些价值观。

4. 如何支持人类与海洋的各种关系在未来蓬勃发展,使海洋能够为人类幸福感做出可持续的贡献?这是本研究要回应的首要政策问题。政策研究在评估海洋生产经济商品和服务的能力方面取得了重大进展。这里提出的互补观点旨在让人们关注海洋已经发挥并将继续发挥的更广泛的作用,维持和再现其他人类价值观,如社会和文化认同、个人和集体幸福感、地方感和归属感,以及好奇心、灵性、敬畏和冒险感等人类情感。

5. 通过简要调查过去和现在人类与海洋的各种关系及其对人类幸福感的贡献,并通过研究这些关系的经济和政策影响，我们认为，可持续的海洋经济不仅可以促进经济商品和服务的可持续和公平增长，而且还可以更普遍地促进人类幸福感和繁荣。因此，在下一阶段的人类发展中，海洋可以发挥催化作用，增强人类的能力和自由度，从而为实现联合国可持续发展目标做出贡献。

6. 我们此处的目的不是记录人类与海洋相互作用的每一种方式，无论是好是坏。本系列的其他研究详细分析了我们如何维持和发展海洋食品生产；气候变化如何影响海洋，人类如何应对；我们如何才能更好地处理侵犯人权和海上其他犯罪活动及不公平现象；污染如何威胁海洋，我们如何更好地控制污染；有什么机会可以改善海洋经济的融资和治理；等等。这些问题和解决途径都影响到人与海洋的多种关系，并可能破坏某些关系，加强另一些关系，这在一定程度上取决于它们如何影响现有的与海洋有关的经济不平等现象。我们在这里的观点是，人与海洋关系的相关因素和主观因素尚未在政策领域得到充分阐述，因此在应对这些海洋威胁或抓住海洋经济和保护机会的计划中尚未得到充分考虑。

7. 通过简要概述不同海洋社会中发展起来的代表性社会和法律制度，我们确定了不同社会如何治理海洋空间和海洋体量，以及这些治理机制如何反映不同民族所特有的"海洋价值观"的多样性。我们利用这些对人类与海洋关系多样性的概述，基于历史和文化的海洋权属安排的例子，以及围绕"蓝色经济""蓝色正义""蓝色去增长"的当代政策辩论，确定一系列行动机会，以建立可持续的海洋经济和未来人类与海洋的关系，反映当前和未来海洋公民的世界观和价值观的广度和多样性。这也确认了与海洋关系紧密的人们社会身份的多样性。

8. 进行本研究时，世界正遭受新型冠状病毒肺炎大流行的重创，截至 2020 年 8 月 16 日，已有约 2 130 万人感染，761 779 人死亡。我们认为有必要考虑为减缓病毒传播而采取的公共卫生措施，以及该疾病本身和为遏制该疾病而采取的措施所造成的经济和社会后果，会如何影响人与海洋之间的关系。因此，关于疫情对海洋经济和人与海洋关系的影响，我们简要地考虑一些已知情况。

9. 同样不能忽视的是，应对气候变化的行动至关重要，与此同时，人类正着手进行海洋治理转型。海洋为减少温室气体排放和增加碳捕获与储存提供了许多机会。与海洋相关的气候变化影响可能会加剧沿海社区内部现有的不平等现象，易受影响人群包括生活在热带低洼地区、海洋小岛和北极的人群，还有些易受影响人群，他们的生计与受全球环境变化影响的渔业息息相关。海洋经济的大多数部门将受到气候变化的负面影响，而相互关联的气候和经济过程意味着海洋变化也会对内陆产生影响。需要投资建立适应能力，以适应不同民族的价值观，在考虑如何理解、评估和管理人类与海洋的关系时，应牢记这些投资。

Cultural Background Knowledge: The Ocean and Human Civilization
海洋与人类文明文化背景知识

公元前1600—1450年乃至更早的年代,位于爱琴海的克里特岛上的皮拉斯基人创造了程度较高的古代海洋文明,并使当时的克里特成为海上霸主。它很早就能建造高头低舷的快速远航船,拥有一支强大的海军,曾控制过爱琴海的海上贸易,强迫周围的民族向它称臣缴贡。据记载,古代埃及新王国的法老图特摩斯三世曾租用克里特人的船只从黎巴嫩向埃及运输木材。现代考古工作者在克里特岛发掘出新王国时期埃及的黄金、象牙及工艺品,证明了当时克里特海上商业之发达。

在希腊、罗马奴隶制社会繁荣时期,地中海地区曾在人类文明史上大放异彩。以地中海为中心,古代希腊、罗马的文明曾影响到广大的周边地区,南至撒哈拉、东南至红海地区、东北至黑海地区、西出直布罗陀海峡、北至高卢与英国,时间长达几个世纪,因此,被人们称为"地中海时代"。古代希腊的雅典拥有曲折的海岸线和比利犹斯等良港,为其工商业的发展创造了有利的条件。公元前5世纪至公元前4世纪中叶,雅典的比利犹斯港是古代希腊最大的货物集散地。这里有来自地中海各国的商船、操不同语言的商人;港内有旅馆、剧院、仓库、商品陈列室和银钱兑换所等。通过中介贸易,不仅商业奴隶主获得了巨利,雅典政府也抽得2%的关税。历史学家色诺芬(雅典人、苏格拉底的弟子)在谈到雅典商业时说,雅典的航海贸易最使人向往,它有风平浪静的商港,还有到处可以通用的银币,所以它成了地中海乃至世界上最大的商业中心。

以大西洋为中心的近代资本主义文明是人类近代文明的突出代表,被称为"大西洋时代"。航海技术的进步、新航线的开辟是"大西洋时代"的发端。13世纪,中国人民的伟大发明——指南针,经阿拉伯传入欧洲,到14世纪,欧洲人已普遍使用。15世纪,载重千吨的快速多桅帆船制造出来,航海技术进步很快。同时,随着科学技术的发展和地理知识的进步,地圆学说在欧洲日益流行。这些都为新航线的开辟创造了有利条件。15世纪末到16世纪初,葡萄牙人和西班牙人开辟了从欧洲不经过地中海直达东方的航线和横渡大西洋前往美洲的新航线,完成了第一次环球航行。著名航海家哥伦布(1451—1506)是开辟新航线的主要人物。1492年8月3日拂晓,哥伦布率领88名水手,分乘3艘帆船,从西班牙南端的巴罗斯港出发,经过69天的艰苦航行,于10月12日到达巴哈马群岛中的一个小岛(哥伦布命名为"圣萨尔瓦多")。1493年3月16日,哥伦布返回西班牙。在此之后,哥伦布又3次西航至美洲,先后发现了牙买加、波多黎各、多米尼加等岛屿,并到过中美洲的洪都拉斯、巴拿马以及南美洲大陆北岸,为西班牙的殖民扩张打下了基础。继哥伦布等人的探险之后,葡萄牙人麦哲伦(1470—1521)率领的船队经过3年时间的艰苦航行,成功地完成了人类历史上第一次环球航行,使地圆学说得到了证实。

Unit Two
History of the Ocean

海洋历史

Text A

History of the Ocean
By Anonymous Author

Ideological and Political Education：维护海洋和平安宁 构建海洋命运共同体

海洋覆盖了地球70%以上的面积,是人类的生命之源。海洋如同人类社会一样也拥有悠久而丰富的自然形成过程,它的海沟、海床、海底火山和大洋山脊等主要海洋地形随着千万年来地球所经历的沧海桑田的变迁而渐次形成。通过阅读和学习这篇文章,读者可以"目睹"海洋形成、发展和变化的一幅幅美妙奇绝的图景,更加深入地了解这片人类赖以生存的家园。

2019年4月23日,中国人民解放军海军成立70周年多国海军活动举行,国家主席、中央军委主席习近平在青岛集体会见应邀出席中国人民解放军海军成立70周年多国海军活动的外方代表团团长时提出"海洋命运共同体"的重要理念。习近平主席提出我们人类要像关爱自己的生命一样关爱我们的海洋,在我们共同居住的这个世界上,海洋不是将世界分割成孤立的若干孤岛,而是真正将人类连接在一起的纽带。海洋和人类原本就是休戚与共的共同体。习近平主席的这一重要思想高屋建瓴,从全人类生存和发展的高度为未来海洋的探索、研究和开发擘画了蓝图。

中国自古以来就是一个海洋大国,海洋的和平与安宁直接关系到中国的国家安全和发展福祉。习近平主席提出的构建海洋命运共同体的主张,彰显了中国高举多边主义旗帜,推动各方共护海洋和平、共筑海洋秩序、共促海洋繁荣的负责任大国担当。中国坚决维护和支持《联合国海洋法公约》权威和在全球海洋治理中的作用,促进实现海洋环境共同维护、海上安全共同保护、海上争端和平解决。从亚丁湾、索马里海域护航和人道主义行动,到"和平方舟"号医院船的"和谐使命",中方的务实行动让人类命运共同体和海洋命运共同体理念更加深入人心。在国家治理层面上,中国历来高度重视海洋生态文明建设,持续加强海洋环境污染防治,保护海洋生物多样性,实现海洋资源有序开发利用,为子孙后代留下一片碧海蓝天。

1. Geological processes that occur beneath the waters of the sea affect not only marine life, but dry land as well. The processes that mold ocean basins occur slowly, over tens and hundreds of millions of years. On this timescale, where a human lifetime is but the blink of an eye, solid rocks flow like liquid, entire continents move across the face of the earth and mountains grow from flat plains. To understand the sea floor, we must learn to adopt the unfamiliar point of view of geological time. Geology is very important to marine biology. Habitats, or the places where organisms live, are directly shaped by geological processes. The form of coastlines; the depth of the water; whether the bottom is muddy, sandy, or rocky; and many other features of a marine habitat are determined by this geology. The geologic history of life is also called Paleontology.

2. The presence of large amounts of liquid water makes our planet unique. Most other planets have very little water, and on those that do, the water exists only as **perpetually** frozen ice or as vapor in the atmosphere. The earth, on the other hand, is very much a water planet. The ocean covers most of the globe and plays a crucial role in regulating our climate and atmosphere. Without water, life itself would be impossible.

3. Our ocean covers 72% of the earth's surface. It is not distributed equally with respect to the Equator. About two-thirds of the earth's land area is found in the Northern Hemisphere, which is only 61% ocean. About 80% of the Southern Hemisphere is ocean.

4. The ocean is traditionally **classified** into four large basins. The Pacific is the deepest and largest, almost as large as all the others combined. The Atlantic "Ocean" is a little larger than the Indian "Ocean", but the two are similar in average depth. The Arctic is the smallest and **shallowest**. Connected or **marginal** to the main ocean basins are various shallow seas, such as the Mediterranean Sea, the Gulf of Mexico and the South China Sea.

5. Though we usually treat the oceans as four separate **entities**, they are actually interconnected. This can be seen most easily by looking at a map of the world as seen from the South Pole. From this view it is clear that the Pacific, the Atlantic and Indian oceans are large branches of one vast ocean system. The connections among the major basins allow seawater, materials, and some organisms to move from one "ocean" to another. Because the "oceans" are actually one great interconnected system, **oceanographers** often speak of a single world ocean. They also refer to the continuous body of water that surrounds Antarctic as the Southern Ocean.

6. The earth and the rest of the solar system are thought to have **originated** about 4.5 billion years ago from a cloud or clouds of dust. This dust was debris remaining from a huge **cosmic** explosion called the big bang, which **astrophysicists** estimate occurred about 15 billion years ago. The dust particles **collided** with each other, merging into larger particles. These larger

particles collided in turn, joining into pebble-sized rocks that collided to form larger rocks, and so on. The process continued, eventually runs the earth and other planets.

7. So much heat was produced as the early earth formed that the planet was probably **molten**. This allowed materials to settle within the planet according to their density. Density is the weight, or more correctly, the mass, of a given volume of a substance. Obviously, a pound of **styrofoam** weighs more than an ounce of lead, but most people think of lead as "heavier" than styrofoam. This is because lead weighs more than styrofoam if equal volumes of the two are compared. In other words, lead is denser than styrofoam. The density of a substance is calculated by dividing its mass by its volume. If two substances are mixed, the denser material will tend to sink and the less dense will float.

8. During the time that the young earth was molten, the densest material tended to flow toward the center of the planet, while lighter materials floated toward the surface. The light surface material cooled to make a thin crust. Eventually, the atmosphere and oceans began to form. If the earth had settled into orbit only slightly closer to the sun, the planet would have been so hot that all the water would have **evaporated** into the atmosphere. With an orbit only slightly farther from the sun, all the water would be perpetually frozen. Fortunately for us, our planet orbits the sun in a narrow zone in which liquid water can exist. Without liquid water, there would be no life on earth.

9. The earth is composed of three main layers: the iron-rich core, the semiplastic **mantle** and the thin outer crust. The crust is the most familiar layer of earth. Compared to the deeper layers it is extremely thin, like a rigid skin floating on top of the mantle. The composition and characteristics of the crust differs greatly between the oceans and the continents.

10. The geological distinction between ocean and continents is caused by the physical and chemical differences in the rocks themselves, rather than whether or not the rocks happen to be covered with water. The part of earth covered with water, the ocean, is covered because of the nature of the underlying rock.

11. Oceanic crustal rocks, which make up the sea floor, consist of minerals collectively called **basalt** that have a dark color. Most continental rocks are of general type called **granite**, which has a different mineral composition than basalt and is generally lighter in color. Ocean crust is denser than continental crust, though both are less dense than the underlying mantle. The continents can be thought of as thick blocks of crust "floating" on the mantle, much as icebergs float on water. Oceanic crust floats on the mantle too, but because it is denser it doesn't float as high as continental crust. This is why the continents lie high and dry above sea level and oceanic crust lies below sea level and is covered by water. Oceanic crust and

continental **crust** also differ in geological age. The oldest oceanic crust is less 200 million years old, quite young by geological standards. Continental rocks, on the other hand, can be very old, as old as 3.8 billion years!

12. The East Pacific Ocean Rise. In the years after World War II, sonar allowed the first detailed surveys of large areas of the sea floor. These surveys resulted in the discovery of the mid-oceanic **ridge** system, a 40,000 mile continuous chain of volcanic submarine mountains and valleys that encircle the globe like the seams of a baseball. The mid-oceanic ridge system is the largest geological feature on the planet. At regular intervals the mid-ocean ridge is displaced to one side or the other by cracks in the earth's crust known as transform faults. Occasionally the submarine mountains of the ridge rise so high that they break the surface to form islands, such as Iceland and the Azores.

13. The portion of the mid-ocean ridge in the Atlantic, known as the Mid-Atlantic Ridge, runs right down the center of the Atlantic Ocean, closely following the curves of the opposing coastlines. The ridge forms an inverted Y in the Indian Ocean and runs up the eastern side of the Pacific. The main section of ridge in the eastern Pacific is called the East Pacific Rise. Surveys also revealed the existence of a system of deep depressions in the sea floor called trenches. Trenches are especially common in the Pacific.

14. When the mid-ocean ridge system and **trenches** were discovered, geologists wanted to know how they were formed and began intensively studying them. They found that there's a great deal of geological activity around these features. Earthquakes are **clustered** at the ridges, for example, and volcanoes are especially common near trenches. The characteristics of sea floor rocks are also related to the mid-oceanic ridges. Beginning in 1968, a deep-sea drilling ship, the Glomar Challenger, obtained samples of the actual sea floor rock. It was found that the farther rocks are from the ridge **crest** the older they are. One of the most important findings came from the studying the **magnetism** of rocks on the sea floor. Bands of rock alternating between normal and reversed magnetism parallel the ridge.

15. It was the discovery of the magnetic anomalies on the sea floor, together with other evidence, that finally led to an understanding of plate **tectonics**. The earth surface is broken up into a number of plates. These plates, composed of the crust and the top parts of the mantle, make up the lithosphere. The plates are about 100 km thick. As new **lithosphere** is created, old lithosphere is destroyed somewhere else. Otherwise, the earth would have to constantly expand to make room for the new lithosphere. Lithosphere is destroyed at the trenches. A trench is formed when two plates collide and one plate dips below the other and slides back down in to the mantle. This downwards movement of the plate into the mantle is called **subduction**. Because subduction occurs at the trenches, trenches are often called subduction

zones. Subduction is the process that produces earthquakes and volcanoes, also underwater. The volcanoes may rise from the sea floor to create chains of volcanic islands.

16. We now realize that the earth's surface has undergone dramatic alterations. The continents have been carried long distances by the moving sea floor, and the ocean basins have changed in size and shape. In fact, new oceans have been born. Knowledge of the process of plate tectonics has allowed scientists to reconstruct much of the history of these changes. Scientists have discovered, for example, that the continents were once united in a single supercontinent called Pangaea that began to break up about 180 million years ago. The continents have since moved to their present position.

17. The characteristics of seawater are due both to the nature of pure water and to the materials dissolved in it. The solids **dissolved** in seawater come from two main sources. Some are produced by the chemical weathering of rocks on land and are carried to sea by rivers. Other materials come from the earth's interior. Most of these are released into the ocean at **hydrothermal** vents. Some are released into the atmosphere from volcanoes and enter the ocean in rain and snow. Seawater contains at least a little of almost everything, but most of the solutes or dissolved materials, are made up of a surprisingly small group of ions. In fact, only six ions compose over 98% of the solids in seawater. **Sodium** and **chloride** account for about 85% of the solids, which is why seawater tastes like table salt. The salinity of the water strongly affects the organisms that live in it. Most marine organisms, for instance, will die in fresh water. Even slight changes in **salinity** will harm some organisms.

(1,828 words)

https://www.marinebio.org/oceans/history/

New Words

1. perpetually [pəˈpetʃuəli]
   ad. seemingly uninterrupted 永久地；恒久地
   He was perpetually involving himself in this long lawsuit.
   他和这些旷日持久的官司总是纠缠不清。

2. classify [ˈklæsɪfaɪ]
   v. to arrange sth. in groups according to features that they have in common 将……分类；将……归类
   Rather than relying on how others classify you, consider how you identify yourself.
   与其等着别人把你分门别类，倒不如自己考虑如何定义自己。

3. shallow [ˈʃæləʊ]
   a. not having much distance between the top or surface and the bottom 浅的
   These fish are found in shallow waters around the coast.
   这些鱼生长在海边浅水水域。

4. marginal [ˈmɑːdʒɪn(ə)l]

   a. not part of a main or important group or situation 非主体的；边缘的

   For other more marginal languages some measures should be taken.

   对于其他更边缘的语言，应该采取一些措施。

5. entity [ˈentəti]

   n. (formal) something that exists separately from other things and has its own identity 独立存在物；实体

   These countries can no longer be viewed as a single entity.

   这些国家不能再被看成一个单独的实体。

6. oceanographer [ˌəʊʃəˈnɒɡrəfə(r)]

   n. the scholar who makes the scientific research of ocean 海洋学家

   According to research of oceanographers, marine resources are much more than the land.

   根据海洋学家的研究，海洋的资源比陆地多得多。

7. originate [əˈrɪdʒɪneɪt]

   v. to happen or appear for the first time in a particular place or situation 起源；发端于

   The disease is thought to have originated in the tropics.

   这种疾病据说起源于热带地区。

8. cosmic [ˈkɒzmɪk]

   a. connected with the whole universe 宇宙的

   Do you believe in a cosmic plan?

   你相信冥冥中的安排吗？

9. astrophysicist [ˌæstrəʊˈfɪzɪsɪst]

   n. someone who studies astrophysics 天体物理学家

   When I grew up, I became a professional astrophysicist.

   长大之后，我成了一名职业的天体物理学家。

10. collide [kəˈlaɪd]

    v. if two people, vehicles, etc. collide, they crash into each other; if a person, vehicle, etc. collides with another, or with sth. that is not moving, they crash into it 碰撞；相撞

    The car and the van collided head-on in thick fog.

    那辆小轿车和货车在浓雾中迎面相撞。

11. molten [ˈməʊltən]

    a. heated to a very high temperature so that it becomes liquid 熔化的；熔融的

    A nearby volcano erupted violently, sending out a hail of molten rock and boiling mud.

    一座附近的火山猛烈爆发，喷出大量熔岩和沸腾的泥浆。

12. styrofoam [ˈstaɪrəˌfəʊm]

    n. it is a very light, plastic substance, used especially to make containers 泡沫聚苯乙烯

    It's so light weight and styrofoam is 100 times heavier.

    它是这么轻，泡沫聚苯乙烯塑料比它重100倍。

13. lava [ˈlɑːvə]

n. hot liquid rock that comes out of a volcano(火山喷出的)熔岩,岩浆

The lava will just ooze gently out of the crater.

熔岩只会缓缓地从火山口流出来。

14. evaporate [ɪˈvæpəreɪt]

    v. liquid's changing into a gas, especially steam (使)蒸发,挥发

    Crustal movements closed the straits, and the landlocked Mediterranean began to evaporate.

    地壳运动封闭了海峡,内陆的地中海开始蒸发。

15. mantle [ˈmæntəl]

    n. the part of the earth below the crust and surrounding the core 地幔

    More and more of this plate, the ocean floor, would go down under the continent into the mantle.

    越来越多的这种板块,即海床,将在大陆下方进入地幔。

16. crust [krʌst]

    n. a hard layer or surface, especially above or around sth. soft or liquid (尤指软物或液体上面、周围的)硬层,硬表面

    The most ancient parts of the continental crust are 4,000 million years old.

    该大陆地壳最古老的那些部分有40亿年了。

17. basalt [ˈbæsɔːlt]

    n. a type of dark rock that comes from volcanoes 玄武岩

    Basalt is the commonest volcanic rock.

    玄武岩是最普通的火山岩。

18. granite [ˈgrænɪt]

    n. a type of hard grey stone, often used in building 花岗岩;花岗石

    The towers are made of steel cased in granite.

    这些塔楼是钢结构的,花岗岩贴面。

19. ridge [rɪdʒ]

    n. a narrow area of high land along the top of a line of hills; a high pointed area near the top of a mountain 山脊;山脉

    He slowed the pace as they crested the ridge.

    当他们到达山脊时,他放慢了步伐。

20. trench [trentʃ]

    n. a long deep narrow hole in the ocean floor 海沟;大洋沟

    A diver explores a continental trench near Iceland in 2010.

    2020年一位潜水者正在探索冰岛附近的大陆海沟。

21. cluster [ˈklʌstə(r)]

    v. to come together in a small group or groups 群聚;聚集

    The children clustered together in the corner of the room.

    孩子们聚集在房间的角落里。

22. crest [krest]

　　n. the top part of a hill or wave 山顶；顶峰；波峰；浪尖

　　She neared the crest of a hillock and was soon out of sight.

　　她走近一座小山丘的顶部，很快就不见了踪影。

23. magnetism ['mægnətɪzəm]

　　n. a physical property of some metals such as iron, produced by electric currents, that causes forces between objects, either pulling them towards each other or pushing them apart 磁性；磁力

　　Other areas under investigation include magnetism, landmarks, coastlines, sonar, and even smells.

　　其他正在调查的领域包括磁性、地标、海岸线、声呐甚至气味。

24. anomaly [ə'nɒməli]

　　n. a thing, situation, etc. that is different from what is normal or expected 异常事物；反常现象

　　The January warmth turned out to be part of a remarkably persistent weather anomaly.

　　天气状况的持续的显著反常表现之一就是一月时节的温暖如春。

25. tectonics [tek'tɒnɪks]

　　n. geological structural features as a whole 构造

　　With the ocean lost, Venus could not kick-start plate tectonics.

　　随着海洋的消失，金星可能再也不能进行板块重组了。

26. lithosphere ['lɪθəsfɪə(r)]

　　n. the layer of rock that forms the outer part of the earth 岩石圈；岩石层

　　The hydrosphere and the lithosphere together form the Earth's surface.

　　水圈与岩石圈一起形成地球的表面。

27. subduction [səb'dʌkʃən]

　　n. the act of descending 俯冲，向下冲

　　Basically, a subduction zone is where two tectonic plates collide.

　　俯冲带基本上就是两个地壳构造板块碰撞的地方。

28. dissolve [dɪ'zɒlv]

　　v. to mix with a liquid and become part of it 溶解

　　The substance does not dissolve in water whether heated or not.

　　无论加热与否，这种物质都不溶于水。

29. hydrothermal [ˌhaɪdrəʊ'θɜːməl]

　　a. of or relating to the action of water under conditions of high temperature, esp. in forming rocks and minerals 高温水作用的

　　It all began in 1977 with the exploration of hydrothermal vents on the ocean floor.

　　一切始于1977年的一次海底深海热泉的探索。

30. sodium ['səʊdiəm]

　　n. a chemical element 钠

Common salt is a compound of sodium and chlorine.
普通食盐是钠和氯的化合物。

31. chloride [ˈklɔːraɪd]

    n. a chemical compound of chlorine and another substance 氯化物

    The fish or seafood is heavily salted with pure sodium chloride.
    鱼或海产品以厚厚的纯氯化钠盐腌制。

32. salinity [səˈlɪnəti]

    n. the relative proportion of salt in a solution 盐度，盐分，盐性

    There are three basic processes that cause a change in oceanic salinity.
    导致海洋盐度变化有三个基本过程。

Phrases and Expressions

with respect to 相对于……
speak of 谈到，论及
build up 形成，建成
be composed of 由……组成
consist of 由……组成
run up 向上爬升
break up into 分解成，分割成
be due to 因为，由于
account for 占……比例

Terminology

paleontology 古生物学
ocean basin 洋盆
the iron-rich core 富含铁质的地核
the semiplastic mantle 半塑性地幔
the outer crust 外壳
transform faults 转换断层
subduction zones 俯冲带，隐没带

Proper Names

Northern Hemisphere 北半球
South China Sea 南中国海
South Pole 南极点
Mid-Atlantic Ridge 大西洋中脊
East Pacific Rise 东太平洋海隆
Glomar Challenger "格罗玛挑战者"号

Pangaea 泛大陆,盘古大陆

一、翻译策略/方法/技巧:省译法

省译法是一种非常重要且十分常见的翻译技巧。出于译文语法结构、表达习惯、译文文化等的需要,将原文中需要而译文中不需要的词在翻译过程中加以省略,或者在译文中用简洁明了的语言代替原文中烦琐累赘的语言,从而完整准确、通顺流畅、言简意赅地翻译出原文的内容。省译法是与增译法相对应的一种翻译方法。增译法的使用原则是"增词不增意",省译法的使用原则是"省词不省意"。以下是省译法在实践应用中常见的两个角度。

1. 语法角度的省译

语法角度的省译主要是针对英汉两种语言在句法结构、遣词造句上的不同,根据上下文的语境和译文的表达习惯,省略不必要的冠词、代词、介词、连词、动词、范畴词等。如:

(1) An elephant is much superior to a giraffe in strength.

大象比长颈鹿有力量。

英语中有冠词,包括定冠词和不定冠词。而汉语中没有冠词。因此英译汉时,需要根据中文的表达习惯将定冠词 the 和不定冠词 a 或者 an 省略。

(2) It is patently obvious that he is lying.

他显然在撒谎。

英译汉时,常常省略英语中表示泛指的人称代词、用作定语的物主代词、起强调作用的代词 it 和反身代词等。

(3) In September 1939, the Second World War broke out.

1939 年 9 月,第二次世界大战全面爆发。

英语中常用介词表示时间、地点、方位等,而汉语常通过语序和逻辑关系来表示。因此,英译汉时常省略 on、in、at 等介词。

(4) If winter comes, can spring be far behind?

冬天来了,春天还会远吗?

英语重形合,常用 if、and、as 等连词表示上下文逻辑关系,而汉语重意合,结构比较松散,较少使用连词。故英译汉时,常将连词省略掉。

(5) 继续加强基础设施和基础工业,大力振兴支柱产业。

Continuous efforts will be dedicated to strengthening infrastructure, basic and pillar industries.

汉语表达呈动态,一句话中可以出现多个动词,构成"连动式"或"兼语式"结构。而英语的表达呈静态,在一个句子中往往只使用一个实义动词来表达最重要的动作含义,其他动作含义则基本借用含有动作意义的动词不定式、分词、名词等来表达,甚至可以略去。原文中"加强"和"振兴"意义相近,省略后者不译,更符合英语表达。

(6) 中国足球的落后状态必须改变。

The backwardness of the Chinese football must be changed.

范畴词是指用来表示行为、现象、属性等概念所属范畴的词,是汉语常用的特指手段,而英语则少用。因此,译文中省译了"状态"一词。

2. 修辞角度的省译

修辞角度的省译包括汉译英时重叠词的省译如"高高兴兴 happy"、比喻喻体的省译如"鸦雀无声 utterly quiet"、特定文化意象词的省译如"这些人不讲道德,简直猪狗不如。These people are immoral."、成语典故中人名或地名的省译如"南柯一梦 be a dream"。英译汉时,可以根据情况对句子中一些重复出现的、汉语表达习惯里可有可无的词语做适当的省略。如:

(1) The applicants who had work experience would be preferable in getting the position (over those who had not).

有工作经验的人优先录用。

省译"over those who had not",使译文简洁凝练。

(2) He used poetry (as a medium) for writing in prose.

他用诗歌来写散文。

省译"as a medium"作为媒介,避免译文重复啰唆。

二、译例与练习

Translation

1. On this timescale, where a human lifetime is but the blink of an eye, solid rocks flow like liquid, entire continents move across the face of the earth and mountains grow from flat plains.(Para.1)

在这段如此悠长的岁月里,人的生命仅是转瞬,磐石也碎烂成泥,如水一般流淌,整块整块的大陆在地球表面移动,座座高山也从片片平原间拔地而起。

【这个句子的主干是由三个简单句构成的,结构上并不复杂。此句翻译上的难点主要在选词上。一是 the blink of an eye,flow like liquid 和 grow from flat plains 的选词。若是照搬三个表达的词典释义,译文会显得极为生硬呆板,不通顺。因此,必须在原意基础上进行适当引申,同时考虑到三者的并列关系,选词上还要兼顾到译文音韵上的协调,可以考虑使用四字结构,尤其是成语。第一个表达提示我们可以使用"转瞬即逝"类似的表达,考虑原文中有个 but(仅仅),为了搭配顺畅,可以在"转瞬即逝"的基础上改译成"仅是转瞬"。第二个表达中的 solid rock 让我们联想到中文中"海枯石烂"的成语,此处只需要取"石烂"。一般来看,石头再烂,也不会成水,原文中仅是一种明喻而已。译文中可将"水"置换成"泥",这样比较符合常识,也能同"烂"搭配起来。第三个表达中的 grow from 若使用词典中的"增长""成长"的话,显然不通顺。这里可以根据语境做适当引申。前后文中的 mountain 和 plain,结合该文讲述的是地球经历的造山运动,可以想到使用成语"拔地而起"一词,这样可以比较充分地展现出造山运动的剧烈。】

2. This dust was debris remaining from a huge cosmic explosion called the big bang, which astrophysicists estimate occurred about 13.8 billion years ago.(Para.6)

这片尘埃实际上是一次巨大的宇宙爆炸(也叫大爆炸)的残骸,据天体物理学家们估测这次大爆炸大约发生在 138 亿年前。

【这个句子结构稍显复杂,其主体部分是 This dust was debris,其从属结构分别有 remaining 引导的现在分词短语修饰 debris,called 引导的过去分词短语修饰 a huge cosmic explosion，再就是 which 引导的定语从句对 the big bang 进行补充说明。明确了各个部分之间的逻辑关系，下一步就是以此逻辑关系为依据，按照汉语的表达习惯，重新组合成译文。这也是此类句子英译汉时的一般方法。需要注意的是，分词短语如果内容相对简单可以置于被修饰词或短语的汉译之前，此句的译文就是这么处理的，若过于繁杂，则可另起一句进行表述。对于定语从句的汉译，一般需运用分句法，处理成一个单独的句子。】

3. During the time that the young earth was molten, the densest material tended to flow toward the center of the planet, while lighter materials floated toward the surface. (Para. 8)

新土熔化期间，密致物质开始向地心流淌，与此同时那些轻质物质则向地表飘升。

4. If the earth had settled into orbit only slightly closer to the sun, the planet would have been so hot that all the water would have evaporated into the atmosphere. (Para. 8)

如果地球离太阳稍近，那么地球就会无比炽热，炽热到所有的水都会蒸发不见。

5. The geological distinction between ocean and continents is caused by the physical and chemical differences in the rocks themselves, rather than whether or not the rocks happen to be covered with water. (Para. 10)

海洋和陆地间地理学上的差异在于岩石的物理和化学构成，而非岩石是否浸于水中。

【这个句子的主要信息实际上集中在一个并列结构上，该结构表达的是取舍含义，以 rather than 连接。这个并列结构稍显复杂的地方在于后项是一个以 whether 引导的宾语从句。翻译上另一个难点在于 is caused by 的处理上。有些人习惯按照原文将其译成"由……引起"。这样处理的话，英文在通顺度上很差，因为中文使用被动式的频率远无英文那样高，只有在一些比较极端的状况下才使用。因此，这里需要通过"转换法"将原文中的被动语态转换成主动语态，同时还需将动词 cause 转换成名词"原因"。如此处理，整个译文既通顺又达意，达到了较为理想的翻译效果。】

6. Oceanic crustal rocks, which make up the sea floor, consist of minerals collectively called basalt that have a dark color. (Para. 11)

海床由大洋地壳岩构成，这种岩石乃是由一种统称为玄武岩的矿物质组成的，这种岩石通体为黑色。

【这个句子在结构上稍显复杂的地方在于句末的过去分词结构。跟一般的所不同的是，该结构中还含有一个 that 引导的定语从句，与英文长句中的"套句"十分类似。翻译中处理这样结构的一般方法是"分译法"。也就是将结构复杂的"套句"按照自身的逻辑关系拆分成若干部分，之后按照译语表达习惯进行重新组合。例句中的 that 定语从句修饰的 basalt。鉴于 basalt 前面修饰成分较多，尽管该定语从句并不长，但是为了避免译文表达的累赘琐碎，可以将其彻底拆分为一个单独的句子。】

7. This is why the continents lie high and dry above sea level and oceanic crust lies below sea level and is covered by water. (Para. 11)

这可用来诠释为何大陆高于海平面且干燥少水，而大洋地壳则低于海平面且浸入水中。

8. These surveys resulted in the discovery of the mid oceanic ridge system, a 40,000 mile continuous chain of volcanic submarine mountains and valleys that encircle the globe like the seams of a baseball. (Para. 12)

这些调查发现了一系洋中山脊。山脊长 40 000 英里,连绵不断,乃海底火山和峡谷,它如同一条锁链环绕着地球,就像一个棒球上的线缝。

【这个句子在翻译上的难点主要在 mountains and valleys 前后诸多修饰语的处理上。这个名词短语语法上看是 the mid-oceanic ridge system 的同位语,用来详尽描述"洋中山脊"的各个特性,包括其长度、形状、分布的状况。翻译其长度时,可将原文直接对译成一个中文名词词组——"长 40 000 英里";翻译其形状时,考虑到下文在描述其分布状况时,作者用了一个明喻,将其喻成棒球上的线缝,为了让该比喻在近距对照中更加形象生动,势必要述及洋中山脊的形状,因此可将原文中 continuous chain of 挪到此处进行翻译。这样处理既可以更加直接地凸显明喻的形象性,还可以避免语言上的赘述,两全其美。】

9. The portion of the mid-ocean ridge in the Atlantic, known as the Mid-Atlantic Ridge, runs right down the center of the Atlantic Ocean, closely following the curves of the opposing coastlines. (Para. 13)

大西洋底的洋中山脊,也就是为人熟知的大西洋中脊,不偏不斜一直延伸至大西洋的中心,与对面曲折的海岸线紧紧相随。

10. It was the discovery of the magnetic anomalies on the sea floor, together with other evidence, that finally led to an understanding of plate tectonics. (Para. 15)

正是海床磁场异常现象以及其他证据的发现才最终让人类对板块构造有了些许认识。

Exercises

1. Fill in the blanks with the proper given words, and then translate the sentences into Chinese.

  perpetual marginal with respect to originate consist of entity

1) Petroleum, _____ crude oil and natural gas, seems to derive from organic matter in marine sediment.

2) The biographer has to make a dance between two shaky positions _____ the subject.

3) They operate by eroding your trust in your own intellect, gradually convincing you to put your trust into some external _____.

4) How does it compare with the other, seemingly _____ health scares we confront, like panic over lead in synthetic athletic fields?

5) _____ from the ice sheets, icebergs form on land from millions of years of snowfall.

6) The tribunals were established for the well-integrated members of society and not for _____ individuals.

2. Translate the following sentences into Chinese.

1) The majority of pollutants that make their way into the ocean come from human activities along the coastlines and far inland.

2) The circulation of deep water is mostly controlled by water flowing from the Atlantic Ocean, the Red Sea, and Antarctic currents.

3) Most variation in the ocean floor consists of steep-sided, flat-topped submarine peaks called seamounts.

4) In addition to affecting the weather, the Kuroshio also likely affects the climate, although its impact on thousand- and million-year time scales is still unclear.

5) Deep sea creatures have evolved some fascinating feeding mechanisms because food is scarce in these zones.

6) Ironically, the material that has caused so much environmental concern now originally started as a tool to preserve the natural world from depletion and overexploitation.

7) Changes in ocean color are primarily due to changes in the type and concentration of organisms suspended in the water.

8) Evaporation and precipitation determine the salinity of the ocean in any given region and both of these processes depend on energy from the sun.

9) Considering the volume of water above the deepest parts of the ocean, it's no wonder that hydrostatic pressure is one of the most important environmental factors affecting deep sea life.

10) This is because cold water can dissolve more oxygen than warm water, and the deepest waters generally originate from shallow polar seas.

3. Translate the following sentences into English.
1) 深海具备如下特点:高压、无光、水温低、盐度高、氧含量高、沉积物多。
2) 海洋新知中传递的第一重信息是关于海洋世界的无休止变化。
3) 这次海底火山喷发并不是单一事件造成的,而是地下能量的持续聚集的结果。
4) 当今人类已探索的海底只有5%,还有95%大海的海底是未知的。
5) 各个地质时代都发生过多次大规模的地壳变动,形成了许多海岸山脉。
6) 虽然时至今日深潜技术仍然十分不尽如人意,但已经能够瞥见海洋最深处的神秘风景了。
7) 当人们确定海床深度后,便想看看海床是怎样的一番景象。
8) 绝大部分海底火山位于构造板块运动的附近区域,被称为中洋脊。
9) 太平洋,南北最长约15 900 km,东西最宽约19 000 km,总面积为18 134.4万 km^2。
10) 大洋洋脊在大西洋位置居中,走向与大西洋东西两岸大体平行,呈S形展布。

4. Choose the best paragraph translation. And then answer why you choose the first translation or the second one.
　　印度洋在我国古代被称为"西洋"。在古希腊时期,著名地理学家、历史学家希罗多德曾称之为"厄立特里亚海",意为"红海"。到古罗马时期,印度洋被罗马人称为"鲁都姆海",但这个名字只不过是希腊语"厄立特里亚"的意译,也是"红海"的意思。直到15世纪末,葡萄牙著名航海家达·伽马为了寻找通往印度的航线,绕过非洲南端的好望角进入这个大洋后,才开始使用印度洋这个名称。这个名称逐渐为人们所接受,便成为通用的名称。

译文一：

Indian Ocean has been named by "Xi Hai" in ancient China, by "Erythraean Sea" referring to "Red Sea" by Herodotus, a well-known geologist and historian in ancient Greek period, and by "Rudum Sea" by ancient Romans, which just served to be literal translation of "Erythraean Sea" in Greek and had the identical meaning with "Red Sea". Not until the end of 15th century did Da Gama, the well-known Portuguese navigator, start to put it to use in order to seek for the navigating route heading for India after he bypassed the Cape of Good Hope at Africa and steered into the ocean. Since then this name has been widely accepted as the universal title.

译文二：

Indian Ocean has been named by "Xi Hai" in ancient China. It was named by "Erythraean Sea" by Herodotus, a well-known geologist and historian in ancient Greek period, and referred to "Red Sea". It was named by "Rudum Sea" by ancient Romans, and served to be literal translation of "Erythraean Sea" in Greek and had the identical meaning with "Red Sea". Till the end of 15th century, Da Gama, the well-known Portuguese navigator, started to put it to use in order to seek for the navigating route heading for India after he bypassed the Cape of Good Hope at Africa and steered into the ocean. Since then this name has been widely accepted as the universal title.

5．Translate the following passage into Chinese.

The characteristics of seawater are due both to the nature of pure water and to the materials dissolved in it. The solids dissolved in seawater come from two main sources. Some are produced by the chemical weathering of rocks on land and are carried to sea by rivers. Other materials come from the earth's interior. Most of these are released into the ocean at hydrothermal vents. Some are released into the atmosphere from volcanoes and enter the ocean in rain and snow. Seawater contains at least a little of almost everything, but most of the solutes or dissolved materials, are made up of a surprisingly small group of ions. In fact, only six ions compose over 98％ of the solids in seawater. Sodium and chloride account for about 85％ of the solids, which is why seawater tastes like table salt. The salinity of the water strongly affects the organisms that live in it. Most marine organisms, for instance, will die in fresh water. Even slight changes in salinity will harm some organisms.

6．Translate the following passage into English.

原始海洋就其规模而言,远没有现代海洋这么大。据估算,它的水量大约只有现代海洋的10％。后来,贮藏在地球内部的结构水的加入,才使其逐渐壮大,形成了蔚为壮观的现代海洋。原始海洋中的水不像现代海水那样又苦又咸。现代海洋海水中的无机盐,主要是通过自然界周而复始的水循环,由陆地带入海洋而逐年增加的。可是,原始海洋中的有机大分子要比海洋中的丰富得多。原始大气化学演化过程中所形成的有机分子都随着雨水冲进了原始海洋,并迅速地下沉到原始海洋的中层,从而避免了因原始大气缺乏臭氧层而造成的紫外线伤害。

三、翻译家论翻译

刘宓庆论翻译中的"形合"与"意合"——"所谓'合',就是组合(syntagma)。组合是语言符号由'散'(个体的词)到'集'(词组,即语符列)的基本表现手段,也就是线性组织手段。形合和意合两种表现手段通常都是并存于一种语言中,我们称之为'相对性'(relativity)。形合与意合虽然并存于一种语言中,但二者的作用绝不是等量齐观的。很多语言都表现出各有侧重。在汉英双语中,汉语重意合,英语重形合。英语之所以重形合,是因为英语具有比较多样的形式组合手段(syntagmatic devices)。形合的优势是视觉分辨率(visual differentiation)高,语法关系属于显性表现式。汉语有'意定于笔'的传统。更早以前庄周提出了'舍象求意'(《庄子·物外》);'象'就是语言的外在表象,认为最重要的是内在的'意',力求做到'意余象外'。因此汉语的意合不仅有语言结构上的原因,还有哲学美学上的渊源。意合的优点是文句也有意序相连,形成音乐的'意脉',使'意'成为连接、贯通、流迁、缀合的逻辑线索,而不受'形'的框驭。在汉语意合的地方,英语总是要以各种连接手段将它形合化,不允许留下结构上的'脱漏'"。试看以下译例:

例1:You try to persuade him now, I talked to him all last night, therefore I've done my part.

译文:你去说服他吧,我昨晚和他谈了一夜,尽力了。

原文可以看出,第一句、第二句和第三句之间存在着明显的因果关系。前两句是在陈述原因,最后一句是在说明结果或结论。对于英语这种重形合的语言来说,一般要在外在语言形式上用某些连词将这种因果关系表现出来,因此原文中使用了表示因果关系的连词"therefore"。当转译成重意合的汉语时,这种因果关系是否用表示因果关系的中文连词来表现就在两可之间。所以,在译文中,第二句和第三句之间就没有使用"因此"这样的连词。当然,如果一定要使用,整个句子也是通顺的。但是对于英语来说,若没有"therefore"的话,整个句子是不恰当的。

例2:往者不可谏,来者犹可追。(《论语·微子》)

译文:Although the past cannot be recovered, the future still can be pursued.

这句《论语·微子》中的原文意思是"过去的事情已然无法挽回,但是将来的事情依然可以追求"。相较于现代汉语,古代汉语重意合的倾向更加明显。某种程度上说,意合是古代汉语表达观念中的主流。这两句古文之间虽然存在着转折逻辑关联,但是一般是不会将"虽然……但是……"写出来的,而是暗含在句子背后。但是若将其翻译成英语的话,这种暗含的转折的逻辑关联就必须用"although"或"but"表现在译文中,否则这句英语是不符合语法规范的。

(摘自:刘宓庆.新编汉英对比与翻译[M].北京:中国对外翻译出版公司,2010:403-404,407.)

Text B

A History of the Study of Marine Biology

By Anonymous Author

Ideological and Political Education：推进海洋研究探索 建设"海洋强国"

浩瀚的海洋是人类赖以生存和发展的家园。对于这片广袤的天地,人类自诞生之日起就一直怀着强烈的好奇之心。海洋探索和研究几乎涵盖了人类文明整个形成和发展的历史。在数千年的历史流变中,人类社会在海洋探索和研究方面也涌现出了一大批重要代表人物和成果,这些都为人类未来更为科学合理地研究和探索海洋指明了发展方向,也更加凸显了研究和探索海洋与人类命运的息息相关。在中共中央政治局第八次集体学习时,习近平总书记强调:"21世纪,人类进入了大规模开发利用海洋的时期。海洋在国家经济发展格局和对外开放中的作用更加重要,在维护国家主权、安全、发展利益中的地位更加突出,在国家生态文明建设中的角色更加显著,在国际政治、经济、军事、科技竞争中的战略地位也明显上升。"习近平总书记的这一重要思想高屋建瓴,从国家发展和民族振兴的高度为中国海洋的探索、研究和开发指明了方向。

我国是一个陆海兼备的发展中大国,建设海洋强国是全面建设社会主义现代化强国的重要组成部分。党的十八大做出了建设海洋强国的重大部署。实施这一重大部署,对推动经济持续健康发展,对维护国家主权、安全、发展利益,对实现全面建成小康社会目标、进而实现中华民族伟大复兴都具有重大而深远的意义。建设海洋强国是实现中华民族伟大复兴的重大战略任务。要推动海洋科技实现高水平自立自强,加强原创性、引领性科技攻关,把装备制造牢牢抓在自己手里,努力用我们自己的装备开发油气资源,提高能源自给率,保障国家能源安全。进入21世纪,海洋在国家经济发展格局和对外开放中的作用更加重要,在维护国家主权、安全、发展利益中的地位更加突出,在国家生态文明建设中的角色更加显著,在国际政治、经济、军事、科技竞争中的战略地位也明显上升。经过多年发展,我国海洋事业总体上进入了历史上最好的发展时期,海洋作为高质量发展战略要地的地位日益凸显。在实现第二个百年奋斗目标的新征程上,我们必须进一步关心海洋、认识海洋、经略海洋,协同推进海洋生态保护、海洋经济发展和海洋权益维护,推动我国海洋强国建设不断取得新成就。

1. It wasn't until the writings of Aristotle from 384—322 BC that specific references to marine life were recorded. Aristotle identified a variety of species including **crustaceans, echinoderms, mollusks**, and fish. He also recognized that cetaceans are mammals, and that marine **vertebrates** are either **oviparous** (producing eggs that hatch outside the body) or **viviparous** (producing eggs that hatch within the body). Because he is the first to record observations on marine life, Aristotle is often referred to as the father of marine biology.

2. The Early Expeditions. The modern day study of marine biology began with the exploration by Captain James Cook (1728—1779) in 18th century Britain. Captain Cook is most known for his extensive voyages of discovery for the British Navy, mapping much of the world's uncharted waters during that time. He **circumnavigated** the world twice during his lifetime, during which he logged descriptions of numerous plants and animals then unknown to most of mankind. Following Cook's explorations, a number of scientists began a closer study of marine life including Charles Darwin (1809—1882) who, although he is best known for the Theory of Evolution, contributed significantly to the early study of marine biology. His expeditions as the resident naturalist aboard the HMS Beagle from 1831 to 1836 were spent collecting and studying specimens from a number of marine organisms that were sent to the British Museum for cataloging. His interest in geology gave rise to his study of coral reefs and their formation. His experience on the HMS Beagle helped Darwin formulate his theories of natural selection and evolution based on the similarities he found in species specimens and fossils he discovered in the same geographic region.

3. The voyages of the HMS Beagle were followed by a 3-year voyage by the British ship HMS Challenger led by Sir Charles Wyville Thomson (1830—1882) to all the oceans of the world during which thousands of marine specimens were collected and analyzed. This voyage is often referred to as the birth of oceanography. The data collected during this trip filled 50 volumes and served as the basis for the study of marine biology across many disciplines for many years. Deep sea exploration was a **benchmark** of the Challenger's voyage disproving British explorer Edward Forbes' theory that marine life could not exist below about 550 m or 1,800 feet.

4. The *Challenger* was well equipped to explore deeper than previous expeditions with laboratories aboard stocked with tools and materials, microscopes, chemistry supplies, trawls and dredges, thermometers, devices to collect specimens from the deep sea, and miles of rope and **hemp** used to reach the ocean depths. The end product of the *Challenger's* voyage was almost 30,000 pages of oceanographic information compiled by a number of scientists from a wide range of disciplines.

5. The Institutions. These expeditions were soon followed by marine laboratories established to study marine life. The oldest marine station in the world, Station Biologique de Roscoff was

established in Concarneau, France founded by the College of France in 1859. Concarneau is located on the northwest coast of France. The station was originally established for the cultivation of marine species, such as Dover sole, because of its location near marine **estuaries** with a variety of marine life. Today, research is conducted on molecular biology, biochemistry, and environmental studies.

6. In 1871, Spencer Fullerton Baird, the first director of the US Commission of Fish and Fisheries (now known as the National Marine Fisheries Service), began a collection station in Woods Hole, Massachusetts because of the abundant marine life there and to investigate declining fish stocks. This laboratory still exists now known as the Northeast Fisheries Science Center, and is the oldest fisheries research facility in the world. Also at Woods Hole, the Marine Biological Laboratory (MBL) was established in 1888 by Alpheus Hyatt, a student of Harvard naturalist Louis Agassiz who had established the first seaside school of natural history on an island near Woods Hole. MBL was designed as a summer program for the study of the biology of marine life for the purpose of basic research and education. The Woods Hole Oceanographic Institute was created in 1930 in response to the National Academy of Science's call for "the share of the United States of America in a worldwide program of oceanographic research" and was funded by a $3 million grant by the Rockefeller Foundation.

7. An independent biological laboratory was established in San Diego in 1903 by University of California professor Dr. William E. Ritter, which became part of the University of California in 1912 and was named the Scripps Institution of Oceanography after its **benefactors**. Scripps has since become one of the world's leading institutions offering a **multi-disciplinary** study of oceanography.

8. The Explorers. The **advent** of scuba diving introduced other pioneers to the study of marine biology. Jacques Cousteau (1910—1997) was determined to safely breathe **compressed** air underwater in order to lengthen dive times. His work with Emile Gagnan ultimately led to the invention of the regulator which releases compressed air to divers "on demand" (as opposed to a continuous flow). The combination of the Cousteau-Gagnan regulator with compressed air tanks allowed Cousteau the freedom to film underwater, and by 1950 he had produced the Academy Award winning "*The Silent World.*" By the 1970s he was bringing the underwater realm into millions of homes with his PBS series "*Cousteau Odyssey.*" Cousteau's television documentaries won 40 Emmy Awards. Like other oceanography pioneers, Cousteau was criticized for his lack of scientific **credentials**, however his legacy **fostered** a greater knowledge and understanding of the **devastation** caused by threats to ocean health such as pollution of marine resources and resource exploitation.

9. Cousteau's Austrian counterpart, Dr. Hans Hass (1919—), also helped introduce the wonders

of the underwater world to the public. Hass and his wife Lotte were both passionate about underwater exploration and protection of the marine environment, and together they produced numerous documentaries and wrote a variety of books on their underwater experiences. During his career as an underwater explorer, Hass also made significant contributions to diving technology. He invented one of the first underwater flash cameras and contributed to the development of the Drager oxygen rebreather which he and Lotte used in 1942 to film *"Men Amongst Sharks"* and continued to use on diving expeditions aboard their research vessel *"Xarifa"* in the Red Sea and Caribbean. Hass is also known as one of the first humans to interact with a **sperm** whale underwater which helped him become a pioneer in the study of marine animal behavior.

10. The Future. Today, the possibilities for ocean exploration are nearly infinite. In addition to scuba diving, rebreathers, fast computers, remotely-operated vehicles (ROVs), deep sea submersibles, reinforced diving suits, and satellites, other technologies are also being developed. But **interdisciplinary** research is needed to continue building our understanding of the ocean, and what needs to be done to protect it. In spite of ongoing technological advances, it is estimated that only 5% of the oceans have been explored. Surprisingly, we know more about the moon than we do the ocean. This needs to change if we are to ensure the longevity of the life in the seas—and they cover 71% of the earth's surface. Unlike the moon, they are our backyard. Without a detailed collective understanding of the **ramifications** of pollution, overfishing, coastal development, as well as the long-term **sustainability** of ocean oxygen production and carbon dioxide and monoxide **absorption**, we face great risks to environmental and human health. We need this research so that we can act on potential problems—not react to them when it is already too late.

11. Fortunately, thanks to the work of past and present ocean explorers, the public is increasingly aware of these risks which encourage public agencies to take action and promote research. The efforts of public agencies using a multi-disciplinary approach, together with the efforts provided by numerous private marine conservation organizations that work on issues such as **advocacy**, education, and research, will help drive the **momentum** needed to face the challenges of preserving the ocean.

(1,241 words)

https://www.marinebio.org/creatures/marine-biology/history-of-marine-biology/

New Words

1. crustacean [krʌˈsteɪʃ(ə)n]

   n. any mainly aquatic arthropod usually having a segmented body and chitinous exoskeleton 甲壳纲动物

   They use these to capture and tear apart their crustacean prey.

   他们用它捕捉甲壳类的猎物,并且用它们去掉猎物的甲壳。

2. echinoderm [ɪˈkaɪnəʊˌdɜːm]

   n. an animal characterized by tube feet, a calcite body-covering (test), and a five-part symmetrical body. The group includes the starfish, sea urchins, and sea cucumbers 棘皮动物门海洋生物

   It is called an echinoderm, which means it is spiny-skinned.

   它是一种棘皮类动物,身上皮肤长满刺。

3. mollusk [ˈmɒləsk]

   n. an animal such as a snail, clam, or octopus which has a soft body. Many types of mollusc have hard shells to protect them 软体动物

   The mollusc phylum includes all soft-bodied animals without backbones.

   软体动物门包括所有无脊椎的软体动物。

4. vertebrate [ˈvɜːtɪbrət]

   n. any animal with a backbone, including all mammals, birds, fish, reptiles and amphibians 脊椎动物

   The abundance of vertebrate species fell by a third between 1970 and 2006.

   在1970年到2006年之间,本来丰富的脊椎动物物种减少了三分之一。

5. oviparous [əʊˈvɪpərəs]

   a. producing eggs rather than live babies 卵生的

   Every bird is oviparous, and it is viviparity however.

   凡禽鸟都是卵生,而它却是胎生的。

6. viviparous [vɪˈvɪpərəs]

   a. an animal producing live babies from its body rather than eggs 胎生的

   The whale is viviparous, the baby to eat mother's milk to grow up.

   鲸是胎生的,幼鲸靠吃母鲸的奶长大。

7. circumnavigate [ˌsɜːkəmˈnævɪɡeɪt]

   v. to sail all the way around sth., especially all the way around the world 环绕……航行;(尤指)环绕地球航行

   He is preparing for his sixth solo attempt to circumnavigate the globe.

   他准备独自环球一周,这已是第六度挑战了。

8. benchmark [ˈbentʃmɑːk]

n. something which can be measured and used as a standard that other things can be compared with 基准

Tests at the age of seven provide a benchmark against which the child's progress at school can be measured.

七岁时进行的测试为孩子在学校中的学习发展提供了一个衡量基准。

9. hemp [hemp]

n. a plant which is used for making rope and cloth, and also to make the drug cannabis 大麻

Fibers of seed plants such as cotton, flax, and hemp are woven into cloth.

棉花、亚麻和大麻等种子植物的纤维都可被织成布。

10. estuary ['estʃuəri]

n. the wide part of a river where it flows into the sea（入海的）河口；河口湾

The river opens up suddenly into a broad estuary.

河面突然变宽，形成了一个宽阔的河湾。

11. benefactor ['benɪfæktə(r)]

n. a person who gives money or other help to a person or an organization such as a school or charity 施主；捐款人；赞助人

An anonymous benefactor stepped in to provide the prize money.

一位匿名捐助人参与进来提供了奖金。

12. multi-disciplinary [ˌmʌltɪ'dɪsɪˌplɪnərɪ]

a. of or relating to the study of one topic, involving several subject disciplines 多种学科的；与多种学科有关的

There are a large number of multi-disciplinary centers at the University.

另外，该大学设有许多跨领域的研究中心。

13. advent ['ædvent]

n. the coming of an important event, person, invention, etc.（重要事件、人物、发明等的）出现，到来

The advent of the computer has brought this sort of task within the bounds of possibility.

电脑的出现使这种任务的完成成为可能。

14. compress [kəm'pres]

v. to press or squeeze sth. together or into a smaller space; to be pressed or squeezed in this way（被）压紧，压缩

The bin can also compress the waste so it will take up less space.

垃圾桶还可以压缩垃圾，这样就会占用更少的空间。

15. credential [krə'denʃ(ə)l]

n. a document attesting to the truth of certain stated facts 证明，资格证书

A credential can be anything from simple password to a complex cryptographic key.

凭证可以是从简单的密码到复杂的加密密钥的任何东西。

16. foster [ˈfɒstə(r)]

　　v. to encourage sth.to develop 促进；助长；培养；鼓励

　　The club's aim is to foster better relations within the community.

　　俱乐部的宗旨是促进团体内部的关系。

17. devastation [ˌdevəˈsteɪʃn]

　　n. great destruction or damage, especially over a wide area（尤指大面积的）毁灭，破坏，蹂躏

　　The war brought massive devastation and loss of life to the region.

　　战争给该地区造成巨大的破坏以及生命的丧失。

18. sperm [spɜːm]

　　n. a cell that is produced by the sex organs of a male and that can combine with a female egg to produce young 鲸蜡油；精子 sperm whale, 巨头鲸，抹香鲸

　　For a long time, we assumed that all whales that had teeth including sperm whales and killer whales were closely related to one another.

　　在很长一段时间里，我们认为所有有牙齿的鲸鱼，包括抹香鲸和虎鲸，彼此都是密切相关的。

19. interdisciplinary [ˌɪntədɪsəˈplɪnəri]

　　a. involving different areas of knowledge or study 多学科的；跨学科的

　　The new century is an interdisciplinary century that needs to address this problem.

　　新世纪是个跨学科的世纪，是一个能找出解决问题办法的世纪。

20. ramification [ˌræmɪfɪˈkeɪʃn]

　　n. one of the large number of complicated and unexpected results that follow an action or a decision（众多复杂而又难以预料的）结果，后果

　　These changes are bound to have widespread social ramifications.

　　这些变化注定会造成许多难以预料的社会后果。

21. sustainability [səˌsteɪnəˈbɪləti]

　　n. the property of being sustainable 可持续性

　　I'll pose some of the questions and the challenges we are facing in the area of access in sustainability.

　　我将提出我们在可持续性这方面所面临的问题和挑战。

22. absorption [əbˈzɔːpʃ(ə)n]

　　n. the process of a liquid, gas or other substance being taken in（液体、气体等的）吸收

　　Vitamin D is necessary to aid the absorption of calcium from food.

　　从食物中吸取钙需靠维生素 D 的帮助。

23. advocacy [ˈædvəkəsi]

　　n. the giving of public support to an idea, a course of action or a belief（对某思想、行动方针、信念的）拥护，支持，提倡

　　Health advocacy groups have hinted that even more might be coming.

　　健康倡导组织已经暗示了未来要面临的可能更多。

24. momentum [mə'mentəm]
   n. the ability to keep increasing or developing 推进力；动力；势头
   The vehicle gained momentum as the road dipped.
   那辆车顺着坡越跑冲力越大。

Phrases and Expressions

a variety of 多种多样的
give rise to 使发生，引起
serve as 起到……的作用
a wide range of 许多各种不同的
for the purpose of 为了……，出于……的目的
in response to 对……做出回应
in addition to 除……之外，还有

Terminology

molecular biology 分子生物学
biochemistry 生物化学

Proper Names

Station Biologique de Roscoff（法国）罗斯科夫生物站
US Commission of Fish and Fisheries (now known as the National Marine Fisheries Service) 美国渔业委员会
Northeast Fisheries Science Center（美国）东北渔业科学中心
Marine Biological Laboratory (MBL) （美国）海洋生物实验室
Woods Hole Oceanographic Institute（美国）伍兹霍尔海洋研究所
National Academy of Science （美国）国家科学院
Rockefeller Foundation 洛克菲勒基金会
Scripps Institution of Oceanography（美国）斯克里普斯海洋研究所

一、翻译简析

1. Following Cook's explorations, a number of scientists began a closer study of marine life including Charles Darwin (1809—1882) who, although he is best known for the Theory of Evolution, contributed significantly to the early study of marine biology. (Para. 2)

紧随库克海外探险的步伐，许多科学家都纷纷对海洋世界投以青睐，详加考察，甚至就连查理·达尔文(1809—1882)也厕身其间。一提到达尔文，大多数人立即会想到进化论，殊不知他在海洋生物学的发轫上也贡献良多。

【这个句子的结构稍显复杂之处乃是 who 引导的定语从句中内含一个转折关系的状语从句,这也是英文句子中常见的"套句"模式。细心的读者可能会发现,译文中加上了原文中没有对应词的"一提到"和"殊不知"。运用增词法如此处理的主要考量在于原文中的 although 一词内含了因对达尔文身份的固见而造成了对其海洋生物学先驱身份的忽略。为了更为充分地展现出这层隐含的意思,便在译文中加上了上述两个表达。整个处理过程也体现了增词法乃是"有中生有",而非"无中生有"的特性。】

2. His experience on the HMS Beagle helped Darwin formulate his theories of natural selection and evolution based on the similarities he found in species specimens and fossils he discovered in the same geographic region. (Para. 2)

达尔文在"皇家小猎犬号"上的探险经历给予其自然选择和生物进化论的形成颇有助益,因为在相同的地理环境中他发现了物种标本和化石间的颇多相似之处。

【这个句子句式结构不太复杂,主语带有介词短语,宾语补语是由 and 连接的并列成分组成,并列成分又由定语从句修饰。翻译时,将定语修饰成分"he found……"及"he discovered……"译为汉语的动宾结构"发现了……",在其前面翻译成汉语的状语句式"因为……",使译文通顺自然。】

3. The voyages of the HMS Beagle were followed by a 3-year voyage by the British ship HMS Challenger led by Sir Charles Wyville Thomson (1830—1882) to all the oceans of the world during which thousands of marine specimens were collected and analyzed. (Para. 3)

"皇家小猎犬号"探险之旅后,紧随其后的是查理·威伟尔·汤姆森(1830—1882)爵士率领的探险活动。三年间,他们乘坐"皇家挑战者号"遍游世界各大洋,先后搜集和研究了数千份海底标本。

【这个句子在结构上十分紧密,毫无间断地一气呵成。处理这样结构极为紧凑的句子需要按照句子本身的语法结构一点点地拆解。这样一来,整个句子可以拆解成六个部分:"The voyages of the HMS Beagle were followed""by a 3-year voyage""by the British ship HMS Challenger""led by Sir Charles Wyville Thomson (1830—1882)""to all the oceans of the world""during which thousands of marine specimens were collected and analyzed"。化整为零之后,无论是在理解上还是在翻译上,难度都大为降低。接下来便是按照原文的语义构成进行重新划分。整个句子的语义其实分作三个单元:一是点出汤姆森率领的海外探险,对应的是第一部分;二是介绍这次探险活动的持续时间、所用舰船和活动区域,对应的是第二、三、四、五部分;三是介绍这次探险活动的成果,对应的是第六部分。以此为顺序,按照汉语的表达习惯,重新组合,便可得到参考译文。】

4. The Challenger was well equipped to explore deeper than previous expeditions with laboratories aboard stocked with tools and materials, microscopes, chemistry supplies, trawls and dredges, thermometers, devices to collect specimens from the deep sea, and miles of rope and hemp used to reach the ocean depths. (Para. 4)

较之之前的海外探险,"挑战者号"装备精良,探险活动也更为深入。船上的实验室备有各种工具和材料,还有显微镜、各类化学试剂、拖网、挖泥船、气压计以及搜集深海标本的工具,还有长达数英里的麻绳,专做下潜海底之用。

【这个句子的句式结构不太复杂,"stocked with""to collect""used to"都是定语修饰各自前面的名词。翻译时,先将被动语态采用转态译法,译为汉语的主动语态"装备精良……",然后将几个定语修饰语译为"……的"。】

5. Like other oceanography pioneers, Cousteau was criticized for his lack of scientific credentials, however his legacy fostered a greater knowledge and understanding of the devastation caused by threats to ocean health such as pollution of marine resources and resource exploitation. (Para.12)

尽管人们通过库斯托进一步认识到了危害海洋生态健康带来的严重破坏,包括对海洋生态资源的污染和过度使用,但是如同其他海洋科学的先行者一样,他也常遭人诟病为缺乏专业素养。

【这个句子结构上以 however 可以划分为两个相对独立的语义单元,两者之间其实是转折关系。按照汉语的表达习惯,需要先陈述"尽管"或"虽然"后面的内容,再陈述"但是"之后的内容。因此,翻译时需要从后往前翻,这也是英译汉中常用的所谓"倒序法"。】

二、参考译文

海洋生物学研究史

1. 人类关于海底生物的具体记载最先出现在亚里士多德的著作中,他生活在公元前 384 到 322 年间。书中亚里士多德为很多种生物定了名字,像甲壳类动物、棘皮动物、软体动物和鱼类。他将鲸类生物定为哺乳动物,还认为海洋脊椎动物要么卵生(产生体外孵化的卵),要么胎生(产生体内孵化的卵)。因为亚里士多德是第一个观察和记录海洋生物的人,因此常被称为海洋生物学之父。

2. 早期的探险活动:现代海洋生物学研究肇始于詹姆斯·库克船长(1728—1779)18 世纪的探险活动。他最广为人知的乃是为英国皇家海军所做的全球航海探险,发现并记录了多块不为人知的水域。终其一生,库克完成了两次环球航行,其间他曾记录了不计其数的动植物,这些都为时人所未闻未见。紧随库克海外探险的步伐,许多科学家都纷纷对海洋世界青睐有加,详加考察,甚至就连查理·达尔文(1809—1882)也厕身其间。一提到达尔文,大多数人马上会想到进化论,殊不知他在海洋生物学的发轫上也贡献良多。作为"皇家小猎犬号"上一位驻船博物学家,从 1831 到 1836 年达尔文花了五年时间搜集和研究了大量海洋生物标本,这些标本最后都送到了大英博物馆分类收藏。达尔文在地理学上的兴趣引发了他对珊瑚礁及其形成的研究。他在"皇家小猎犬号"上的探险经历给予其自然选择和生物进化论的形成颇有助益,因为在相同的地理环境中他发现了物种标本和化石间的颇多相似之处。

3. "皇家小猎犬号"探险之旅后,紧随其后的是查理·威伟尔·汤姆森(1830—1882)爵士率领的探险活动。三年间,他们乘坐"皇家挑战者号"环游世界各大洋,先后搜集和研究了数千份海底标本。这次探险活动标志着海洋学的诞生。搜集的科学数据最后集成 50 大册,为多年来许多领域中的海洋学研究奠定了基础。"皇家挑战者号"完成的深海探险也成为验证英国探险家爱德华·福布斯理论正确性的标尺,该理论认为 550 米或 1 800 英尺以下的海水不适于生

4. 较之之前的海外探险,"皇家挑战者号"装备精良,探险活动也更为深入。船上的实验室备有各种工具和材料,还有显微镜、各类化学试剂、拖网、挖泥船、气压计以及搜集深海标本的工具,还有长达数英里的麻绳,专做下潜海底之用。最终"皇家挑战者号"搜集到的海洋学方面的信息长达 30 000 页,多领域的多名科学家先后参与到了信息编辑工作中。

5. 众多机构:紧随这些探险活动之后的乃是多家专门研究海洋生物的海洋实验室的建立。世界上历史最悠久的海洋实验室是创建于法国西北沿海孔卡诺的"罗斯科夫生物站"。这所实验室最初是为了培育海洋生物(比如多弗比目鱼)而建,因为孔卡诺位于河流入海口附近,海洋生物种类极为丰富。目前这所实验室主要从事的是分子生物学、生物化学和环境研究。

6. 1871 年,斯宾塞·富乐顿·巴德,"美国渔业委员会"(现名为"国家海洋渔业局")首任负责人,在马萨诸塞州的伍兹霍尔建立了一处采集站以调查数量逐年减少的鱼类生存状况,因为当地有着丰富的海洋生物资源。这家实验室现名为"东北渔业科学中心",乃是世界上存在时间最久的鱼类研究机构。在伍兹霍尔还有一家实验室,这就是 1888 年由阿尔菲俄斯·凯特创建的"海洋生物实验室"。凯特是哈佛博物学家路易斯·阿加西的门生,阿加西曾在伍兹霍尔附近的一个小岛上建立了美国沿海地区首所专授博物史的学校。海洋生物实验室还为海洋生物学研究者专设夏令营以促进基础研究和教育。1930 年,为了响应美国国家科学院"同世界分享美国海洋学研究成果"的号召,在"洛克菲勒基金会"300 万美元资助下,"伍兹霍尔海洋研究所"得以建立。

7. 1903 年,加州大学教授威廉·E.瑞特尔在圣地亚哥创建了一所独立的生物学实验室,到了 1912 年成为加州大学的一部分,并以其资助者姓名重新命名为"斯克里普斯海洋研究所"。目前该研究所是世界上主要的海洋学研究机构之一,尤其是在跨学科海洋学研究上。

Cultural Background Knowledge: History of the Ocean
海洋历史文化背景知识

早些年科学家们为追溯海洋的形成,提出了大胆的设想,设想 60 亿年前地球表面温度比较低,其他物质之间经过亿万年变化混合,紧接着地球表面温度急剧增高,经过热压力作用,使地球大部分形成类似火球的东西,之后又经过亿万年地球表面又冷却下来,这里所用到的知识就是热胀冷缩的物理知识,从此地理表面开始凹凸不平,形成了一个个盆地。

盆地形成以后地球最初的水存在于地球各个地方,后来经过重力作用使其互相挤压,被挤出来的水蒸气越来越多,最终形成大规模的地震、火山爆发等诸多情况,使得大量水蒸气迅速集结,最终形成了江河大海雏形。

有人提出设想就有人想推翻设想,还有一部分人认为在地球漫长演化过程中,地球的重力作用导致轻重物质分离,质量较轻的水分从质量较重的岩石、矿物中分离出来,最终形成了地球初期的水资源。

随着提出设想的人越来越多,各派系纷纷提出设想,还有人认为地球上的水并非地球本身存在的,是由其他行星带来的。当然这种设想并没有确切的科学依据。还有一部分人提出设想,认为地球上的水是在地球形成之前就产生了,地球后期由于重力作用经不起挤压,将自身分离出去形成月球,由于月球在分离的时候和地球产生了巨大的震动,形成了巨大的空隙,各大海洋系就此诞生!

科学家们分为多派系,每个派系都保持各自的观点。虽然科学越来越进步,越来越先进,但终究是无法探索广阔海洋是怎么形成的,海水究竟来自哪里。目前对于海洋的理解我们仍处在探索阶段,但可以明确的是两个派系,第一种则是水是地球本身固有的,第二种就是水是其他地方的。

地球区别于其他星球,是因为地球的水资源丰富,其他已知的行星都是贫水的。海洋是地球的生命之源,也是文明起源之地,对于我们整个人类的发展有着十分重要的意义。近来随着科学的发展,人类对于海水的污染也在急剧增加,希望随着我们环保意识的觉醒,能够很好地保护我们的生命之本。虽然现在的我们无时无刻都在使用着地球上的水,但是海洋究竟是怎样形成的,海水又从哪里来,还是一个谜题。要想揭开谜题,还需要进一步的探索发现。

Unit Three
Cultural Exchanges along the Maritime Road

海路文化交流

Text A

The Maritime Silk Road
By Kelly Pang

Ideological and Political Education：开创新型海洋文明 建设海上丝绸之路

21世纪海上丝绸之路,是2013年10月习近平总书记访问东盟时提出的构想。海上丝绸之路自秦汉时期开通以来,一直是沟通东西方经济文化交流的重要桥梁,而东南亚地区自古就是海上丝绸之路的重要枢纽和组成部分。在我国着眼于与东盟建立战略伙伴十周年这一新的历史起点,为进一步深化我国与东盟的合作,提出"21世纪海上丝绸之路"的构想。这一构想,是我国在世界格局发生复杂变化的今天,主动创造合作、和平、和谐、共生的国际合作的有力保障,为我国全面深化改革开放创造良好的机遇与国际环境。

海上丝绸之路见证了沿线各国人民通过海洋交流融通、互利合作的悠久历史。共建"21世纪海上丝绸之路",不仅有深厚的历史渊源,也具有坚实的现实基础,对促进海上丝绸之路沿线各国经济发展、文化交流有着重要价值。

海上丝绸之路开创了新型海洋文明,我国有渤海、黄海、东海、南海四大海域,蔚蓝的海洋孕育了开放、多元、包容、共享的蓝色文化,为构建守望相助、休戚与共、和谐共生、持续发展,与沿线各国共建人类命运共同体提供了重要的思想启迪。海上丝绸之路促进海路文化交流,帮助人类更好地认识海洋、开发海洋、利用海洋、保护海洋、管控海洋,共建蓝色伙伴关系,推动构建合作共赢的海洋命运共同体,实现人与自然的和谐发展。

建设海上丝绸之路,我国的优势体现在基础设施建设上,比如高铁建设,中国在温带、寒带甚至热带地区都具有丰富的经验;我国的载人潜水器在马里亚纳海沟下潜突破1万米等。我国积极推进海洋治理体系和治理能力现代化,全面推动海洋管理体制创新、海洋权益维护、海洋生态文明建设和海洋法律法规制定。同时,国际海洋合作尤其是同沿线国家之间的海洋合作空间不断拓展,合作成效日益显著。构建21世纪海洋命运共同体成为全球海洋治理的正道,倡导新型的海洋治理观。

我国坚持与沿海丝路国家共同倡导共商、共建、共享的海洋开发原则,聚焦发展、治理和安全问题进行深化合作,不断推动海洋文化交融和增进海洋福祉,为构建海洋命运共同体奠定更加坚实的基础。

1. The Maritime Silk Road was a **conduit** for trade and cultural exchanges between China's southeastern coastal areas and foreign countries. There were two major routes: the East China Sea Silk Route and the South China Sea Silk Route.

2. Starting from Quanzhou Fujian Province, the Maritime Silk Road was the earliest voyage route that was formed in the Qin and Han dynasties, developed from the Three Kingdoms Period to the Sui Dynasty, flourished in the Tang and Song dynasties, and fell into decline in the Ming and Qing dynasties. Through the Maritime Silk Road, silks, china, tea, and brass and iron were the four main categories exported to foreign countries; while spices, flowers and plants, and rare treasures for the court were brought to China. Therefore, the Maritime Silk Road was also known as "the maritime China road" or "the maritime spices road".

3. East China Sea Silk Route. The East China Sea Silk Route mainly went to Japan and Korea. It dates back to the Zhou Dynasty (1112 BC) when the government sent some Chinese people to Korea to teach its people farming and **sericulture**, departing from the port of Bohai Bay, Shandong Peninsula. From that time, the skills and techniques of raising silkworms, silk **reeling** and **weaving** were introduced into Korea slowly via the Yellow Sea. When Emperor Qin Shi Huang united China (221 BC), many people from the States of Qi, Yan and Zhao fled to Korea and took with them silkworms and raising technology, which sped up the development of silk spinning in Korea.

4. During the Sui and Tang dynasties, Japanese envoys and monks travelled to China frequently. They brought back the blue **damask** silks that they got in Taizhou, Zhejiang Province, which served as samples. Since the Tang Dynasty, silk products from Jiangsu and Zhejiang provinces were directly transported to Japan by sea, and the silk products formally became the commodities. In the Song Dynasty, lots of silk products were exported to Japan. In the Yuan Dynasty, the government set up Shi Bo Si (市舶司) in many ports, such as Ningbo, Quanzhou, Guangzhou, Shanghai, Ganpu (澉浦), Wenzhou, and Hangzhou, in order to export the silk products to Japan. Shi Bo Si's Oceangoing and Marketing Department, was set up in each port to administrate foreign economy-related affairs by sea during the dynasties of Tang, Song and Yuan, and the early part of the Ming Dynasty.

5. The Maritime Silk Road fell into decline because of the Haijin policy of the Qing Dynasty. The vast trade networks of the Silk Roads carried more than just merchandise and precious commodities. In fact, the constant movement and mixing of populations brought about the widespread transmission of knowledge, ideas, cultures and beliefs, which had a profound impact on the history and civilizations of the Eurasian peoples. Travellers along the Silk Roads were attracted not only by trade but also by the intellectual and cultural exchange taking place in cities along the Silk Roads, many of which developed into **hubs** of culture and learning.

Science, arts and literature, as well as crafts and technologies were thus shared and **disseminated** into societies along the lengths of these routes, and in this way, languages, religions, and cultures developed and influenced one another.

6. Silk is a textile of ancient Chinese origin **fabric** from the protein fibre produced by the silkworm as it makes its **cocoon**. The cultivation of silkworms for the process of making silk, known as sericulture, was, according to Chinese tradition, developed sometime around the year 2700 BCE. Regarded as an extremely high value product, silk was reserved for the exclusive usage of the Chinese imperial court for the making of cloths, **drapes, banners**, and other items of prestige. Its production technique was a fiercely guarded secret within China for some 3,000 years, with imperial **decrees** sentencing to death anyone who revealed to a foreigner the process of its production. Tombs in Hubei province dating from the 4th and 3rd centuries BCE contain the first complete silk garments as well as outstanding examples of silk work, including **brocade, gauze** and embroidered silk.

7. At some point during the 1st century BCE, silk was introduced to the Roman Empire, where it was considered an exotic luxury that became extremely popular, with imperial **edicts** being issued to control prices. Silks popularity continued throughout the Middle Ages, with detailed Byzantine regulations for the manufacture of silk clothes, illustrating its importance as a quintessentially royal fabric and an important source of revenue for the crown. Additionally, the needs of the Byzantine Church for silk garments and hangings were substantial. This luxury item was thus one of the early **impetuses** for the development of trading routes from Europe to the Far East.

8. Knowledge about silk production was very valuable and, despite the efforts of the Chinese emperor to keep it a closely guarded secret, it did eventually spread beyond China, first to India and Japan, then to the Persian Empire and finally to the west in the 6th century CE. This was described by the historian Procopius, writing in the 6th century:
"About the same time [*circa.* 550 CE] there came from India certain monks; and when they had satisfied Emperor Justinian Augustus that the Romans should no longer buy silk from the Persians, they promised the emperor in an interview that they would provide the materials for making silk so that never should the Romans seek business of this kind from their enemy the Persians, or from any other people whatsoever. They said that they were formerly in Serinda, which they call the region frequented by the people of the Indies, and there they learned perfectly the art of making silk. Moreover, to the emperor who plied them with many questions as to whether he might have the secret, the monks replied that certain worms were manufacturers of silk, nature itself forcing them to keep always at work; the worms could certainly not be brought here alive, but they could be grown easily and without difficulty; the eggs of single hatchings are innumerable; as soon as they are laid men cover them with **dung**

*and keep them warm for as long as it is necessary so that they produce insects. When they had announced these tidings, led on by liberal promises of the emperor to prove the fact, they returned to India. When they had brought the eggs to Byzantium, the method having been learned, as I have said, they changed them by **metamorphosis** into worms which feed on the leaves of **mulberry**. Thus began the art of making silk from that time on in the Roman Empire."*

9. Beyond Silk: A Diversity of Routes and Cargos. These routes developed over time according to shifting **geopolitical** contexts throughout history. For example, merchants from the Roman Empire would try to avoid crossing the territory of the Parthians, Rome's enemies, and therefore took routes to the north instead, across the Caucasus region and over the Caspian Sea. Similarly, whilst extensive trade took place over the network of rivers that crossed the Central Asian **steppes** in the early Middle Ages, their water levels rose and fell, and sometimes rivers dried up altogether, and trade routes shifted accordingly.

10. The history of maritime routes can be traced back thousands of years, to links between the Arabian Peninsula, Mesopotamia and the Indus Valley Civilization. The early Middle Ages saw an expansion of this network, as sailors from the Arabian Peninsula forged new trading routes across the Arabian Sea and into the Indian Ocean. Indeed, maritime trading links were established between Arabia and China from as early as the 8th century CE. Technological advances in the science of navigation, in astronomy, and also in the techniques of ship building, combined to make long-distance sea travel increasingly practical. Lively coastal cities grew up around the most frequently visited ports along these routes, such as Zanzibar, Alexandria, Muscat, and Goa, and these cities became wealthy centres for the exchange of goods, ideas, languages and beliefs, with large markets and continually changing populations of merchants and sailors.

11. The great variety of routes were illustrated, which were available to merchants transporting a wide range of goods and travelling from different parts of the world, by both land and sea. Most often, individual merchant caravans would cover specific sections of the routes, pausing to rest and replenish supplies, or stopping altogether and selling on their cargos at points throughout the length of the roads, leading to the growth of lively trading cities and ports. The Silk Roads were dynamic and porous; goods were traded with local populations throughout, and local products were added into merchants' cargos. This process enriched not only the merchants' material wealth and the variety of their cargos, but also allowed for exchanges of culture, language and ideas to take place along the Silk Roads.

12. Routes of Dialogue. Despite the Silk Roads history as routes of trade, the man who is often credited with founding them by opening up the first route from China to the West in the 2nd

century BC, General Zhang Qian, was actually sent on a diplomatic mission rather than one motivated by trading. Sent to the West in 139 BCE by the Han Emperor Wudi to ensure alliances against China's enemies the Xiongnu, Zhang Qian was ultimately captured and imprisoned by them. Thirteen years later he escaped and made his way back to China. Pleased with the wealth of detail and accuracy of his reports, the emperor then sent Zhang Qian on another mission in 119 BCE to visit several neighbouring peoples, establishing early routes from China to Central Asia.

13. These routes were also fundamental in the **dissemination** of religions throughout Eurasia. Buddhism is one example of a religion that travelled the Silk Roads, with Buddhist art and **shrines** being found as far apart as Bamiyan in Afghanistan, Mount Wutai in China, and Borobudur in Indonesia. Christianity, Islam, Hinduism, Zoroastrianism and Manicheism spread in the same way, as travellers absorbed the cultures they encountered and then carried them back to their homelands with them. Thus, for example, Hinduism and subsequently Islam were introduced into Indonesia and Malaysia by Silk Roads merchants travelling the maritime trade routes from the Indian Subcontinent and Arabian Peninsula.

14. Travelling the Silk Roads. As trade routes developed and became more **lucrative, caravanserais** became more of a necessity, and their construction intensified across Central Asia from the 10th century onwards, continuing until as late as the 19th century. This resulted in a network of caravanserais that stretched from China to the Indian subcontinent, Iranian Plateau, the Caucasus, Turkey, and as far as North Africa, Russia and Eastern Europe, many of which still stand today.

15. Maritime traders had different challenges to face on their lengthy journeys. The development of sailing technology, and in particular of ship-building knowledge, increased the safety of sea travel throughout the Middle Ages. Ports grew up on coasts along these maritime trading routes, providing vital opportunities for merchants not only to trade and **disembark**, but also to take on fresh water supplies, as one of the greatest threats to sailors in the Middle Ages was a lack of available drinking water. Pirates were another risk faced by all merchant ships along the Maritime Silk Roads, as their lucrative cargos made them attractive targets.

16. The Legacy of the Silk Roads. Today, many historic buildings and monuments still stand, marking the passage of the Silk Roads through caravanserais, ports and cities. However, the long-standing and ongoing legacy of this remarkable network is reflected in the many distinct but interconnected cultures, languages, customs and religions that have developed over **millennia** along these routes. The passage of merchants and travellers of many different nationalities resulted not only in commercial exchange but in a continuous and widespread process of cultural interaction. As such, from their early, exploratory origins, the Silk Roads

developed to become a driving force in the formation of diverse societies across Eurasia and far beyond.

(2,010 words)

https://www.chinahighlights.com/travelguide/maritime-silk-road.htm

New Words

1. conduit ['kɒndjuɪt]

    n. a passage (a pipe or tunnel) through which water or electric wires can pass(水或电线的)管道,导管;渠道,通道;(保护线路的)导线管,电缆沟

    The organization had acted as a conduit for money from the arms industry.

    那家机构充当了从军火工业向它处周转资金的渠道。

2. sericulture ['serɪˌkʌltʃə]

    n. raising silkworms in order to obtain raw silk; the production of raw silk by raising silkworms 养蚕;蚕丝业

    China was the first country in the world to cultivate silkworms and develop silk weaving. More than 3,000 years ago, sericulture and silk weaving were already significantly developed.

    中国是世界上最早发明养蚕和丝织的国家,早在3 000多年以前,蚕桑、丝织业已相当发达。

3. reel [riːl]

    v. wind onto or off a reel; revolve quickly and repeatedly around one's own axis 卷,绕(reel sth.in)

    She is reeling the silk thread off cocoons.

    她在从茧中抽出丝。

4. weave [wiːv]

    v. interlace by or as it by weaving(用织布机)编,织

    They would spin and weave cloth, cook and attend to the domestic side of life.

    他们纺纱、织布、做饭,还料理家务。

5. damask ['dæməsk]

    a. incarnadine 缎子的;粉红色的

    The duck dropped the dirty double damask dinner napkin.

    那只鸭子弄掉了肮脏的双缎料晚餐餐巾。

6. transmission [trænz'mɪʃ(ə)n]

    n. communication by means of transmitted signals; the act of sending a message; causing a message to be transmitted(无线电、电视等信号的)播送,发送;(电台或电视等的)信息,广播;传递,传播,传染

    The transmission of the programme was brought forward due to its unexpected topicality.

    该节目提前播送是因有出人意料的热门话题。

7. hub [hʌb]

    n. a center of activity, interest, commerce or transportation; a focal point around which events revolve; the central part of a car wheel (or fan or propeller etc.) through which the shaft or axle passes(活动的)中心;枢纽机场,枢纽城市;(轮)毂,(推进器、风扇等的)旋翼叶毂

This grand building in the centre of town used to be the hub of the capital's social life.

这座位于城镇中央的宏伟建筑曾经是首都社交生活的中心。

8. disseminate [dɪˈsemɪneɪt]

v. cause to become widely known 散布,传播(信息、知识等)

Political advocacy groups have begun to use information services to disseminate information that is then accessed by the public via personal computer.

政治宣传团体已经开始使用信息服务来传播信息,公众可以通过个人电脑获取这些信息。

9. fabric [ˈfæbrɪk]

n. artifact made by weaving or felting or knitting or crocheting natural or synthetic fibers 布料,织物

The fabric was red, flecked with gold.

织物是红色的,带有金黄色的斑点。

10. cocoon [kəˈkuːn]

n. silky envelope spun by the larvae of many insects to protect pupas and by spiders to protect eggs 茧,卵袋;(尤指保护性或安慰性的)遮盖(或包裹)物;安全宁静的地方,避风港;(防止金属设备腐蚀的)防护膜

The cocoon-to-butterfly theory only works on cocoons and butterflies.

茧变蝴蝶理论只适用于茧和蝴蝶。

11. drape [dreɪp]

n. hanging cloth used as a blind (especially for a window); a sterile covering arranged over a patient's body during a medical examination or during surgery in order to reduce the possibility of contamination 窗帘;褶裥

Anna twists herself in a velvet drape like it's a gown. She looks gorgeous, but she looks ridiculous.

安娜卷进天鹅绒窗帘里,就像它是礼服一样。她表现得很华丽,但她看上去很可笑。

12. banner [ˈbænə(r)]

n. long strip of cloth or paper used for decoration or advertising; a newspaper headline that runs across the full page 横幅标语;(网页上的)横幅广告;信仰,准则;旗帜

I felt the banner rip as we were pushed in opposite directions.

当我们被挤向相反的方向的时候,我感到横幅撕裂了。

13. decree [dɪˈkriː]

n. a legally binding command or decision entered on the court record (as if issued by a court or judge) 法令,政令;裁定,判决

The decree imposed strict censorship of the media.

这个法令强制实行严格的媒体审查制度。

14. brocade [brəˈkeɪd]

n. thick heavy expensive material with a raised pattern 织锦;锦缎

Brocade is a class of richly decorative shuttle-woven fabrics, often made in colored silks with gold and silver threads.

锦是一种华美的织物,由染色的丝绸和金银线织成。

15. gauze [gɔːz]

n. a net of transparent fabric with a loose open weave;(medicine) bleached cotton cloth of plain weave used for bandages and dressings 薄纱;薄雾;纱布

It's almost like a green gauze veil.

它几乎就像一块绿色的薄纱。

16. embroider [ɪmˈbrɔɪdə(r)]

v. decorate with needlework;add details to 刺绣;装饰;镶边;绣花;刺绣

Only with many embroidery techniques is it possible to embroider such wonderful patterns.

绣法多,绣出来的图案才会多姿多彩。

17. edict [ˈiːdɪkt]

n. a formal or authoritative proclamation;a legally binding command or decision entered on the court record (as if issued by a court or judge) 法令;布告

He issued an edict that none of his writings be destroyed.

他下了一道命令:他写的所有东西都不得毁掉。

18. quintessentially [ˌkwɪntɪˈsenʃəli]

ad. typically 典型地;标准地

It is a familiar, and quintessentially Russian ritual.

它是个熟悉而又典型的俄罗斯仪式。

19. impetus [ˈɪmpɪtəs]

n. a force that moves something along;the act of applying force suddenly 动力,促进;动量,冲力

By 2022, she was restless and needed a new impetus for her talent.

到 2022 年,她焦躁不安,需要一个新的对其才华的推动力。

20. circa [ˈsɜːkə]

prep. at approximately, in approximately, or of approximately, used especially with dates 大约,左右(主要用于日期前)

ad. in about 大约

The story tells of a runaway slave girl in Louisiana, circa 1850.

这个故事讲述的是路易斯安那州一个年轻的逃跑女奴,时间大约在 1850 年。

21. dung [dʌŋ]

n. fecal matter of animals(动物的)粪,粪肥;肮脏的东西

She's cooking on a stove fuelled by dried animal dung.

她正在一个以动物干粪便做燃料的火炉上做饭。

22. metamorphosis [ˌmetəˈmɔːfəsɪs]

n. a complete change of physical form or substance especially as by magic or witchcraft 变形;变质

By what invisible power has this surprising metamorphosis been performed?

是什么无形的力量完成了这一惊人的蜕变?

23. mulberry ['mʌlbəri]

n. sweet usually dark purple blackberry-like fruit of any of several mulberry trees of the genus Morus 桑树;桑葚,桑椹;深紫红色

The room was so still that I could hear mockingbirds quarreling in the fruitless mulberry trees.

房间里非常安静,我能听到知更鸟在不结果实的桑树上嘹亮地对唱。

24. geopolitical [ˌdʒiːəʊpə'lɪtɪkl]

a. of or relating to geopolitics 地理政治学的

Early resolution of geopolitical issues would be beneficial.

尽早解决地缘政治学议题是有益的。

25. steppes [steps]

n. extensive plain without trees (associated with eastern Russia and Siberia) 草原;干草原

The foraging communities of the cultures in the region of the Don and Dnieper rivers took up stock breeding and began to exploit the neighboring steppes.

顿河和第聂伯河流域的游牧部落开始从事家畜饲养,并开始开发邻近的大草原。

26. dissemination [dɪˌsemɪ'neɪʃ(ə)n]

n. the opening of a subject to widespread discussion and debate; the act of dispersing or diffusing something 宣传,散播

He actively promoted the dissemination of scientific ideas about matters such as morality.

他积极地推动诸如道德之类的科学观念的传播。

27. shrine [ʃraɪn]

n. a place of worship hallowed by association with some sacred thing or person 圣地;神殿;神龛;圣祠

Wimbledon is a shrine for all lovers of tennis.

温布尔登是所有网球爱好者的圣地。

28. lucrative ['luːkrətɪv]

a. producing a sizeable profit 获利多的,赚大钱的

Thousands of ex-army officers have found lucrative jobs in private security firms.

成千上万的退役军官在私人保安公司找到了薪水丰厚的工作。

29. caravansary [ˌkærə'vænsəri]

n. an inn in some eastern countries with a large courtyard that provides accommodation for caravans 驿站;商队旅馆;大旅舍

The whole caravansary had fallen in like a card house.

这座大旅舍就像纸牌搭的房子一样整个坍塌了。

30. disembark [ˌdɪsɪm'bɑːk]

v. go ashore 登陆,下车;上岸;使……登陆;使……上岸

Now moored in West Java, its passengers are refusing to disembark.

这艘船现在停靠在西爪哇岛,船上的乘客拒绝上岸。

31. millennia [mɪˈleniə]

n. one thousand years 千年期；千周年纪念日（millennium 的复数）

Helium is the by-product of millennia of radioactive decay from the elements thorium and uranium.

氦是钍和铀元素数千年放射性衰变的副产品。

Phrases and Expressions

date back to 追溯到；从……开始
depart from 离开；开出；从……出发
speed up 加速；使加速
cover with 覆盖
feed on 以……为食；以……为能源
a lack of 缺乏，缺少
as such 同样地；本身；就其本身而论
far beyond 远远超出

Terminology

the Maritime Silk Road 海上丝绸之路

Proper Names

the Three Kingdoms Period 三国时期
BCE 公元前（Before Common Era）
the Middle Ages 中世纪
Byzantine 拜占庭人，拜占庭派的建筑师；拜占庭式的；东罗马帝国的；错综复杂的；暗中的
the Far East 远东
CE 公元（Common Era）
Emperor Justinian Augustus 查士丁尼·奥古斯都皇帝
the Persians 波斯人
the Caucasus region 高加索地区
the Caspian Sea 里海
the Indus Valley Civilization 印度河流域文明
Zoroastrianism 琐罗亚斯德教
Manicheism 摩尼教

一、翻译策略/方法/技巧：分译法

在翻译时，为了适应目标语的表达习惯，可以改变原文的结构，即：把原文的某个成分分离出来，译成一个独立成分，这个过程叫作分译。分离出来的成分可以与原单位大小相等，也可以不等，即：小单位可以分译为大成分，大单位也可以分译成小成分。按照分离出来的原文单

位,大体可以分为单词分译、短语分译、小句分译和复句分译。

1. 单词分译

下面的例子中,如果按照英语的结构去翻译,就会产生蹩脚的译文。但是,如果把画线的部分从句子中分离出来,与其他成分结合翻译,就会译出高质量的译文。

例1:The computer can give Mary the right lesson for her, neither too fast nor too slow.

译文:计算机能给玛丽上课,上得恰到好处,既不太快也不太慢。

例2:He tried vainly to talk us into agreement with the unrealistic proposal.

译文:他试图劝说我们同意接受这项不切实际的建议,但还是白费了力气。

2. 短语分译

与单词分译相比,短语分译更常见。在语义结构上,短语的语义更完整,更接近于小句,因此,分离出来的短语,常常可以译成小句。比如下面的例子:

例3:Several theories of atomic structure were produced when the electron was discovered, only to be discarded when more information about the atom was obtained.

译文:电子被发现后,人们提出了好几种原子结构理论,但是随着对原子的了解越来越深入,这些理论就被抛弃了。

3. 小句分译

英语的后置修饰语有时很长,而汉语的修饰语一般前置,不宜过长。因此,英语中的定语从句等从句,如果可能的话,尽量译成短语或词,作为定语置于中心词的前面。如下面的例子:

例4:Elizabeth was determined to make no effort for conversation with a woman, who was now more than usually insolent and disagreeable.

译文:伊丽莎白不肯再和这个极其无礼、非常讨厌的女人说话。

4. 复句分译

有些英语句子很长,结构比较复杂;而汉语的句子往往短小。如果按照英语的表形结构去翻译,汉语的译文会让读者不好理解,甚至感觉莫名其妙。如下面的例子:

例5:And all this was done in the heart of a crowded city with very little storage space, so that the arrival of each part had to be scheduled to the minute, and the construction work so organized that there was no delay and no one got in another's way.

译文:一切都在某个拥挤的城市中心展开。储料空间非常小,所以每次进料的时间分秒不差;工地进度也是如此,不容拖延,互不相碍。

原文是主从复合句,状语情况复杂。状语"in the heart of a crowded city"与主干部分关系紧密,构成了完整的语义;而短语"with very little storage space"与后续部分逻辑关系很强。译文用了分译法,依据各层次内容之间的紧密程度,把原句处理成两个单独的小句,而且译文第二句还使用了分号,以区分细小层次。

(摘自:黄忠廉.翻译方法论[M].北京:中国社会科学出版社,2019:122-130.)

二、译例与练习
Translation

1. Starting from Quanzhou Fujian Province, the Maritime Silk Road was the earliest voyage route that was formed in the Qin and Han dynasties, developed from the Three Kingdoms Period to the Sui Dynasty, flourished in the Tang and Song dynasties, and fell into decline in the Ming and Qing dynasties. (Para.2)

　　海上丝绸之路始于福建泉州。这条航路最早形成于秦汉时期、发展于三国至隋代、兴盛于唐宋、衰落于明清时期。

　　（这里主要是指 2013 年 9 月习近平总书记提出建设"新丝绸之路经济带"构想和 2015 年 3 月 28 日国家发展改革委、外交部、商务部联合发布了《推动共建丝绸之路经济带和 21 世纪海上丝绸之路的愿景与行动》的相关文章。）

2. Travellers along the Silk Roads were attracted not only by trade but also by the intellectual and cultural exchange taking place in cities along the Silk Roads, many of which developed into hubs of culture and learning. (Para.5)

　　丝绸之路沿线的旅行者不仅被贸易活动所吸引，还被丝绸之路沿线城市发生的知识和文化交流所吸引，许多沿线城市已发展成为文化和学习的中心。

3. Similarly, whilst extensive trade took place over the network of rivers that crossed the Central Asian steppes in the early Middle Ages, their water levels rose and fell, and sometimes rivers dried up altogether, and trade routes shifted accordingly. (Para.9)

　　同样地，在中世纪早期，人们在中亚草原的河流网络上进行广泛的贸易，水位忽高忽低，有时候河流会完全干涸，贸易路线也随之发生改变。

4. Lively coastal cities grew up around the most frequently visited ports along these routes, such as Zanzibar, Alexandria, Muscat, and Goa, and these cities became wealthy centres for the exchange of goods, ideas, languages and beliefs, with large markets and continually changing populations of merchants and sailors. (Para.10)

　　在这些海上丝绸路线上，人们经常访问的港口附近，如桑给巴尔港、亚历山大港、马斯喀特港和果阿港，出现了生机勃勃的沿海城市。这些城市成为众多商品贸易、不同观点与语言交流、信仰表达的中心，拥有着巨大的市场和人数不断变化的商人与水手们。

5. Most often, individual merchant caravans would cover specific sections of the routes, pausing to rest and replenish supplies, or stopping altogether and selling on their cargos at points throughout the length of the roads, leading to the growth of lively trading cities and ports. (Para.11)

　　通常情况下，个体商队会在特定的路线上停留、休息和补充物资，或完全停下来，在整个丝绸之路的商品销售点上出售货物，这促进了城市贸易的发展和港口的不断增加。

6. Despite the Silk Roads history as routes of trade, the man who is often credited with founding them by opening up the first route from China to the West in the 2nd century BC, General Zhang Qian, was actually sent on a diplomatic mission rather than one motivated by trading. Sent to the West in 139 BCE by the Han Emperor Wudi to ensure alliances against China's enemies the Xiongnu, Zhang Qian was ultimately captured and imprisoned by them. (Para.12)

尽管丝绸之路历史上是贸易之路,在公元前2世纪,张骞将军开辟了中国通往西域的第一条路线,但实际上张骞是被派去执行外交任务,而不是以贸易为主。公元前139年,汉武帝派张骞出使西域,以联盟对抗匈奴人的入侵,张骞最终被匈奴人俘虏并被监禁起来。

【在翻译这个长句中的"in the 2nd century BC"时,采用换序译法,对原文的时间状语按照中文的习惯放在前面进行翻译。这种改变原文词语的前后词序,根据语言习惯,对原文的词序进行调整,使译文通顺,这就是换序译法。在翻译"Han Emperor Wudi"时采用合词译法,在原文基础上把词汇 Emperor 与 Wudi 表达同一意思的"皇帝"合起来进行翻译,这是合词译法。无论是英译汉还是汉译英,我们需要注意意思相近或同义词采用合词译法进行准确翻译。】

7. Thus, for example, Hinduism and subsequently Islam were introduced into Indonesia and Malaysia by Silk Roads merchants travelling the maritime trade routes from the Indian Subcontinent and Arabian Peninsula. (Para.13)

例如,通过丝绸之路,商人通过印度次大陆和阿拉伯半岛的海上贸易路线把印度教和随后的伊斯兰教传入印度尼西亚和马来西亚。

8. As trade routes developed and became more lucrative, caravanserais became more of a necessity, and their construction intensified across Central Asia from the 10th century onwards, continuing until as late as the 19th century. (Para.14)

随着贸易路线的发展和利润的增加,商队客栈变得越来越必要,从10世纪一直到19世纪,人们在中亚各地加强了商队客栈建设。

9. Ports grew up on coasts along these maritime trading routes, providing vital opportunities for merchants not only to trade and disembark, but also to take on fresh water supplies, as one of the greatest threats to sailors in the Middle Ages was a lack of available drinking water. (Para.15)

这些海上贸易路线沿岸的港口不断发展,不仅为商人们提供了贸易和登上陆地的重要机会,还为他们提供了淡水供应,因为在中世纪,水手们面临的最大威胁之一就是缺乏饮用水。

10. However, the long-standing and ongoing legacy of this remarkable network is reflected in the many distinct but interconnected cultures, languages, customs and religions that have developed over millennia along these routes. (Para.16)

然而,这一非凡网络拥有长期与可持续发展的遗产。许多数千年来相互联系的文化、语言、习俗和宗教反映出了沿着这些路线发展的情况。

【这个长句是个被动句。翻译时将被动语态直接转换成汉语的主动语态。原文中后面的定语从句部分的词与词组采用对等译法,在翻译过程中把原文中常用的词进行对等翻译,是名副其实的对等表达。一般来说,英汉两种语言中单词的对等率最高,而词组、成语、谚语的对等率就要相对低一些。因此,英译汉时,根据具体情况运用转态译法进行翻译。】

Exercises

1. Fill in the blanks with the proper given words, and then translate the sentences into Chinese.

oceangoing   navigation   route   maritime   encounter   merchant

1) There's been a big reduction in the size of the _____ fleet in recent years.
2) Today we live in a world where GPS systems, digital maps, and other _____ APPs are available on our smartphones.
3) Both of these activities were based on seafaring, an ability the Phoenicians developed from the example of their _____ predecessors, the Minoans of Crete.
4) In international commerce about 5,000 _____ vessels dock at New Orleans annually, and more than 40 nations have consular offices in the city.
5) Researchers are trying to get at the same information through an indirect _____.
6) Why do some students give up when they _____ difficulties, whereas others who are no more skilled continue to strive and learn?

2. Translate the following sentences into Chinese.

1) It was one of the most important trade routes during this time. Caravans loaded with silk often travelled through this passage, making it historically and economically important structure.
2) As the marine channel for silk became more and more famous, Silk Road trade began becoming redundant. It was during the Sui dynasty that new silk ports like Nanhai came into existence.
3) Ever since the inception of a marine link between China and rest of the world for trade of silk, several countries began taking interest in silk trade.
4) The hottest of the silk trade routes started from Chinese ports of southern regions including Wu, Wei, Qi and Lu regions. Owing to closeness of marine channels, the export and import of silk through these ports was always very easy.
5) Silks popularity continued throughout the Middle Ages, with detailed Byzantine regulations for the manufacture of silk clothes, illustrating its importance as a quintessentially royal fabric and an important source of revenue for the crown.
6) Knowledge about silk production was very valuable and, despite the efforts of the Chinese emperor to keep it a closely guarded secret, it did eventually spread beyond China, first to India and Japan, then to the Persian Empire and finally to the west in the 6th century CE.
7) Moreover, to the emperor who plied them with many questions as to whether he might have the secret, the monks replied that certain worms were manufacturers of silk, nature itself forcing them to keep always at work.
8) The early Middle Ages saw an expansion of this network, as sailors from the Arabian Peninsula forged new trading routes across the Arabian Sea and into the Indian Ocean.

9) The great variety of routes were illustrated, which were available to merchants transporting a wide range of goods and travelling from different parts of the world, by both land and sea.

10) Christianity, Islam, Hinduism, Zoroastrianism and Manicheism spread in the same way, as travellers absorbed the cultures they encountered and then carried them back to their homelands with them.

3. Translate the following sentences into English.

1) 通过海上丝绸之路,丝绸、瓷器、茶叶、黄铜和铁是出口到国外的四大品类,香料、花草和宫廷的珍奇珍宝也被带到中国。

2) 从中华人民共和国成立初期开始,中国的内部贸易和商业化高度发展,其总值一直大大超过国际贸易。

3) 这是第二条丝绸之路。它的水域和岛屿海峡就像中亚的沙漠和山口,它的港口就像商队旅馆。

4) 泉州(位于福建省)被许多专家认为是古代海上丝绸之路的终点站,至少在元朝之前,它可以说是南海贸易中最重要的港口。

5) 在传统航运技术下,没有现代地图和导航辅助设备,在南海水域航行极其危险。两千多年来,中国船只一直在南海的危险海域上航行。

6) 从罗马帝国衰落到至少在16世纪时期,中国的商业化、城市化、技术和文化水平远远领先于欧洲。

7) 中国政府的"陆海新丝绸之路"政策以发展基础设施和商务关系为核心。

8) 在许多重要方面,中国宣布的新政策建立在中国与中亚和东南亚之间古代贸易网络和文化互动的历史基础之上。

9) 为了支持商业和促进社会稳定,基础设施建设是中国两千多年来长期繁荣的基石。

10) 早在汉代,南海周边国家和海上丝绸之路沿线的国家就向中国统治者派遣了朝贡使团。

4. Choose the best paragraph translation. And then answer why you choose the first translation or the second one.

值得注意的是,这条路线最初是为了一般贸易目的,但后来因为这条特殊通道上发生的大量丝绸贸易而被称为"海上丝绸之路"。海上丝绸之路的流行在后来的许多朝代都得到了延续,包括唐朝和宋朝。这也促进了丝绸之路的建设,丝绸之路贸易也从这条路线的建设中受益匪浅。它连接了中国西安、玉关、阳关、丛岭、库尔勒、阿克苏、和田等几个重要城市,并进一步与中亚和西亚的城市相连。

译文一:

It is important to note that this route was started for general trade purposes but later got the name of "Maritime Silk Road" owing to the vast silk trade that occurred over this particular channel. The popularity of Maritime Silk Road for the trade was continued by many later dynasties including the Tang and Song dynasties. This also promoted building of Silk Road route and the Silk Road trade benefitted immensely from construction of this route. It connected several important cities starting from the Chinese town of Xian, Chang'an, Jade Pass, Yangguan, Cong Range, Korla, Aksu and Hetian and going further to meet cities in Middle and West Asia.

译文二：

It is worthy of this road being a common purpose at first, but later it is a special route that emerges lots of silk trade, thus called as "the Maritime Silk Road". The popular way to remain the later Tang and Song dynasties, which promotes the construction of the Maritime Silk Road. It is valuable that the Silk Road is on the way, which connects several cities, such as Xian, Chang'an, Jade Pass, Yangguan, Cong Range, Korla, Aksu and Hetian, and further connect the cities in Middle and West Asia.

5. Translate the following passage into Chinese.

Silk is a textile of ancient Chinese origin fabric from the protein fibre produced by the silkworm as it makes its cocoon. The cultivation of silkworms for the process of making silk, known as sericulture, was, according to Chinese tradition, developed sometime around the year 2700 BCE. Regarded as an extremely high value product, silk was reserved for the exclusive usage of the Chinese imperial court for the making of cloths, drapes, banners, and other items of prestige. Its production technique was a fiercely guarded secret within China for some 3,000 years, with imperial decrees sentencing to death anyone who revealed to a foreigner the process of its production. Tombs in Hubei province dating from the 4th and 3rd centuries BCE contain the first complete silk garments as well as outstanding examples of silk work, including brocade, gauze and embroidered silk.

6. Translate the following passage into English.

最古老的海上丝绸之路路线位于中国南部与印度洋和南太平洋沿岸的岛屿之间。到目前为止，这些最古老的丝绸贸易路线已经到达朝鲜、新罗、日本、印度和波斯，覆盖了亚洲北部、南部和东南部的大部分地区。其他最古老的海上丝绸贸易路线存在于中国与波斯湾和红海海岸之间，包括印度半岛的奎龙湾、苏门答腊岛、奥尔岛和暹罗湾和越南海岸。到元朝末年，已有220多个国家以这条海上丝绸之路为起点，成为贸易路线的一部分。清朝初年时期，在明代诸多航线的基础上又开辟了北美航线、俄罗斯航线、大洋洲航线等。

三、翻译家论翻译：林纾论译

林纾（1852—1924年），光绪八年举人，原名群玉，字琴南，号畏庐、畏庐居士，笔名冷红生，自号践卓翁，福建闽县（今福州）人。近代著名翻译家、文学家、画家、诗人。林纾的成就在他的"林译小说"中体现出来,他的诗文和画都为其译著所掩。

林纾少孤,喜欢读书,深爱中国传统文学,与文学结下不解之缘。

林纾翻译小说始于清光绪二十三年(1897年),与精通法文的王寿昌合译法国小仲马《巴黎茶花女遗事》,1899年在福州由畏庐刊行。这是中国第一部介绍西洋小说的译著,为国人见所未见,一时风行全国,备受赞扬。林纾受商务印书馆的邀请专译欧美小说,先后共译作品180余种。介绍来自美国、英国、法国、俄国、希腊、德国、日本、比利时、瑞士、挪威、西班牙的作品。单行本主要由商务印书馆刊行,未出单行本的多在《小说月报》《小说世界》上刊载。

"译才并世数严林,百部虞初救世心",是康有为赠林纾某诗的首联。林纾译得最多的是英国的哈葛德作品,有《迦因小传》《鬼山狼侠传》等20种;其次为英国柯南道尔的《歇洛克奇案开场》等7种。林纾译世界名作家和世界名著的小说,有俄国托尔斯泰的《现身说法》等6种,法国小仲马《巴黎茶花女遗事》等5种,大仲马《玉楼花劫》等2种,英国狄更斯的《贼史》等5种,莎士比亚的《恺撒遗事》等4种,司各特的《撒克逊劫后英雄略》等3种,美国欧文的《拊掌录》等3种,希腊伊索的《伊索寓言》,挪威易卜生的《梅孽》,瑞士威斯的《口巢记》,西班牙塞万提斯的《魔侠传》,英国笛福的《鲁滨孙漂流记》,斯威夫特的《海外轩渠录》,美国斯托夫人的《黑奴吁天录》,法国巴尔扎克的《哀吹录》,雨果的《双雄义死录》,日本德富健次郎的《不如归》,等等。这些翻译小说使国人认识了狄更斯、司各德、哈葛德、雨果、大小仲马等欧美作家,开始认识西洋文学,并认识到西洋文学的灿烂新鲜,在某种程度超越了我国文学。可以说是林纾开创了文学翻译的局面,使外国文学的翻译成为自觉。

林纾不懂外文,选择原本之权全操于口译者之手,因而也产生了一些疵误,如把名著改编或删节的儿童读物当作名著原作,把莎士比亚和易卜生的剧本译成小说,把易卜生的国籍误译成德国等。即使这样,林纾仍然译了40余种世界名著,这在中国,不曾有过第二个。

林纾在《译林·叙》和《剑底鸳鸯·序》中清楚地表达了自己的翻译思想:"吾谓欲开民智,必立学堂,学堂功缓,不如立会演说,演说又不易举,终之唯有译书……不能著书以勉我国人,则但有多译西产英雄之外传,俾吾种亦去其倦敝之习,追踪于猛敌之后,老怀其以此少慰乎!"可见,林纾认为只有发展翻译事业,才能"开民智",从而抵御欧洲列强,强我中国。林纾强调译者应投入主观感情,与原文或原作品的作者与作品中的人物进行心灵交流,才能达到原作的艺术效果,提出译者在翻译时应忠实于原著,并提出译者应"存其文不至踬其事"。

林纾在《鹰梯小豪杰》的序中这样表述:"或喜或愕,一时颜色无定,似书中之人,即吾亲切之戚畹。遭难为悲,得志为喜,则吾身直一傀儡,而著书者为我牵丝矣。"这段序言言简意赅地体现出林纾的翻译反映了他所处时代的文化心态和翻译理想,开创了中国人译介外国文学作品的先河,推动了中国新文学的革新。

(摘自:曾诚.实用汉英翻译教程[M].北京:外语教学与研究出版社,2002:280.)

Text B

Maritime Rules of the Road
By Tony Bessinger

Ideological and Political Education:鉴航海进步发展史 展大国风采

中国航海历史悠久。一个国家的航海史能够折射出一个国家的发展历程。早在距今 7 000 年前的新石器时代,我们的祖先已经能用火与石斧"刳木为舟、剡木为楫",到了春秋战国时期开始制造木帆船,出现了大规模的海上运输与海上战争。三国魏晋南北朝时期,中国商船远航到了波斯湾。从隋唐五代到宋元时期,中国航海业繁荣发展,海上丝绸之路远及红海与东非之滨。到了明朝永乐年间,大航海家郑和先后七次下西洋,遍访亚非各国,将中国航海业推向顶峰,也竖起了人类航海史上前无古人的历史丰碑。中华人民共和国成立后,我国高度重视领海主权,在开发、利用、管理领海等方面进入了一个新的历史阶段,颁布了一系列涉及领海问题的规章制度,尤其是 12 海里领海制的确定。

向海而生,是一种探索的勇气,一种开放的胆魄,一种通达的梦想。在世界航海史上,中华民族不仅为世界航海业贡献了指南针、舵、水密舱壁、车轮舟等影响世界的重要发明,也在人类海洋文明发展史上刻上了中华民族的深深烙印。

面向未来,我们要在习近平总书记提出的建设"丝绸之路经济带"的倡议下,不断与世界各国在"和平合作、开放包容、互学互鉴、互利共赢"的丝路精神和"不畏艰险、勇于开拓、同舟共济、尚新图变"的航海精神下,不断加强航海和海洋事业发展,薪火相传,使之成为新时代中国航海人的精神谱系。

21 世纪的中国,逐梦深蓝,坚持陆海统筹,走依海富国、人海和谐、合作共赢的发展道路。我们要坚守初心使命,勇做新时代的奋进者、追梦人,加快建设海洋强国、交通强国、造船强国,为实现中华民族伟大复兴的中国梦,推动构建人类命运共同体而努力奋斗。

21 世纪海上丝绸之路的伟大实践,尽管道路曲折艰难,但动摇不了中国建设强大海洋力量并以此维护自己海洋主权不受侵犯的坚定信念,动摇不了中国引领世界海洋命运共同体的意识担当。

1. Let's take a simplified, basic look at the Maritime Rules of the Road, also known as the International Regulations for Preventing Collisions at Sea. Rule 1: the basic premise that the rules apply to all vessels. Rule 2: the idea that nothing in the rules **exonerates** you from common sense. In other words, you cannot blindly follow the rules to the point of colliding with another vessel not following the rules. Rule 3: make it clear that you can depart from the rules if the need arises, i.e. **imminent** collision. Rule 4: lay out the **hierarchy** of the waterways, also known as the pecking order. It seems complicated at first-read, but makes complete sense. The rule first defines what makes a vessel, and makes it clear that even seaplanes and Wing-In-Ground (skimming-type) vessels are subject to the rules.

2. Next up are Fishing Vessels, and we'll start this rule up by correcting a common misconception. Fishing Vessels are vessels that use: "nets, lines, **trawls**, and other fishing **apparatus**," but do not include vessels that are **trolling, aka**, sports fishermen. One step up from fishing vessels are vessels restricted in their ability to **maneuver**. These are vessels that, "through the nature of their work are restricted in their ability to maneuver." These vessels include **dredges, minesweepers**, cable-layers, **buoy tenders**, survey vessels, aircraft carriers conducting flight operations, and even (sometimes) tugs and tows. These vessels show shapes or lights and make Securite calls on the VHF radio about their situation.

3. The following are vessels **constrained** by draft, but this is a tricky rule, because it applies only in International Waters. The bays and sounds we mostly use are not in international waters, and therefore this rule does not apply. However, deepwater vessels that enter bays and sounds must stick to the marked channels, and there is a different section of the rules that deals with vessels wanting to cross or use marked channels that don't need the depth available and are interacting with vessels that do. A simple way to think about this? Don't interfere with **deep-draft** vessels in marked channels.

4. At the top of the **heap** are vessels not under command. By "command", the rules don't mean there's no captain, and they mean that commands to the engine room or steering station can't be followed; therefore the vessel is either close to incapable or completely incapable of following the rules. The strict wording is quite clear: "...... a vessel which through some exceptional circumstance is unable to maneuver as required by the rules". These vessels show shapes or lights and make Securite calls about their situation, much the same as vessels restricted in their ability to maneuver.

5. The next section of the Rules deals with steering and sailing and the rules are for any visibility.

6. First, the rules define Safe Speed. Basically this means: don't go faster than the situation

allows. In other words, if it's crowded, you have poor visibility, or are driving a vessel with limited **maneuverability**, you must take all these factors into account.

7. The next rule says that all vessels must maintain a proper lookout at all times, using any and all available means of doing so. That means if you have a VHF radio, you are required to monitor Channel 16, and if you have RADAR or AIS, you must monitor them as well. You must also, of course, look around to determine if a risk of collision exists. The rule also states that any action that must be taken be positive, obvious, and made in good time. Slight course changes aren't obvious, so make it clear to the other vessel what your intentions are, and do them before the situation becomes extreme. Although unspoken, this rule also means that distracted driving, such as using your mobile phone to call or text, is not OK, as it means you are not maintaining a proper lookout.

8. The next section applies to vessels that are in sight of one another, and deals with sailing vessels first. For sailors, these are rules we know well. Port tack gives way to **starboard** tack, **windward** gives way to **leeward**, and if you're on port tack and unsure of the other vessel's point of sail, you must give way.

9. In all the situations there are two types of vessels, the Stand On, and the Give Way. The Give Way vessel must take early and substantial action to keep clear. The Stand On vessel must maintain her course and speed until it becomes clear that the Give Way vessel isn't following the rules, and even when that happens the rules state that the Stand On vessel should avoid turning to port.

10. Overtaking is one of the most misunderstood rules. Quite simply, it does not matter what type of vessel you are. If you are overtaking, you must stay clear of the other vessel. Even if you're a sailing vessel overtaking a power-driven vessel, you must stay clear. If you are the Overtaken vessel, you must maintain your course and speed, otherwise how can the other vessel stay clear?

11. Crossing Situations are also often misunderstood. Simply put, the vessel that's coming from your starboard side has the right of way, and you are required to either alter course or slow down so as to pass **astern** of that vessel.

12. Dealing with Commercial Vessels: Most commercial boaters have at least some idea about how to deal with commercial vessels (see the "Tonnage Rule"). However, they sometimes have a hard time knowing how to do that, and how to properly communicate with the commercial operators. By knowing the rules above and being aware of your surroundings, you can usually avoid issues with commercial vessels in the first place. Sometimes, especially in

crowded waterways, you're going to have to talk directly to the commercial vessel and make your intentions clear, or he is going to call you up and ask what your intentions are. All commercial vessels monitor VHF channels 13 and 16; 13 is for bridge-to-bridge communications, and is the channel that commercial vessels use to speak to one another regarding their crossing, passing, or overtaking situations. The conversations are usually brief and the agreements are laid out very clearly, and often repeated twice. Like all vessels, including yours, commercial vessels monitor 16 so as to hear any Securite, Pan Pan, Mayday, or Coast Guard communications. We often hear commercial vessels trying to **hail** recreational vessels on 16; sadly, very few answer. If you are anywhere near a commercial vessel that's underway, you should be listening very carefully to 16 and answer promptly if hailed. Chances are he's not calling to chat, but to warn you that what you're doing is putting you in danger.

13. If you feel the need to hail a commercial vessel, hail them on 13, and be as professional and brief as you can. Call them up and wait for their answer before launching into your **spiel**. Then, once communications have been established, tell them what you're doing to avoid them, or ask whatever question you may have about what their intentions are. Things not to do include saying: "**Tugboat** off my starboard bow, this is the white sailboat, is it OK if I go ahead of you?" The answer you'll get will go something like this: "Uh, Cap, which vessel are you, exactly? I have five white sailboats in sight right now and don't want to steer you wrong." This is where AIS is invaluable to both the recreational and commercial **mariner**. You'll be able to contact the **tug** by name, and he'll be able to see exactly which sailboat you are.

(1,282 words)

https://www.boats.com/how-to/maritime-rules-of-the-road-a-primer-for-boaters/

New Words

1. exonerate [ɪɡˈzɒnəreɪt]

    v. pronounce not guilty of criminal charges 使免罪

    Under specific circumstances, the lawyer is able to exercise the counterplead right to exonerate from part or all of the compensation liability.

    在特定的情形下,律师可以行使抗辩权,全部或部分免除赔偿责任。

2. imminent [ˈɪmɪnənt]

    a. close in time; about to occur 即将发生的,临近的

    Everyone feels that a disaster is imminent, as if a catastrophe is about to come.

    谁都感到大事不好,似乎有什么极大的灾难即将来临。

3. hierarchy [ˈhaɪərɑːki]

    n. a series of ordered groupings of people or things within a system; the organization of people at different ranks in an administrative body 等级制度；统治集团；等级体系

Like most other American companies with a rigid hierarchy, workers and managers had strictly defined duties.

像大多数其他等级制度森严的美国公司一样，工人和管理人员都有严格界定的职责。

4. peck [pek]

v. eat by pecking at, like a bird; hit lightly with a picking motion; eat like a bird 啄，啄食；匆匆轻吻

When the Prince eventually realizes Cinderella is the one for him, as a punishment for evil, birds peck out the stepmother's and sisters' eyes.

当最后王子发现灰姑娘才是他的命中注定时，作为对邪恶的惩罚，鸟啄去了继母和姐姐们的眼睛。

5. trawl [trɔːl]

n. a conical fishnet dragged through the water at great depths; a long fishing line with many shorter lines and hooks attached to it (usually suspended between buoys) 拖网；排钩；用拖网捕鱼

In the previous study of trawl fishery resources of the Yellow sea, the number of hauls is usually taken as the unit of fishing effort.

在以往的黄海底拖网渔业资源研究中，投网次数常作为捕捞量的度量单位。

6. apparatus [ˌæpəˈreɪtəs]

n. equipment designed to serve a specific function; (anatomy) a group of body parts that work together to perform a given function 设备，器具；机构，组织；器官

One of the boys had to be rescued by firemen wearing breathing apparatus.

其中的一个男孩得由戴着呼吸设备的消防队员们营救。

7. troll [trɒl]

v. to try to catch fish by pulling a baited line through the water behind a boat; cause to move round and round; circulate, move around 拖捞，拖钓；搜寻，寻找；随便浏览，随便翻翻

They were trolling the colder waters of the Channel.

他们在海峡的寒冷水域采用曳绳钓鱼。

8. aka [ˌeɪ keɪ ˈeɪ]

abbr. used when someone has another name 又名，亦称 (also known as)

Exotropia is the term used to describe outward turning of the eyes (aka "wall-eyed").

外斜视的症状是眼球向外转（又名"白眼"）。

9. maneuver [məˈnuːvə(r)]

v. direct the course; determine the direction of travelling; perform a movement in military or naval tactics in order to secure an advantage in attack or defense 巧妙地移动；操纵，使花招；部署（军队、船只等）；演习

She can then maneuver the 3D image on the computer screen to map the shortest, least invasive surgical path to the tumor.

然后,她可以操纵电脑屏幕上的3D图像,以绘制出最短、切口最小的通往肿瘤的手术路径。

10. dredge [dredʒ]

n. a power shovel to remove material from a channel or riverbed 挖泥船,疏浚机;拖捞网

Hammer dredge is a technical improvement of the existing dredge sampler.

重锤式拖网取样器是对现行常规拖网的技术改进。

11. minesweeper ['maɪnswiːpə(r)]

n. ship equipped to detect and then destroy or neutralize or remove marine mines 扫雷舰;扫雷艇

The loss of a minesweeper, at a time of rising tension in the Persian Gulf over Iran's nuclear programme, was questioned by opposition MPs.

在波斯湾因伊朗核项目而日益紧张之际,一艘扫雷舰的损失受到了反对派议员的质疑。

12. buoy [bɔɪ]

n. a float attached by rope to the seabed to mark channels in a harbor or underwater hazards 浮标,航标;救生圈

The buoy floated back and forth in the water.

浮标在水里漂来漂去。

13. tender ['tendə(r)]

n. a boat for communication between ship and shore; car attached to a locomotive to carry fuel and water; someone who waits on or tends to or attends the needs of another (在大船和口岸之间载运人或货物的)供应船,补给船;(装有特别设备的)辅助救火车;服务车,辅助车辆;照料人,照管人

A supply ship carrying people or goods between a large ship and a port is a tender.

供应船在大船和口岸之间载运人或货物的船只。

14. constrain [kən'streɪn]

v. hold back; restrict 限制,约束;强迫,迫使

Policies not only constrain decisions, but also may require decisions to be made.

政策不仅约束决策,而且还可能要求制定决策。

15. deep-draft ['diːpˌdrɑːft]

a. 深水的;载重吃水深的

Yang Shan deep-draft port is an important part of Shanghai international shipping center, so it attracts worldwide attention whether the project could move forward fast, safely and steadily.

上海洋山深水港作为上海国际航运中心的重要组成部分,工程项目能否快速、安全、稳步推进,举世瞩目。

16. heap [hiːp]

n. a collection of objects laid on top of each other(凌乱的)一堆;许多,大量

She piled the papers in a heap on her desk, just anyhow.

她把文件在桌上随便搁成一堆。

17. maneuverability [məˌnuvəˈbɪlətɪ]

    n. the quality of being maneuverable 可操作性;机动性

    Maneuverability is one of the most important tactical and technical performances for fighter planes.

    飞机的机动性是战斗机重要的战术、技术指标之一。

18. starboard [ˈstɑːbəd]

    n. the right side of a ship or aircraft to someone facing the bow or nose(船舶或飞机的)右舷,右侧

    The hull had suffered extensive damage to the starboard side.

    船体右舷遭到大面积损坏。

19. windward [ˈwɪndwəd]

    n. the direction from which the wind is coming 迎风面;上风方向

    Each rise correlates with the windward-facing slope of the terrain, while declines in cloudiness are found in valleys on the leeward side.

    每一处云量的上升都与迎风面斜坡相关,同时,云量的下降出现在背风面的山谷。

20. leeward [ˈliːwəd]

    n. the side of something that is sheltered from the wind 背风,下风

    A leeward side of the box is provided with at least one air outlet having a vent check valve.

    箱体的背风面至少具有一个带有出气单向阀的出气口。

21. astern [əˈstɜːn]

    ad. at or near or toward the stern of a ship or tail of an airplane; stern foremost or backward 向船尾,在船尾;向后

    Ship's emergency astern is an urgent measure to prevent collision between ships.

    船舶在船尾紧急制动,是船舶航行中避碰而采取的应急措施。

22. hail [heɪl]

    v. call for; be a native of; precipitate as small ice particles 呼喊,招呼;赞扬,欢呼;下冰雹;(大量物体)像雹子般落下(或击打)

    As they smashed up neat Georgian barracks, the Russians cursed their own poverty and hailed their victory at the same time.

    当他们将整洁的格鲁吉亚兵营夷为平地,俄罗斯人在诅咒自己贫困的同时,却又在为胜利欢呼。

23. spiel [ʃpiːl]

    n. plausible glib talk (especially useful to a salesperson) 流利夸张的讲话,招揽生意的

言辞

Once they've done their spiel, you can ask them plenty of difficult questions.

一旦他们开始喋喋不休口若悬河,你就可以扔出一堆难题来考他们。

24. tugboat ['tʌgbəʊt]

   n. a powerful small boat designed to pull or push larger ships 拖船(等于 towboat 或 tug)

   The tugboat is small and dark, and it's making a lot of smoke. But it's very strong, because it's pulling the much bigger ship.

   拖船又小又黑,还冒着很多烟,但它非常有力,因为它在拖着比它更大的船。

25. mariner ['mærɪnə(r)]

   n. a man who serves as a sailor 水手;船员

   When a storm arises, the hardy mariner doesn't turn off steam and drift helplessly before the wind.

   当船舶在海上遇上风暴时,勇敢的水手是不会关掉引擎,任由自己绝望地迎风漂流的。

26. tug [tʌg]

   n. a powerful small boat designed to pull or push larger ships; a sudden abrupt pull 拖船;猛拉,拽

   The tug crossed our stern not fifty yards away.

   那艘拖船在不到 50 码的地方与我们的船尾错过。

Phrases and Expressions

collide with 冲突
depart from 离开;开出;从……出发
be subject to 受支配,从属于;常遭受……;有……倾向的
next up 接下来
stick to 坚持;粘住
interact with 与……相互作用
at all times 一直;始终
give way 让路;撤退;倒塌;失去控制
astern of 在……后面
lay out 展示;安排;花钱;为……划样;提议

Proper Names

international waters 公海;国际水域
VHF 甚高频(very high frequency)
AIS 航空情报处(Aeronautical Information Service)

一、翻译简析

1. However, deepwater vessels that enter bays and sounds must stick to the marked

channels, and there is a different section of the rules that deals with vessels wanting to cross or use marked channels that don't need the depth available and are interacting with vessels that do. (Para. 3)

然而，进入海湾和海峡的深水船只必须航行在有标记的通道里。对于想要穿越或使用标记航道无须可通行深度的船只，与需要达到通行深度的深水船只相互关联，需要遵守不同的规则条款。

【这个句子是一个较长的复杂句。首先查词典，确定单词 sound 的词义是"海峡，海湾"的意思。句子原封不动地进行翻译，往往会使读者感到费解。把这个长句断开来翻译，先找出句子的主语、谓语、宾语和其他成分再进行翻译。主语是"deepwater vessels"，谓语是"stick to"，宾语是"the marked channels"，"and there is……"作为并列结构补充说明。翻译时如果把这种结构直接转换成汉语，译文会显得臃肿不堪。这时，我们需要考虑采用断句译法，将"there is……"进行断句翻译。理清句子的语法层次后，按照译文的语言习惯调整搭配，恰如其分地断开长句进行翻译。】

2. At the top of the heap are vessels not under command. By "command", the rules don't mean there's no captain, and they mean that commands to the engine room or steering station can't be followed; therefore the vessel is either close to incapable or completely incapable of following the rules. (Para. 4)

在顶端的是不接受管辖的船只。所谓"管辖"，规则并不意味着没有船长，而是指对船舱或操舵舱的命令不能执行的情况。因此，这艘船要么几乎没有能力，要么完全没有能力遵守规则。

【翻译时首先遇到 heap 这个词，不要直接译为"一堆"，根据上下文译为"顶端的"，原文后面的三个词汇作为定语放在船只的前面。理清句子的语法层次后，按照译文的语言习惯调整搭配，再进行翻译。】

3. Port tack gives way to starboard tack, windward gives way to leeward, and if you're on port tack and unsure of the other vessel's point of sail, you must give way. (Para. 8)

左舷航向变成右舷航向，迎风航向变成下风航向，如果你在左舷航向，不确定对方的航行方位点时，你就必须转向。

【这是个并列复合句，句式结构不是很复杂，适合于对句式进行直译。"port tack"译为"左舷航向"，"starboard tack"译为"右舷航向"。句末的"give way"有"让路；撤退；倒塌；失去控制"的意思，但翻译时根据上下文句意译为"转向"。】

4. The Stand On vessel must maintain her course and speed until it becomes clear that the Give Way vessel isn't following the rules, and even when that happens the rules state that the Stand On vessel should avoid turning to port. (Para. 9)

待命船必须保持航向和航速，直到发现让路船没有遵守规则，即使出现这种情况，规则也规定待命船应避免转向左舷。

【翻译"Stand On vessel"时，首先要确定"Stand On"的词义为"持续向同一方向航行"，根据句意，译为"待命船"。port 作为名词是（船的）舷门舷窗的意思，作为动词有"（船，舵）向左转"的词意，这里译为"左舷"。】

5. Crossing Situations are also often misunderstood. Simply put, the vessel that's coming

from your starboard side has the right of way, and you are required to either alter course or slow down so as to pass astern of that vessel. (Para.11)

关于超船驾驶规则也经常被误解。简单地说，从右舷驶来的船只有先行权，其他船只必须改变航向或减速，以便从那艘船的后面驶过。

【这个较长的复杂句中的定语从句部分"that's coming from your starboard side"，将其翻译到所修饰的先行词的前面，用"的"来连接。这种翻译方法称为"前置翻译法"。】

二、参考译文

海上交通规则

1. 让我们简单地了解一下《海上交通规则》，也就是《国际海上避碰规则》。规则1：规则的基本前提是适用于所有船舶。规则2：在一般意义上没有任何情形可免受规则约束。换句话说，你不能盲目地遵守规则，以至于与另一艘不遵守规则的船只相撞。规则3明确规定，如果有必要，即将发生碰撞时你可以不遵守规则。规则4：排列出水道的等级，也被称为啄食顺序。乍一看似乎很复杂，但完全有道理。该规则首先定义了船只的构成，并明确表示即使是水上飞机和地面翼(撇油型)船只也受该规则的约束。

2. 接下来是渔船，我们将通过纠正一个常见的错误来讲述这个规则。渔船是使用"网、线、拖网和其他捕鱼设备"的船只，但不包括拖网渔船，又称"运动渔民"。比渔船高一级的是机动操纵受限船只。这些船只"由于工作性质使操纵能力受到限制"，包括挖泥船、扫雷舰、电缆敷设船、浮标船、测量船、执行飞行任务的航空母舰，甚至(有时候)是拖船和牵引船。这些船只显示外形特征或光源，并通过甚高频无线电汇报其所在位置，进行安全呼叫。

3. 然后是受吃水深度限制的船只，但这是一个复杂的规则，因为它只适用于国际水域。我们主要航行的海湾和海峡不在国际水域，因此这条规则不适用。然而，进入海湾和海峡的深水船只必须航行在有标记的通道里。对于想要穿越或使用标记航道无须可通行深度的船只，与需要达到通行深度的深水船只相互关联，需要遵守不同的规则条款。

4. 在顶端的是不接受管辖的船只。所谓"管辖"，规则并不意味着没有船长，而是指对船舱或操舵舱的命令不能执行的情况。因此，这艘船要么几乎没有能力，要么完全没有能力遵守规则。严谨的措辞十分明确："……在某些特殊情况下无法按照规则要求操纵的船只"。这些船只能够显示外形特征或光源，并发出安全呼叫，汇报其所在位置，这与操纵能力受限的船只大致相同。

5. 下面的规则内容涉及船舶驾驶和航行规则及任何可见性规则。

6. 首先，规则规定了安全航速。基本上来说：不要超出规则允许的速度。换句话说，要考虑以下的因素，如船上很拥挤、能见度很低、驾驶船只操纵能力有限等。

7. 下一条规则规定，所有船只必须使用各种可用手段，全程保持正规的瞭望。这意味着如果有甚高频无线电，就必须监控第16频道；如果有雷达或航空情报设备，也必须监控第16频道。当然，还必须环顾四周，以确定是否存在碰撞风险。该规则还规定，必须采取积极的、明显的、及时的各项措施。细微的航向改变并不明显,所以要让其他船只清楚行驶船只的意图,并在极端情况出现之前采取行动。虽然这条规则没有明确说明,但它也意味着不能分神驾驶,比如不允许用手机打电话或发短信,因为这意味着你没有保持正规的瞭望。

8. 下面的规则适用于彼此视线范围内的船只,首先涉及帆船。对于水手来说,这些规则都很熟悉。左舷航向变成右舷航向,迎风航向变成下风航向,如果你在左舷航向,不确定对方的航行方位点时,你就必须转向。

9. 在所有情形下,有两种类型的船只,持续向同一方向航行或避让时,避让船只必须尽早采取实质性的行动保持航线畅通。待命船必须保持航向和航速,直到发现避让船只没有遵守规则,即使出现这种情况,规则也规定避让船只应避免转向左舷。

10. 超船规则是最容易被误解的规则之一。很简单,什么类型的船只并不重要。如果超船,必须与其他船只保持适当的距离。即使是一艘帆船,超越了一艘电力驱动的船只,也必须保持清醒。如果是被追赶的船只,必须保持航向和速度,否则其他船只怎么能保持清醒?

11. 关于超船驾驶规则也经常被误解。简单地说,从右舷驶来的船只有先行权,其他船只必须改变航向或减速,以便从那艘船的后面驶过。

12. 与商船打交道:大多数商船船东至少对如何与商船打交道有一些了解(参见"吨位规则")。然而,有时候很难知道如何做到这一点,如何正确地与商务运营商沟通。通过了解上述规则与周围环境,通常可以在一开始就避免与商船发生碰撞。有时,特别是在拥挤的水域,必须直接与商船商谈,表明你的意图,否则他会打电话问你的意图是什么?所有商船都监测甚高频频道13和16,频道13是驾驶台与驾驶台之间的通信联络,是商船用于相互交流的通道,涉及穿越、通过或超船情况。通话通常很简短,协议也写得很清楚,而且经常重复两遍。像所有船只一样,包括你的船只,商船只会监控频道16,以便能够听到任何关于安全、紧急警报、求救或海岸警卫队的通信联络。我们经常听到商船在频道16号航道上试图与休闲船只打招呼;遗憾的是,很少有人回应。如果在一艘正在行驶的商船附近,应该非常仔细地收听频道16的广播,如果有人打电话,应立即回答。对方打电话很有可能不是为了聊天而是为了警告,警告你正在做的事情处于危险之中。

13. 如果需要呼叫某一商船,请用频道13进行呼叫,并尽可能说得具有专业性并简短表述。打电话呼叫时,待回答后再开始讲述船只的情况。接下来,一旦建立了沟通,告诉对方你正在做什么来避免与对方发生碰撞,或者问一些涉及对方意图的问题。不要做的事情包括:"拖船在我的右舷船头,这是白色的帆船,我可以走在你前面吗?"你会得到这样的答案:"呃,船长,你到底在哪艘船上?我现在看到了五艘白色的帆船,我不想让你误入歧途。"这就是船舶自动识别系统对休闲船和商船船员都非常宝贵的地方。可以通过船名联系拖船,就能准确地知道你在哪艘帆船上。

Cultural Background Knowledge: Cultural Exchanges along the Maritime Road
海路文化交流文化背景知识

浩瀚广袤的海洋赋予了人类精神财富与物质财富,海路文化因其与海洋深深相吸又博大精深变得源远流长。海上与陆上丝绸之路并肩启航,有力地促进了东西方的经济文化交流,对促进区域繁荣、推动全球经济发展具有重要意义,同时将大大拓展中国经济发展战略空间,为中国经济持续稳定发展提供有力支撑。

海路文化交流离不开航海活动。轮船是工业革命的产物,1807年出现了有实际应用价值的木制蒸汽轮船,1820年铁轮船"艾伦蔓比"号出现,自此,航海活动给人类带来了巨大影响。古罗马时期的伽太基(在今天的突尼斯共和国境内,迦太基坐落在非洲大陆北海岸)航海家希米尔康于公元前520年为了找锡探索了大西洋,挪威人埃里克劳德在982年发现了格陵兰岛,他的儿子在1003年发现了纽芬兰岛和美洲大陆。15世纪初,著名的航海家郑和,自永乐三年(1405年)至宣德八年(1433年)的28年时间里,七次下西洋。据《明史》记载,郑和第一次下西洋时,率领众部27 000多人,船只长达44丈、宽18丈、9桅12帆的就有62艘,每艘船载重量为1 500~2 500吨,规模之大前所未有。1492年,哥伦布的旗舰载重量为250吨,他在1492—1502年期间进行了环球航行,发现了美洲大陆。达伽马开辟了连接大西洋和印度洋的航线。1906年,挪威人阿蒙森驾船穿越了加拿大北极地区,到达阿拉斯加西海岸的诺姆港,实现了打通"西北航线"的目标。此后的百年间,海上航路不断开拓,到了21世纪,海路文化随着各个航路的开辟不断开创了各国之间的海上路线,带来了全球的经济文化发展。

自此,海路文化交流不断发展,海上丝绸之路已经成为我国与各国之间文化交流之路,海上丝绸之路对于我国以及其他国家的重要性,它带动的不仅仅是不同文化的交流碰撞,更是推动了世界的进步与发展,是东西方友好交往的一个重要通道。

Unit Four
Maritime Folklife

海洋民俗生活

Text A

Pirate Cultural Relics

By Heather E. Hatch

Ideological and Political Education：海洋民俗傍海而生 因海而兴

傍海而生，因海而兴。

民俗文化是由特定区域劳动人民所创造、传承、分享、发展的，具有良好的群众基础、深厚的文化内涵及生活化和仪式化特征的一种特殊的传统文化形态。我国拥有绵长的海岸线，特殊的地理环境构成了具有鲜明个性和丰富内容的海洋民俗文化。它反映了沿海居民对海洋的认识经历，折射出沿海居民的生活方式、娱乐习惯和原始信仰崇拜等，体现了沿海居民创造海洋文化的历史。

我国海洋民俗文化具有四个主要特性：

海陆交融性。由于地理环境的特殊性，海洋型民俗文化既包括了海洋民俗文化属性，又包括了内陆民俗文化属性，比如沿海地区特有的"闹海""赛泥马""人龙舞""开渔节""攻淡菜"等活动都体现出内陆生产和海洋渔业活动的交织性，并在此过程中将生产劳动、舞蹈和音乐进行结合。

历史传承性。沿海地区在悠久的历史中形成了世代传承、共同信奉的文化传统习俗。

休闲娱乐性。海洋民俗文化活动的形成，与当地原生态的海洋型生产生活密不可分。大量的渔业生产劳动技能被渔民应用到休闲娱乐活动之中，如放海灯、妈祖出巡、船舞、吃普度等，体现了劳动人民吃苦耐劳、拼搏向上的精神。

功能多样性。随着沿海经济的发展，海洋民俗文化活动将逐渐集社会、经济、教育、娱乐和文化等多种功能于一体。

海洋民俗文化是海洋文化中一个非常重要的组成部分，它反映了海岛居民的日常生活和思想情感，表现了当地渔民的审美观念和艺术情趣。

习近平总书记指出："我们人类居住的这个蓝色星球，不是被海洋分割成了各个孤岛，而是被海洋连结成了命运共同体，各国人民安危与共。"海洋民俗文化连接着世界各国，是共同打造人类命运共同体的基石。同时，海洋的和平安宁关乎世界各国的安危利益，需要我们共同维护。与各国携手构建海洋命运共同体，促进合作共赢，是我们建设海洋强国的必由之路。

1. Pirates are an interesting group-a somewhat liminal, and definitely maritime, **offshoot** of the larger European **colonial** culture of the time. This research focuses specifically on pirates and piracy in the period known as the Golden Age of Piracy, from 1680 to 1730, through analyzing the archaeological record of artifacts of the time exploring how past peoples have used and transformed their landscapes, maritime and otherwise.

2. Behaviors that might be represented by the Barcadares **assemblage** can aid in understanding the social **dynamics** of the logwood cutters at the Barcadares, and by extension, other members of their subculture. Comparison with the Ridge Complex and Port St. George highlighted some of the aspects of the logwood camp that might suggest a distinct piratical or maritime orientation in the material culture.

3. The categories used in this analysis are Food Consumption (subdivided into drinking, tableware, and utilitarian ware), Architecture, Weaponry, Tobacco Pipes, Clothing, Food Production, and Industrial Activity. Artifacts within the groups are also subdivided by class, material, ware, and type. The analysis **disregarded** material of a clearly modern nature where possible. For the Nevisian sites, this included materials that clearly fell outside the temporal boundaries of the early- to mid- eighteenth century. These comparisons are not perfect by any means-apart from the biases introduced by different site formation and collection processes, Meneketti's information did not always allow the material from his projects to be arranged into directly **analogous** categories. Nevertheless, the results are fairly informative and indicate the potential of this approach. This analysis is not intended to provide figures for direct comparison, but rather to give a more general impression of the differences visible in the site assemblages. A brief examination of each category follows, along with a final summary of findings and potential of this experiment.

4. Food Consumption. There are several notable differences between the Food Consumption wares at the Barcadares and the two Nevisian sites. The first is the difference in the variety of forms. The Ridge Complex and Port St. George assemblages both contain a wider spectrum of **vessel** forms in all categories, including mugs, cups, platters, and serving dishes all completely absent from the Barcadares. With the exception of one plate and one **saucer** sherd, all Barcadares forms are rounded forms capable of holding liquid, while the forms from the other two sites serve much more **diversified** functions.

5. In the Drinking subgroup, glass bottles dominate the Barcadares assemblage but are much less significant at the two Nevisian sites. While bottles still comprise the majority of the drinking assemblage from Port St. George, this figure may be overly weighted with later period bottles. Nevertheless, the site does not exhibit the same

disproportionate representation as the logwood pirate camp. Cups and mugs dominate the Ridge Complex drinking forms, but there are none of these common forms at the Barcadares. The Drinking subgroup at the Barcadares comprises nearly double the percentage of the total assemblage in comparison to the two Nevisian sites, and comprises a much larger percentage of the Food Consumption Group. If the **baymen** used drinking vessels of perishable material, such as horn or wood, this would only increase the **disparity**.

6. The lack of Utilitarian Wares at the Barcadares is notable, as they make up a significant percentage of **identifiable** wares at the Ridge Complex and Port St. George. If bowls and the **porringer sherds** were included as a drinking form at the Barcadares, there would be almost no Tablewares. This comparison of vessel forms appears to reinforce the conclusions drawn previously about the baymen's focus on communal foodways.

7. This same discrepancy of diversity is evident in the ware types as well. Delftware dominates at the Barcadares, but other wares make a good showing on Nevis, including white stoneware and **porcelain** at the Ridge Complex and various stoneware types at Port St. George. The Ridge Complex also has some colonoware, another ware type absents from the logging camp which points to differences in social attitudes between the baymen and colonial sugar planters-the loggers did not typically keep slaves. It is impossible to **extrapolate** back to the decisions made by the baymen in their pottery selection, but it is interesting to note that tin enameled wares, while readily available in this period, had a much softer paste and were generally less durable than stonewares. They were also often made to imitate more expensive porcelain and higher status wares, unlike the sturdier and cheaper utilitarian wares largely absent from the logging camp.

8. Architecture. Construction methods and housing structures account for much of the difference between the logwood camp and the two sites on Nevis. The baymen did not **erect** permanent structures, but lived in tents made of sailcloth on raised platforms constructed of thin logs. The planters, on the other hand, erected permanent brick structures for habitation and also for industrial purposes. These differences in construction are represented in the types of architectural artifacts found on the sites as well as their numbers. On the Nevisian sites, bricks, tile, **mortar**, nails, and even standing foundations were common. At the Barcadares, nails were the only architectural artifact recovered.

9. Weaponry. This is not a strongly represented group at any of the three sites. It is unsurprising that it is largest at the Barcadares, considering that documentary evidence

emphasizes the popularity of hunting and recreational shooting among the baymen.

10. Tobacco Pipes. This is the largest group of artifacts at the Barcadares at 36% of the total site assemblage. Smoking was an important part of maritime culture, and this is clear when the Barcadares is compared to other sites. South gave a percentile range of 1.9%-14.0% for his Tobacco Pipes Group in the Frontier Pattern and 0-20.8% for the Carolina Pattern. The Ridge Complex assemblage is just outside this range at 22.35%, but Port St. George falls well inside with 15.69%. South asserted that the pattern should hold for most British colonial sites. The percentage at all these sites is high compared to the mean, but it is clear that the percentage of the Tobacco Pipe Group from the Barcadares is exceptional. Interestingly, Meniketti notes that an exceptionally high proportion (65%) of the Port St. George pipe assemblage comes from the older Zone 3, which is potentially associated with the use of the site as a port, and likely indicates that sailors, slaves, or both account for most of the smokers from that site.

11. Clothing. Only the Barcadares produced any material for this category, and there it consisted of a single buckle.

12. Food Preparation. This is a fairly significant category at the Barcadares, partly because of the decision to include the fragments of clay **hearths** in the analysis. The lack of representation of artifact in this group on Nevis is notable but misleading. At the Ridge Complex, Meniketti's landscape focus meant that sampling methods were not designed to detect areas with Food Production remains such as trash pits. In the case of the Port St. George materials, none of the South's original categories included faunal remains, and this comparison relies on Meniketti's use of South's analytical model. Meniketti referred to faunal remains in the text of his report, and noted that there were not many-some egg shells inside the main complex, one sheep, one bird, one large mammal (probably a cow), fish, and some goats remain that were probably intrusive. Many untended goats roam the site and the complex, which Meniketti described as being "where goats come to die". The fact that the Barcadares Food Preparation Group is so **conspicuous** in comparison is an important marker of the different lifestyles of the baymen and the Nevisian colonists and planters.

13. Although Finamore used the vessel forms to suggest that the baymen engaged in some kind of communal food sharing, the multiple small hearths recovered suggest that at the least, the "community" involved in the sharing of prepared food may have been relatively small.

14. Industrial Activity. The Barcadares, Ridge Complex, and Port St. George are all associated with some form of industry: logging in Belize, and sugar production at the two sites on Nevis. The Ridge Complex has by far the largest industrial assemblage, which is logical as it represents a connected industrial and domestic complex. Port St. George is also an industrial site, and it is surprising that more small industrial remains are not represented. Some of the unidentified iron fragments categorized by Meniketti as architectural remains may represent industrial materials.

15. Interpretation. There are clear differences in the assemblages of these three sites. The Nevisian sites have a greater diversity of both forms and wares in the Food Consumption category. Although Nevis was a peripheral outpost of English culture, middle-class planters still had access to a wide variety of **ceramic** goods in the early eighteenth century. As well as having access to a broader range of forms and ware types, they also acquired large amounts of expensive wares such as porcelain. The other end of the social spectrum is evident at these sites in the form of slave-made colonoware. There is no record of slavery in either the documentary or archaeological record from the Barcadares-another major difference between these two sites, and between the groups that created them. The presence of utilitarian wares at both Port St. George and the Ridge Complex suggests that their absence at the Barcadares may also be part of the pirate model.

16. It is possible that the logwood cutters and pirates had a more limited selection of wares from which to choose their ceramics, but the presence of porcelain at the Barcadares suggests that they had some access to higher status items. It is less likely that they only had access to bowls. Rather, they selected only those vessel forms best suited to their lifestyle and community.

17. The other notable difference between the sites is in the Tobacco Pipes Group. Smoking was important at all sites, but most **prevalent** at the Barcadares-smoking was clearly an important social activity for the baymen.

18. The Ridge Complex has slightly higher percentages in both the Architecture and Industrial Activities Groups, but these are only significant because they seem low for what might be expected from an industrial site. This is probably a result of the sampling methods used. Construction methods used at the sites varied widely, from **elevated** canvas huts to multistorey brick structures, and the types of remains recovered reflect this variation. The absence of material in the Food Production category may in part be the result of the sampling strategies used, and also indicates differences in related behaviors between baymen and planters.

19. Comparison of the Barcadares with the Ridge Complex and Port St. George sites from Nevis reinforces the distinctiveness of several characteristics of the pirate encampment. These include the high percentage of tobacco pipes, the dominance of bowls, the lack of drinking forms and utilitarian wares, and the generally limited diversity of ceramic wares and forms. These differences represent different choices made by the baymen and the Nevisians in the selection of their material culture. While additional work is still required to determine a true pirate pattern, the model suggested here is clearly capable of highlighting features of pirate activity in the archaeological record that can be tested in the future.

20. The Barcadares is the only clearly pirate-associated site from this period **excavated** to date, and there is little work being done specifically on this topic. There are few pirate sites to examine. The few pirate shipwrecks that have been excavated are embroiled in debates over ethics and identification. A bigger problem, however, is one that has been mentioned before when looking at "pirate" materials-how can they be distinguished from other maritime groups? There is no coherent basis for comparison at this stage-no pattern established for understanding maritime communities in terms of their artifact assemblages, and consequently no way of judging how a maritime population may internalize its interactions with the maritime landscape. While pirates may show some further variation in terms of material culture, sufficient data do not currently exist to answer questions about the reciprocal relationship of influence between culture and landscape in this context. The approach presented here is valid for looking at a broader definition of maritime groups-not just pirates-leaving room for future research.

(2,057 words)

https://link.springer.com/chapter/10.1007/978-1-4419-8210-0_12

Extracted from *Material Culture and Maritime Identity: Identifying Maritime Subcultures Through Artifacts in The Archaeology of Maritime Landscapes*, 217-232, 2011.

New Words

1. offshoot [ˈɒfʃuːt]

    n. a thing that develops from sth., especially a small organization that develops from a larger one 衍生物；分支

    Psychology began as a purely academic offshoot of natural philosophy.

    心理学最初只是自然哲学的一个学术分支。

2. colonial [kəˈləʊniəl]

    a. connected with or belonging to a country that controls another country 殖民地的；殖民主义的

I love the colonial fabrics, all the silver work, the furnishings, the combination of elegance and simplicity.

我喜欢殖民时期的织物、银器、家具，又优雅，又简洁。

3. assemblage [əˈsemblɪdʒ]

n. (formal) a collection of things; a group of people (人、物的)集聚；装配；集合；聚集；集会；集合物；聚集的物或人

He had an assemblage of old junk cars filling the backyard.

他在后院堆满了废旧汽车。

4. dynamics [daɪˈnæmɪks]

n. the branch of mechanics concerned with the forces that change or produce the motions of bodies 动力学

I don't think he understood the dynamics of how the police and the city administration relate.

我认为他不理解警方与市政部门之间的相互作用关系。

5. disregard [ˌdɪsrɪˈɡɑːd]

v. bar from attention or consideration; give little or no attention to refuse to; acknowledge 忽视，轻视，不尊重

She shows a total disregard for other people's feelings.

她丝毫不顾及别人的感受。

n. the act of treating sb./sth. as unimportant and not caring about them/it 忽视

Whoever planted the bomb showed a total disregard for the safety of the public.

不论谁安置了炸弹都是对公众安全的全然漠视。

6. analogous [əˈnæləɡəs]

a. similar or equivalent in some respects though otherwise dissimilar; corresponding in function but not in evolutionary origin 相似的，类似的；(器官)同功的

The mantle's motions, analogous to those in a pot of boiling water, cool the mantle by carrying hot material to the surface and returning cooler material to the depths.

地幔的运动类似于一壶沸水的运动，它把热的物质带到表面，再把较冷的物质带到深处，从而使地幔冷却。

7. vessel [ˈves(ə)l]

n. (technical) (old use) a container used for holding liquids, such as a bowl, cup, etc. (盛液体的)容器，器皿

These algae now release their oil, which floats to the surface of the culture vessel.

现在，这些藻类可以释放出藻类油，藻类油能漂浮到培养容器的水面上。

8. saucer [ˈsɔːsə(r)]

n. a small shallow round dish that a cup stands on; an object that is shaped like this 茶碟

Rae's coffee cup clattered against the saucer as she picked it up.

蕾把咖啡杯拿起来的时候，杯子碰响了茶碟。

9. diversify [daɪˈvɜːsɪfaɪ]

v. make (more) diverse; vary in order to spread risk or to expand(使)多样化,(使)不同;扩大经营范围,增加……的品种

As demand has increased, so manufacturers have been encouraged to diversify and improve quality.

制造商们受需求增加的刺激而扩大生产品种,提高产品质量。

10. disproportionate [ˌdɪsprəˈpɔːʃənət]

a. too large or too small when compared with sth. else 不成比例的

The Whigs, however, enjoyed disproportionate strength among the business and commercial classes.

然而,辉格党在交易和商业阶层中享有不成比例的优势。

11. baymen [ˈbeɪmən]

n. a person and especially a fisherman who lives or works on or about bay 港湾居民

Fish stands thrived, said Dale Parsons, a fifth-generation bayman, who runs Parsons Seafood, a company based in Tuckerton, N.J., that grows clams and oysters.

戴尔·帕森斯说,鱼摊生意兴隆了起来。他已经是第五代海湾人了,经营着帕森斯海鲜公司(位于新泽西州塔克顿),主要养殖蛤蜊和牡蛎。

12. disparity [dɪˈspærəti]

n. (formal) a difference, especially one connected with unfair treatment 明显差异

Many organizations struggle with the disparity and distribution of information.

信息的不一致和分散让很多组织十分烦恼。

13. identifiable [aɪˌdentɪˈfaɪəb(ə)l]

a. capable of being identified 可辨认的;可认明的;可证明是同一的

For ecological studies, the most important factor is collecting identifiable samples of as many of the different species present as possible.

对于生态学研究来说,最重要的是收集尽可能多的不同物种的可识别样本。

14. porringer [ˈpɒrɪn(d)ʒə]

n. a small dish, often with a handle, for soup, porridge, etc. 汤碗

The child was eating pottage from a porringer.

那个孩子正在喝一小碗浓汤。

15. sherd(=shard) [ʃɑːd]

n. pieces of broken glass, pottery, or metal(玻璃、陶瓷或金属的)碎片

Eyewitnesses spoke of rocks and shards of glass flying in the air.

目击者们提到了在空中飞舞的石子和玻璃碎片。

16. porcelain [ˈpɔːsəlɪn]

n. a hard white shiny substance made by baking clay and used for making delicate cups, plates and decorative objects; objects that are made of this 瓷;瓷器

There were lilies everywhere in tall white porcelain vases.

到处是插在高高的白瓷花瓶里的百合花。

17. extrapolate [ɪkˈstræpəleɪt]

v. to estimate sth. or form an opinion about sth. , using the facts that you have now and that are valid for one situation and supposing that they will be valid for the new one 推断

Extrapolating from his latest findings, he reckons about 80% of these deaths might be attributed to smoking.

从最近的发现推断,他估计大约 80% 的这类死亡可能归因于吸烟。

18. erect [ɪˈrekt]

    v. to build sth. 建造

    Opposition demonstrators have erected barricades in roads leading to the parliament building.

    反对派示威者在通往议会大厦的路上设置了路障。

19. mortar [ˈmɔːtə(r)]

    n. a mixture of sand, water, lime and cement used in building for holding bricks and stones together 砂浆,灰浆

    Several soldiers mix cement on the street, lifting it up to their colleagues who use it as mortar, placing a concrete block on top of concrete block.

    几名士兵在街道上混合水泥,举给同事们当砂浆,用来垒砌混凝土块。

20. hearth [hɑːθ]

    n. the area around a fireplace or the area of floor in front of it 壁炉炉床;壁炉边

    It was winter and there was a huge fire roaring in the hearth.

    时值冬天,壁炉里的火在熊熊燃烧。

21. conspicuous [kənˈspɪkjuəs]

    a. without any attempt at concealment; completely obvious 显眼的

    Most people don't want to be too conspicuous.

    大多数人不愿过于显眼。

22. ceramic [səˈræmɪk]

    n. a pot or other object made of clay that has been made permanently hard by heat 陶瓷

    We'll now look at another ceramic which is made from mixing sand with minerals and heating to over 600 degrees Celsius.

    现在我们来看看另一种陶瓷,它是由沙子和矿物质混合,再加热到 600 多摄氏度制成的。

23. prevalent [ˈprevələnt]

    a. that exists or is very common at a particular time or in a particular place 盛行的;普遍存在的

    Trees are dying in areas where acid rain is most prevalent.

    在酸雨非常严重的地区,树木正面临枯死。

24. elevate [ˈelɪveɪt]

    v. to make someone or something more important or to improve something 提升;提高;改进

    These factors helped to elevate the town to the position of one of the most beautiful in the country.

这些因素提高了该镇的声望,使之成为该国最具魅力的小镇之一。

25. excavate ['ekskəveɪt]

v. to dig in the ground to look for old buildings or objects that have been buried for a long time; to find sth. by digging in this way 发掘,挖出(古建筑或古物)

A new Danish archaeological team is again excavating the site in annual summer digs.

一支新的丹麦考古队又在对那个遗址进行一年一度的夏季挖掘。

Phrases and Expressions

be represented by 由……代理;用……来表示
apart from 远离,除……之外;且不说;缺少
be arranged into 分门别类
be intended to do……是用来(做)……
give an……impression of sth.……使得人们对某事物有一个……的印象
make up 构成;编造;补足;与某人和好;铺(床)
be associated with 和……联系在一起;与……有关
have access to 有使用/进入……的权利
in part 部分地;在某种程度上
be embroiled in sth. 卷入某事

Terminology

multistoried brick structures 多层砖石建筑
standing foundation 永久性地基
faunal remains 动物遗骸

Proper Names

Belize 伯利兹(中美洲加勒比地区的国家)
Barcadares 巴卡达瑞斯(位于伯利兹的伯利兹河上,是一个非法砍伐原木的地方,运营于18世纪)
Ridge Complex Site 利奇群落遗址(位于纳维斯一个小岛的西南部,这里的早期居民种植制糖作物)
Port St. George 圣乔治港(与利奇群落遗址相邻,该港口的功能主要是将岛上的制糖作物运送出去)

一、翻译策略/方法/技巧:合译法

合译法又叫合句译法(combination),是汉译英中常用的翻译技巧。一般说来,英语句子要比汉语句子长,一个英语句子要比一个汉语句子具有更大的容量。因此,我们在汉译英时有必要也有可能把两个或两个以上的汉语句子翻译成一个英语句子。比较口语化的英语句子也比较短,我们在英译汉时也要经常使用合句法,把两个或两个以上的英语句子翻译成

一个汉语句子。需要特别指出的是,采用合句法时,都是根据原文各句之间的逻辑关系,在译文的句与句之间加上连接词语,如汉语译文中的"因为""但是""又……又……",以及英语译文中的 and、while、so、but、because、for、as、who 等。试看以下译例:

例1:旧历新年快来了。这是一年中的第一件大事。除了那些负债过多的人以外,大家都热烈地欢迎这个佳节的到来。

译文:The Spring Festival was on the way, which has been the first event within one year traditionally. Everyone but those who were overindebted was enthusiastically yearning for her final arrival.

原文是巴金先生的名著《家》中的文字。整段文字共包含三个句子。前两个句子逻辑关联较为紧密,第三个句子相比之下无论是在语义上还是在逻辑关联上都较为独立。这样来看,英语译文大体上可作为两个句子。原文中的前两个句子考虑到有一个指示代词"这"连接,可以运用合句法并作一句,将其译成一个非限定性定语从句,用来修饰前句中的"旧历新年"。原文中的第三个句子本身即为一个简单句,因此可直接译成一个英语简单句。这样处理的话,原文中三个中文句子通过运用合句法译成了两个英文句子,整个英语译文显得主次分明,集中体现了英文在构造句子时句子长、容量大的特点。

例2:讲动武,祥子不能打个老人,也不能打个姑娘。他的力量没地方用。

译文:Considering resorting to force, it was impossible for Xiangzi to hit the elder and a girl, so there was no place for him to exhaust his energy.

这是老舍先生的名著《骆驼祥子》里的原文。整段文字包含四个小句子。从句意来看,前三个句子在逻辑关联上更加紧密一些,第四个句子相对独立一些,它与前三句是因果关系。这是英语译文需要表现的主逻辑关系,可用 so 等表示因果关联的连词来表现。然后再看前三个小句子。三句中的后两句很显然可以利用 it be……for……to do 句型合为一句,而"讲动武"可以处理成一个现在分词短语,置于句首,作为整句的背景性描述。最终原文中的四个小句子被合并成了一个以 so 连接的表示前因后果的主从复合句。合句法在这一译例中得到了淋漓尽致的展现。

例3:She is very busy at home. She has to take care of the children and do the kitchen work.

译文:她在家很忙,又要照顾孩子,又要做饭。

原文中尽管包含三个简单句,但是三个动作由同一个施动者发出的,三者是时间上的顺承关系。这种逻辑关系恰好可以用汉语中的连动句来表现。所谓连动句,是指由两个或两个以上的动词充当谓语,连续说明同一个主语的动词谓语句,可以表示目的、因果、方式和先后等逻辑关系。这样一来,英语原文中的三个句子就合并成汉语译文中的一个连动句。必须指出的是,相较于汉译英,合句法在英译汉中的应用较少,这也是由英语重"形合"、汉语重"意合"的语言特点决定的。

二、译例与练习
Translation
1. These comparisons are not perfect by any means-apart from the biases introduced by different site formation and collection processes, Meneketti's information did not always

allow the material from his projects to be arranged into directly analogous categories. (Para. 3)

无论从哪个方面看,这些比较都不是完美的,且不论由于不同的遗址信息和不同的收集过程所导致的偏差,梅内克缇也并不总是能够根据他所获得的信息将其项目中的材料归类到恰好对应的范畴中。

2. The Drinking subgroup at the Barcadares comprises nearly double the percentage of the total assemblage in comparison to the two Nevisian sites, and comprises a much larger percentage of the Food Consumption Group. (Para. 5)

巴卡达瑞斯的饮酒器具在小类别组合总数中所占的比例几乎是尼维斯遗址的两倍,其食物消费类所占的比例也更大。

3. The Ridge Complex also has some colonoware, another ware type absents from the logging camp which points to differences in social attitudes between the baymen and colonial sugar planters—the loggers did not typically keep slaves. (Para. 7)

利奇群落还有一些殖民时期的器具,这是另一个伐木营地没有的器具。这表明海湾人和殖民地甘蔗种植园主之间的社会态度不同——伐木者一般不养奴隶。

4. It is impossible to extrapolate back to the decisions made by the baymen in their pottery selection, but it is interesting to note that tin enameled wares, while readily available in this period, had a much softer paste and were generally less durable than stonewares. (Para. 7)

我们不可能推断出海湾人在选择陶器时所做的决定,但有趣的是在这一时期已经存在的锡制搪瓷器皿涂层硬度很低,不如石器耐用。

【此句穿插了各种修饰成分和插入语,如后置定语"made by the baymen"和插入语"while readily available in this period",其逻辑结构非常复杂。如果直接按照原文的结构翻译复杂长句可能造成译文的表述不符合汉语习惯。可以考虑把插入语"while……this period"译为定语修饰主语,使之脱离句子主干之间的逻辑关系,成为一种补充说明成分,保证了译文结构的完整性。】

5. This is not a strongly represented group at any of the three sites. It is unsurprising that it is largest at the Barcadares, considering that documentary evidence emphasizes the popularity of hunting and recreational shooting among the baymen. (Para. 9)

这三个遗址都不是很有代表性。据文献记载,海湾居民非常喜欢狩猎和休闲射击,如此说来,在巴卡达瑞斯挖掘出来的武器数目最多,也就不足为奇了。

【此处采用了"逆译法"。英文的表达习惯是先果后因、先结果后条件,而中文相反,先因后果、先条件后结果。例如,此例中放在句首的"It is unsurprising that……"就是一个结果描述。在汉译中,如果按照原文的表达顺序,把"不足为奇"放在句首,后面的小句就很难与整体协调。此外,为了符合汉语的语用习惯,译文还添加了"如此说来",这样读者就会有一种心理预期,知道后面的文字是对某个结果的解释了。】

6. Meniketti referred to faunal remains in the text of his report, and noted that there were not many-some egg shells inside the main complex, one sheep, one bird, one large mammal (probably a cow), fish, and some goats remain that were probably intrusive. (Para. 12)

梅内克缇在他的报告中提到了动物遗骸,并指出这些遗骸并不是很多:一些来自主群落的蛋壳、一只绵羊、一只鸟、一只大型哺乳动物(可能是牛)、鱼和某些侵入的山羊遗骸。

7. Although Finamore used the vessel forms to suggest that the baymen engaged in some kind of communal food sharing, the multiple small hearths recovered suggest that at the least, the "community" involved in the sharing of prepared food may have been relatively small. (Para.13)

菲纳莫雷根据容器外形特征,得出海湾居民曾经进行过某种公共食物分享活动的结论。但找到的多个小灶台表明,参与分享成品食物的"团体"活动可能非常少。

8. There is no record of slavery in either the documentary or archaeological record from the Barcadares-another major difference between these two sites, and between the groups that created them. The presence of utilitarian wares at both Port St. George and the Ridge Complex suggests that their absence at the Barcadares may also be part of the pirate model. (Para.15)

在巴卡达瑞斯的文献或考古记录中都没有奴隶制的记录,这是这两个遗址以及留下这些遗址的群体之间的另一个主要区别。在圣乔治港和利奇群落都出现了实用器皿,而在巴卡达瑞斯时期并未出现,这也可能是海盗模式的一部分。

9. It is possible that the logwood cutters and pirates had a more limited selection of wares from which to choose their ceramics, but the presence of porcelain at the Barcadares suggests that they had some access to higher status items. (Para.16)

供伐木者和海盗选择的物品种类可能非常有限,可从中选择的陶器种类就更有限了。但巴卡达瑞斯出土的瓷器,表明他们有权使用那些一般只有更高地位的人才能使用的物品。

(在世界各地的文化中,人们使用物品的等级受到地位、阶级的限制。一般来说,平民不允许使用贵族的器物。)

10. While pirates may show some further variation in terms of material culture, sufficient data do not currently exist to answer questions about the reciprocal relationship of influence between culture and landscape in this context. (Para.20)

虽然海盗可能在物质文化方面表现出进一步的变化,但目前还没有足够的数据来回答在这一背景下文化和景观之间相互影响。

Exercises

1. Fill in the blanks with the proper given words, and then translate the sentences into Chinese.

renovate   coexist   increase   colonial   surmount   fragment

1) After a careful investigation, the committee decided to _____ the old house in which the great writer used to live.
2) A capacity _____ of that magnitude would be extraordinary for a country that has never produced more than 11 million barrels a day.
3) One of the aims of cross-cultural research is to _____ the obstacles, to seek for the cultural differences and cultural identification.
4) She reads everything, digesting every _____ of news.

5) The swift growth of this merchant marine diversified the northern _____ economy and made it more self-sufficient.

6) Newspaper websites can _____ with the newspapers and still bring in healthy profit margins.

2. Translate the following sentences into Chinese.

1) Work on Documenting Maritime Folklife commenced in July 1986, when I tested documentation techniques in the fishing village of Mayport, on the northeast coast of Florida.

2) Although the bulk of the examples of maritime traditions used here are taken from Florida, the techniques for documenting cultural resources can be applied and adapted to many other maritime settings.

3) Documenting Maritime Folklife has two main purposes: to promote understanding of maritime cultural heritage—the body of distinctive traditional knowledge found wherever groups of people live near oceans, rivers, lakes, and streams; and to provide laymen with a basic guide for the identification and documentation of common maritime traditions.

4) The historical museum of southern Florida's folklife program completed a project to survey the folklife of the Florida Keys in 1989.

5) Archeology also tells us, the original aesthetic object is often useful object, or a production tool, or by hunting animals.

6) The significance of maritime occupational folklife stems from the historical role of maritime enterprise, along with the distinctive character of the local culture that produced and continues to sustain it.

7) Florida boasts teeming coastal waters and inland lakes and rivers that have spawned both recreational and commercial fishing traditions.

8) Maritime industries have helped transform Florida into a strategically important region in navigation and shipping, as well as in seafood production.

9) I was to conduct an ethnographic survey of Miami-Dade County, with a special emphasis on Latin American culture, for the Florida Folklife Program.

10) Catering culture, which has always been a symbol of social progress and civilization, has unique aesthetic value.

3. Translate the following sentences into English.

1) 为了合理开发和利用海洋文化资源，首先应对海洋文化资源的价值予以评估。

2) 中国的海洋文化历史悠久，且潮汐文化占有很大比重。

3) 美国沿海地区的民俗，既有职业传统，也有娱乐风俗，有从捕捞到烹饪的一系列餐饮文化，也有民间艺术。

4) 舟山渔民的画有着浓厚的地域民间艺术风格，体现了与众不同的海洋文化风情。

5) 整个城市就是一份历史遗产，鹅卵石铺就的小巷，以及带中世纪风格的尖顶房屋。

6) 毫无疑问,凡是去过欧洲旅游的人,无不对它的地标建筑、博物馆和历史遗迹印象深刻。
7) 今年年初以来,约有100条船只遭到索马里海盗的袭击,其中有7艘中国船只。
8) 我说应该禁止吸烟,但她反对。她说烟草是政府税收收入的一个重要来源。
9) 似乎制陶技术是不同地域的人们在不同时期独自发明的,而这项技术遍布世界各地。
10) 在采购易腐食品时,我会挑那些到保质期还有很长时间的,尽管我打算立即把它们吃掉。

4. Choose the best paragraph translation. And then answer why you choose the first translation or the second one.

作为最具影响力的海洋女神,妈祖是中国沿海地区和世界无数华人社区的一系列信仰和习俗的核心。妈祖,字面意思是"母亲的祖先",在她的家乡福建省被称为妈祖婆。在澳门特别行政区被称为阿妈,澳门的葡萄牙语名称"澳门"来自"妈祖阁"的发音。妈祖信仰和习俗于2009年9月30日被联合国教科文组织列入《人类非物质文化遗产代表作名录》。妈祖信仰和习俗主要包括祭祀仪式、民俗和民间故事三个部分,已深深融入中国沿海地区及其子孙后代的生活中。

译文一:

As the most influential goddess of the sea in China, Mazu is at the heart of a host of beliefs and customs, throughout the country's coastal areas and countless Chinese communities around the world. Mazu or Ma-Tsu, literally referring to "maternal ancestor," is also known as "Mazupo" in east China's Fujian Province, where her hometown is located, and "A-Ma" in Macao Special Administrative Region (SAR), whose Portuguese name of "Macao" came from the pronunciation of the word "Mazu Pavilion." The Mazu belief and customs, inscribed on the Representative List of the Intangible Cultural Heritage of Humanity by UNESCO on September 30, 2009, primarily contains three parts-sacrificial and worship ceremonies, folk customs and folk tales, which have been deeply integrated into the lives of coastal Chinese and their descendants.

译文二:

As the most influential goddess of the sea, Mazu is at the heart of a range of beliefs and practices in coastal China and countless Chinese communities around the world. Mazu, literally meaning "mother's ancestor", is known as "Mazupo" in her home province of Fujian and "A-Ma" in the Macao Special Administrative Region, where the Portuguese name "Macao" comes from the pronunciation of "Mazu Pavilion". Mazu beliefs and customs were included in the Representative List of the Intangible Cultural Heritage of Humanity by UNESCO on September 30, 2009. Mazu beliefs and customs mainly include sacrificial ceremonies, folk customs and folk stories, which have been deeply integrated into the life of China's coastal areas and their descendants.

5. Translate the following passage into Chinese.

Comparison of the Barcadares with the Ridge Complex and Port St. George sites from

Nevis reinforces the distinctiveness of several characteristics of the pirate encampment. These include the high percentage of tobacco pipes, the dominance of bowls, the lack of drinking forms and utilitarian wares, and the generally limited diversity of ceramic wares and forms. These differences represent different choices made by the baymen and the Nevisians in the selection of their material culture. While additional work is still required to determine a true pirate pattern, the model suggested here is clearly capable of highlighting features of pirate activity in the archaeological record that can be tested in the future. The Barcadares is the only clearly pirate-associated site from this period excavated to date, and there is little work being done specifically on this topic. There are few pirate sites to examine, period. The few pirate shipwrecks that have been excavated are embroiled in debates over ethics and identification. A bigger problem, however, is one that has been mentioned before when looking at "pirate" materials-how can they be distinguished from other maritime groups?

6. Translate the following passage into English.

郑和(1371—1433年)是明朝(1368—1644年)时期的伟大航海家和外交家。这位有影响力的历史人物出生在中国西南部的云南省。其父亲是穆斯林"哈吉",曾去过麦加朝圣。据说他的家族是云南早期蒙古统治者的后裔,也是布哈拉国王穆罕默德的后裔。他本姓马,来自中文穆罕默德的译法。当他10岁时,云南被新建立的明朝重新征服。年轻的马和(他当时被称为马和)在被俘虏后送进了军队当勤务兵。1390年,马和受命于燕王,作为一名初级军官,他精通战争和外交。燕王反抗他的侄子建文帝,于1402年即位。在燕王成为永乐皇帝后,饱受战争摧残的明朝经济很快得到恢复。于是皇帝想要在海外展示其海上的力量。郑和被皇帝选为西海使团的总司令,指挥62艘船,带领27 800人于1405年启航。

三、翻译家论翻译

汪榕培论中国典籍英译。学者汪榕培先生是著名英语教育家、翻译家、教授,先后在大连外国语大学(校长)、苏州大学、大连理工大学担任博士生导师,发表论文六十余篇,出版论著几十部,译著十几部,有《英译诗经》《英译老子》《英译庄子》《英译汉魏六朝诗三百首》《英译孔雀东南飞·木兰诗》《英译陶诗》《陶渊明集》《英译汤显祖戏剧全集》等,为中国文化走向世界做了大量工作。

在中国典籍英译领域,汪榕培先生的译著,无论是数量还是质量,都是成绩斐然。关于为什么要将中国典籍翻译成英语,他认为"中国典籍是中国古老文明的结晶,是世界文明的一个组成部分,也是应该让世界各国人民都能够共同分享的一笔财富"。

那么这些典籍应该由谁来翻译呢?瑞典汉学研究者、翻译家、诺贝尔文学奖评委马悦然(Goran Malmqvist,1924—2019年)认为,"中国人英语不够好,因此不应该由中国人将中国文学作品翻译为英文,而应由一个文学修养很高的英国人来翻译"。汪先生非常反对他的观点,并反驳道:"外国译者翻译中国的典籍,最致命的弱项是难以理解中国人的思维"。汪先生认为,"就文学经典的翻译而论,主要是特定译者的语言和文学修养决定了作品的翻译水平"。例如,陶渊明《饮酒》中的诗句:"结庐在人境,而无车马喧。问君何能尔,心远地自偏。"下面的几种译文均是由母语为英语的人所翻译,虽然语法通顺,但是完全没有传达出诗歌语言的素美,以及诗人内心的隐逸。这里的"心远"是一种恬淡自得,不为尘世所扰的悠然心态,而非遥远、心不在焉,而"心远地自偏"的真正意义是"any place is calm for a peaceful mind"。

(1) When the heart is far the place of itself is distant. (William Acker)
当心遥远了,地方也远了。
(2) When my heart is absent the place itself is absent. (Andrew Boyd)
当我心不在这个地方,它就心不在焉了。
(3) With the mind detached, one's place becomes remote. (James Robert Hightower)
随着思想的超脱,一个人的位置变得遥远。
(4) With a mind remote, the region too grows distant. (Robert Watson)
心灵遥远,这个地区也会变遥远。
(5) When the heart is remote, earth standing aloof. (R. H. Kotewall)
心远则地远。

对于中国典籍英译译文的评价,仁者见仁智者见智,似乎永远是一个哲学问题。一方面,它要求译者对原文有深刻的理解,而原文的真正含义,已经散落在各个历史时代的不同解读之中,无论采用其中的哪一个,都只能是"片面"。另一方面,它又要求译者对译语文化有全面的了解,以免"译者无意,读者有心",产生不必要的误解。对于这个问题,汪先生认为"译可译,非常译"。所谓"译可译"是说人类的思想感情都是互通的,一切皆可译。所谓"非常译"是说"原始语的文本是固定不变的,但是内容可能会有因人而异、因时而异,乃至因地而异的不同理解,从而产生不同的译入语文本;译入语所采用的形式和措辞也会因人、因时、因地而做出不同的选择"。但是,不同的译本之间总会有一种"更加接近原著的精髓,比较接近'常译'"。汪先生还将"非常译"与"常译"之间的关系比作相对真理和绝对真理。他认为,"非常译"诚然可能有错译、误译,但它们不妨是通往"常译"的渠道。

对于如何尽可能地接近"常译",尽可能地让译文体现原著的精髓,汪先生提出了"传神达意"的翻译原则。该原则一直贯穿于汪先生所有译著的翻译过程。在翻译中国典籍的过程中,既要考虑中国人的思维特点,又要兼顾西方人的表达习惯,要找到一种平衡,就要"传神地达意",即通过一定的手段或方法达到传递原著意蕴的目的。拿诗歌来说,汪先生认为只有把诗歌翻译成诗的形式,才能达到最佳效果,而诗歌在翻译过程中丢失的原有特点,可以"通过补偿的方式使其获得新的生命"。

(摘自:汪榕培.汪榕培学术研究文集[M].上海:上海外语教育出版社,2017:13-37.)

Text B

Goan Houses: Memory of Home
By Pedro Pombo

Ideological and Political Education：关注海洋民俗 保护海洋民俗文化

我国拥有数千年的海洋文明史，在16世纪之前相当长的一段时期，中国人所创造的海洋文明遥遥领先于世界，为人类海洋文明史谱写了辉煌的篇章。同时，历史悠久的中华海洋文明积淀了灿若星海的海洋文化遗产。然而，中国海洋文化遗产的状况令人担忧，特别是随着沿海开发开放步伐的日益加快以及工业化、现代化和城市化进程的加速推进，许多传承久远的海洋民俗文化遗产正濒临消失。因此，海洋民俗文化遗产的抢救性保护不仅十分必要，而且刻不容缓。

严格意义上说，民俗文化与文化遗产之间存在一定区别，不能简单画等号。然而，民俗文化与文化遗产特别是非物质文化遗产之间的关系又是十分密切的，二者之间难以截然分开。"一方面，非物质文化遗产在很大程度上体现了民俗文化的精粹，把传统的难登大雅之堂的民间生活文化引入现代人关注的领域和研究视野之中；另一方面，非物质文化遗产的形成与传承离不开民俗文化的滋养和内涵化。"海洋民俗与海洋文化遗产之间的这种特殊关联性，决定了海洋民俗研究不应绕开海洋文化遗产保护传承问题，而应通过扎实的研究成果为更好保护传承海洋文化遗产提供科学支撑。同样道理，海洋文化遗产保护传承也不能将海洋民俗弃置一旁，而要将其作为重要对象加以保护传承。

我国海洋民俗研究意识淡薄的状况已经引起学术界的关注。要想改变我国海洋民俗研究薄弱的状况，补上我国海洋民俗研究的"短板"，最终落脚点还是要广泛地开展实地调查研究，在此基础上形成更多的研究成果，不断丰富和发展我国海洋民俗研究。通过深入研究，系统挖掘海洋民俗文化遗产中蕴含的文化价值，把握海洋民俗文化遗产保护传承的客观规律，做到科学保护和有序传承。

1. Goa, India's smallest state is situated along the west coast of India, which has been deeply influenced by the Catholic Portuguese, who ruled from 1510 to 1961. During the nineteenth century, Goans started to migrate other countries.

2. The growing **migratory** movements to South Asian and East African cities at the turn of the twentieth century would provoke a significant, and still understudied, architectural activity in coastal Goa's villages, translating flows of **remittances** and social changes sustained by the diverse influences that **diaspora** communities would bring to their ancestral villages. The monetary benefits brought back home by the diaspora would allow many Goan families to slowly increase their economic status. Remittances from relatives settled across the Indian Ocean colonial territories would allow a wide spread construction activity in ancestral homes, visually marking social and economic changes brought by migration to the state.

3. Many, or even most, of the houses we see while traversing this region of Goa were built or **renovated** in this period. This fact translates to an architectural culture of the growing migratory movements, while the ancestral house, as it is commonly mentioned in Goa, would be not only a central place of personal belonging but also a privileged location of showing to the village society the success of the migrants in their new life.

4. This is a perceptible cultural feature of the Goan diaspora and the Goan built landscape: the house, its aesthetics and material qualities, would stay as fundamental spatial, social and emotional reference not only in life experiences but equally in cultural production and heritage attachments.

5. Selma Carvalho, a dedicated researcher on Goan literatures and diasporas, mentions this particular aspect, affirming that there has been a "consistent compulsion of the Goan Catholic writer to employ 'home' as the centripetal location", reinforcing the place that home occupies in the sensorial, emotional and imaginary attachments to the home-land and family histories. The ancestral home, thus, is central to those who migrate as it is to the ones who stay, and the house, understood in its material and aesthetical elements, is, of course, the repository of concepts of home and family history. Decorative and architectural influences brought from **cosmopolitan** port cities across the Indian region would be integrated in the Goan built heritage, turning the houses in repositories of geographic and cultural circulations.

6. Large landlord houses would grow in scale and richness, in a continuation of what

succeeded in the late eighteenth century, becoming locally mentioned as palaces or mansions. New objects and technologies, decorative **motifs** and materials would fill large reception halls: Venetian or Belgian **chandeliers** suspended from high ceilings, Chinese porcelain from Macau invading the halls in blue and white dining sets or coloured monumental vases surmounting chest of drawers or side tables, **gramophones** filling the space with new music, hot water pipe systems serving modern toilet spaces, European made mosaic floor under the feet of dancing parties. The **exuberance** and **ostentation** of the nineteenth-century European art history would reach Goan villages and transform their architectural compositions. In much modest scales, smaller homes would equally reflect these influences. Typical architectural elements of Goa houses, as the balcão (**balcony**, a space at the entrance of the **veranda** with seating arrangements that serves both as a pleasure space and as a reception antechamber) and the veranda, would systematically incorporate small details that denote this period: the changing profile of the oyster-shell windows, the enlargement of proportions or the increasing relevance of the public façade of the house. In their inner spaces, we can usually observe the development of more formal **reception** room, often with large windows opening to the main veranda, creating a sense of scale and decoration that many times was not present in the more private areas of the house. The house would be enlarged in relation to the roads, in visual and material representations of family histories spread across vast distances. This would be accompanied by the visit of the migrants to their ancestral homes and villages, sustaining the incorporation of a whole universe of music, furniture, fashion senses, habits and cultural references that were mostly from outside a Portuguese sphere of influence, originated in globalizing cultural languages enabled by technologies as the steamship and the increasing colonial appropriation of the African continent.

7. The house, then, while integrating local social hierarchies and material elements of a "traditional" Goan life and heritage, would also embody traces of social mobility, economic and cultural changes, allowed by maritime migratory routes. The centrality of oceanic circulations in the **coeval** migratory routes and the development of diasporic cultures is perceived in the way they leave marks, physical and cultural, in these rural landscapes. We may not see the ocean from these balconies, but at the same time that we feel the **diaphanous** light coming from the traditional oyster-shell windows, we can still perceive a certain cosmopolitan modernity while seating among furniture and objects brought from other continents, observing old photographs and reading letters sent by those who left to settle in new cities, faraway shores and different landscapes.

8. Another layer where oceanic crossings are perceptible in Goan houses and their importance as literary locations is in the still blurred history of the slave trade to Goa during the long colonial period and the contemporary presence of Afro-descendant populations in the state.

9. While the role of the Portuguese in the Atlantic and Indian oceans slave trade for centuries is widely acknowledged, the presence of African slaves in Goa, as well in the other Portuguese colonial territories in the Indian subcontinent is still to be recognized in its full dimension. The fact that the economic power of trading families and businesses in Portuguese India was, among other activities, sustained by the slave trade is an historical element that has been slowly but steadily **obliterated** from the local social memory and cultural references. It is known that before the abolition many **maroon** slaves settled in the forested hills just outside the Portuguese border, in the actual state of Karnataka, where among the Siddi communities (communities of Afro-descendants) we find many Christians with Portuguese origin names. The settling, however, of African descendants in Goa after the abolition is, on the contrary, an unclear history which we can perceive through diverse archival dimensions connected with the home as simultaneously a family space and a location of historical processes. The contemporary existence of Afro-Goans, perceptible when crossing south Goa, can be curiously observed in photographic and literary dimensions, directly connected with the space of the house and its familiar dimension.

10. Research projects using family photographic archives, as the one developed by Savia Viegas, a writer, educationist and researcher or the book on Goan fashion by the late Wendrell Rodricks, are important to unveil untold stories. In both projects we can see Goan extended families posing in front of their homes, translating the senses of self-fashioning and family atmosphere of the time. The peculiarity of these images is that they are accompanied by their **Mozambican** servants. At the turn of the twentieth century, Goan families with ties in **Mozambique**, which had been definitely conquered by Portugal in 1885, would often bring children as house servants to live in their Goan ancestral homes. These two examples of photographic **archival** research constitute important testimonies of this less known layer of the long maritime dimensions in Goa and its domestic spaces.

(1,230 words)

https://link.springer.com/book/10.1007/978-3-030-99347-4

Extracted from *Tales of Ocean, Migration and Memory in Ancestral Homespaces in Goa* in *The Palgrave Handbook of Blue Heritage*, 31-40, 2022.

New Words

1. migratory [ˈmaɪɡrət(ə)ri]

   a. connected with, or having the habit of, regular migration 迁移的；迁徙的；移栖的

   Some residents voiced concern that the rice could threaten certain species of rare migratory birds.

   一些居民担心这种大米会威胁到某些稀有候鸟的生存。

2. remittance [rɪˈmɪt(ə)ns]

   n. a sum of money that is sent to sb in order to pay for sth. 汇款

   Please enclose your remittance, making checks payable to Valley Technology Services.

   请随信附上您的汇款，支票收款方为瓦利科技服务公司。

3. diaspora [daɪˈæspərə]

   n. the movement of people from any nation or group away from their own country（任何民族的）大移居

   Then India relied on soft loans and deposits from its diaspora to fill the hole.

   然后，印度采取了软贷款和从离散的犹太人中吸取存款的措施来弥补缺口。

4. renovate [ˈrenəveɪt]

   v. to repair and paint an old building, a piece of furniture, etc. so that it is in good condition again 修复，翻新

   Nevertheless, this does not mean that modernization and new building should be discouraged in order to renovate and protect heritage buildings.

   然而，这并不意味着为了翻新和保护历史建筑而打压现代化与新型建筑。

5. cosmopolitan [ˌkɒzməˈpɒlɪtən]

   a. growing or occurring in many parts of the world; composed of people from or at home in many parts of the world; especially not provincial in attitudes or interests 世界主义的，四海为家的；国际性的，国际化的；见过世面的，见识广的；广布的，遍生的

   No dynasty had been so cosmopolitan and outward-looking.

   没有哪个朝代是如此的世界性和外向型。

6. motif [məʊˈtiːf]

   n. a design that consists of recurring shapes or colors 图案

   He then stenciled the ceiling with a moon and stars motif.

   他接着用模板在天花板印上月亮和星星图案。

7. chandelier [ˌʃændəˈlɪə(r)]

   n. a large round frame with branches that hold lights or candles 垂吊灯；枝形吊灯

   The 23rd-floor suite is decked with marble floors, a grand piano and a 22-carat gold chandelier.

   该套间位于23层，铺有大理石地板，里面配备了一架大钢琴和22 K金做成的吊灯。

8. gramophone [ˈɡræməfəʊn]

   n. an antique record player 留声机

   The gramophone has been displaced by the tape recorder.

留声机已被磁带录音机所取代。

9. exuberance [ɪgˈzjuːbərəns]

   n. joyful enthusiasm 情感洋溢的言行

   Her burst of exuberance and her brightness overwhelmed me.

   她的活力勃发和她的聪明智慧征服了我。

10. ostentation [ˌɒstenˈteɪʃ(ə)n]

    n. an exaggerated display of wealth, knowledge or skill that is made in order to impress people 卖弄；炫耀

    On the whole, she had lived modestly, with a notable lack of ostentation.

    总体来说，她生活俭朴，不事张扬。

11. balcony [ˈbælkəni]

    n. a platform that is built on the upstairs outside wall of a building, with a wall or rail around it 露台，阳台；楼座，楼厅

    Each has a snug mezzanine area and an antler-fenced balcony about eight meters above ground.

    每个房间都有一个舒适的夹层区域和一个离地约八米的鹿角围栏阳台。

12. veranda [vəˈrændə]

    n. a platform with an open front and a roof, built onto the side of a house on the ground floor (房屋底层有顶半敞的) 走廊，游廊

    There is a wide veranda under the overhang of the roof.

    悬吊的屋顶下是一条宽阔的游廊。

13. reception [rɪˈsepʃ(ə)n]

    n. the area inside the entrance of a hotel, an office building, etc. where guests or visitors go first when they arrive 接待处；接待区

    All visitors must report to the reception desk on arrival.

    所有参观者到达后务必在接待处报到。

14. coeval [kəʊˈliːv(ə)l]

    a. of or belonging to the same age or generation 同时代的

    The industry is coeval with the construction of the first railways.

    这一产业和初期铁路的建造相伴而生。

15. diaphanous [daɪˈæfənəs]

    a. so light and fine that you can almost see through it 薄得几乎透明的

    Clothiers sell only the most flattering of outfits cut from the most diaphanous of silks.

    服装商只出售用最透明的丝绸裁剪的最性感的衣服。

16. obliterate [əˈblɪtəreɪt]

    v. to remove all signs of sth., either by destroying or covering it completely 摧毁

    Their warheads are enough to obliterate the world several times over.

    他们的弹头足以摧毁这个世界好几次。

17. maroon [məˈruːn]

n. dark brownish-red in color 褐红色

She opened the tie box and looked at her purchase. It was silk, with maroon stripes.

她打开领带盒子,看着自己买的东西。那是真丝的,有栗色的条纹。

18. Mozambican [ˌməuzæmˈbiːkən]

a. of or relating to Mozambique or its inhabitants 莫桑比克的;莫桑比克人的

They walked through the dense Mozambican bush for thirty-six hours.

他们步行穿过茂密的莫桑比克灌木丛区,走了36个小时。

19. Mozambique [ˌməuzæmˈbiːk]

n. Portuguese East Africa country, a republic, until 1975 a dependency of Portugal 莫桑比克(东非国家)

Mozambique became independent in 1975.

莫桑比克于1975年获得独立。

20. archival [ɑːˈkaɪvəl]

a. of or relating to or contained in or serving as an archive 档案的

They need computer access to the archival material.

他们需要获准用电脑连线检索该档案材料。

Phrases and Expressions

a diversity of 多种多样的
at the turn of 在……的转弯处;即将进入……阶段
originate in 起源于
on the contrary 正相反;反而
employ……as…… 把……当作……来使用
emotional attachment 情感依恋
be accompanied by 伴随;同时发生
a sense of 一种……的感觉
link……with…… 把……和……联系起来

Terminology

Façade 建筑物正面

Proper Names

Goa 果阿(印度城邦)
Karnataka 卡纳塔克邦(印度城邦)
Macau 澳门(中国地名)

一、翻译简析
Translation

1. This fact translates to an architectural culture of the growing migratory movements, while the ancestral house, as it is commonly mentioned in Goa, would be not only a central place of personal belonging but also a privileged location of showing to the village society the success of the migrants in their new life. (Para. 3)

　　这一时期的建筑活动,最终转变为由日益壮大的移民运动所引发的一种建筑文化,而果阿邦经常被提及的祖屋,不仅是个人归属的中心场所,也是向乡村社会展示移民新生活的荣耀之地。

　　【原文句子较长,成分复杂。从句中插入语"as it is commonly mentioned in Goa"和"the success of the migrants in their new life"是"showing"的宾语,简化一下就可以明白,主干是"show the success to the village society"。相比之下,汉语的表达习惯是简短明快。如果按照原文结构翻译,其译文将混乱不堪,无法阅读。因此,这里将原文长句拆解为两个相对独立的短句,并调整各个成分的语序,同时尽量保持原文各成分之间的逻辑关系,这样翻译出来的译文更符合汉语的表达习惯。】

2. Selma Carvalho, a dedicated researcher on Goan literatures and diasporas, mentions this particular aspect, affirming that there has been a "consistent compulsion of the Goan Catholic writer to employ 'home' as the centripetal location", reinforcing the place that home occupies in the sensorial, emotional and imaginary attachments to the home-land and family histories. (Para. 5)

　　塞尔玛·卡瓦略,一位专门从事果阿文学和流散人口研究的学者,曾提到这一特殊事实。她证实"果阿天主教作家始终有一种把'家'作为向心作用的强烈欲望"。而这种向心作用进一步加强了"家"在他们对家乡和家族历史的感知和情感依恋中的地位。

　　【原文句子较长,主语与谓语之间嵌入了一个插入成分"a dedicated researcher on Goan literatures and diasporas",造成主语与谓语的距离被拉长。如果按照原文结构进行翻译,会使得汉语译文的逻辑不明晰,很难理解。因此,此处采用拆译法,将原文的插入成分与主语结合变成一个完整独立的主谓结构小句,然后,再另起一句,给原句后面的谓语添加一个新的主语,用代词"she",除了避免语义冗赘外,还可以利用代词的回指功能,让第二句与第一句产生关联。这种关联不但可以使译文流畅,还部分地保持了原文的逻辑结构。】

3. Typical architectural elements of Goa houses, as the balcão (balcony, a space at the entrance of the veranda with seating arrangements that serves both as a pleasure space and as a reception antechamber) and the veranda, would systematically incorporate small details that denote this period: the changing profile of the oyster-shell windows, the enlargement of proportions or the increasing relevance of the public façade of the house. (Para. 6)

　　果阿房屋的典型建筑元素,如凉台(阳台,门廊处的一个空间,安设座位后,既可作为娱乐空间,也可作为接待前厅)和门廊,有机地融入了一些体现这一时期特点的小细节:牡蛎壳窗户的轮廓变化了,房屋各个部分的比例扩大了,房屋正面公共区域的实用性增强了。

【此句的翻译难点在于冒号后面的三个名词短语:"the changing profile of the oyster-shell windows""the enlargement of proportions""the increasing relevance of the public façade of the house"。但是按照原文的逻辑结构分别翻译为"……的改变""……的扩大""……的增加",这样的译文既不符合汉语的习惯,其形式也与前面的句子主干不协调。这时,我们需要采用"换译法",将原文的静态表达转换为汉语的动态表达。】

4. Another layer where oceanic crossings are perceptible in Goan houses and their importance as literary locations is in the still blurred history of the slave trade to Goa during the long colonial period and the contemporary presence of Afro-descendant populations in the state.(Para.8)

从果阿房屋和一些文学描述中,我们可以看到海洋跨越的另一个层面。在漫长的殖民时期,殖民者向果阿贩卖奴隶的那段历史仍是模糊的,而该邦现存的非洲裔人口似乎可以证明这段历史。

【原文中的"slave trade to Goa"是一个抽象、静态的名词短语,加之读者可能对奴隶买卖的历史背景不了解,如果完全按照字面翻译为"果阿的奴隶贸易"就会造成不解。因此,这里采用了"增译"和"分译"的方法,将该名词短语从原结构中分离出来,翻译为主谓结构的小句,并增加了"就是……证明",从而使得原文中隐含的信息凸显出来。】

5. The fact that the economic power of trading families and businesses in Portuguese India was, among other activities, sustained by the slave trade is an historical element that has been slowly but steadily obliterated from the local social memory and cultural references. (Para.9)

除其他活动外,葡属印度的贸易家族和企业的经济实力,是由奴隶贸易维持的,而这一段历史已从当地的社会记忆和文化参照中逐渐消失了。

【原文的句子结构复杂,除了主语带有同位语从句外,还有插入成分"among other activities",这些都导致主语与谓语的距离被拉长,因此如果按照原文直接翻译,会造成译文的可读性变差。对于这种情况,首先我们可以将原句拆解为两个相对独立的简单句,这样不仅使插入成分脱离了句子主干成分之间的逻辑关系,让译文更加简洁明快,同时凸显了原文作者真正想要强调的"经济实力"这一侧面。】

二、参考译文

果阿的房屋:家的记忆

1. 果阿邦,印度最小的城邦,位于印度西海岸,深受信奉天主教的葡萄牙人的影响。葡萄牙人于1510—1961年统治此地,19世纪,果阿人开始向其他国家移民。

2. 20世纪初,不断涌向南亚和东非一些城市的移民在果阿沿海村庄引发了一场重大的建筑活动(还未得到充分研究)。那些移民不仅为这些建筑活动带回了资金,同时还给祖先的村庄带来了社会变化和各种影响。流散在海外的移民带回来的金钱使许多果阿家庭的经济地位慢慢提高。在印度洋殖民领土上定居的亲属的汇款将使祖屋的建筑活动得以广泛开展,形象化地标志着移民给该邦带来的社会和经济变化。

3. 当我们穿越果阿地区时,我们看到的许多甚至大部分房屋都是在这一时期建造或翻修的。这一时期的建筑活动,最终转变为由日益壮大的移民运动所引发的一种建筑文化,而果阿邦经常被提及的祖屋,不仅是个人归属的中心场所,也是向乡村社会展示移民新生活的荣耀之地。

4. 这是果阿海外移居和果阿建筑景观的一个可感知的文化特征:房子。它的艺术审美和材料品质,不仅在生活经验中,而且在文化生产和观念传承中,都是基本的空间、社会和情感参照,因而被保留下来。

5. 塞尔玛·卡瓦略,一位专门从事果阿文学和流散人口研究的学者,曾提到这一特殊事实。她证实"果阿天主教作家始终有一种把'家'作为向心作用的强烈欲望"。而这种向心作用进一步加强了"家"在他们对家乡和家族历史的感知和情感依恋中的地位。因此,祖屋对于迁居者和留居者都是至关重要的,而房子,从其材料和美学元素上理解,当然就是家和家族历史的宝库。果阿的古老建筑受到印度国际港口城市的建筑装饰风格的影响,使房屋成为地理和文化流通的丰富资源。

6. 大地主住宅的规模在18世纪晚期之后继续不断增长,其外表也更加华美,在当地被称为宫殿或宅第。接待大厅里到处可以看到高级的装饰材料和新颖的装饰图案:威尼斯或比利时的枝形吊灯悬挂在高高的天花板上,来自澳门的中国瓷器以蓝色和白色的餐具或彩色纪念性花瓶的形式摆进大厅,留声机用新音乐填充空间,热水管道系统为现代厕所空间服务,欧洲制造的马赛克地板在跳舞的人脚下。19世纪欧洲艺术史的繁荣和奢华将影响整个果阿村庄,并改变它们的建筑构成,小一点的房子也同样反映了这些影响。果阿房屋的典型建筑元素,如凉台(阳台,门廊处的一个空间,安设座位后,既可作为娱乐空间,也可作为接待前厅)和门廊,有机地融入了一些体现这一时期特点的小细节:牡蛎壳窗户的轮廓变化了,房屋各个部分的比例扩大了,房屋正面公共区域的实用性增强了。在房屋的内部,我们通常可以观察到更正式的接待室,通常有大窗户通向主阳台,创造出一种规模感和装饰感,这在房子的私人区域中是不存在的。相比于街道的狭窄,房子所扩大的程度,足以在外观和材料上体现出家族历史跨越巨大的时空距离。每当那些海外移民回家探亲的时候,就会从世界各地带回音乐、家具、时尚、习惯以及带有文化标志的东西。这些东西大部分都来自葡萄牙影响范围之外,起源于技术的文化语言全球化,如蒸汽船和非洲大陆日益增长的殖民占有。

7. 因此,这座房子在整合当地社会等级和"传统"果阿生活和遗产的物质元素的同时,也体现了海上迁徙路线所允许的社会流动、经济和文化变化的痕迹。海洋环流在同时代迁徙路线和流散文化发展中的中心地位,可以从它们在这些乡村景观中留下的物质和文化印记中感受到。从这些阳台上我们可能看不到大海,但在我们感受到传统牡蛎壳窗透出的透明光线的同时,我们仍然可以感受到某种国际化的现代性,坐在从其他大陆带来的家具和物品之间,看着旧照片,阅读那些离开这里的人们发来的信件,他们或定居在新的城市或在遥远的海岸与不同景观地带。

8. 从果阿房屋和一些文学描述中,我们可以看到海洋跨越的另一个层面。在漫长的殖民时期,殖民者向果阿贩卖奴隶的那段历史仍是模糊的,而该邦现存的非洲裔人口似乎可以证明这段历史。

9. 虽然葡萄牙人几个世纪以来在大西洋和印度洋奴隶贸易中的作用已被广泛承认,但存在于果阿以及印度次大陆其他葡萄牙殖民领土上的非洲奴隶仍有待全面认识。除其他活动外,葡属印度的贸易家族和企业的经济实力,是由奴隶贸易维持的,而这一段历史已从当地的社会记忆和文化参照中逐渐消失了。众所周知,在废除奴隶制之前,许多黑人奴隶定居在葡萄牙殖民地边境外的森林山丘上。在卡纳塔克邦,在非洲后裔的西迪社区,我们发现许多基督徒的名字来自葡萄牙语。然而,恰恰相反,奴隶制废除后,非洲后裔在果阿的定居是一段不明确的历史。我们可以通过与家相关的不同的档案维度来感知,家不仅是一个家庭的空间,也是历史过程的发生地。当穿越果阿南部时,我们可以直接从与房子空间及其他方面相关的摄影和文学作品来观察现存非洲裔果阿人的当代存在。

10. 利用家庭照片档案的研究项目,如作家、教育家和研究员萨维亚·维加斯开发的项目,或已故的温德尔·罗德里克斯关于果阿时尚的书,对于揭开不为人知的故事非常重要。在这两个项目中,我们可以看到果阿大家庭在他们的家门前摆拍,体现了一种刻意的塑造感和当时的家庭氛围。这些照片的特点是,他们都有莫桑比克仆人跟随左右。在20世纪初,与莫桑比克有联系的果阿家族(莫桑比克在1885年被葡萄牙占领)经常把儿童带到他们的果阿祖屋做家仆。这两个摄影档案研究的例子构成了重要的证据,揭示了果阿及果阿人的家庭空间在长期海洋活动中不太为人所知的一面。

Cultural Background Knowledge: Marine Folklife
海洋民俗生活文化背景知识

曲金良先生曾写道:"海洋文化,作为调整人与海洋的关系,是人类文化的重要组成部分,是人类开发利用海洋过程中形成的精神成果和物质成果。具体表现为人类对海洋的认识、观念、思想、意识和心态,以及由此而生成的生活方式,包括经济结构、法规制度、衣食住行习俗和语言文学艺术等形态。海洋文化的本质,就是人类与海洋的互动关系及其产物。"相比其他文化亚群,海洋文化具有如下特征。

漂流性与传播性:一个民族的民俗文化,会随着某个群体的迁徙而传播。族人漂流到哪里,他们的民俗文化也就漂流到哪里。海洋民俗文化亦是如此。这方面最典型的就是妈祖文化。起源于中国福建的"妈祖"信仰,随着福建沿海一些渔民的迁徙,传播到了20多个国家和地区,全世界有超过2亿的信徒、5 000多座妈祖寺庙。除了人口迁徙,促进海洋文化传播的还有商业。伴随着商业开发的进程贯穿在世界历史中,由商业需求带来的航海探险、国际贸易、港口储运、观光旅游等,极大地促进了沿海国家经济和文化的发展,古代海上丝绸之路更是成为海上商业文化发展的典型象征。

开放性与变异性：在相当长的历史时期，我国盛有发达的海洋文化。以郑和下西洋为代表，在国际贸易、文化交流等方面，中国的灿烂文明通过海洋逐步传播到直达非洲的广阔海域内，在东北亚、东南亚地区形成了汉文化圈。但是，这种影响往往随着时间的推移和空间改变，会产生不同形式的变异。例如，原来居住在福建省沿海地区的一些渔民，为了生存，漂洋过海，到了泰国。他们还带去了中国福建沿海特有的风俗。但是，这些风俗受到当地自然与人文环境的影响，与当地的风俗融合，产生了新的形式。

包容性与融合性：相较于世界其他国家，中国的海洋文化有很强的包容性和融合性。开放性使我们传播中华文明，更好地促进本国文化的发展；包容性使我们吸收一国文明，取其精华，更好地促进国与国的交流与文化融合。这种包容性是有益的、积极的，对各种异国文化都能主动接纳、与之和谐共处，并在共处中吸取精华，不断完善和充实自我。

功利性与神秘性：海洋民俗文化的功利性是显而易见的，特别是其中的海神信仰。人们无非是出于功利目的而祈求海神保佑出海安全、渔业丰收。在海洋民俗文化中，许多信仰都有巫术掺杂，充满神秘色彩。相传中国台湾的高山族赛夏人保留着一个神秘的祭祀活动。据说他们的祖先，在古代曾经消灭过矮人，为了避免矮灵作祟，每隔两年逢阴历十月十五举行矮灵祭。赛夏人的文化中，矮灵是灵，也是神，充满了神秘感。

Unit Five
Ocean Ambience

海洋风情

Text A

Native Hawaiian Culture

By Anonymous Author

Ideological and Political Education：依自然风情 展经略海洋

经略海洋,放眼国际、纵观全局,立足国内现状和长远发展,统筹协调海洋和陆地、中央和地方、维权和维稳、资源和环境、行业与管理等关系,在涉及海洋发展的重要问题上做出具有战略性的谋划,从而推动海洋强国战略目标的实现。建设海洋强国是中国特色社会主义事业的重要组成部分。党的十八大做出了建设海洋强国的重大部署。实施这一重大部署,对推动经济持续健康发展,对维护国家主权、安全、发展利益,对实现全面建成小康社会目标、进而实现中华民族伟大复兴都具有重大而深远的意义。海洋对于人类社会生存和发展具有重要意义。海洋孕育了生命、联通了世界、促进了发展。我们人类居住的这个蓝色星球,不是被海洋分割成了各个孤岛,而是被海洋连结成了命运共同体,各国人民安危与共。海洋的和平安宁关乎世界各国安危和利益,需要共同维护,倍加珍惜。

当前,以海洋为载体和纽带的市场、技术、信息、文化等合作日益紧密,中国提出共建21世纪海上丝绸之路的倡议,就是希望促进海上互联互通和各领域务实合作,推动蓝色经济发展,推动海洋文化交融,共同增进海洋福祉。在新时代的征程上,在实现中华民族伟大复兴的奋斗中,建设强大的人民海军的任务从来没有像今天这样紧迫。要深入贯彻新时代党的强军思想,坚持政治建军、改革强军、科技兴军、依法治军,坚定不移加快海军现代化进程,善于创新,勇于超越,努力把人民海军全面建成世界一流海军。

1. Ocean **stewardship** is deeply **embedded** in Native Hawaiian culture. **Humpback** whales, or koholā, are an important part of history, legend, and connection to the sea. The cultural and maritime heritage of Hawaii plays a role in management decisions and how the **sanctuary**'s daily activities are carried out.

2. Native Hawaiian Culture. **Indigenous** peoples have profound connections with their native lands, territories, and resources. For hundreds of years, Native Hawaiians have **coevolved** with the natural environment of these islands and have accumulated deep knowledge and understanding of their **ancestral** lands and seas. Today, Native Hawaiians continue to rely on the environment as a primary source and foundation for Hawaiian culture and worldview. According to Hawaiian **cosmology**, Native Hawaiians have a unique kinship relationship with the natural world. The Kumulipo, a Hawaiian creation chant, depicts the power of the cosmos erupting into motion and heat that causes natural elements to inspire creation as evolving out of the night by the gradual accumulation of life forms. Beginning with the coral **polyp** in the first wā or era, the Kumulipo announces the existence of the whale in the second wā. "Hānau ka palaoa noho i kai", born is the whale living in the ocean. Eventually, humans are born from this common origin.

3. Native Hawaiian culture acknowledges the whale as an ancient being. They have been honored as aumākua which are family or personal gods, **deified** ancestors who might assume the shape of either animals, plants, or natural phenomena. Native Hawaiian families and individuals have **reciprocal** relationships with their aumākua. Aumākua are prayed to and sometimes fed while they warn and **reprimand** mortals through dreams, visions, and other omens. These traditions were common practice when Hawaiian religion and spirituality were more prevalent. Today, there are families and individuals that continue to **perpetuate** these beliefs and practices.

4. Humpback whales and several other large whale species are known as koholā in the Hawaiian language. There are several place names in Hawaii that are connected to the koholā. Koholālele, which translates as leaping whale, is the name of a **fishpond** in the land division of Kualoa on the island of Oʻahu. Kualoa is also historically known for the **ivory** of the palaoa or sperm whale that drifts ashore. In ancient times, an ivory tooth or niho was carved into a large tongue-shaped **pendant** and fastened to **braided** hair of ancestors to create a sacred object known as a lei niho palaoa or ivory pendant necklace. It represents a chief's status and authority to speak which is based on his or her **genealogy**. Whales have also long been revered as physical manifestations of Kanaloa, god of the sea and open-ocean voyaging. Kanaloa is also another name for the island of Kahoʻolawe, known in pre-contact times as a training center for the art and skill of open

ocean navigation. After the attack on Pearl Harbor, the U.S. military used Kaho'olawe as a bombing range.

5. In the 1970s, the Hawaiian Cultural Renaissance began as a **resurgence** of a distinct cultural identity that drew upon traditional Kānaka Maoli (Native Hawaiian) culture. During the Hawaiian Cultural Renaissance, Native Hawaiians protested the bombing of Kanaloa/Kaho'olawe by the U.S. military, and built the first Polynesian voyaging canoe seen in 200 years. Today, Kanaloa/Kaho'olawe has become a center of teaching and training in environmental restoration, and the Native Hawaiian community has **revitalized** non-instrument voyaging and navigation. For over four decades now, the double-hull voyaging canoe, Hōkūle'a, has sailed thousands of miles across the world and is a proud symbol for the Native Hawaiian community and all of Hawaii.

6. Native Hawaiian culture is grounded in traditional values and principles that are based on ancestral relationships between people and places. The koholā is part of this immense cultural heritage found within oral histories, storied place names, and material culture. Although there have been **drastic** changes to the social, cultural, environmental, economic, and political qualities of Hawaii for over two hundred years, Native Hawaiians have endured and adapted. Their culture continues to thrive and evolve as they navigate modern times because they rely on their traditional values and cultural principles that were forged in their ancestral homeland of Hawaii in the presence of the whale.

7. Committed to Heritage. We at Hawaiian Islands Humpback Whale National Marine Sanctuary recognize that as sanctuary managers, we are public servants. We, therefore, have developed meaningful relationships within our island communities and with Native Hawaiians who are committed to perpetuating Hawaiian culture. We cultivate these relationships through a strong network of partners, including the community-based Sanctuary Advisory Council, and through extensive public participation in education and outreach venues on different islands. We learn from organizations and entities that are committed to perpetuating cultural heritage, and we strive to create opportunities to integrate place-based knowledge into all aspects of our marine resource stewardship.

8. Voyaging Canoe on the Ocean. The Hawaiian **archipelago** has a long history of continuous and intensive maritime activity. Many historic shipwrecks and other types of submerged archaeological sites can be found in and around Hawaiian Islands Humpback Whale National Marine Sanctuary.

9. Polynesian Voyaging. Over hundreds of years, double-hull voyaging canoes, guided by

specialists raised in the art and science of non-instrument navigation, moved eastwards into the Pacific. The canoes carried the men and women and all necessary provisions needed for remote settlement. Navigators used solar and **celestial** observations of the sea, and swell direction and ocean temperature, to maintain their course beyond the sight of land. The successful colonization of remote islands and **atolls** also relied on the technical capabilities of double-hulled oceanic voyaging canoes, or wa'a. These were advanced vessels, well designed for their environments. In a series of bold exploratory voyages, all of Eastern Polynesia appears to have been inhabited by around 700 CE. It is then from Eastern Polynesia, from Tahiti and the Marquesan Islands, that the remote corners of the Polynesian triangle, Rapa Nui (Easter Island), Aotearoa (New Zealand), and Hawaii were originally discovered and settled. The speed and carrying capacity of many types of Pacific canoes impressed 18th- and 19th-century observers such as Captain James Cook. The voyaging canoe design was clearly the **preeminent** Pacific vessel of ancient exploration and migration, a capable platform for colonizing remote islands and therefore of immense cultural importance.

10. Ocean Connection. Sunset over a Fishpond. Following Polynesian settlement, advanced aquaculture techniques flourished in the Hawaiian Islands, where there may have been between 400 and 500 stone fishponds producing around two million pounds of fish annually. Fishponds are still some of the most significant traditional cultural resources in Hawaii. They demonstrate traditional capacity in sustainability, food sovereignty, and natural resource management. There is a continued interest in the repair and operation of traditional Hawaiian fishponds for their cultural, economic, and ecological value. Shell fishing hooks, as well as scattered artifacts such as **octopus** lure weights, fish trap weights, and canoe anchors, are found on some near shore reefs, even in developed areas near Honolulu on the island of O'ahu. Multiple ways of fishing continue to this day, including traditional methods like throw nets, torch fishing, and spears.

11. Western Contact. The wreck site of the brig Cleopatra's Barge highlights the interactions between Native Hawaiians and foreign cultures in the decades following Western contact. The luxurious 1816 vessel was sold to King Kamehameha II (Liholiho) in 1820 and renamed Ha'aheo o Hawaii (Pride of Hawaii), to serve as the royal yacht. Lost in 1824 in Hanalei Bay off Kauai, the remains of Ha'aheo o Hawaii were excavated in 1995. The collection is now on display at the Kaua'i Museum in Lihu'e. These and other Native Hawaiian, Asian, and Western artifacts in Hawaii tell the story of social and economic change among the islands.

12. Whaling Ships. Soon after the crews of Balaena and Equator harpooned the first whale

off Kealakekua Bay in 1819, Hawaii won its place on the maps of the whalers. Pacific whaling grounds became dominated by American vessels in the mid 19th century, as whale oil became a major economic component of economic expansion in both New England and the Hawaiian Islands. Some residents in Hawaii today can trace their lineage to the frequent deserter from a whaling ship. There are at least 18 documented whaling ship losses in and around the Hawaiian island chain, five of these reported historically within the sanctuary. The whalers Drymo (1845), Paulina (1860), and Young Hero (1858) were lost near Maui, Jefferson (1842) in Hanalei Bay on Kaua'i, and Helvetius (1834) near O'ahu. These shipwrecks testify to the once active and bloody American involvement in Pacific whale hunting.

13. Shipwreck Beach. The sanctuary's boundaries also include "Shipwreck Beach" on the north shore of Lāna'i. During the 19th and early 20th centuries, inter-island navigation companies used Shipwreck Beach as an area for the intentional abandonment of vessels, a "rotten row" of old ships. Other vessels were also lost on the coast's treacherous reefs by accident. Several of these now historic sites have been surveyed, such as the Pearl Harbor survivor YO-21, the schooner Mary Alice and the Hawaiian steamship Hornet, but many other wrecks along the eight-mile stretch have yet to be identified. Shipwreck Beach is also the location of a Hawaiian battleground. Seeking to strike against the political satellite of Maui, Kalani'ōpu'u, a war chief from the Island of Hawaii, landed his warriors along the north shore and raided Lāna'i in 1778.

14. World War II. Unprecedented naval activity took place among the Hawaiian Islands during World War II, in the skies as well as on and under the sea. Hundreds of Navy fighter aircraft and pilots took part in intensive training activities in preparation for combat operations in the Pacific. Over 1,500 naval aircraft sank in the vicinity of the Hawaiian Islands, and of these some 39 are known to have been lost in sanctuary waters. Some of these submerged aircraft crash sites are war graves. Other submerged archaeological sites include landing craft and AMTRACKS (**amphibious assault** vehicles) lost during massive training exercises in the islands. These protected resources, property of the federal government, bear witness to our nation's commitment and sacrifice during the war, a period which changed the shape of the entire Pacific region.

15. Dive Sites. Shipwreck and aircraft sites within the sanctuary also function as sport diving destinations, enjoyed by local and visiting divers. The steamship Maui, lost on the Island of Hawaii, the F4U-1 Corsair in Oahu's Maunalua Bay, the PB4Y-1 Liberator near Maui, numerous U.S. Navy landing craft lost during training operations on beaches throughout the main Hawaiian Islands, and Carthaginian II, the **replica** whaling supply **brig** that once welcomed Lahaina visitors to tour on board, now **entice**

divers to share a bit of the maritime past. The sport diving industry plays an important role in sharing the connection of Hawaii to the sea. You can read more about Maui's World War Ⅱ Legacy.

16. Resources. Heritage resources within the sanctuary reflect the historical phases of past maritime activity: Native Hawaiian aquaculture and fishing, Pacific whaling, inter-island sail and steam navigation, and naval activity among the islands. Many of the maritime heritage resources within the sanctuary fall within state waters. The state agency for preservation management of these heritage resources is the State Historic Preservation Office under the Department of Land and Natural Resources, co-manager of the sanctuary.

17. Native Hawaiian Fishpond. The historical fishpond named Kō'ie'ie fronts the visitor center on Maui and provides a natural classroom for educational efforts. The three-acre pond is one of the last remaining intact traditional fishponds along the south Maui coastline. The sanctuary partners with Ao'ao O Nā Loko I'a O Maui (Association of the Fishponds of Maui), which is working to protect and restore the ancient fishpond in order to educate the public about traditional Native Hawaiian values, practices, and traditions. A major restoration of the ancient Hawaiian fishpond by Ao'ao O Nā Loko I'a O Maui is currently taking place. Stone by stone, the fishpond association workers and volunteers are rebuilding the wall by hand using traditional methods.

(2,011 words)

https://hawaiihumpbackwhale.noaa.gov/heritage/native-culture.html

New Words

1. stewardship [ˈstjuːədʃɪp]

    n. the position of steward 管理工作;管事人的职位及职责

    The UK's stewardship code is actively encouraging institutional shareholders to co-operate with companies.

    英国的管理法规也积极提倡机构股东与公司合作。

2. embed [ɪmˈbed]

    v. fix or set securely or deeply; attach to, as a journalist to a military unit when reporting on a war(使)嵌入,把……插入;深信,使深留脑中;(计算机中)内置

    In fact, the process can embed small flaws in the genes of clones that scientists are only now discovering.

    事实上,这个过程可能会在克隆的基因中嵌入了一些小的缺陷,而科学家们直到现在才发现这些缺陷。

3. humpback [ˈhʌmpbæk]

n. large whalebone whale with long flippers noted for arching or humping its back as it dives;an abnormal backward curve to the vertebral column 座头鲸;驼背

Even humpback whales prefer to use the right side of their jaws to scrape sand eels from the ocean floor.

即使是座头鲸也喜欢用它们的右颚从海底刮沙鳗鱼。

4. sanctuary [ˈsæŋktʃuəri]

n. a consecrated place where sacred objects are kept;a shelter from danger or hardship 避难权,庇护;避难所,庇护所;动物保护区,禁猎区;(某种受伤或遭遗弃动物的)收养所;宗教圣所,圣殿;(教堂内的)圣坛

A friend is a sanctuary. A friend is a smile.

朋友是你心灵的庇护所,朋友是一个向你笑容满面的人。

5. indigenous [ɪnˈdɪdʒənəs]

a. originating where it is found 本土的,固有的

The tropical island of new guinea is home to over nine hundred languages, Russia, twenty times larger, has 105 indigenous languages.

热带岛屿新几内亚是900多种语言的家园,而比它面积大20倍的俄罗斯,有105种土著语言。

6. coevolve [ˌkəʊɪˈvɒlv]

v. 共同进化(发生于两种或多种生物)

Satisfying these conflicting demands requires that the organization and process coevolve in order to deliver a solution that adds maximum value and delivers competitive advantage.

满足这些相互冲突的要求,需要组织和过程共同改进,以交付一个增加最大价值和巩固竞争优势的解决方案。

7. ancestral [ænˈsestrəl]

a. inherited or inheritable by established rules (usually legal rules) of descent 祖先的;祖传的

Perhaps that is why we explore the starry skies, as if answering a primal calling to know ourselves and our true ancestral homes.

也许这就是我们探索星空的原因,仿佛在回应一种原始的召唤,去了解我们自己和祖先真正的家园。

8. cosmology [kɒzˈmɒlədʒi]

n. the metaphysical study of the origin and nature of the universe 宇宙论,宇宙学

Cosmology, geology, and biology have provided a consistent, unified, and constantly improving account of what happened.

宇宙学、地质学和生物学对所发生的事情提供了一致的、统一的和不断改进的解释。

9. cosmos [ˈkɒzmɒs]

n. everything that exists anywhere;popular fall-blooming annuals 宇宙;完整和谐的一统体系

A team of astronomers working at the European Southern Observatory in Chile, led by

Simon Clark, spotted a flyer in the cosmos.

在智利的欧洲南方天文台,由西蒙·克拉克领导的天文学家团队发现了宇宙中的一个飞行物。

10. polyp ['pɒlɪp]

n. a small vascular growth on the surface of a mucous membrane; one of two forms that coelenterates take, e.g. a hydra or coral: usually sedentary and has a hollow cylindrical body usually with a ring of tentacles around the mouth 息肉;珊瑚虫;水螅虫

As to the allergic rhinitis with sinusitis and nasal polyp, both may be the reason for smell disturbance.

变应性鼻炎合并鼻窦炎和鼻息肉者,其嗅觉障碍的原因可能为机械阻塞加炎症。

11. deify ['deɪɪfaɪ;'diːɪfaɪ]

v. consider as a god or godlike; exalt to the position of a God 把……奉若神明;把……神化;崇拜

Under the influence of many factors, people, on the one hand, have high expectation of and show great concern to school education, and even exaggerate and deify its role.

在众多因素的影响下,人们一方面对学校教育寄予了高度的期待和积极的关注,甚至将学校教育功能夸大、神圣化。

12. reciprocal [rɪ'sɪprək(ə)l]

a. concerning each of two or more persons or things; especially given or done in return; of or relating to the multiplicative inverse of a quantity or function 相互的,互惠的,报答的;(路线,方向)反向的;(代词,动词)互相的;(量,函数)倒数的

The two colleges have a reciprocal arrangement whereby students from one college can attend classes at the other.

两所学院有一项互惠协定,允许学生在院际间选课。

13. reprimand ['reprɪmɑːnd]

v. rebuke formally; censure severely or angrily 谴责,训斥

I once had a manager who would reprimand his subordinates in public.

我曾经有这样一个上司,他会在公众场合批评自己的下属。

14. omen ['əʊmən]

n. a sign of something about to happen 预兆,征兆;预兆性

Her appearance at this moment is an omen of disaster.

她此时的出现是灾难的预兆。

15. perpetuate [pə'petʃueɪt]

v. cause to continue or prevail 使持续,使长久(尤指不好的事物)

Comics tend to perpetuate the myth that "Boys don't cry".

连环画往往在延续着"男儿有泪不轻弹"的神话。

16. fishpond ['fɪʃpɒnd]

n. a freshwater pond with fish 鱼池;养鱼塘

Aerator should be generally placed in the center of fishpond, or like windward on the drawing.

增氧机一般应放置在鱼池中央,或略偏上风的位置。

17. ivory ['aɪvəri]

n. a hard smooth ivory colored dentine that makes up most of the tusks of elephants and walruses 象牙,(其他动物的)长牙;象牙制品;象牙色,乳白色

The Ivory Coast became the world's leading cocoa producer.

象牙海岸成了世界上领先的可可粉生产地。

18. pendant ['pendənt]

n. an adornment that hangs from a piece of jewelry (necklace or earring); branched lighting fixture; often ornate; hangs from the ceiling 垂饰;坠饰;吊灯;短索;补充作品;三角旗

We envision the camera becoming so small that it integrates into clothing, such as the button of a shirt, a brooch, or a pendant.

我们想象一下,如果这个摄影机非常小巧,我们就可以把它放进衣服里,像衬衫扣子、胸针或是项链的坠子一样。

19. braid [breɪd]

v. make by braiding or interlacing; decorate with braids or ribbons; form or weave into a braid or braids 把……编成绳子(或穗带);把(头发)编成辫子;(用饰带等)镶边,镶缀;(河或小溪)迂回流动,交叉往来

Can you braid my hair?

可以帮我编下头发吗?

20. genealogy [ˌdʒiːniˈælədʒi]

n. successive generations of kin; the study or investigation of ancestry and family history 宗谱;血统;家系;系谱学

Books, movies, family genealogy, environmental living programs and travel can all be vehicles to exploring history.

书籍、故事片、族谱、现实生活事件、旅游等,都是孩子探索历史的好材料。

21. revere [rɪˈvɪə(r)]

v. regard with feelings of respect and reverence; consider hallowed or exalted or be in awe of 尊敬,崇敬

If we translated Sanskrit into Chinese, we would revere Sakyamuni Buddha as "Lord Sakyamuni".

如果从梵文翻译成中文的话,我们也称呼释迦牟尼为"主释迦牟尼"。

22. resurgence [rɪˈsɜːdʒəns]

n. bringing again into activity and prominence 复兴,复苏,再次兴起

The revival of the language, particularly among young people, is part of a resurgence of national identity sweeping through this small, proud nation.

语言的复兴,尤其是在年轻人当中的复兴,是席卷这个自豪的小国的民族认同感复苏的一部分。

23. revitalize [ˌriːˈvaɪtəlaɪz]

  v. restore strength; give new life or vigor to 使恢复生机,使复兴

  Private courier companies, which have more dollars to spend, use their expertise in logistics to help revitalize damaged areas after a disaster.

  私人快递公司可支配的资金更多,它们利用自身在物流方面的专业技术来助力灾区的重振。

24. drastic [ˈdræstɪk]

  a. forceful and extreme and rigorous 极端的,激烈的

  Against a backdrop of drastic changes in economy and population structure, younger Americans are drawing a new 21st-century road map to success, a latest poll has found.

  一项最新民调发现,在经济和人口结构发生巨变的背景下,美国的年轻一代正在绘制一幅 21 世纪通向成功的地图。

25. archipelago [ˌɑːkɪˈpeləgəʊ]

  n. a group of many islands in a large body of water 群岛,列岛;多岛的海区

  Herpetologist Robert Mertens later argued that Varanus probably originated in the archipelago.

  爬虫学家罗伯特·默滕斯后来提出,巨蜥属可能起源于这个群岛。

26. celestial [səˈlestiəl]

  a. of or relating to the sky; relating to or inhabiting a divine heaven 天空的,天上的;天国的,天堂的;精美绝伦的

  Humans have sent many missions, both manned and robotic, beyond our planet to explore our neighboring celestial bodies.

  人类已经向太空发射了许多载人的和遥控的飞行器,用来探索地球之外的临近天体。

27. atoll [ˈætɒl]

  n. an island consisting of a circular coral reef surrounding a lagoon 环礁,环状珊瑚岛

  Although the atoll lost many corals, it kept an abundance of grazing fish that help reefs recover by keeping them clean.

  尽管这里的环礁损失了很多珊瑚,但依然有大量的草食鱼类通过清理珊瑚礁来帮助它们复苏。

28. preeminent [ˌpriːˈemɪnənt]

  a. greatest in importance or degree or significance or achievement 卓越的;超群的

  The story tells about a group of guests from all over the world in a Tong Fu preeminent experience something legendary.

  故事讲述的是一群来自五湖四海的江湖人士在一个名叫"同福客栈"中所经历的传奇故事。

29. artifact [ˈɑːtɪfækt]

  n. a man-made object taken as a whole(尤指有文化价值或历史价值的)人工制品,历史文物;非自然存在物体,假象

  This view of the artifact will be updated when the source artifact is modified and saved.

工件的视图将会在源工件更改和保存时得到更新。

30. octopus ['ɒktəpəs]

  n. bottom-living cephalopod having a soft oval body with eight long tentacles; tentacles of octopus prepared as food 章鱼；章鱼肉

  The nocturnal animals hide in underwater caves by day, then venture out at night, feeding on small fish, squid, and octopus.

  这种夜行性动物白天时隐藏在水底洞穴里，晚上才出来冒险，它们以小鱼、乌贼和章鱼为食。

31. schooner ['skuːnə(r)]

  n. sailing vessel used in former times; a large beer glass 纵帆船；大酒杯；大篷车

  He would build a patriarchal grass house like Tati's, and have it and the valley and the schooner filled with dark-skinned servitors.

  他要把海湾和峡谷当作大本营，要修建一幢塔提家的那种草屋，让那草屋、峡谷和大帆船里满是皮肤黝黑的仆人。

32. amphibious [æm'fɪbiəs]

  a. operating or living on land and in water; relating to or characteristic of animals of the class Amphibia 两栖的，水陆两用的；具有双重性的

  Left atrial appendage has become a research hotspot for its special anatomic structure and function characteristics.

  左心耳特殊的解剖结构、功能特点使其成为目前研究的热点。

33. assault [ə'sɔːlt]

  n. close fighting during the culmination of a military attack; a threatened or attempted physical attack by someone who appears to be able to cause bodily harm if not stopped（军事）袭击，攻击；人身攻击，侵犯人身（罪）；抨击，闯关，冲击（难关）

  The paper's assault on the president was totally unjustified.

  这份报纸对总统的攻击纯属无稽之谈。

34. replica ['replɪkə]

  n. copy that is not the original; something that has been copied 复制品，仿制品

  The bedroom was an exact replica of the original, perfect right down to the patterns on the wallpaper and the hairbrushes on the dressing table.

  这间卧室精确地复制了原来的房间，模仿之细甚至连墙纸的图案和梳妆台上的发刷都一模一样。

35. brig [brɪg]

  n. two-masted sailing vessel square-rigged on both masts; a penal institution (especially on board a ship) 双桅横帆船；禁闭室；警卫室

  Despite its cuddly name, the Beagle was a naval brig outfitted with 10 guns.

  尽管它的名字很是逗人喜爱，但小猎犬号是一艘装备有10支枪的海军双桅横帆船。

36. entice [ɪn'taɪs]

  v. provoke someone to do something through (often false or exaggerated) promises or persuasion 诱使，引诱

They'll entice thousands of doctors to move from the cities to the rural areas by paying them better salaries.

他们将通过支付更高的薪水怂恿成千上万的医生从城市迁往农村。

Phrases and Expressions

have a relationship with 与……有关系；与……有关联
begin with 以……开始；开始于……
draw upon 利用；开出；总结
in the presence of 在……面前；有某人在场
in preparation for 为……做准备
in the vicinity of 在……附近；在……上下
on board 在船(火车、飞机)上
take place 发生；举行

Terminology

double-hull 双层壳
economic expansion 经济扩张
amphibious assault vehicles 两栖攻击车

Proper Names

sperm whale 巨头鲸；抹香鲸(等于 Physeter catodon)
Pearl Harbor 珍珠港(美国夏威夷港口)
Polynesian 波利尼西亚的；波利尼西亚人的；波利尼西亚人
Hawaiian Islands Humpback Whale National Marine Sanctuary 夏威夷群岛座头鲸国家海洋保护区
Sanctuary Advisory Council 保护区咨询委员会
Tahiti 塔希提岛(位于南太平洋，法属波利西亚的经济活动中心)
the Marquesan Islands 马克萨斯群岛
Hanalei Bay 哈纳莱伊湾
Maunalua Bay 冒纳卢阿湾(美国地名)

一、翻译策略/方法/技巧：重组法

重组法：科技英语为了表示严谨、精确的含义，长句使用较多。长句主要是由基本句型扩展而成，其方式有增加修饰成分如定语、状语，用各种短语如介词、分词、动名词或不定式短语充当句子成分，也可能是通过关联词将两个或两个以上的句子组合成并列复合句或主从复合句。英译汉时，往往要先分析句子结构、形式，才能确定句子的功能、意义。在遇到长句时，宜通过形态识别突显主、谓、宾、表等主干成分，了解其"骨架含义"及次要成分的含义,理解这些成分之间的逻辑关系和修饰关系,然后通过适当的方法翻译出来。英语利用形态变化、词序和虚词三大语法手段可构成包含许多修饰成分或从句的长句或复合句,句中各部分顺序灵活多样。通常英语句中的表态部分,如判断和结论在先,而叙事部分,如事实和描写在后,汉语则正好相反;英语句先短后长,显得"头轻脚重",而汉语也正好相反;英语借助形态变化和连接手段将句中成分灵活排列,汉语则常按时间和逻辑顺序由先到后、由因到果、由假设到推论、由事实到结论这样排列。此外,汉语少用长句,多用短句、分句、流水句,按一定的时间或逻辑顺序,分层叙述。因此,长句翻译时应根据叙述层次和顺序对句子进行拆分重组。重组法是指在进行翻译时,为了使译文流畅和更符合译入语的叙事论理习惯,在捋清长句的结构、弄懂英语原意的基础上,彻底摆脱原文语序和句子形式,对句子进行重新组合,有时甚至会从文章全篇入手进行重组。重组主要包括句子在长度上的调整和结构上的调整。

例1:Decision must be made very rapidly; physical endurance is tested as much as perception, because an enormous amount of time must be spent making certain that the key figures act on the basis of the same information and purpose.

必须把大量时间花在确保关键人物均根据同一情报和目的行事,而这一切对身体的耐力和思维能力都是一大考验。因此,一旦考虑成熟,决策者就应迅速做出决策。

例2:The isolation of the rural world because of distance and the lack of transport facilities is compounded by the paucity of the information media.

因为距离远,交通工具缺乏,造成了农村社会与外界的隔绝,这种隔绝又由于信息媒介不足,而变得更加严重。

例3:The poor are the first to experience technological progress as a curse which destroys the old muscle-power jobs that previous generations used as a means to fight their way out of poverty.

对于以往的几代人来说,旧式的体力劳动是一种用以摆脱贫困的手段,而技术的进步则摧毁了穷人赖以为生的体力劳动。因此,首先体验到技术进步之害的是穷人。

例4:The big problem of comprehension of the English text and the bigger problem of how to express it in rich, present-day Chinese which ranges from the classical to the colloquial both have to be solved in the course of translation.

了解英语原义是一大问题,现代汉语既然是文言口语兼收并蓄,那么怎么用这样丰富多彩的文字来表达英文原文是一个更大的问题,这两个问题在翻译中都要得到解决。

二、译例与练习
Translation
1. According to Hawaiian cosmology, Native Hawaiians have a unique kinship relationship with the natural world. The Kumulipo, a Hawaiian creation chant, depicts the power of the cosmos erupting into motion and heat that causes natural elements to inspire creation as evolving out of the night by the gradual accumulation of life forms. (Para. 2)

根据夏威夷宇宙学,夏威夷原住民与自然界有着独特的亲缘关系。库穆里波是夏威夷的一首（上帝）创造天地的圣歌，描绘了宇宙力量爆发为运动和热量，导致自然元素激发创造，如同由黑夜逐渐积累进化而来的生命形态。

【夏威夷州（Hawaii State）是美国唯一的群岛州，由太平洋中部的132个岛屿组成。首府位于瓦胡岛上的火奴鲁鲁（檀香山）。】

2. There are several place names in Hawaii that are connected to the koholā. Koholālele, which translates as leaping whale, is the name of a fishpond in the land division of Kualoa on the island of O'ahu. (Para. 4)

夏威夷有好几个地名都与鲸鱼（夏威夷语）有关。拉海纳，翻译过来就是跳跃的鲸鱼，是瓦胡岛的古兰尼境内地区一个鱼塘的名字。

3. Today, Kanaloa/Kaho'olawe has become a center of teaching and training in environmental restoration, and the Native Hawaiian community has revitalized non-instrument voyaging and navigation. (Para. 5)

今天，卡纳洛阿/卡荷拉维已成为环境恢复的教学和培训中心，夏威夷土著群体已经使不用仪器航海重新焕发了生机。

4. Their culture continues to thrive and evolve as they navigate modern times because they rely on their traditional values and cultural principles that were forged in their ancestral homeland of Hawaii in the presence of the whale. (Para. 6)

在现代社会中，文化继续繁荣发展，因为文化依赖于传统的价值观和文化原则，这些价值观和文化原则是在祖先们的家乡夏威夷的鲸鱼面前形成的。

5. We cultivate these relationships through a strong network of partners, including the community-based Sanctuary Advisory Council, and through extensive public participation in education and outreach venues on different islands. (Para. 7)

我们通过包括以群体为基础的保护区咨询委员会在内的强大的合作伙伴网络，以及在不同岛屿上广泛参与教育和拓展活动，来培养这些体系。

6. In a series of bold exploratory voyages, all of Eastern Polynesia appears to have been inhabited by around 700 CE. It is then from Eastern Polynesia, from Tahiti and the Marquesan Islands, that the remote corners of the Polynesian triangle, Rapa Nui (Easter Island), Aotearoa (New Zealand), and Hawaii were originally discovered and settled. (Para. 9)

在一系列大胆的探险航行中，整个东波利尼西亚似乎在公元700年左右就有人居住了。然后，人类从东波利尼西亚、塔希提岛和马克桑群岛上最初发现了波利尼西亚三角的偏远角落、拉帕努伊（复活节岛）、奥特罗阿（新西兰）和夏威夷岛屿，并在此定居下来。

7. There is a continued interest in the repair and operation of traditional Hawaiian fishponds for their cultural, economic, and ecological value. Shell fishing hooks, as well as scattered artifacts such as octopus lure weights, fish trap weights, and canoe anchors, are found on some near shore reefs, even in developed areas near Honolulu on the island of O'ahu. (Para. 10)

由于夏威夷传统鱼塘的文化、经济和生态价值，人们对其修复和运营一直很感兴趣。在一些近岸珊瑚礁上，甚至在奥胡岛檀香山附近的发达地区发现了贝壳鱼钩以及分散的文物，如章鱼诱饵重量、鱼饵重量和独木舟锚。

【这个长句在翻译时,对 There be 句型进行具体化翻译,没有译为"有……",而是译为了原因状语从句。后面句子的状语"even in......of O'ahu"翻译时采用了换序译法,对原文进行调整,使译文符合汉语的表达习惯,自然通顺。】

8. Several of these now historic sites have been surveyed, such as the Pearl Harbor survivor YO-21, the schooner Mary Alice and the Hawaiian steamship Hornet, but many other wrecks along the eight-mile stretch have yet to be identified. (Para. 13)

其中一些历史遗迹已经被调查过了,比如珍珠港幸存者 YO-21 号、双桅帆船玛丽·爱丽丝号和夏威夷大黄蜂号蒸汽船,但沿着这条 8 英里长的海域还有许多其他残骸尚未确认。

【这个长句含有两个被动语态。翻译时将第一个被动语态译为汉语的被动语态"被调查"。第二个被动语态 be identified,直接转换成汉语的主动语态"未确认",译文通顺流畅能够为读者所接受。因此,英译汉时,可根据具体情况运用转态译法进行翻译。】

9. The steamship Maui, lost on the Island of Hawaii, the F4U-1 Corsair in Oahu's Maunalua Bay, the PB4Y-1 Liberator near Maui, numerous U.S. Navy landing craft lost during training operations on beaches throughout the main Hawaiian Islands, and Carthaginian II, the replica whaling supply brig that once welcomed Lahaina visitors to tour on board, now entice divers to share a bit of the maritime past. (Para. 15)

无论是在夏威夷岛上丢失的毛伊号蒸汽船、在瓦胡岛的毛那鲁阿湾丢失的 F4U-1 海盗号、在毛伊岛附近丢失的 PB4Y-1 解放者号、在整个夏威夷群岛海滩上训练行动中丢失的众多美国海军登陆艇及迦太基 II 号,还是曾经欢迎拉海纳游客到船上游览的仿冒捕鲸供应禁猎艇,现在都吸引潜水员分享过去的海事历史。

10. The sanctuary partners with Ao'ao O Nā Loko I'a O Maui (Association of the Fishponds of Maui), which is working to protect and restore the ancient fishpond in order to educate the public about traditional Native Hawaiian values, practices, and traditions. (Para. 17)

保护区合伙人与 Ao'ao O Nā Loko I'a O(夏威夷语)毛伊(毛伊岛鱼塘协会)合作,该协会致力于保护和恢复古老的鱼塘,以培养公众了解夏威夷原住民的传统价值观、习俗与传统。

Exercises

1. Fill in the blanks with the proper given words, and then translate the sentences into Chinese.

　　worldview　　involvement　　flourish　　ground　　sustainability　　harpoon

1) The Foreign Ministry has issued a flat denial of any _____.
2) Energy and _____ experts say the answer to our future energy needs will likely come from a lot of comprise, both traditional and alternative.
3) He _____ the gears on the car.
4) As your plants _____, you'll need to repot them in bigger pots.

5) The _____ rope had to be very long, and very carefully arranged in the boat so that it would run out freely when the whale dived.
6) When they were first published, Lu Xun's short stories were unorthodox in their language, as well as their _____.

2. Translate the following sentences into Chinese.
1) The deep sea typically has a sparse fauna dominated by tiny worms and crustaceans, with an even sparser distribution of larger animals.
2) This deep sea creepy-crawly gave oil workers a fright, after the unexpected visitor hitched a ride on a submarine sent from a rig in the Gulf of Mexico.
3) Some futurists envision nanotechnology also being used to explore the deep sea in small submarine, or even to launch finger-sized rockets packed with micro miniature instruments.
4) In Scotland, the number of wild salmon have been reduced because of uncontrolled deep-sea and coastal netting, by pollution, and by various other threats to the fish's habitat.
5) Most deep-sea faunas rely for food on particulate matter, ultimately derived from photosynthesis, falling vent would dwarf any contribution from advection.
6) The sample consisted of pebbles of hardened sediment that had once been soft, deep-sea mud, as well as granules of gypsum and fragments of volcanic rock.
7) So why has it taken so long to move towards the commercial exploitation of deep sea massive sulphide deposits?
8) The authors point out that global climate change could affect the food supply to the deep sea in many ways.
9) Beyond the sea water seeping into the soil, there is also acidification of the ocean, which is leading to coral bleaching.
10) Because deep sea trawling is a recent phenomenon, the damage that has been done is still limited.

3. Translate the following sentences into English.
1) 夏威夷在吹过椰子树的微风中，海浪的轻拍声中，甜蜜的爱情颂歌中，低低吟唱着浪漫。
2) 我们知道，浮游生物会吸收大气中的二氧化碳，当它们死亡时会把二氧化碳带入洋底，并在那里停留数千年。
3) 他们综合日射日光、海洋和大气的成分及流通的有关数据，模拟影响地球气候变化的过程。
4) 诸如2004年触发印度洋海啸的地震这样的自然灾难也会导致珊瑚礁损失。
5) 加利福尼亚州被山区和海洋所限制，环境学家都强烈要求保持该州的自然美景。
6) 只是在2005年发生印度洋海啸后，东盟才达成了一项让该组织的成员国在遭受自然灾害侵袭时彼此帮助的协议。
7) 导致温和气温的其中一个重要因素是强大的厄尔尼诺现象，这是太平洋表面温度的自然上升现象,厄尔尼诺对2015—2016年的气候造成了影响。

8) 因其岛屿被深蓝色的大海所环绕,新西兰的很多城市便坐落于海湾沿岸,并拥有天然的深水良港。

9) 当设计师第一次参观这个地方的时候,他就知道这个项目的重点是如何交织、如何分解、剪切和吸收周围的自然元素,如海洋、绿植、蓝天和风。

10) 岛国的周围是蔚蓝色的海洋,新西兰许多城市都建在海湾上,拥有天然的深水港。

4. Choose the best paragraph translation. And then answer why you choose the first translation or the second one.

中国的海岛,星罗棋布,形态各异,如同一颗颗璀璨的明珠,镶嵌在波光粼粼的蓝色海洋上。一些适合开发旅游的岛屿已成为热门的旅游目的地。悠闲舒适的海岛生活令人向往。迷人的热带海洋风情让海南独具魅力。海南的一些小岛逐渐进入游客的视野。蜈支洲岛海水清澈透明,被誉为"中国第一潜水基地"。南湾猴岛森林覆盖率达95%,生活着近1 500只猕猴,是中国也是世界唯一的热带岛屿型猕猴自然保护区。

译文一:

China's islands, dotted with various forms, are like a bright pearl inlaid in the sparkling blue ocean. Some islands suitable for tourism development have become popular tourist destinations. The leisure and comfortable island life is attractive. The charming tropical ocean makes Hainan unique. Some small islands in Hainan are gradually coming into the view of tourists. Sea water in Wuzhizhou Island is clear and transparent, known as "China's first diving base". Nanwan Monkey Island has a forest coverage rate of 95%, and nearly 1,500 macaques live on it, which is the only tropical island type macaque nature reserve in China and the world.

译文二:

China's islands are dotted with different shapes, like bright pearls, embedded in the sparkling blue sea. Some islands suitable for tourism development have become popular tourist destinations. Leisure and comfortable island life is desirable. The charming tropical marine customs make Hainan unique. Some islands in Hainan are gradually coming into the sight of tourists. Wuzhizhou Island is known as the "First Diving Base in China" for its clear and transparent water. Nanwan Monkey Island, with 95% forest coverage and nearly 1,500 macaques, is the only tropical island type macaque nature reserve in China and the world.

5. Translate the following passage into Chinese.

Native Hawaiian Culture. Indigenous peoples have profound connections with their native lands, territories, and resources. For hundreds of years, Native Hawaiians have coevolved with the natural environment of these islands and have accumulated deep knowledge and understanding of their ancestral lands and seas. Today, Native Hawaiians continue to rely on the environment as a primary source and foundation for Hawaiian culture and worldview. According to Hawaiian cosmology, Native Hawaiians have a unique kinship relationship with the natural world. The Kumulipo, a Hawaiian creation chant, depicts the power of the cosmos erupting into motion and heat that causes natural elements to inspire creation as evolving out of the night by the gradual accumulation of life forms. Beginning with the coral polyp in the first wā or era, the Kumulipo announces the existence of the whale in the second wā. "Hānau ka palaoa noho i kai", born is the whale living in the ocean. Eventually, humans are born from this common origin.

6. Translate the following passage into English.

深海中有着多样化的生境，如深海平原、海山、深渊、热液、冷泉、鲸落等。不同生境孕育了不同的生态系统，其中海山生态系统中的生物多样性最高，栖息着几乎所有门类的动物，从最原始的微生物到最高等的哺乳动物都有。海山被认为是深海生物的聚居地，目前人类从海山发现了约2 000种生物，而实际"居住"在海山的物种数可能远不止这些。由于海洋生物有分层分布的特点，即每一种生物都分布在特定水层，海山的立体结构可使其容纳不同水层的生物。

三、翻译家论翻译

许渊冲的优化论：许渊冲把自己的翻译经验总结为"美化之艺术、创优似竞赛"。"美"是指"意美、音美、形美"，提出译诗要尽量传达原诗的"意美、音美、形美"；"化"，即"等化、浅化、深化"，源于钱钟书的"化境说"；"之"即"知之、好之、乐之"，源于孔子"知之者不如好之者，好之者不如乐之者"；"艺术"即"翻译是艺术，不是科学"；"创"即"文学翻译等于创作"；"优"即"翻译要发挥译文语言优势"；"似"，即"意似音似、形似"或"意似、形似、神似"；"竞赛"即"翻译是两种语言的竞赛，文学翻译更是两种文化的竞赛"。优化论的核心是重实践、重创造、重艺术；优化论的灵魂是忠实与创造的统一；优化论的特点是：不是从理论到理论的"空对空"的研究，或照搬外国"全盘西化"，而是要贴上中国标签。许渊冲说："翻译时找不到对等词，译文不是优于原文，就是劣于原文，劣不如优，所以应该发挥译语的优势，也就是用最好的译语表达方式，这可以简称作优化法。"在比较英汉语法结构差异的基础上，许渊冲提出了优化论。其中，英语与汉语在语法结构上的主要差异是形合与意合的差异，它们是语言连词成句的内在依据。

1. 四字结构的使用

汉语语言优美，其最重要的体现是四字结构的使用。除了使用固有成语之外，译者可以主动发挥主观能动性，创造性地创造四字结构。

例1：He is determined to turn over a new leaf.

他决定洗心革面。

上面这个例句中用"洗心革面"这个四字结构，不仅形式上美观，而且读起来也朗朗上口。原文与译文对比，可见四字成语恰当运用，没有滥用，可以说是尽情发挥了汉语语言的优势。

2. 叠字的运用

叠字（音）是"把两个相同的字重叠在一块，中间无间隔，以加重语言的感情色彩或性状程度，增强表达效果"。叠字的运用，可以更好地传达感情，恰到好处的使用叠字，可以为译文加分，有在精彩的原文上锦上添花的效果。

例2：Wish all of you safe and sound.

祝大家平平安安。

上面的例句用"平平安安"这一叠词，使句子自然且没有故意叠字的嫌疑。通过叠字的使用不仅恰到好处，而且还给原文增添了色彩。

3. 对偶的使用

对偶是一种富有浓郁民族文化色彩的修辞方法。对偶是"用字数相等，句法相似的两句，成双成对排列成功的辞格"。构成对偶的前后两个语言单位可以表达相反或相关的意思。

例3：A blare of sound, a roar of life, a vast array of human hives, appeal to the astonished senses in equivocal terms.

嘈杂的声音、熙攘的生活、蜂窝的人群，足以让感官震惊，为之吸引，感觉难以名状。

以上对偶修辞手法的使用，使得句子读起来铿锵有力、抑扬顿挫、口齿留香。在英译汉的过程中，如果能够恰当地使用汉语的对偶修辞手法，会给译文添彩，发挥译文语言的优势。而且对偶的使用，使译文产生了结构上的视觉美和声律上的听觉美。

4. 重复的使用

重复即字词反复。字词的反复，与叠字不同。叠字叠词，指字词的连续重复使用；字词的反复，则主要指字词的间隔重复使用。使用字词反复的手法可以有效地表达作者的感情，这一修辞格的直接效果就是说服、激励、感染读者听众，引起共鸣。在英译汉时，可以根据翻译的需要，适当使用这一修辞格，从而更好地表达原文的深层含义。

例4：Studies serve for delight, for ornament, and for ability.

读书足以怡情，足以博彩，足以长才。

（摘自：陈宏薇.高级汉英翻译[M].北京：外语教学与研究出版社，2020：56.）

Text B

Oceania's Indigenous Peoples
By Joshua Cooper

Ideological and Political Education：推动海洋发展 依海富国

21世纪以来,世界各国逐渐将开发和利用海洋的着力点放在提高海洋资源的开发效能上。习近平总书记强调,要提高海洋资源开发能力,着力推动海洋经济向质量效益型转变。多年来,我国从提高海洋资源开发能力、加快优化海洋产业结构步伐、促进海洋新兴产业发展、强化陆海联动的沿海港口建设五个方面,将推动海洋经济发展作为海洋强国战略的一条主线,立足海洋发展视野全面提升海洋战略价值认识,加强对海洋领域的开发建设,努力实现海洋经济成为中国经济新的增长点,有效落实"依海富国"战略。《中国海洋经济发展报告2020》显示,2019年,我国海洋生产总值超过8.9万亿元,同比增长6.2%。海洋经济对国民经济增长的贡献率达到9.1%,拉动国民经济增长0.6个百分点。

未来,海洋的产业发展仍然潜力巨大、竞争新优势明显,市场需求广阔。在重视海洋资源开发基础上要进一步优化海洋产业布局,不断壮大海洋新兴产业。对此,要加快沿海港口建设。一个经济大国必须是一个海洋和航运大国,要依托港口的建设、管理和发展,进一步提升我国对外开放的水平。海洋港口作为连接陆海的关键位置,是海洋经济发展的重要引擎,要实现陆海交通运输连接畅通,加快发展大物流和临港工业,不断提高临港经济对国民经济和社会发展的贡献率。要提升海洋科技创新能力,为海洋经济发展方式的转变提供前进动力。发挥海洋科技的先导作用和重要地位,一是要加强海洋科技成果转化,把海洋科技的创新发展与海洋产业转型升级紧密结合起来,增加海洋科技研发投入,大力提升海洋科技的研发水平,促进科技成果直接转化为生产力,实现海洋科技与实体产业的协同发展;二是增强海洋科技创新能力,以完善海洋科技创新体制机制为保障,实现海洋科技创新进步与发展,打破传统海洋科技创新的无序局面,构建适合我国国情的海洋科技创新体制机制,推动海洋科技创新进一步向前发展。

1. As the climate change crisis continues to crash on the shores of the atoll nations, every day Pacific Islanders are challenged to exercise their rights of self-determination. While the lack of defined determinism by developed states could **doom** the small island developing states to disappear, **indigenous** communities, their countries, and global civil society continue to create innovative initiatives illustrating indigenous island traditional knowledge. They are answering the moral call to action by sharing these sustainable models with one another.

2. The communities of Oceania face severe, **imminent** consequences connected with climate change. Because of rising sea levels, low-lying atoll areas are already being flooded and coastal shores **eroded**, along with salt water **intrusion**. The results are violations of the residents' fundamental human rights to water, food, housing, and health. Yet there are many more consequences impacting the cultures and livelihoods of Pacific Islanders. Beyond the sea water seeping into the soil, there is also **acidification** of the ocean, which is leading to coral **bleaching**. The death of the coral **equates** to the loss of an important natural barrier to king tides and rising sea levels, as well as loss of a food source. Shell fish cannot survive and fish no longer inhabit the dying reefs.

3. Indigenous Peoples of the Pacific aren't waiting to be saved, however. Instead, they are showing significant strength in sharing knowledge and **strategic** initiatives of the islands. Kiribati is an example of education, empowerment, and engagement at all levels of society. At the community level, the Otin Taai Declaration of 2004 outlines basic economic, social, and cultural rights. Beyond mere words, the work is impressive with over 37,000 **mangroves** planted under the Kiribati Environment and Conservation Division Kiribati Adaptation Program Phase II. What is vital is the planting with youth groups and school students to start and manage the mangroves. Helene Jacot Des Combes of the University of the South Pacific's Pacific Centre for Environment and Sustainable Development explains the importance of the project: "To **replant** native plants contributes to the renewal of the ecosystem. When planting is successful it has impacts both on the protection of the coast and also for food security by providing more fish and crabs for the community."

4. The world is aware of the actions taken in Kiribati due to President Anote Tong's leadership in international negotiations such as the UN Framework Convention on Climate Change. Tong argues and advocates from a human rights perspective, demanding dignity and equality for the citizens of Kiribati. He also coordinated a visit of UN Secretary-General Ban Ki-moon to his country. Beyond bearing witness as the first

Secretary-General ever to step foot in Kiribati, Ki-moon literally stood in solidarity with the youth, planting his feet in the sinking sands of Stewart Causeway while planting mangroves to protect the beachfront area from rising sea levels.

5. Another climate change adaptation action combines education and cultural exchange to combat the climate crisis. The Climate Challenger Voyage was launched with 10 crew in one canoe for a voyage of 4,000 kilometers, with the purpose of connecting 25 Pacific communities to share climate change adaptation practices. Manuai Matawai, **initiator** of the Climate Challenger Voyage, summarizes his vision: "Five years ago I dreamt of building an ocean voyaging canoe to sail the Pacific and unite with other Pacific Island communities on what we can do about climate change".

6. The Titan tribe of Manus Province, Papua New Guinea, coordinated the traditional voyage to connect culture, conservation, and climate change adaptation in the Pacific. On its initial voyage, it visited various communities to share educational material about climate change and to serve the community with an exchange of initiatives by fellow Pacific Islanders to realize their human rights. The Climate Challenger Voyage shared innovative initiatives that are already proven in the neighbor nations. The different models of locally managed marine areas continue to increase and prove successful; there is **seaweed** farming to assist with **reefs**, along with mangrove planting and the addition of crabs to increase life along the mangroves and provide food. With the community, the crew actually built a 7-meter long dry stone wall on Buala for coastal protection, leaving behind not just a living model, but training material and the traditional knowledge for construction.

7. While the Climate Challenger Voyage is important in the Pacific region, another traditional voyaging community is aiming to teach about climate change on a worldwide voyage from 2014—2017. The Polynesian Voyaging Society will sail the entirely traditional Ho-ku-le'a and solar powered escort safety canoe, Hikianalia, around the globe visiting groups with a message of sustainability and peace. As Jenna Ishii, an education specialist, said, "We know there are a lot of issues we are facing. But as a people we can find solutions." The World Wide Voyage will teach a curriculum of climate change during its travels and share examples of adaptation accumulated while sailing. There will also be Google **hangout** group that will connect directly from canoe to classroom and communities around the world.

8. These climate change adaptation practices provide a **prolonged** and **principled** pursuit of sustainable development and saving our planet. A voice from the Pacific represents the

ripple that must crash on the conscience of citizens of the world. The poem "Tell Them," by Kathy Jetnil-Kijiner from the Republic of the Marshall Islands, makes sure that the existence of Indigenous Peoples is never forgotten, and indeed remains at the forefront of policymakers' minds.

9. Beyond writing poems to share with humanity's soul, Jetnil-Kijiner was also **pivotal** in the creation of Jo-Jikum, an environmentalist group empowering youth to maintain positive changes for people and land in the Marshall Islands. "I want the world to know of our resilience and our strength that we are in no way going to just give up, pack up, and leave our islands. Our culture is rooted in our land, and our land is our life," she says. Climate change is an **insidious** form of colonialism denying the fundamental freedoms of Indigenous Peoples. It is no longer an **ominous** threat of the future but a dawning, deadly reality. Indigenous Peoples of the Pacific are responding to this new reality with **fortitude** for the continued existence of the rich cultures of Oceania. Their resistance is rooted in human rights **advocacy** and is resulting in a renewed sense of self and commitment to sustainable development not only to save one's homeland, but all of humanity. Their actions and adaptations are the difference between life and death.

(1,085 words)

https://www.culturalsurvival.org/publications/cultural-survival-quarterly/oceanias-indigenous-peoples-rising

New Words

1. doom [duːm]

    v. decree or designate beforehand 注定

    n. an unpleasant or disastrous destiny 厄运，劫数；悲观，沮丧

Policymakers who refocus efforts on improving well-being rather than simply worrying about GDP figures could avoid the forecasted doom and may even see progress.

如果政策制定者将注意力放在改善福祉上，而不是单纯担心国内生产总值的数据，就能避免预测中的厄运，甚至可能看到进展。

2. indigenous [ɪnˈdɪdʒənəs]

    a. originating where it is found 本土的，固有的

Over 75% of Mexico's forests, which range from temperate spruce and fir to tropical rainforest, are controlled by ejidos or indigenous groups.

墨西哥的森林既有温带的云杉和冷杉，也有热带雨林，其中75%以上由合作农场或原住民团体控制。

3. initiative [ɪˈnɪʃətɪv]

　　n. readiness to embark on bold new ventures;the first of a series of actions 措施,倡议;主动性,积极性;主动权;(美国某些州的)公民立法提案程序

　　This universal consistency among education experts indisputably demonstrates an immutable principle of learning: initiative and correct methods are fundamental to academic success.

　　教育专家的这种普遍共识无可争辩地证明了一个不可改变的学习原则:主动性和正确的方法是学术上成功的基础。

4. imminent [ˈɪmɪnənt]

　　a. close in time;about to occur 即将发生的,临近的

　　The mere mention of the words "heart failure" can conjure up, to the layman, the prospect of imminent death.

　　一提到"心力衰竭"几个字,外行人就会想到立即死亡。

5. erode [ɪˈrəʊd]

　　v. become ground down or deteriorate;remove soil or rock 侵蚀,腐蚀;削弱,降低

　　Usually, price wars wreak havoc because they erode the pricing power of an entire business.

　　价格大战通常损失严重,因为它们破坏了整个行业的定价机制。

6. intrusion [ɪnˈtruːʒ(ə)n]

　　n. entrance by force or without permission or welcome; any entry into an area not previously occupied 扰乱,侵犯;闯入,侵入;(对周围岩层的)火成岩侵入

　　They claim the noise from the new airport is an intrusion on their lives.

　　他们声称新机场的噪声侵扰了他们的生活。

7. acidification [əˌsɪdɪfɪˈkeɪʃən]

　　n. the process of becoming acid or being converted into an acid 酸化;成酸性;使……发酸

　　However, their effects are probably accentuated by climatic factors, such as drought and hard winters, or soil imbalances such as soil acidification, which damages the roots.

　　然而,它们的影响可能因气候因素而加剧,如干旱和严冬,或因土壤酸化等土壤不平衡而损害到根系。

8. bleach [bliːtʃ]

　　v. emove color from;make whiter or lighter 漂白,使褪色;晒白;使失去活力,使失去实质内容;用漂白剂给……清洗,消毒

　　The tree's roots are stripped and hung to season and bleach.

　　这些树根被剥去皮并挂起来风干晒白。

9. equate [ɪˈkweɪt]

　　v. consider or describe as similar, equal, or analogous; be equivalent or parallel, in mathematics;make equal, uniform, corresponding, or matching (使)等同;使(两个或两个以

上事物)相等,使平衡

Simplicity in website design doesn't necessarily equate with a minimalist design aesthetic.

在网页设计中,简洁并不一定等同于极简主义者的设计美学。

10. strategic [strə'ti:dʒɪk]

a. relating to or concerned with strategy; highly important to or an integral part of a strategy or plan of action especially in war 战略(性)的,策略(上)的;(计划或战略上)有用的,重要的;(武器、战争或地方)战略性的,有战略优势的

According to the Centre for Strategic and International Studies, about three quarters of energy we use to move things, including ourselves, accomplish no useful work.

根据战略与国际研究中心的数据,我们用于移动包括我们自己的物体的能量中,约有四分之三没有完成任何有用的工作。

11. mangrove ['mæŋgrəʊv]

n. a tropical tree or shrub bearing fruit that germinates while still on the tree and having numerous prop roots that eventually form an impenetrable mass and are important in land building 红树林

Mangrove root systems have the ability to absorb and well trap sediments and pollutants in water that flows through them before they enter the ocean.

红树林根系有能力吸收并很好地截留水中的沉积物和污染物,这些沉积物和污染物在进入海洋之前流经这些根系。

12. replant [ri:'plɑ:nt]

v. plant again or anew 改种;移植;移居

If you are using a potted pine or spruce tree, you can replant it after using it which will preserve the tree's life and beautify where ever you decide to replant it.

如果你正在使用一棵盆栽的松树或云杉,你可以在使用后再种它,这样可以保存它的寿命并且美化你所决定的再植它的地方。

13. initiator [ɪ'nɪʃieɪtə(r)]

n. a person who initiates a course of action 发起人,创始人

The initiator of this program is Hou Haiyang, a 23-year-old graduate from Dalian Ocean University.

这个计划的发起人叫侯海洋,23岁,是大连海洋大学的毕业生。

14. seaweed ['si:wi:d]

n. plant growing in the sea, especially marine algae 海藻,海草

Like seaweed, sea grasses are very susceptible to human pollution.

和海藻一样,海草很容易受到人类污染的影响。

15. reef [ri:f]

n. a submerged ridge of rock or coral near the surface of the water; a rocky region in the southern Transvaal in northeastern South Africa; contains rich gold deposits and coal and manganese 住,礁,暗礁,礁脉;(尤指含金的)矿脉;缩帆部,帆的可收缩部

A quarter of all sea species spend at least a part of their life in a reef and many reefs are in cold or temperate waters.

四分之一的海洋物种在珊瑚礁中度过了它们一生中至少一半的时间,而且很多珊瑚礁处于寒冷或温和的水中。

16. hangout ['hæŋaʊt]

n. a frequently visited place(非正式)常去的地方(社交或娱乐场所)

It was a cloudy Wednesday night in Chicago and their regular hangout had turned into an institution of stools.

那是个多云的星期三的晚上,在芝加哥,他们常去的消遣之地俨然已成为一个被工具所统治的机构。

17. prolong [prə'lɒŋ]

v. lengthen in time;cause to be or last longer 拉长,拖长(感觉或活动)

In government laboratories and elsewhere, scientists are seeking a drug able to prolong life and youthful vigor.

在政府实验室和其他地方,科学家们正在寻找一种能够延长寿命、让人保持年轻活力的药物。

18. principled ['prɪnsəpld]

a. based on or manifesting objectively defined standards of rightness or morality 有原则的;有操守的

Lovelock's principled nonconformity can be traced to his childhood.

洛夫洛克自成一套的非主流表现可以追溯到他的童年时期。

19. ripple ['rɪp(ə)l]

n. a small wave on the surface of a liquid;(electronics) an oscillation of small amplitude imposed on top of a steady value 涟漪;(外观或运动)如波纹的东西;(声音的)传播起伏,(感情的)逐渐扩散;连锁反应;彩条冰激凌;表面张力波;脉动

The air was so still that there was hardly a ripple on the pond's surface.

没有风,池塘的水面上几乎看不到波纹。

20. pivotal ['pɪvət(ə)l]

a. being of crucial importance 中枢的,关键的;有枢轴的,像枢轴的

The elections may prove to be pivotal in Colombia's political history.

这些选举也许会证明其在哥伦比亚政治史上是至关重要的。

21. insidious [ɪn'sɪdɪəs]

a. beguiling but harmful;intended to entrap;working or spreading in a hidden and usually injurious way 潜伏的,隐袭的;阴险的,狡猾的

The very reason that clichés so easily seep into our speech and writing, their insidious memorability, is exactly why they played such an important role in oral storytelling.

陈词滥调之所以如此容易渗透到我们的演讲和写作中，是由于它们潜在的可记忆性，这正是它们在口述故事中扮演如此重要角色的原因。

22. colonialism [kəˈləunɪəlɪzəm]

n. exploitation by a stronger country of weaker one; the use of the weaker country's resources to strengthen and enrich the stronger country 殖民主义，殖民政策

In a 1984 book, Claire Robertson argued that, before colonialism, age was a more important indicator of status and authority than gender in Ghana and in Africa generally.

在1984年的一本书中，克莱尔·罗伯逊认为，在殖民主义出现之前，在加纳和整个非洲，年龄是地位和权威的一个更重要的指标，而不是性别。

23. ominous [ˈɒmɪnəs]

a. threatening or foreshadowing evil or tragic developments; presaging ill fortune 预兆的，不吉利的

I can't get enough of them, especially around Halloween when the shadows from the leafless trees take on ominous shapes.

对于它们我百看不厌，尤其是在万圣节期间，那些光秃秃的树的阴影呈现出不详的形状。

24. fortitude [ˈfɔːtɪtjuːd]

n. strength of mind that enables one to endure adversity with courage 刚毅；不屈不挠；勇气

Humor, albeit of a particularly hard-won variety, suggests Kennedy (a sometime stand-up comic), is necessary for fortitude.

他的幽默，即使带着来之不易的独特的捉摸不定感，暗示了肯尼迪（他曾经是一位单人喜剧表演者）必定有着坚韧不拔的毅力。

25. advocacy [ˈædvəkəsi]

n. active support of an idea or cause etc.; especially the act of pleading or arguing for something 拥护，提倡；辩护，辩护术；游说（团体或组织）；律师职业（或工作）

Borlaug's advocacy of intensive high-yield agriculture came under severe criticism from environmentalists in recent years.

博洛格提倡集约化的高产农业，近年来遭到环保人士的严厉批评。

Phrases and Expressions

the lack of 缺乏；没有

share with one another 彼此分享

along with 沿（顺）着；连同……一起；与……一道；除了……

dreamt of 梦想；梦见

assist with 帮助（照料，做）；在……给予帮助

leave behind 留下；遗留；超过

aim to 目的在于
know of 知道,了解……;听说过……
pack up 整理;把……打包

Terminology
indigenous peoples 土著民族
sea level 海平面
human rights 人权

Proper Names
the Otin Taai Declaration 奥廷塔伊宣言
the UN Framework Convention on Climate Change 联合国气候变化框架公约
The Climate Challenger Voyage 气候挑战者号航行
Papua New Guinea 巴布亚新几内亚

一、翻译简析

1. While the lack of defined determinism by developed states could doom the small island developing states to disappear, Indigenous communities, their countries, and global civil society continue to create innovative initiatives illustrating Indigenous island traditional knowledge. (Para.1)

虽然发达国家缺乏明确的决定论,可能会导致发展中国家小岛屿的消失,但土著群体、土著地域和全球公民社会继续创造创新举措,展示土著岛屿传统知识。

【这是个较长的让步状语从句,句式结构不是很复杂。在汉译时,While 译为"虽然……但……",然后更换原文的词语的前后词序,将 developed states 放在 the lack of 的前面,译为"发达国家缺乏……",根据汉语的语言习惯,使译文在最大程度上通顺,这是换序译法的翻译方法。their countries 的 countries 有"国,国家;全国人民,全体国民(the country);乡下,乡村;地区,区域;乡村音乐;故乡"的词意,前面有"群体"、后面有"公民",这里就翻译为"地域"。在翻译过程中,用具体的词进行翻译,从而降低语言差别,使译文产生与原文同样的效果。】

2. Indigenous Peoples of the Pacific aren't waiting to be saved, however. Instead, they are showing significant strength in sharing knowledge and strategic initiatives of the islands. Kiribati is an example of education, empowerment, and engagement at all levels of society. At the community level, the Otin Taai Declaration of 2004 outlines basic economic, social, and cultural rights. (Para.3)

然而,太平洋的土著民并没有等着被拯救。相反,他们在分享这些岛屿知识和战略倡议方面表现出巨大的力量。基里巴斯是在社会各阶层开展教育、赋权和参与的典范。在群体层面,2004 年的《奥廷塔伊宣言》概述了基本的经济、社会和文化权利。

【这个句子中的 Indigenous Peoples 译为"土著民或土著居民",这是翻译中的省词译法,将民族译为"(土著)民"。At the community level 中的 level 译为"在群体层面",翻译时需要用具体的词汇进行翻译的具体译法。最后的小句谓语动词 outlines 有"概述,略述;勾勒,描画……的轮廓"词意,翻译时译为"概述了"的增词译法多了个"了"时译文符合汉语的表达习惯,使译文被读者接受。】

3. Manuai Matawai, initiator of the Climate Challenger Voyage, summarizes his vision: "Five years ago I dreamt of building an ocean voyaging canoe to sail the Pacific and unite with other Pacific Island communities on what we can do about climate change". (Para.5)

气候挑战者航行的发起人马努埃·马塔瓦伊总结了他的愿景:"五年前,我梦想建造一艘远洋独木舟,航行太平洋,并与其他太平洋岛屿群体联合起来,共同为气候变化做些什么"。

【这个句子结构不复杂。在汉译时,按照汉语的语序,把句子断开来翻译,先将翻译主语部分的"Manuai Matawai"放在"……发起人"的后面。然后用断句译法翻译引号内的原文,使译文简洁自然。】

4. The Titan tribe of Manus Province, Papua New Guinea, coordinated the traditional voyage to connect culture, conservation, and climate change adaptation in the Pacific. On its initial voyage, it visited various communities to share educational material about climate change and to serve the community with an exchange of initiatives by fellow Pacific Islanders to realize their human rights. (Para.6)

巴布亚新几内亚马努斯省的泰坦部落协调了传统的航行,将太平洋的文化、保护和气候变化适应联系起来。在第一次航行中,它访问了各个群体,分享有关气候变化的教育材料,并与太平洋岛民交流实现其人权的倡议,为群体服务。

【这个句子在翻译时,按照汉语的语序习惯,将状语部分的国家放在译文的句首。整个句子进行可以直译,后面的目的状语 to serve……human rights 运用断句译法,放在句后译为"为群体服务"。】

5. The different models of locally managed marine areas continue to increase and prove successful; there is seaweed farming to assist with reefs, along with mangrove planting and the addition of crabs to increase life along the mangroves and provide food. (Para.6)

当地不同的管理海洋区域模式继续增加,并证明是成功的;这里有海藻养殖来帮助珊瑚礁,还有红树林的种植,以及螃蟹的加入增加了红树林沿线的生命并提供了食物来源。

【译文按照汉语语序与习惯,采用断句译法,前半句的 and 并列结构翻译时进行断句,译为"并证明……"。将"along with mangrove planting"分开原文的结构进行断句,并把动词"plant"译为名词"的种植"的转性译法,断句为"还有红树林的种植"。英译汉时,恰当使用翻译策略,使译文为读者接受。】

二、参考译文

大洋洲的土著民

1. 随着气候变化危机继续冲击环礁国家的海岸,太平洋岛民每天都面临着行使自决权的挑战。虽然发达国家缺乏明确的决定论,可能会导致发展中国家小岛屿的消失,但土著群体、土著地域和全球公民社会继续创造创新举措,展示土著岛屿传统知识。他们通过彼此分享这些可持续模式,响应了行动的道义号召。

2. 大洋洲群体面临着严重的与气候变化相关的迫在眉睫的后果。由于海平面上升,低洼的环礁地区已经被淹没,海岸被侵蚀,同时还有盐水入侵。其结果是侵犯了居民享有水、食物、住房和健康的基本人权。然而,影响太平洋岛民文化和生计的后果还有很多。除了海水渗入土壤之外,海洋也在酸化,导致珊瑚白化。珊瑚的死亡等于失去了抵御潮汐和海平面上升的重要天然屏障,同时也失去了一种食物来源。贝壳类、鱼类无法生存,鱼类也不再栖息于濒临死亡的珊瑚礁上。

3. 然而,太平洋的土著民并没有等着被拯救。相反,他们在分享这些岛屿知识和战略倡议方面表现出巨大的力量。基里巴斯是在社会各阶层开展教育、赋权和参与的典范。在群体层面,2004年的《奥廷塔伊宣言》概述了基本的经济、社会和文化权利。除了文字之外,这项工作令人印象深刻,在基里巴斯环境和保护司,基里巴斯在适应计划的第二阶段下种植了3.7万多棵红树林。至关重要的是与青年团体和学校学生一起开始种植和管理红树林。南太平洋大学太平洋环境与可持续发展中心的海伦·杰科特·德·库姆斯解释了该项目的重要性:"重新种植本地植物有助于生态系统的更新。当种植成功时,它既可以保护海岸,也可以为群体提供更多的鱼和螃蟹,从而对粮食安全产生影响"。

4. 全世界都知道基里巴斯总统唐安诺在《联合国气候变化框架公约》等国际谈判中发挥的领导作用。董(人名)从人权的角度进行论证和倡导,要求基里巴斯土著民享有尊严和平等。他还协调了联合国秘书长潘基文对该国的访问。作为历史上首位踏足基里巴斯共和国的秘书长,潘基文不仅见证了这一事件,还与年轻人站在一起,在斯图尔特堤道下沉的沙滩上踏足,同时种植红树林,保护海滨地区不受海平面上升的影响。

5. 另一项气候变化适应行动将教育和文化交流结合起来,以应对气候危机。"气候挑战者"号由10名船员乘坐一艘独木舟起航,航行4 000千米,目的是连接25个太平洋群体,分享适应气候变化的实践行动。气候挑战者航行的发起人马努埃·马塔瓦伊总结了他的愿景:"五年前,我梦想建造一艘远洋独木舟,航行太平洋,并与其他太平洋岛屿群体联合起来,共同为气候变化做些什么"。

6. 巴布亚新几内亚马努斯省的泰坦部落协调了传统的航行,将太平洋的文化、保护和气候变化适应联系起来。在第一次航行中,它访问了各个群体,分享有关气候变化的教育材料,并与太平洋岛民交流实现其人权的倡议,为群体服务。"气候挑战者"号航行分享了已经在邻国得到验证的创新举措。当地不同的管理海洋区域模式继续增加,并证明是成功的;这里有海藻养殖来帮助珊瑚礁,还有红树林的种植,以及螃蟹的加入增加了红树林沿线的生命并提供了食物来源。在群体的帮助下,船员们在布阿拉(所罗门群岛地名)上建造了一堵7

米长的干石墙,用于保护海岸,留下的不仅是一个活的模型,还有培训材料和传统的建筑知识。

7. 虽然"气候挑战者"号航行在太平洋地区很重要,但另一个传统航海者正致力于在2014—2017年的全球航行中教授气候变化的知识。波利尼西亚航海协会将驾驶传统的呼库勒船(Ho-ku-le'a的音译)和太阳能护航安全独木舟希基亚拉尼(夏威夷语)环游世界,为可持续发展与和平传递信息。正如教育专家珍娜·石井所说:"我们知道面临着很多问题。但作为一个民族,我们可以找到解决方案"。"环球航行"将在航行过程中教授有关气候变化的课程,并分享航行过程中积累的适应气候变化的例子。还有谷歌聚会小组,将直接从独木舟连接到世界各地的教室与社区。

8. 这些适应气候变化的做法为可持续发展与拯救我们的地球提供了长期和原则的追求。来自太平洋的声音代表着一定会冲击世界公民良知的涟漪。马绍尔群岛共和国的凯西·杰特尼尔-柯基妮(年轻女诗人)的诗歌《告诉他们》确保土著民的存在永远不会被遗忘,而且确实仍然是政策制定者需考虑的首要问题。

9. 除了写诗分享人类的灵魂,杰特尼尔-柯基妮创建的乔基姆发挥了关键作用。乔基姆是一个环保组织,赋予年轻人力量,为马绍尔群岛的人民和土地赋予了积极的变化。"我想让世界知道我们的韧性和力量,我们绝不会就此放弃,收拾行李,离开我们的岛屿。我们的文化根植于我们的土地,我们的土地就是我们的生命"。气候变化是一种阴险的殖民主义形式,剥夺了土著民的基本自由。它不再是不祥的未来威胁,而是一个黎明与致命的现实。太平洋土著民正以坚定的态度对这一新的现实做出反应,以使大洋洲丰富的文化继续存在下去。他们的反抗植根于倡导的人权,并导致了一种新的自我意识和对可持续发展的承诺,不仅是为了拯救自己的家园,而且是为了拯救全人类。他们的行动和适应是生与死的区别。

Cultural Background Knowledge: Ocean Ambience
海洋风情文化背景知识

中国的海岛,星罗棋布,形态各异,如同一颗颗璀璨的明珠,镶嵌在波光粼粼的蓝色海洋上。一些适合开发旅游的岛屿已成为热门的旅游目的地。悠闲舒适的海岛生活令人向往。

迷人的热带海洋风情让海南独具魅力。海南的一些小岛逐渐进入游客的视野。分界洲岛是中国首家海岛型国家5A级景区,处在海南热带和亚热带的分界线上。蜈支洲岛海水清澈透明,被誉为"中国第一潜水基地"。南湾猴岛森林覆盖率达95%,生活着近1 500只猕猴,是中国也是世界唯一的热带岛屿型猕猴自然保护区。

除了海南,沿海岸线自南向北,分布着特色各异的海岛。位于广西北海南面的涠洲岛是中国最年轻的火山岛。炽热的熔岩与冰冷的海水孕育奇特的火山遗迹,塑造出海蚀崖、海蚀洞、海蚀平台等地质奇观。福建的崳山岛集大海、沙滩、草场、湖泊于一身,两个水清如镜的湖泊周边是被誉为"南国天山"的万亩草场。浙江舟山群岛以海、渔、城、岛、港、航、商为特色,集海岛风光、海洋文化和佛教文化于一体。山东长岛由32个岛屿组成,山水相依,如诗如画,是"京津之门户""渤海之锁钥"。

  中国海岛资源优势明显,但旅游开发仍处在初级阶段。不同的海岛旅游应该有不同的创意,因地制宜,构建差异化的目的地品牌。同时,在自然和历史文化资源既定的情况下,商业环境和生活品质也是提升海岛游品质的重要因素。海岛游将成为高品质国内游的重要供给。海南、平潭和横琴相继建设国际旅游岛,引领着中国海岛游的发展。

Unit Six
Marine Tourism

海洋旅游

Text A

The Challenges to Sustainability in Island Tourism

By Pauline J. Sheldon

Ideological and Political Education：探索海洋蓝色治理 拓展海洋旅游新业态

2021年,我国立足新发展阶段,贯彻新发展理念,加快构建新发展格局,持续推进海洋经济高质量发展,海洋经济强劲恢复,产业结构调整步伐加快,自主创新能力不断提升,供给保障能力持续增强,国际竞争优势进一步巩固,实现"十四五"良好开局。《2021年中国海洋经济统计公报》数据显示：全国海洋生产总值2021年首破9万亿,国民经济增长贡献率为8%。海洋经济结构得到不断优化,海洋第一、二、三产业占比为5.0：33.4：61.6。作为全球97%的水和约80%动植物的家园,海洋是地球"最后的边疆"。如何进一步挖掘海洋的经济潜力,设法在"蓝色增长"和"蓝色治理"方面处于领先地位,使"蓝色增长"真正成为社会经济发展引擎,实现中国建设海洋强国目标,是新时期发展的重中之重。2021年,旅游产业已在海洋经济发展中占据相当大的比重,但到目前为止,海洋旅游经济产能还没得到充分释放,还可以通过进一步将海洋旅游作为抓手,通过拓展海洋旅游新业态,推进"旅游+消费""旅游+工业""旅游+农业""旅游+会展""旅游+体育""旅游+医疗""旅游+康养"等新业态,进一步拓展海洋旅游消费空间,促进海洋旅游"蓝色增长"。

同时,通过探索海洋旅游市场"云监管"方式,加强海洋旅游"蓝色治理",在质量标准化建设的同时,加强海洋旅游产品特色化建设,通过大力弘扬发展当地文化、民族文化以及历史文化等,推动其与旅游深度融合发展,从而进一步有效整合资源,优化配套体系,提高旅游资源要素配置效率,推动从数量扩张到质量提升转变,从而进一步促进海洋旅游可持续发展。

1. Islands are special places with a natural attraction for tourists and a special challenge to **sustainability**. The thousands of islands on the face of the earth include some of the finest and most sought after destinations, such as the Balearic Islands, the Hawaiian Islands, the Galapagos Islands, the Canary Islands, the French Polynesian Islands, and the Caribbean islands. The mystique associated with islands is dependent on a **blend** of different lifestyles, **indigenous** cultures, unique land formations, flora and fauna, and ocean and coastal resources. To keep that mystique alive and **thriving**, islands must **implement** sustainable tourism policies in all areas including environmental, economic and socio-cultural.

2. Types of Island Tourism Destinations. Islands vary in many ways, and understanding the various types clarifies for the decision-makers the policies that need to be used. One classification is islands' climate which can be cold, temperate or tropical. Even though tropical islands (Caribbean, Hawaii, French Polynesia) tend to have most **allure** for tourists, cold and temperate islands also have environmental or cultural features and lifestyles that attract tourism—for example the Shetland islands off the coast of Scotland. Baum, in his book entitled *"The Fascination of Islands: A Tourist Perspective in Island Tourism: Trends and Prospects"* describes the general attractivity of North Atlantic islands, including their remoteness, their small size, the slower pace of life, the chance to go back-in-time, the wilderness environment, the water-focused society and the sense of difference yet familiarity. Very cold islands such as Iceland and Greenland offer unique landscapes and flora and fauna and are **alternate** destinations often attracting scientists, photographers and other specialized travelers. Another island classification is the **proximity** to the related mainland and also its size. Islands that are more remote and distant face more challenging **accessibility** and transportation issues due to their isolation. Visitors will tend to stay longer in islands that are remote and larger, whereas those close to the mainland and smaller may experience more excursionist tourism. For example, Cousin Island in the Seychelles, hosts only day visitors that leave the island at the end of each day. The island's choice to host excursionists versus stay-over visitors requires a careful evaluation of the strengths and weaknesses of each type of tourism. A third classification is whether an island is a single island or part of an archipelago. Multi-destination travel in island chains may be an added attraction for tourists, whereas the peace, or 'sun, sand, sea' experience of a single island vacation may be the choice for others. Cooperative marketing and **complementary** product development is important for archipelago islands. This will create a diverse touristic experience giving archipelagos an advantage over single islands, particularly if they are small. A fourth classification is the governance of the island destination. Some islands have **autonomous** governments and others are part of the mainland government system. Those with autonomy have more control over the direction of sustainable development of the island. They are also more likely to reap the maximum economic benefits from tourism, without any revenue being leaked to the mainland through taxes and other means. Islands under the **jurisdiction** of the mainland need to ensure adequate

representation in the government decision-making. Fifth, some island destinations have growing resident populations (which may be due in part to tourism) and others with weak economies are experiencing declining populations. In the latter case, there is a special need to ensure economic **viability** to prevent the out-migration of residents, especially young ones who look for opportunities elsewhere. Some islands, of course, have no human population and are simply nature-reserves, and others are privately owned with their own policies. The last classification relates to the **homogeneity** of the population and the socio-cultural sustainability of island destinations. Islands with homogeneous, indigenous populations are particularly vulnerable to tourism development since they have different cultures with different values than the source markets. The close interaction that islands create between hosts and guests must be managed. Islands with more **heterogeneous** populations may be more **resilient** to socio-cultural impacts. It is clear from the categories above that islands differ in many ways from the mainland and from each other. Each island has its uniqueness and that uniqueness needs to be **nurtured** and strengthened through sustainable tourism policies.

3. All islands must address issues of economic impact, environmental consequences and those relating to the social, cultural and political fabric of the island, all of which are affected by the density of tourism on the island. High tourist and resident densities in islands such as Malta are the source of many sustainability problems and carrying capacity needs to be considered.

4. Measures of tourism density are important for policy makers to assess possible growth scenarios. One measure of tourism density or saturation, which considers all three areas of impact, is the Tourism Penetration Index (TPI). TPI includes three variables: ① visitor spending per capita of population (economic measure), ② average daily visitors per 1,000 population (social measure), and ③ hotel rooms per square kilometer of land (environmental measure). McElroy and Albuquerque in their research used this to cluster Caribbean islands into different clusters depending on whether their TPI is low, intermediate or high value. McElroy notes that for islands with low TPI, the most important challenges are establishing profitability and international recognition, for those with intermediate TPI, controlling growth is the most important, and for those islands with the highest TPI, the greatest challenge is to sustain vacation quality.

5. Economic Issues. A challenge to the economic health of an island is the often limited economic resource base. Islands may have few resources or viable industries other than tourism to provide revenue and employment for the local population. The value of agricultural and mining commodities on the international markets is declining and fishing is less reliable as fish populations are being **depleted**, and global warming is changing the nature of coastlines and fish movements. Tourism can be an economic catalyst for small island development. In fact, Croes, in as early as 2005, suggests tourism as a tool for small islands to enlarge their economies and

overcome the disadvantages of smallness. The extra market demand produces economies of scale and increases efficiency and decreases costs of production. Tourism also increases competition, encourages new start-up businesses, democratizes market structure, and **deters** rent-seeking behaviors and corruption. He also argues that this competition can provide greater consumer choice, trade openness and increase the quality of life for residents. Despite this, the revenue from tourism must remain in the island economy as much as possible. Policies of import substitution to ensure minimum economic leakages, and 'buy local' policies to maximize linkages are essential. Taxation policies, entrepreneurial **subsidies**, and investment **incentives** are all useful to strengthen the economy. If development strategies are such that the incoming wealth is leaving the island economy, tourism needs to be re-designed. Islands under the governance of the mainland need to ensure a fair share of tax **revenue** due. Islands under their own governance will gain most economically.

6. Seasonality in island tourism is another challenge to the economic sustainability of the island and the well-being of the island people. **Fluctuations** in visitor arrivals must be understood and **mitigated** through product and market diversification so that employment stabilizes and tourism infrastructures and superstructures are well utilized. Escalating land prices represent another economic concern in islands forcing local residents out of the housing market. This promotes out migration, leading to a possible dissolution of the culture, and second home ownership by foreigners. These trends if unmitigated can generate a serious chain of problems for the island economy.

7. Environmental Issues. Environmental issues of sustainability on islands are multi-faceted, since islands have diverse land formations, coastal areas, and wildlife species. Tourism often contributes to the environmental degradation (pollution, erosion, etc.) in small, island states which are host to fragile eco-systems rich in biodiversity. The isolation of the island environment created the biodiversity, and by opening to tourism, some of that sheltered biodiversity is endangered. Islands' prime tourist environmental resource is often the coastal regions (beaches, sand dunes, coral reefs) that are easily damaged, heavily used, and requiring careful visitor management. The large amount of waste (solid and liquid) created by tourism is a problem since space for its disposal on islands is limited.

8. Socio-Cultural Issues. Islands face complex socio-cultural issues, particularly those with indigenous populations. Tourism on islands, particularly small ones, brings hosts and guests into closer contact than on mainland destinations, creating a more vulnerable situation for social disruption. Crime, commoditization of culture, and loss of traditional lifestyles, moral standards and family life impact islands more than mainland destinations. Studies of resident sentiment and response to tourism in the islands of Malta and Hawaii show the importance of this component. Community integration is the key to successful and sustainable tourism

development, meaning that all islanders affected by tourism must be involved in the planning process. Stakeholders on islands are not only in closer proximity but also have long histories of conflict, making it even more important to involve them in the decision-making process.

9. Approaches to Overcome the Challenges to Island Tourism. The experience of many island destinations over the years has provided a rich source of policies that can assist with sustainability on all levels.

10. Long-Term, Stakeholder-Involved Planning. Long term planning, developed with comprehensive community and stakeholder input is becoming an important foundation for tourism on islands. Plans also need to be values-based plans and reflect the indigenous culture and traditions. Long range planning must consider the balance of supply and demand of tourism, both quantitatively and qualitatively. A study of tourism in the Canary Islands showed that when these two growth patterns were out of balance, the industry is not healthy.

11. Once the plans have been put in place, methodologies to measure and monitor impacts of tourism are essential. This requires the assignment of government agencies to the task of ongoing monitoring of effects. The need for stakeholder-driven planning and indicator development is essential. In Hawaii for example, a process that covered almost two years brought together stakeholders to define their vision, goals and indicators for sustainable tourism. Community involvement to guide tourism planning, development, management, research and evaluation of community-based tourism projects has been implemented in Taquile Island in Peru. On this island, decision-making powers, local control and ownership, and type of employment patterns were measure of community involvement in the planning process.

12. Empowerment of the Island Community and Culture. Empowerment of the island community and culture is a necessary part of planning. Frameworks to protect and conserve the social and cultural structure are important also. A building of cultural pride through story-telling and memory of traditions, and a sense of identity are paramount. This may involve the re-enlivenment of festivals, arts, language, folk lore and policies to encouraging local people to engage in entrepreneurial activities. Efforts towards sustainable tourism in French Polynesian islands found the meaningful integration of culture into the tourist experience difficult to accomplish. Achieving a balance of respect for the culture and providing tourists with the opportunity to learn about and appreciate the culture is the core of the challenge.

13. Tourist and resident education are a critical part of island sustainability. To empower the residents, education and training programs are needed for meaningful careers in the

industry. This may involve distance education since islands do not always have comprehensive tertiary education programs in tourism. Residents also need to learn about the impact tourism is having on their community, through the sharing of statistics and facts. The receptivity and openness to change and innovation is also needed. Education for tourists is also important. They need to learn about the unique cultural and environmental features of the island and appropriate behaviors.

14. Environmental Management. Given the challenges to the island's ecosystems, environmental management is crucial for island sustainability. The paucity of land causes land usage issues, and the trade-off of land for tourism versus agriculture and other industries, or preservation and conservation needs to be addressed. The environmental resources are a main visitor attraction and tourist interfaces with those resources need to be planned and cared for. Conlin, in a study of Tasmania tourism, gives nine different types of parks or reserves that can be created (national park, state reserve, nature reserve, game reserve, conservation area, nature recreational area, regional reserve, historic site and private **sanctuary**). This may include designation of zones that are off-limits to tourists, and those that are only visitable with guides and interpreters. Islands with unique wildlife must also take steps to sustain those populations. Tourism can actually assist as an anti-poaching mechanism and an engine for conservation management when residents realize the economic value of the wildlife as in the Seychelles. Policies to keep the land and ocean unpolluted are also necessary. Waste management and recycling programs are essential, particularly on small islands. Also the use, through incentive programs if necessary, for alternative fuel sources (wind, solar, geo-thermal etc.) will make the destination more sustainable. The shortage of land for landfills may need waste to be sent to the mainland for some islands. Recycling programs for all types of waste are essential, and the use of alternative energy sources (wind, solar, geothermal) is an important consideration since tourists use much higher per capita user of energy than locals and rarely can islands survive on fossil fuels. Water shortages also are common on islands, limiting the amount of tourism development.

15. Environmental management includes recovery from natural disasters to which islands are so vulnerable. Disasters such as tidal waves, volcanic eruptions, cyclones, drought and rise in sea level are all natural hazards that islands face. Funds for conservation and disaster management are needed and can be gained through taxation, visitor fees or other mechanisms.

(2,284 words)

https://www.researchgate.net

New Words

1. sustainability [səˌsteɪnə'bɪləti]

    n. the quality of causing little or no damage to the environment and therefore able to continue for a long time 可持续性

    When choosing what products to buy and which brands to buy from, more and more consumers are looking into sustainability.

    在选择购买产品和品牌时，越来越多的消费者开始关注可持续性。

2. blend ['blend]

    n. a mixture of different things or styles 混合物

    Their music is a blend of jazz and African rhythms.

    他们的音乐融合了爵士乐和非洲音乐。

3. indigenous [ɪn'dɪdʒənəs]

    a. naturally existing in a place or country rather than arriving from another place 本土的

    Are there any species of frog indigenous to the area?

    这个地区是否有本地品种的青蛙？

4. thrive [θraɪv]

    v. to grow, develop, or be successful 茁壮成长，兴旺，繁荣

    His business thrived in the years before the war.

    战前那些年，他的生意很兴隆。

5. implement ['ɪmplɪment]

    v. to start using a plan or system 执行，贯彻

    The changes to the national health system will be implemented next year.

    国民医疗保健制度的改革将于明年实行。

6. allure [ə'lʊə(r)]

    n. the quality of being attractive, interesting, or exciting 吸引力，魅力

    This is a sign of the continued allure of creating scientific coalitions across borders.

    这是一个迹象，表明跨国界建立科学联盟具有持续吸引力。

7. alternate [ɔːl'tɜːnət]

    a. with first one thing, then another thing, and then the first thing again 轮流的，交替的

    Private cars are banned from the city on alternate days.

    该市私家车限隔日出行。

8. proximity [prɒk'sɪməti]

    n. the state of being near in space or time（时间、空间的）靠近，亲近

    Yunnan Province is located in southwest China, enjoying geographical proximity to Myanmar, Laos and Vietnam.

    云南省地处中国西南部，地理上邻近缅甸、老挝和越南。

9. accessibility [əkˌsesə'bɪləti]

    n. the fact of being able to be reached or obtained easily 易得到；易触及；可访问性

    Two new roads are being built to increase accessibility to the town center.

两条新路正在修建中，以便更快、更便捷地到达市中心。

10. complementary [ˌkɒplɪˈment(ə)ri]

    a. useful or attractive together 补充的；互补的；补足的

    My family and my job both play an important part in my life, fulfilling separate but complementary needs.

    家庭和工作都是我生活中重要的组成部分，两者互为补充，满足我不同的需要。

11. autonomous [ɔːˈtɒnəməs]

    a. independent and having the power to make your own decisions 自主的，有自主权的

    They proudly declared themselves part of a new autonomous province.

    他们自豪地宣布自己是新自治省的一部分。

12. jurisdiction [ˌdʒʊərɪsˈdɪkʃn]

    n. the authority of a court or official organization to make decisions and judgments 司法权；管辖权；审判权

    The court has no jurisdiction in/over cases of this kind.

    该法庭无权审判此类案件。

13. viability [ˌvaɪəˈbɪləti]

    n. ability to work as intended or to succeed；(of living things) capable of normal growth and development 可以实施；可行性；生存能力，发育能力

    Rising costs are threatening the viability of many businesses.

    不断上涨的成本正在威胁许多企业的生存。

14. homogeneity [ˌhəʊməʊdʒəˈniːəti]

    n. the quality of consisting of parts or people that are similar to each other or are of the same type 同种；同质（等于 homogeneousness）

    We're seeing chefs abandoning homogeneity in favor of innovation.

    我们看到厨师们放弃了千篇一律，而选择了创新。

15. heterogeneous [ˌhetərəˈdʒiːniəs]

    a. consisting of parts or things that are very different from each other 各种各样的；混杂的

    Switzerland is a heterogeneous confederation of 26 self-governing cantons.

    瑞士是一个由 26 个不同的自治州组成的联邦。

16. resilient [rɪˈzɪliənt]

    a. able to quickly return to a previous good condition 有弹性的；能复原的；有复原力的；适应性强的

    This rubber ball is very resilient and immediately springs back into shape.

    这种橡胶球很有弹性，可以立即恢复原来的形状。

17. nurture [ˈnɜːtʃə(r)]

    v. to take care of, feed, and protect someone or something, especially young children or plants, and help him, her, or it to develop 养育，培育

    As a record company executive, his job is to nurture young talent.

作为唱片公司主管,他的职责是培养有才华的年轻人。

18. deplete [dɪˈpliːt]

　　v. to reduce something in size or amount, especially supplies of energy, money, etc. 消耗;耗费(资源、金钱、精力等)

　　If we continue to deplete the earth's natural resources, we will cause serious damage to the environment.

　　如果我们继续消耗地球上的自然资源,将对环境造成严重的破坏。

19. deter [dɪˈtɜː(r)]

　　v. to prevent someone from doing something or to make someone less enthusiastic about doing something by making it difficult for that person to do it or by threatening bad results if they do it 阻挠,阻止;威慑;使不敢

　　These measures are designed to deter an enemy attack.

　　这些措施旨在阻止敌人的进攻。

20. subsidy [ˈsʌbsədi]

　　n. money given as part of the cost of something, to help or encourage it to happen 补贴,津贴,补助金

　　The government is planning to abolish subsidies to farmers.

　　政府正计划取消对农民的补贴。

21. incentive [ɪnˈsentɪv]

　　n. something that encourages a person to do something 激励,刺激

　　Tax incentives have been very effective in encouraging people to save and invest more of their income.

　　税收措施在鼓励人们将收入多储蓄和多投资方面很有效。

22. revenue [ˈrevənjuː]

　　n. the income that a government or company receives regularly (企业、组织的)收入,收益;(政府的)税收

　　Government revenues fell dramatically.

　　政府财政收入大幅下降。

23. fluctuation [ˌflʌktʃuˈeɪʃ(ə)n]

　　n. constant change; vacillation; instability 波动,起伏

　　Changes in climate before the Cretaceous period caused severe fluctuation in sea level, resulting in the extinction of the dinosaurs.

　　白垩纪以前的气候变化导致了海平面的剧烈波动,引发了恐龙的灭绝。

24. mitigate [ˈmɪtɪɡeɪt]

　　v. to make less severe or intense; moderate or alleviate 减轻,缓和

　　The leases come with painstaking stipulations to mitigate any possible environmental harm to species like the polar bear.

　　这些租约附带了细致的条款,以减轻对北极熊等物种的任何可能的环境伤害。

25. sanctuary [ˈsæŋktʃuəri]

n. a place of refuge or asylum 避难所，庇护所
His church became a sanctuary for thousands of people who fled the civil war.
他的教堂成了数千逃避内战的人们的一个避难所。

Phrases and Expressions

associate......with...... 与……关联
due to 由于；应归于
tend to 趋向；易于；有……的倾向
gain......an advantage over 使某人占优势；使某人处于更有利的地位
reap the benefits 获利；获得好处；获得收益
under the jurisdiction of 在……管辖下；治下
in part 部分地；在某种程度上
be vulnerable to 易受……的伤害
other than 除……之外
put......in place 到位；落实到位
seek after 追求；探索

Terminology

flora and fauna 动植物
sense of difference 差异感
excursionist tourism 短途旅行
stay-over visitors 过夜游客
autonomous government 自治政府
economic viability 经济活力；经济供给能力；经济可靠性
out-migration 迁出

Proper Names

Balearic Islands 巴利阿里群岛
the Hawaiian Islands 夏威夷群岛
the Galapagos Islands 加拉帕戈斯群岛
the Canary Islands 加那利群岛
the French Polynesian Islands 法属波利尼西亚群岛
the Caribbean islands 加勒比群岛
Tourism Penetration Index 旅游渗透指数

一、翻译策略/方法/技巧:褒贬译法

翻译过程中,词汇的翻译经常在原文中某个单词、词组、成语和常用表达等意义上能够找到对等的表达。汉语词汇的词类分为实词、虚词、叹词、拟声词四个大类,包括名词、动词、形容词、副词、数词、量词、助词和语气词等。英语词汇的词性可分成名词、动词、代词、形容词、副词、连词、介词、叹词、冠词等。词汇按照感情色彩可分为褒义、贬义和中性。在词汇的翻译过程中,词汇的词性并不能原封不动地从原文直接译成译文,需要根据译文的语言习惯进行翻译。对于词汇的感情色彩的翻译一般应严格按照原文的思想与精神进行翻译,不能按照词汇的词性在原文中随意译出。也就是说,原文中的褒义词需要翻译成褒义词,原文中的贬义词需要翻译成贬义词,同样,中性词也应翻译成中性词或译为恰当的褒贬义词。翻译中需要注意的要点是,原文措辞的语义特征、原文作者的表达意图、作者的写作手法与风格、原文描写的人物关系、话语的语用意图等信息内容。

英语中的一些词汇有明显的褒义特征,翻译时应译出褒义特征,包括整个句子中的与此相关的信息单位。

例 1:He was polite and always advised willingly.他很有礼貌,总是诲人不倦。

句子中的 willingly 是褒义词,又通过上下文挖掘出词语的感情色彩,在翻译时斟酌后译为"诲人不倦"。

英语中有些词有明显的贬义特征,翻译时同样要译出贬义的特征,同样包括整个句子中相关的语言单位。

例 2:You'll be sorry if you try anything funny in class.你要是在课上玩什么鬼花样,肯定要后悔的。

句中的 funny 是贬义词"奇怪的,不好的"意思,上下文中的 sorry 从整个句意来看,可译为"后悔"。

英语中有中性语义特征的词汇,翻译时在一定的上下文中可翻译为中性词,亦可译为恰当的褒贬义词。

例 3:由于党的农业政策,我国千千万万农民走上了致富的道路。Thanks to the policies of the Communist Party of China on agriculture, millions of peasants in China are getting rich.

中文句中的词汇都是中性词,译文也应由中性词汇组成。

有的时候,原文句子并不能表明感情色彩,可以将中性词汇根据整个句意和相关的信息单位翻译为褒义词或贬义词。

例 4:He is a man of integrity, but unfortunately he had a certain reputation. I believe the reputation was not deserved.他是位正直诚实的人,但不幸有某种坏名声。我认为他不该有这个坏名声。

词汇 reputation 是个中性词,但从整个句子的句意上看,应译为贬义词,所以译为"坏名声"。

在翻译过程中,源语文本中的一个褒义词或贬义词往往会给整个句子或上下文带来感情色彩,所以,在翻译的过程中使用褒贬译法,会增加语句翻译的准确性和句子思想的表

达。不仅要注意句子的感情色彩明显的词语,更要注意原文中一些无感情色彩的词语,准确恰当地使用褒贬译法进行翻译。

(摘自:冯庆华.实用翻译教程[M].上海:上海外语教育出版社,2010:78-79.)

二、译例与练习

1. The mystique associated with islands is dependent on a blend of different lifestyles, indigenous cultures, unique land formations, flora and fauna, and ocean and coastal resources. (Para.1)

到这里的海岛旅游会给游客带来神秘的体验,这种神秘感是多种因素共同作用的结果,首先,岛上原住民别出一格的生活方式以及独特的土著文化,外加岛上奇异的地形地貌、动植物种类以及海洋和沿海资源都让这种神秘体验有迹可循。

2. To keep that mystique alive and thriving, islands must implement sustainable tourism policies in all areas including environmental, economic and socio-cultural. (Para.1)

为了继续保持海岛特有的神秘感,海岛管理者就必须在多个层面、不同领域(环境、经济和社会文化方面)全面制定和实施有利于促进海岛旅游可持续发展的政策。

【英汉两种语言有一个共同的特点:表达某些概念既可运用概略化的手法,也可采用具体化的方式。但就某一概念的表达而言,英语可能采用概略化的表达方式,而汉语却采用具体化的表达方式,以上这个句子就是如此,英文句子的主语是海岛,该句描述的是为了继续保持海岛特有的神秘感应该出台相关政策,翻译时如果把主语直接转换成海岛,则不符合汉语习惯。英语原文中有不少词语乃至整个句子的字面意义非常抽象、笼统、概括和空泛,如对等译成汉语,不但译文难以表达出原作者的真正意图,而且译文语义也会含糊不清,很难给读者一个准确的概念。遇到这种情况时,译者可以甚至必须根据具体的语言环境和上下文的事理联系,把这些字面意义捉摸不定的词句用汉语里含意明确具体的词句加以表达,使原文比较抽象的表达方式变得比较实在、比较空泛的变得比较具体、比较笼统的变得比较明确。这时,我们需要考虑采用具体化策略,把主语译成"海岛管理者"。】

3. All islands must address issues of economic impact, environmental consequences and those relating to the social, cultural and political fabric of the island, all of which are affected by the density of tourism on the island. (Para.3)

岛上旅游业密度加大,必然会带来种种问题,如对海岛经济的影响,给海岛环境造成的压力,以及海岛旅游对岛上居民的社会、文化和政治结构生态造成的冲击等各种问题,这些问题都亟待解决。

4. Baum, in his book entitled *"The Fascination of Islands: A Tourist Perspective in Island Tourism: Trends and Prospects"* describes the general attractivity of North Atlantic islands, including their remoteness, their small size, the slower pace of life, the chance to go back-in-time, the wilderness environment, the water-focused society and the sense of difference yet familiarity. (Para.2)

鲍姆在他的《岛屿的魅力——岛屿旅游的旅客视角：趋势和前景》一书中描述了这些北大西洋岛屿所普遍具有的吸引力，包括：地理位置相对偏远、人迹罕至、海岛面积小、岛上生活节奏慢、游客在享受荒野美景的同时体验时光倒流的感觉，还有"以水为中心"的社会关系体系给人带来的差异感和熟悉感，都给岛屿旅游增加了无限魅力，让游客流连忘返。

5. Visitors will tend to stay longer in islands that are remote and larger, whereas those close to the mainland and smaller may experience more excursionist tourism. (Para. 2)

游客倾向于在偏远并较大的岛屿上停留时间更长，而对于那些靠近大陆以及面积较小的岛屿，游客则可能选择短途旅行的方式。

6. The island's choice to host excursionists versus stay-over visitors requires a careful evaluation of the strengths and weaknesses of each type of tourism. (Para. 2)

海岛则需要做出仔细评估，来决定选择接待短途游客还是留宿游客，毕竟每种旅游类型都有其各自的优缺点。

7. Multi-destination travel in island chains may be an added attraction for tourists, whereas the peace, or "sun, sand, sea" experience of a single island vacation may be the choice for others. (Para. 2)

对有些游客来说，群岛旅游可能具有额外的吸引力，主要体现在可以安排多目的地旅行路线，而选择在单个岛屿上度假这种旅游方式的也有一定的拥趸，后者更喜欢享受孤岛的安宁或"阳光、沙滩、大海"的体验。

8. They are also more likely to reap the maximum economic benefits from tourism, without any revenue being leaked to the mainland through taxes and other means. (Para. 2)

他们也更有可能从本地旅游业中获得最大的经济利益，而不用担心大陆管理者会以税收和其他方式将任何收入拿走。

9. In the latter case, there is a special need to ensure economic viability to prevent the out-migration of residents, especially young ones who look for opportunities elsewhere. (Para. 2)

在后一种情况下，特别需要做的是为海岛经济注入新活力，以防止岛上居民迁出岛外，特别是岛上年轻人到其他地方寻找更好的工作机会所造成的居民外移。

10. This may involve the re-enlivenment of festivals, arts, language, folk lore and policies to encouraging local people to engage in entrepreneurial activities. (Para. 12)

这可能涉及让岛上传统节日重新活跃起来，利用艺术、语言、民间传说来打造海岛文化魅力，并通过制定政策，鼓励当地居民从事文化创业活动来促进文化传承。

【翻译时，由于汉语和英语表达上的差异，很难做到总按原文的顺序一句一句翻，经常出现这样一种情况，即原文一句，不一定是译文的一句。原文冗长的复句，可以包含主句、分句、形容词组、副词组等。而按照汉语语法，一个句子里容纳不下那么多分句和词组。如果一定要按原文一句翻译一句，就翻译不出原文的意思，所以断句和重组是免不了的。可是如果断句不当，或断成的一句句排列次序不当，译文还是翻译不出原文的意思。那么怎样断句，怎么重新组合（即排列）断成的一句，没有一定的规律，不过还是有个方法，也有个原则。

方法是分清这一句里的主句、分句以及各种词组,并认清以上各部分的从属关系。在这个基础上,把原句断成几句,重新组合,重新按照汉语的思维方式整理出叙述层次。原则是一定要突出主句,并衬托出各部分之间的从属关系。主句没有固定的位置,可在前、可在后、可在中间,甚至也可切断。从属的各分句、各词组都要安放在合适的位置,使这一组重新组合的断句读起来和原文的那一句是同一个意思,也是同样的说法。】

Exercises

1. Fill in the blanks with the proper given words, and then translate the sentences into Chinese.

  sustainable implement generate vulnerable perspective mitigate

1) A large international meeting was held in Paris with the aim of promoting _____ development in all countries.

2) The discussion has been on for decades, but till now, it is still unclear how to _____ the effects of tourism on the island.

3) The wind farm constructed last year may be able to _____ enough electricity/power for 2,000 homes.

4) The government informed the public in the press conference that the changes to the national health system will be _____ next year.

5) Because of its geographical position, Germany's _____ on the situation in Russia is very different from Washington's.

6) Tourists are more _____ to attack, because they do not know which areas of the city to avoid.

2. Translate the following sentences into Chinese.

1) The ocean, which covers 71 percent of the earth's surface, is a basic component of the global bio-support system. It is also a treasure house of resources and an important regulator of the environment.

2) As a major developing country with a long coastline, China attaches great importance to marine development and protection, and takes it as the state's development strategy. It is constantly strengthening comprehensive marine management, steadily improving its marine-related laws, and actively developing science, technology and education pertaining to the oceans.

3) The 180-page documents, with more than 300 articles and eight annexes, definitely covers every conceivable issue dealing with the seas, from the definition of what constitutes an island to the jurisdiction over fish that live in fresh water but spawn in the ocean.

4) The Tribunal did not accept Bangladesh's argument that the natural prolongation was the primary criterion in establishing the argument that Myanmar had no entitlement of CS

because of discontinuity between Myanmar's landmass and the seabed of the Bay of Bengal due to plate boundaries.

5) Recently, the EEZ has been extended by an additional 396,000 km^2 on the Mascarene Plateau through the joint submission by Mauritius and Seychelles to the United Nations Convention on the limits of continental shelf.

6) The people believe that they are at a stage of economic growth and political maturity which, combined with opportunities presented by global and regional developments, will propel them to the next stage of development if they play their cards right.

7) The Indian Ocean is very different from the western Pacific and the seas near China due to its relatively open geography, the fact that there are multiple powers and no local super power that has enforced dominance, strategic issues and the balance of forces.

8) The environmental resources are a main visitor attraction and tourist interfaces with those resources need to be planned and cared for.

9) Hainan Island, which is almost as big as Taiwan, has abundant natural resources, such as rich iron ore, oil and natural gas, as well as rubber and other tropical and subtropical crops. When it is fully developed, the result should be extraordinary.

10) The ocean's resources sustain the livelihoods of about 3 billion people worldwide, the vast majority of whom live in developing countries. But those livelihoods are under threat, as the ocean and its ability to sustain life are in grave danger due to human activities, such as pollution and overfishing.

3. Translate the following sentences into English.

1) 该计划有助于促进国家和组织之间的对话和协调,也有助于两者之间的伙伴关系,可以通过互相借鉴经验吸取教训。

2) "海洋可持续发展倡议"发起各种活动,比如与区域海洋组织和区域渔业机构进行"海洋可持续发展倡议"全球对话,再比如打造"海洋可持续发展倡议"培训师培训计划,这些活动都有一个目的,即在全球层面、区域之间、国家和国家、国家和地方、地方与地方之间实现协同机制,从而实现海洋和沿海生物多样性的保护和可持续资源利用之间的平衡。

3) 本次活动将概述有效实施生态系统管理的方法或途径,重点是展示各国在恢复珊瑚礁方面所做的各种尝试和建立的机制,借此获得可持续融资,从而实现对珊瑚礁的有效养护、保护到最后的恢复。

4) 在这个具有里程碑意义的海洋之年,海洋行动从多个方面得以推进。在里斯本举行的2022年联合国海洋大会上,产生了数百项新的自愿承诺和一项政治宣言,旨在扩大基于科学和创新的海洋联合行动,以落实设定的第14项可持续发展目标。

5) 自20世纪70年代后期以来,在国际海底区域进行的环境研究以及矿物勘探活动使得人们对深海的海洋生物多样性有了更深的认识。

6) 可持续海洋行动旨在促进生物多样性。"海洋小组"共同议程中设定的目标与生物多样性缔约方大会上拟定的"2020年后全球生物多样性框架"中设定的85%的目标保持一致。

7)本次活动将探讨在2020年后框架的背景下,如何将人类掌握的海洋和沿海生态系统方面的知识转化为海洋治理行动,以及在此程中将会面临的挑战和机遇。

8)这个跨机构、跨部门的小组将着重阐述美国将如何从多个层面来实施海洋多样性解决方案,从而减轻气候变化对海洋生物多样性造成的负面影响。

9)在过去的三十年中,由于污染、发展、过度开发和气候变化等人为因素,生物多样性已经急剧下降,海洋也不能幸免,海洋物种也出现了急剧减少、珊瑚礁等标志性景观退化的现象。

10)我们需要立刻采取全方位的措施和行动来制止海洋生物多样性的丧失,通过恢复和保护海洋中重要的生态系统,如珊瑚礁、红树林和盐沼,来确保生态系统可持续性。

4. Choose the best paragraph translation. And then answer why you choose the first translation or the second one.

南麂岛:惬意的海岛生活

　　南麂岛,别名海山,古代又写作"南已山",位于浙江省平阳县,列岛海岸线曲折,岬角丛生,岛上大沙岙是国内唯一罕见的贝壳沙,远看也很美。这里海岸线曲折,岬角丛生,海湾众多,岛上不仅有着突兀的滨海悬崖,还有着绵延的贝壳沙海滩。赤脚漫步在大沙岙沙滩上,犹如踏在地毯上,最是惬意。最美不过于落日下的小岛,余晖相印,毫不吝啬地铺在海面上,金光闪闪,波光粼粼,远处欣赏之时,海平面与天空接壤,渔船与归鸟似乎成为马祖岙渔港的点缀。

译文一:

Nanji Island: An Unexpected Weekend Getaway from Cities

Nanji Island, nicknamed "Haishan" or literally "Sea Mountain", is historically known as "Nanji Mountain" in ancient times. Located in Pingyang County, Zhejiang Province, the archipelago, which consists of exposed bedrock and sharp cliffs, bays and islets, boasts of its rugged and winding coastline. A significant part of its charm is a stretch of inviting beaches. The most famous one, which is called "Da Sha Ao", is the only shell sand beach in China, which is even visible from afar. A walk on the beach barefoot offers you unique experience and makes your trip all the more special. Feeling powder sand massaging your feet, you can at the same time appreciate and admire the sun slowly setting under the horizon by enveloping yourself in the last rays of beautiful sunlight shimmering on the seawater. That grandeur is enough to make you forget where you are and who you are, then suddenly you will be called back to reality by gliding and soaring seabirds against the azure sky and occasionally you may spot fishing boats breaking the landscape, which will put you again at ease and forget all the hustle and bustle of the human society.

译文二:

Nanji Island: A Comfortable Island Life

Nanji Island, alias Haishan, also written "Nanji Mountain" in ancient times, is located in Pingyang County, Zhejiang Province, the coastline of the archipelago is tortuous, the headland is thick, the island is large sand, is the only rare shell sand in China, and it is also beautiful from afar. With a winding coastline, headlands and bays, the island is not only abrupt coastal cliffs, but also stretches of shell sand beaches. Walking barefoot on the beach is like stepping on a carpet, which is the most pleasant. The most beautiful is the island under the sunset, the sunset is imprinted, unsparingly spread on the sea, shining gold, sparkling, when admired from afar, the sea level borders the sky, fishing boats and returning birds seem to become the embellishment of Matsu Fish Port.

5．Translate the following passage into Chinese．

Given the challenges to the island's ecosystems, environmental management is crucial for island sustainability. The paucity of land causes land usage issues, and the trade-off of land for tourism versus agriculture and other industries, or preservation and conservation needs to be addressed. The environmental resources are a main visitor attraction and tourist interfaces with those resources need to be planned and cared for. Conlin, in a study of Tasmania tourism, gives nine different types of parks or reserves that can be created (national park, state reserve, nature reserve, game reserve, conservation area, nature recreational area, regional reserve, historic site and private sanctuary). This may include designation of zones that are off-limits to tourists, and those that are only visitable with guides and interpreters. Islands with unique wildlife must also take steps to sustain those populations. Tourism can actually assist as an anti-poaching mechanism and an engine for conservation management when residents realize the economic value of the wildlife as in the Seychelles. Policies to keep the land and ocean unpolluted are also necessary. Waste management and recycling programs are essential, particularly on small islands. Also the use, through incentive programs if necessary, for alternative fuel sources (wind, solar, geothermal etc.) will make the destination more sustainable.

6. Translate the following passage into English.

海洋旅游是以海洋为旅游场所，以探险、观光、娱乐、运动、疗养为目的的旅游活动形式。海洋面积辽阔，开发潜力很大。海洋空气中含有相当量的碘、大量的氧、臭氧、碳酸钠和溴，灰尘极少，有利于人体健康，适于开展各种旅游活动。在海上旅行具有与陆地迥然不同的趣味，海上传统项目通常是海上观看日出日落，开展划船、海水浴，以及各种体育和探险项目，如游泳、潜水、冲浪、钓鱼、驰帆、赛艇等。游船是海洋旅游的主要交通工具。当今世界拥有数百艘豪华型游船，不仅可为游客提供食宿，而且具有各种服务项目和娱乐设施。

三、翻译家论翻译

早在2004年，黄友义就在《中国翻译》上发表了一篇名为《坚持"外宣三贴近"原则，处理好外宣翻译中的难点问题》的文章。在文章中，黄友义指出了外宣翻译最应该注意的是要潜心研究外国文化和外国人的心理思维模式，善于发现和分析中外文化的细微差异和特点，时刻不忘要按照国外受众思维习惯去把握翻译。黄友义还指出，最好的外宣翻译不是逐字逐句机械地把中文转换为外文，而是根据国外受众思维习惯，对中文原文进行适当的加工，有时要删减，有时要增加背景内容。在文章中，黄友义提出了"外宣三贴近"原则，即贴近中国发展的实际，贴近国外受众对中国信息的需求，贴近国外受众的思维习惯的"外宣三贴近"原则。

1. 充分考虑文化差异，努力跨越文化鸿沟

中西文化上的差异，导致思维习惯和表达方式上的明显不同。许多中文里约定俗成的词句原封不动地翻译成英文后，不但难以达到忠实地传达中文原意的目的，反而会引起不必要的误解。有时非但不能回答外国人对中国某一方面的疑问，反而引出新的问题。

如果机械地翻译涉及我国政治体制的内容，很可能引起国外受众的误解。比如有的文章说，"在中国，共产党是执政党，此外还有八个民主党派。"由于西方长期的反共宣传，那里的很多人错误地把共产党等同于不讲民主的专政党，这时，他们会很自然地把共产党和民主党派这两个字当作反义词来解读。因此，当他们读到"In China, the Communist Party is the party in power. Besides, there are also eight democratic parties."（他们对此产生的理解却是：中国实施的是共产党的一党专制，同时中国存在着八个追求民主的反对党。）因此，只是按照字面直译造成的后果就是这样严重。其实，如果动动脑筋，把这句话中的"democratic parties"译为"other political parties"，就会淡化外国人头脑中国共产党和民主党派"完全对立"的错误印象。

2. 熟知外国语言习俗，防止落入文字陷阱

有许多词句在外国文化里，已经被赋予了特定的一些含义，若使用不当，就会产生误会。这就要求翻译人员特别留心语言的发展变化。

比如，在奥维尔写的《一九八四》一书中，大洋国有个无处不在、战无不胜、永远正确、心狠手辣的统治者，他的名字叫"老大哥"（BigBrother）"Big Brother is watching you."（老大哥正在看着你）。这是一个骇人听闻、令人毛骨悚然的场景。然而在中文里，老大哥则是具有亲切感的词汇。因此，如果翻译直接处理成 big brother，就会出国际笑话。

因此，可以说外宣翻译有其特有的规律，对翻译工作者有着特殊的、更高的要求。而外宣翻译工作者每一张口或每下一笔，都要以沟通为目的，需要具备厚重的中外政治、经济、文化背景知识功底，而不能仅仅为了完成翻译而翻译，真可谓难度大、要求高。

（摘自：黄友义.坚持"外宣三贴近"原则，处理好外宣翻译中的难点问题[J].中国翻译，2004(6)：27-28.）

Text B

Sustainable Travel: Eight Best Islands for Ecotourism
By Gina Vercesi

Ideological and Political Education：讲好海洋文化故事 打造旅游品牌效应

中共中央办公厅、国务院办公厅印发《关于建立健全生态产品价值实现机制的意见》（以下简称《意见》）指出：建立健全生态产品价值实现机制，是贯彻落实习近平生态文明思想的重要举措，是践行绿水青山就是金山银山理念的关键路径，是从源头上推动生态环境领域国家治理体系和治理能力现代化的必然要求，对推动经济社会发展全面绿色转型具有重要意义。《意见》还指出，应该依托优美自然风光、历史文化遗存，引进专业设计、运营团队，在最大限度减少人为扰动前提下，打造旅游与康养休闲融合发展的生态旅游开发模式。

当下，文旅产业正在经历深刻变革。《2021—2022年中国文旅景区发展总结与趋势报告》指出旅游发展的几个新特征：① 文化和旅游融合进一步深化，文旅新产品、新业态、新形式不断涌现。② 横向跨界与纵向联动愈加频繁，"文旅＋"的融合特性更加多元化，沉浸式业态蓬勃兴起。在此背景下，如何重塑文旅产业生态，驱动文旅产业发展新浪潮，是我们新阶段应该优先考虑的问题。

然而，纵观当前的海洋旅游，普遍缺乏的是人文内涵。旅游的商业化严重、品牌效应不明显、旅游项目同质化严重、传统的海洋观光和休闲度假产品难以满足游客需求，以及海洋旅游资源挖掘不够、海洋文化资源转化困难、海洋文化的旅游开发与真实性保护存在矛盾、对海洋文化本身认知具有局限性等。具体还表现在走马观花式的旅游，品牌效应不明显，海洋文化特色不突出；海洋旅游产品的海洋文化特色缺乏；以消费为目的的同类型旅游项目在不同滨海城市重复出现；文化旅游项目以"表演"替代文化的"参与感""真实感"。

促进海洋文化与旅游融合发展应该从以下三方面进行：首先，应深度挖掘海洋文化，依托海洋非物质文化遗产，讲好海洋文化的"故事"，突出海洋文化特色，形成品牌效应。其次，要注重海洋文化资源的空间规划和统筹协调。最终，将文旅融合理念与社区治理理念结合起来，重塑"生活化"的社区，打造嵌入式、体验式、深层次的文旅体验空间。

1. In honor of Earth Day, we're **highlighting** ecotourism destinations that double as **sublime** vacation spots. Climate change and the **subsequent** impact to the global environment stands at the fore of many people's consciousness, and the United Nations has declared 2022 the Year of Sustainable Tourism. That's why making responsible travel choices has become all the more crucial. These coastal and island communities have all made environmental stewardship and ethical tourism practices a **priority**, allowing visitors to enjoy their beauty while minimizing human impact.

2. Dominica. Deep in the center of the Caribbean Sea between Martinique and Guadeloupe lies a lush, volcanic island packed with ecological wonders. Self-**proclaimed** the "Nature Island of the Caribbean," Dominica launched its green efforts long before doing so became a trend. (The island has held court as a centerpiece in the world's ecotourism industry for well over a decade.) Known for its incredible biodiversity, hundreds of rare plant, animal, and bird species— including the native Sisserou parrot—call the 290-square-mile island home. Winding trails through **pristine** rainforest lead to stunning waterfalls and natural, **bubbling** hot springs while **secluded** black-sand beaches are **nestled** between rocky cliffs. And as the whale-watching capital of the Caribbean, visitors delight in spotting pods of cetaceans **frolicking** in the waters off Dominica's shores.

3. St. John, U. S. Virgin Islands. With its **crystalline**, turquoise waters that lap gently at flawless, powder-white beaches, St. John, the smallest of the USVI, embodies the archetypal tropical paradise. It also remains **blissfully** untouched in comparison to many of its Caribbean neighbors, largely due to the fact that more than two-thirds of the island falls under the stewardship of the National Park System. Beyond the reaches of land, an additional 12,708 underwater acres comprise Coral Reef National Monument, a **veritable** swim-through aquarium replete with breathtaking marine life. A lush volcanic haven wrapped in laid-back island **ambience**, St. John invites visitors to explore endless **idyllic** beaches or hike to secluded snorkeling spots in sheltered coves where graceful sea turtles munch lazily on the sea grass below.

4. Nevis. Home to more monkeys than people, tiny Nevis, nestled in the northern region of the Lesser Antilles, is working hard to establish itself as "the greenest place on earth," aiming to source 100 percent of its energy from **geothermal** sources within the next year. Because buildings are forbidden to rise higher than 1,000 feet above sea level, only **swaying** palm fronds interrupt views over the dramatic volcanic landscape. Trails through **verdant** rainforest are dotted by natural hot springs, plantation ruins, and forty

varieties of mango wind up towering Nevis Peak, visible from almost everywhere on the island. Down below, inviting beaches, all of which are public, circle the island's sandy shoreline.

5. The Galapagos Islands. As the birthplace of evolutionary biology, the Galapagos archipelago, approximately 600 miles off the Ecuadorian coast, **prevails** as a true **mecca** for ecotourism **aficionados** and nature enthusiasts alike. Home to a remarkable range of animal and plant species, the Galapagos became the first UNESCO Natural Heritage Site for Humanity in 1978 and boasts one of the world's most unique ecosystems. 97 percent of the Galapagos is protected national park or marine biosphere reserve and travel to the islands is closely managed, requiring the company of guides. Each of the mostly uninhabited islands offers its own distinctive landscape and the animal residents are so at ease in their surroundings that it's not unlikely to see them interacting with their human visitors both on land and sea.

6. Grenada. Playing host to waterfalls, volcanic hot springs, and a vast array of indigenous flora and fauna, island locales like Levera National Park and Grand Etang Forest Preserve have helped Grenada's place within the ecotourism niche flourish in recent years. Efforts to protect both natural rainforest and prolific coral reef systems make the island's "Pure Grenada" **moniker** an accurate one, and Grenada's work to construct coral nurseries and promote ecotourism make the island a prime destination for sustainable tourism. Grand Anse Beach, often considered one of the world's loveliest, remains one of the historic Spice Island's stars.

7. Little St. Simons Island, Georgia. The jewel of coastal Georgia's Golden Isles, this privately owned, low-country oasis has a long legacy of environmental stewardship and virtually no development. The historic hunting lodge and five welcoming cabins **accommodate** a total of 32 guests per night who have the island's 11,000 protected acres all to themselves. A nature lover's paradise, Little St. Simons Island boasts a staff of talented onsite naturalists who lead a variety of activities each day, from kayaking through labyrinthine tidal marshlands to creek fishing, bird watching and wildlife hikes. Beach cruisers allow visitors to pedal out to the island's 7-mile long private beach where nesting sea turtles find a safe haven to lay their eggs.

8. Costa Rica. Dubbed the "rich coast" by 16th century Spanish **conquistadors**, the Central American republic of Costa Rica adheres to sustainable tourism practices that emphasize resource efficiency, protection of biodiversity and economic development

within the local community. Abundant wildlife—a half-million species of flora and fauna can be found here—as well as lush jungle, 750 miles of unspoiled beaches, and a proliferation of eco-lodges overlooking the Pacific Ocean are just a few highlights. With its deep passion for la pura vida, Costa Rica endures as an early advocate of wellness travel, encouraging visitors to feed their souls by immersing themselves deeply in the country's culture and nature.

9. Vancouver Island. Known for its lush, emerald rainforest and temperate climate, this 12,000-square-mile island off the Pacific northwest coast of Canada provides a rich habitat for the abundant wildlife who make their home within a variety of diverse ecosystems including salt marsh, freshwater lakes, alpine rivers and sandy ocean shores. Within both the Pacific Rim National Park Reserve and the Clayoquat Sound Biosphere Reserve, visitors can spot black bears, sea lions, pods of whales, and a variety of shorebirds, as well as stunning, old-growth spruce and cedar forest. Wild and windswept, Vancouver Island holds strong as a premier destination for salmon fishing, ocean kayaking, hiking, and whale watching within Canada's unspoiled coastal wilderness.

(1,013 words)

https://www.islands.com

New Words

1. highlight ['haɪlaɪt]

    v. to attract attention to or emphasize something important 使引起注意，强调

    The report highlights the need for improved safety.

    那份报告强调了加大安全力度的重要性。

2. sublime [sə'blaɪm]

    a. extremely good, beautiful, or enjoyable 极好的；极美的；令人极度愉悦的

    The book has sublime descriptive passages.

    这本书中有一些非常优美的描写段落。

3. subsequent ['sʌbsɪkwənt]

    a. happening after something else 随后的，接着的

    The book discusses his illness and subsequent resignation from politics.

    书中讲述了他患病及随后隐退政坛的事。

4. priority [praɪ'ɒrəti]

    n. something that is very important and must be dealt with before other things 优先考虑的事

The management did not seem to consider office safety to be a priority.

管理部门似乎并不认为应优先考虑办公室安全问题。

5. proclaim [prəˈkleɪm]

   v. to announce something publicly or officially, especially something positive 宣布；声明

   Republican party members were confidently proclaiming victory even as the first few votes came in.

   刚刚得到几张选票，共和党人就准备自信地宣布胜利了。

6. pristine [ˈprɪstiːn]

   a. new or almost new, and in very good condition 崭新的；状态良好的

   Washing machine for sale—only two months old and in pristine condition.

   出售洗衣机——只用过两个月，完好如新。

7. bubble [ˈbʌb(ə)l]

   n. a ball of gas that appears in a liquid, or a ball formed of air surrounded by liquid that floats in the air 泡；气泡；泡沫

   I love champagne, and I think it's the bubbles that make it so good.

   我喜欢香槟，我觉得它的美妙之处在于它的气泡。

8. seclude [sɪˈkluːd]

   v. to keep someone or something away from other people or things 不与……接触

   Typically, the bride would seclude herself in another room.

   一般来说，新娘会把自己关在另一个房间里。

9. nestle [ˈnes(ə)l]

   v. to be in, or put something in, a protected position, with bigger things around it（使）位于,（使）坐落于（安全或隐蔽之处）

   Bregenz is a pretty Austrian town that nestles between the Alps and Lake Constance.

   布雷根茨是一个漂亮的奥地利小镇，坐落在阿尔卑斯山和康斯坦茨湖之间。

10. frolic [ˈfrɒlɪk]

    v. to play and behave in a happy way 嬉戏；玩闹

    A group of suntanned children were frolicking on the beach.

    一群晒得黝黑的孩子在海滩上嬉戏。

11. crystalline [ˈkrɪstəlaɪn]

    a. clear and bright like crystal 晶莹剔透的；水晶般清澈透明的

    Her singing voice has a pure, crystalline quality.

    她的嗓音纯净、清澈。

12. blissfully [ˈblɪsfəli]

    adv. in an extremely happy way 极乐地；极幸福地

    It is a place where he is blissfully at ease.

    这是一个让他感到极度幸福自在的地方。

13. veritable ['verɪtəb(ə)l]

a. used to describe something as another, more exciting, interesting, or unusual thing, as a way of emphasizing its character 十足的,不折不扣的;名副其实的

The normally sober menswear department is set to become a veritable kaleidoscope of color this season.

色调通常都比较暗淡的男装部这一季将会成为名副其实的色彩万花筒。

14. ambience ['æmbɪəns]

n. the character of a place or the quality it seems to have 气氛;情调;环境

Despite being a busy city, Dublin has the ambience of a country town.

尽管都柏林是一个繁忙的城市,但颇有乡村城镇的情调。

15. idyllic [ɪ'dɪlɪk]

a. extremely pleasant, beautiful, or peaceful 田园诗般的,田园风光的,恬静的

Till now, he still has a vivid memory of that idyllic childhood he spent in the countryside with his grandfather.

直至今天,他仍然能清晰地记得他和祖父一起在乡下度过的那段恬静愉快的童年。

16. geothermal [ˌdʒiːəʊ'θɜːml]

a. of or connected with the heat inside the earth 地热的,地温的

Eighteen countries now generate electricity using geothermal heat.

现在有18个国家使用地热发电。

17. sway [sweɪ]

v. to move slowly from side to side 摇摆,摆动

The movement of the ship caused the mast to sway from side to side/back and forth.

桅杆随着船的行驶左右/前后摇摆。

18. verdant ['vɜːd(ə)nt]

a. covered with healthy green plants or grass 长满绿色植物的;草木苍翠的

Much of the region's verdant countryside has been destroyed in the hurricane.

这个地区大片草木苍翠的乡村地带都在那次飓风中毁于一旦。

19. prevail [prɪ'veɪl]

v. to get control or influence 占优势,占上风

I am sure that common sense will prevail in the end.

我相信常识最终会获胜。

20. mecca ['mekə]

n. a place to which many people are attracted 胜地,令人向往的地方

His Indiana bookstore became a Mecca for writers and artists.

他在印第安纳州的书店成为作家和艺术家们的一个好去处。

21. aficionado [əˌfɪʃɪə'nɑːdəʊ]

n. someone who is very interested in and enthusiastic about a particular subject

酷爱……者；……迷

I happen to be an aficionado of the opera, and I love art museums.

碰巧我是个歌剧迷,而且喜欢艺术馆。

22. moniker ['mɒnɪkə(r)]

n. a name or nickname 名字；绰号

She's the author of three detective novels under the moniker of Janet Neel.

她是三本署名为珍妮特·尼尔的侦探小说的作者。

23. accommodate [ə'kɒmədeɪt]

v. to provide with a place to live or to be stored in 为……提供住宿；容纳；为……提供空间

New students may be accommodated in halls of residence.

新生可以住在学校宿舍楼里。

24. conquistador [kɒn'kwɪstədɔ:(r)]

n. one of the Spanish people who travelled to America in the 16th century and took control of Mexico and Peru(16世纪前往美洲并占领墨西哥和秘鲁的)西班牙征服者

The Inca Empire was discovered by a Spanish conquistador called Francisco Pizarro accompanied by about 169 Spanish soldiers.

印加帝国被一个名叫弗朗西斯科·皮萨罗的西班牙征服者发现,随同他的还有约169名西班牙士兵。

Phrases and Expressions

take……seriously 认真对待……
all the more 更加
be packed with 挤满；塞满
launch its efforts 展开努力
delight in 因……感到快乐
in comparison to 与……相比
fall under the stewardship of 在……的管理下
replete with 充满
establish itself as 以……身份立足
a vast array of 大量的
adhere to 坚持；拥护,追随
immerse oneself in 专心从事于；埋头

Terminology

la pura vida 纯净生活
ethical tourism 负责任的旅行
ecotourism industry 生态旅游产业
pristine rainforest 原始雨林
marine biosphere reserve 海洋生物圈保护区
labyrinthine tidal marshland 迷宫般的潮汐沼泽
cedar forest 雪松林

Proper Names

UNESCO Natural Heritage Site for Humanity 联合国教科文组织人类自然遗产
Pacific Rim National Park Reserve 太平洋沿岸地区国家公园保护区（位于加拿大卑诗省太平洋洋沿岸）

一、翻译简析

1. A lush volcanic haven wrapped in laid-back island ambience, St. John invites visitors to explore endless idyllic beaches or hike to secluded snorkeling spots in sheltered coves where graceful sea turtles munch lazily on the sea grass below.(Para.3)

此外，圣约翰岛还是一个被悠闲的岛屿氛围所包裹的郁郁葱葱的火山天堂。在这里，游客可以去探索无尽的田园诗般的海滩，或徒步前往隐蔽的海湾浮潜点，在那里，优雅的海龟懒洋洋地咀嚼着下面的海草。

【中文句子强调先出主语原则，本段落都是在介绍该岛的旅游资源，因此对该句进行翻译时，就进行了句子语序调整，把主语圣约翰岛提前，然后以这个主语为出发点，介绍该岛的特色和旅游资源。】

2. Home to more monkeys than people, tiny Nevis, nestled in the northern region of the Lesser Antilles, is working hard to establish itself as "the greenest place on earth," aiming to source 100 percent of its energy from geothermal sources within the next year.(Para.4)

坐落在小安的列斯群岛北部地区的小尼维斯岛，是个猴子比人还多的地方，它正在努力将自己打造成"地球上最绿色的地方"，其建设目标是在明年年内实现能源100%来自地热。

【英语的词语、分句以及句子之间借助语言形式手段（关联词、介词）来实现词语或者句子的连接，通过词汇本身的形态变化来表达语法意义和逻辑关系。汉语重意合，呈"隐形"，按照事情发展的先后顺序以及时序、因果、时空逻辑顺序，以短句的形式行文推进，句子之间无过多的形式连接，叙事从容不迫，层层展开。本句中使用名词短语做同位语，然后一个过去分词，一个现在分词，引导两个状语，翻译时应该按照汉语的逻辑顺序逐层展开，先译该岛位置，然后介绍该岛独特特征，最后介绍该岛发展方向、

建设目标。】

3. Trails through verdant rainforest are dotted by natural hot springs, plantation ruins, and forty varieties of mango wind up towering Nevis Peak, visible from almost everywhere on the island. (Para. 4)

　　游客们穿过葱郁的雨林小径,不仅可以欣赏沿途的天然温泉、庄园遗址,还能看到多达 40 个种类的芒果树,蜿蜒至高耸的尼维斯峰。这样的独特景色在小岛上几乎随处可见。

　　【这个英文句子比较长,使用白描手法对岛上资源景观进行描写,语言静态。旅游宣传文本除了具有信息功能之外,同时还应该具有呼唤功能,如何能吸引读者兴趣、立刻抓住读者眼球,在翻译中扮演着非常重要的作用,因此在遣词造句的同时应该注重营造画面感,具有受众意识,拉近与读者的距离。在翻译本句时,译者有意识地根据汉语动态语言特点,对原文的静态语言进行了处理,着重以游客的视角,以广角镜头推进式动态描述岛上景观,营造了身临其境的画面感,来吸引读者。】

4. A nature lover's paradise, Little St. Simons Island boasts a staff of talented onsite naturalists who lead a variety of activities each day, from kayaking through labyrinthine tidal marshlands to creek fishing, bird watching and wildlife hikes. (Para. 7)

　　作为自然爱好者的天堂,小圣西蒙斯岛拥有一群优秀的野外博物学家,他们每天组织开展各种各样的活动,从迷宫般的潮汐沼泽地划皮划艇到小溪钓鱼,观鸟和漫游海岛辨认岛上各种动植物物种。

　　【英语属于静态语言,名词的作用很大,在句子中充当的成分更多,使用效率较高;汉语属于动态语言,动词的作用举足轻重,串联起了整个句子,使用频率颇高。这里用到词类转译法,英文原文中名词形式介绍的旅游项目汉译时就转化成了符合汉语表述习惯的汉语动词词组。】

5. Known for its lush, emerald rainforest and temperate climate, this 12,000-square-mile island off the Pacific northwest coast of Canada provides a rich habitat for the abundant wildlife who make their home within a variety of diverse ecosystems including salt marsh, freshwater lakes, alpine rivers and sandy ocean shores. (Para. 9)

　　这座岛屿位于加拿大太平洋西北海岸,面积 12 000 平方英里,郁郁葱葱的翠绿雨林和温和的气候非常有名。它为这里丰富的野生动物提供了各种各样的栖息地,让这些动物能够在各种不同的生态系统中安家,包括盐沼、淡水湖、高山河流和沙质海岸。

　　【英语多结构复杂的长句,从句可以充当除了谓语动词之外的所有句子成分。而汉语词汇的黏合力较差,不宜拖带过多的修饰成分,更多擅长使用流水句式。因此,英译汉时往往需要拆译从句或长句。这句的重心就是该岛的雨林和气候为各种不同生物提供不同种类的栖息地,按照这种方式对英语句子进行拆分,然后重新组织,才能符合汉语句子众多、零散、模糊,各句紧密围绕一个中心的语言组织特点。】

二、参考译文

可持续性旅游：追求生态旅游的八大好去处

这些沿海地区和岛屿自然条件本来就得天独厚，当地民众又极重视可持续发展，这就使得以下这些去处理所当然成为生态旅游的绝佳首选，定会给您的旅行增加不一样的色彩。

1. 为了纪念地球日，我们将重点介绍一些秉持生态旅游这一理念的好去处，它们同时也是绝佳的度假胜地。目前，越来越多的人意识到了气候变化及其对全球环境的影响。恰逢联合国宣布2022年为可持续旅游年，因此，做出负责任的旅行选择变得尤为重要。接下来介绍的沿海地区和岛屿都把环境管理和有责任感的旅游作为重中之重，让游客们在享受其美丽风景的同时，能最大限度地减少人类活动对环境和生态带来的影响。

2. 多米尼克岛。在马提尼克岛和瓜德罗普岛之间的加勒比海中心，有一个郁郁葱葱、遍布生态奇迹的火山岛。多米尼克自称"加勒比海的自然之岛"，早在绿色环保成为一种趋势之前就开始践行这种理念。（十多年来，该岛一直是世界生态旅游产业的核心。）多米尼克岛以其令人难以置信的生物多样性而闻名，数以百计的珍稀植物、动物和鸟类（包括本地的帝王亚马逊鹦鹉）都以这个290平方英里的岛屿为家。穿过原始雨林的蜿蜒小径，就能看到令人惊叹的瀑布和天然冒着泡的温泉，隐秘的黑色沙滩则坐落在岩石悬崖之间。此外，作为加勒比海的观鲸之都，游客们也能享受到看鲸鱼群在多米尼克海岸附近水域嬉戏的乐趣。

3. 美属维尔京群岛圣约翰岛。圣约翰岛是美属维尔京群岛中最小的一个，其剔透的青绿色海水轻轻拍打着完美无瑕的白色细软海滩，这里是典型的热带天堂。与加勒比海的许多其他岛屿相比，圣约翰岛仍然保持着纯粹的原生态，想想这一点就让人觉得无比幸福。这主要是因为该岛超过三分之二的面积处于国家公园系统的管理之下。除陆地面积之外，这座岛屿还有12 708英亩的水下面积。水下有珊瑚礁国家纪念碑和一个畅游水族馆，里面住满了令人叹为观止的海洋生物。此外，圣约翰岛还是一个被悠闲的岛屿氛围所包裹的郁郁葱葱的火山天堂。在这里，游客可以去探索无尽的田园诗般的海滩，或徒步前往隐蔽的海湾浮潜点，在那里，优雅的海龟懒洋洋地咀嚼着下面的海草。

4. 尼维斯岛。坐落在小安的列斯群岛北部地区的小尼维斯岛，是个猴子比人还多的地方，它正在努力将自己打造成"地球上最绿色的地方"，其建设目标是在明年年内实现能源100%来自地热。在这里，由于建筑物的高度禁止高于海平面1 000英尺，因此，游客在饱览宏伟的火山景观时，遮挡他们视线的只有风中摇曳的棕榈叶了。游客们穿过葱郁的雨林小径，不仅可以欣赏沿途的天然温泉、庄园遗址，还能看到多达40个种类的芒果树，蜿蜒至高耸的尼维斯峰。这样的独特景色在小岛上几乎随处可见。下面是诱人的海滩，所有海滩都是公共的，这些海滩构成岛屿独特的沙质海岸线。

5. 加拉帕戈斯群岛。加拉帕戈斯群岛距离厄瓜多尔海岸约600英里，是进化生物学的发源地，同时也是生态旅游爱好者和自然爱好者们真正向往的圣地。加拉帕戈斯群岛拥有世界上最独特的生态系统,是众多动植物的家园。1978年,该岛成为联合国教科文组织第一个人类自然遗产。在这里,岛屿面积的97%都处在国家公园以及海洋生物圈保护区保护范围之内,因此,游客去这里旅游会受到严格管理,需要导游陪同。大多数情况下,每一个无人居住的岛屿都有自己独特的景观,动物居民在这样的环境中生活是如此的放松自在,因此,无论是在陆地上还是在海里,都能看到它们与人类游客亲切互动。

6. 格林纳达岛。坐落在格林纳达岛上的莱维拉国家公园和大埃唐森林保护区拥有瀑布、火山温泉和大量本土动植物。近年来,这些保护区的存在帮助格林纳达岛奠定了生态旅游胜地的崇高地位,使其生态旅游业得到了蓬勃发展。人们努力保护天然雨林和富饶的珊瑚礁系统使该岛的称号"纯净格林纳达"实至名归。格林纳达在打造珊瑚培养基地和促进生态旅游方面取得了累累硕果,使得该岛成为可持续旅游的首选站。大安斯海滩,通常被认为是世界上最可爱的海滩之一,也是香料岛历史上的明星岛屿之一。

7. 乔治亚州小圣西蒙斯岛。作为乔治亚州黄金群岛的一颗明珠,小圣西蒙斯岛归私人所有,是一块低地绿洲。长期以来,这里一直实行严格的环境管理,因此几乎没有被开发。这里有历史悠久的狩猎小屋和5间迎客小屋,每晚总共可容纳32名客人,在这里,客人可以独享岛上11 000英亩土地。作为自然爱好者的天堂,小圣西蒙斯岛拥有一群优秀的野外博物学家,他们每天开展各种各样的活动,从迷宫般的潮汐沼泽地划皮划艇到小溪钓鱼,观鸟和漫游海岛辨认岛上各种动植物物种。在这里,游客可以骑着自行车前往岛上7英里长的私人海滩,在那里,筑巢的海龟可以找到一个安全的港湾来产卵。

8. 哥斯达黎加。16世纪的西班牙征服者把哥斯达黎加称为"富庶海岸",这个中美洲共和国一直以来坚持旅游的可持续发展,他们的种种做法都诠释了其强调资源配置效率,对生物多样性保护和促进当地经济发展的理念。这里有丰富的野生动物,还有郁郁葱葱的丛林,在岛上可以找到50万种动植物。还有750英里长的未被破坏的原生态海滩,以及可以俯瞰太平洋的生态旅馆,如雨后春笋般出现,这些只是其中的几个亮点。哥斯达黎加作为健康旅游的早期提倡者,提倡简单纯净的生活方式,鼓励游客通过沉浸在该国文化和美妙的大自然来滋养自己的灵魂。

9. 温哥华岛。这座岛屿位于加拿大太平洋西北海岸,面积12 000平方英里,郁郁葱葱的翠绿雨林和温和的气候非常有名。它为这里丰富的野生动物提供了各种各样的栖息地,让这些动物能够在各种不同的生态系统中安家,包括盐沼、淡水湖、高山河流和沙质海岸。在环太平洋国家公园保护区和克莱奥夸特湾生物圈保护区内,游客可以看到黑熊、海狮、鲸鱼群和各种滨鸟,以及令人惊叹的古老云杉和雪松林。一边享受荒野的美景,一边吹着温哥华岛的海风,这座未受污染的海岛上是游客垂钓鲑鱼、玩皮划艇、徒步旅行和观赏鲸鱼的不二之选。

Cultural Background Knowledge: Marine Tourism
海洋旅游文化背景知识

海洋是世界上最大的水体，约占地球表面总面积的71％。世界上的海滨地区，尤其是气候温暖的中纬度地区，是目前最重要的旅游场所，也是目前世界水休闲旅游的主要场所。早在20世纪80年代，以西班牙为首的地中海沿岸旅游大国，就已经提出了"3S"工程——阳光(sun)、海水(sea)、沙滩(sands)，开创了海滨旅游的新纪元。从此，海滨旅游在世界各地得到迅猛发展。进入21世纪，海滨旅游业已成为沿海国家竞相发展的重点产业及国内外旅游体系的热点，与海洋石油、海洋工程并列为海洋经济的三大新兴产业，其中海滨旅游业是海洋产业的龙头和支柱。海滨旅游收入与海洋产业总产值及当地国民生产总值的关系极其显著，发展海滨旅游业是提高海洋总产值及相应第三产业产值的最佳途径之一，海滨旅游在海洋经济中所起的作用将越来越大。

中国是一个海洋大国，拥有18 000千米的大陆海岸线，14 000千米的海岛海岸线和6 500多个岛屿。根据《联合国海洋法公约》的规定，我国管辖的海域面积达300万平方千米。这片"蓝色国土"是中华民族实现可持续发展的重要战略财富，也是实现民族经济腾飞的重要之处。海洋经济在我国的经济发展中占有重要的地位。20世纪90年代以来，我国的海洋经济取得了高速发展，海洋产业年均增长率在20％以上。1997年全国主要海洋产业产值3 104亿元，海洋产业增加值占全国国内生产总值的2％，以海洋经济为依托的沿海市县国民经济稳步增长。国民生产总值合计达24 472.70亿元，占整个沿海地区国内生产总值的53.1％，占全国国内生产总值的30.8％。2020年全国主要海洋产业总产值达到30 000亿～35 000亿元，占全国GDP的8.5％，海洋经济在整个国民经济中的比重将逐渐加大。随着我国沿海开放战略的实施，沿海成为我国经济最发达的地区，也是我国重要的经济地带，在我国经济发展中起着龙头作用，逐步成为国民经济的新的增长点。沿海地区以占全国13％的土地面积，养活了全国42％的人口，创造了69％以上的国民生产总值。

Unit Seven
Marine Economic Activities

海洋经济活动

Text A

Ocean-based Industries in Developing Countries
By Anonymous Author

Ideological and Political Education: 发展蓝色经济 助力海洋强国建设

浩瀚海洋蕴藏着无限机遇,历来是人类财富的宝库。党十八大报告明确指出,提高海洋资源开发能力,发展海洋经济,保护海洋生态环境,坚决维护国家海洋权益,建设海洋强国。这是新中国历史上第一次将建设海洋强国明确作为国家发展战略。倡议大力发展海洋经济是"海洋强国"战略中的重要一环。我国海洋经济范畴主要包括海洋渔业、海洋交通运输业、海洋船舶工业、海盐业、海洋油气业、滨海旅游业等。

21世纪以来,我国海洋经济发展取得了巨大的进步,呈现出产业规模逐步扩大、产业结构持续优化、新兴产业蓬勃发展的态势,海洋产业体系逐步完善,海洋经济综合实力不断提升。近年来,我国海水养殖产量占全球一半以上,海洋渔获量一直居世界首位。我国沿海港口吞吐量连续多年居世界首位。港口规模位居世界第一,2021年全球港口货物吞吐量和集装箱吞吐量排名前十位的港口中,我国的港口分别占8席和7席。与此同时,我国海运船队的运力规模持续壮大,截至2021年年底,达到3.5亿载重吨,居世界第二位。世界造船大国的地位也进一步巩固。海洋工程总装进入第一方阵,市场份额保持全球领先。在自主研发领域,我国自主研发的海洋药物占全球已上市产品类目的将近30%,海洋糖类药物研发进入国际先进行列,建成了全球规模最大的海洋微生物资源宝藏库。体内植入用超纯度海藻酸钠完成国家药品监督管理局药品审评中心(CDE)登记备案,打破了国际垄断,并完成国产化生产制造。我国自主研发的兆瓦级潮流能发电机组连续运行时间保持世界领先地位。海上风电累计装机容量跃升至全球第一位。

我国在相关海洋产业取得的辉煌成就充分展示了中国智慧、中国方案、中国力量以及中国特色社会主义市场经济的优势,将激励年轻一代以建设海洋强国为己任,认识海洋,经略海洋,守护海洋,发展蓝色经济,助力海洋强国建设。

1. This article explores trends across developing countries in several specific ocean-based industries: seafood production; marine tourism; extractive industries; shipping and passenger transport; shipbuilding; renewable energy; and marine biotechnologies. Some ocean-based sectors account for a considerable share of GDP in developing countries.

2. Seafood production: Fishing, Aquaculture and Fish Processing. Fishing. Capture fisheries and **aquaculture** are important sources of protein in human diets throughout the world. Small-scale fisheries remain the backbone of socio-economic well-being for many coastal communities and especially for developing countries in the tropics, where the majority of fish-dependent countries are located. According to the Organization for Economic Co-operation and Development (OECD) calculations of recent years, the value added from marine fishing alone is highest in the grouping of lower middle-income countries, followed by high-income countries and upper middle-income countries. The lower middle-income countries host largely coastal, artisanal fisheries, which represent very modest value added. The totals are still conservative, since approximations in many countries often do not take into account the informal, small-scale fisheries and subsistence sectors, which are often important in many developing countries. Employment data shows the number of marine fishing jobs is the highest in lower middle-income countries. Upper middle-income countries have the next highest number of marine fishing jobs, followed by high-income countries and lower middle-income countries.

3. Aquaculture. Aquaculture is considered to be one of the sectors with the largest potential for growth and has expanded **substantially** in recent years, driving up total fish production against a more stagnating trend for wild fish catch. More than 50% of aquaculture production is inland and is dominated by freshwater fin fish such as carp species. Aquaculture in coastal areas includes both species farmed in saltwater ponds, such as shrimp, and species produced in cages and man-made structures either adjacent to or on the coast, such as seaweed and molluscs. Developing marine aquaculture could be an opportunity for selected developing countries, although it should not come at the expense of coastal ecosystems. Aquaculture can provide an additional source of income for **vulnerable** coastal populations, who may otherwise rely on farming or fishing. Further, technical improvements in aquaculture systems have greatly increased the feed efficiency of aquaculture in recent years and many systems now achieve a feed conversion ratio similar to poultry systems, albeit with significant variation. More complex and still at a demonstration stage, open ocean farming projects also have potential for more sustainable fish production. In the recent decade, the top countries in seafood production, including both fisheries and aquaculture, have evolved somewhat in terms of live weight tonnes. China has remained the leader. Indonesia rose to second place, followed by the United States, Peru, the Russian Federation, India, Japan, Viet Nam, Norway and Chile.

4. Seafood Processing. Millions of people and particularly women are involved in artisanal fish processing, making it another important ocean-based industry in developing countries but one that faces some common challenges. Post-harvest facilities such as drying equipment, ice plants and cold storage facilities are often lacking. Such installations are needed for adding value to the seafood product and obtaining better prices, but also to reduce post-harvest losses that occur in artisanal fisheries. When no storage facilities are available in the ports with no ice, the fishers sometimes tend to sell their unsold fish at cheaper price or face spoilage of their catches. The Food and Agriculture Organization (FAO) estimates that approximately 35% of the global harvest is either lost or wasted every year. Economic development across the entire fish production system is therefore highly dependent on **enhancing** post-harvest processing, as well as exploring further sustainable fishing practices (e. g. certifications and eco-labels). Looking ahead, the impacts of overfishing, climate change, coastal pollution, biodiversity loss and illegal, unreported and unregulated fishing will take a toll on seafood production, as they add to the inherent challenges of artisanal fisheries. Some countries will increasingly need more effective strategies for marine conservation and sustainable fisheries management to rebuild stocks for nutritional security.

5. Marine Tourism. Tourism is today one of the key sectors in the global economy. Tourism's direct, indirect and induced impact accounted for approximately one in ten of the world's GDP. There are major regional differences in countries' reliance on tourism for their economies. Tourism is the main economic sector in some developing countries and an important source of foreign exchange, income and jobs. In Kenya for example, tourism is an important sector, representing 8.8% of GDP in 2018 and attracting up to two million foreign visitors a year, mainly in national parks and along the coast. In some years, coastal tourism can account for some 60% of the revenues. Small Island Developing States (SIDS) are also particularly dependent on the tourism sector: two out of three SIDS rely on tourism for 20% or more of their GDP. As visitors are frequently concentrated along coastlines, some coastal areas generate up to 80% of total GDP in some countries. In the Maldives, for example, tourism **contributes** up to 40% of national GDP.

6. An increasingly **trenchant** issue for many developing countries is balancing the promotion of commercial activities that cater to foreign demand and the need to address environmental concerns. Global tourism is a case in point, as it has significant adverse environmental impacts by placing pressures on domestic freshwater supplies, food systems and waste disposal systems in particular. Another issue is the impacts of the COVID-19 pandemic on the tourism industry and uncertainly over how long they might last. It is difficult to predict how the industry's recovery will play out, given the extent of the economic damage, the reduced purchasing power of many potential travellers and the possibility that tourists may be reluctant to travel, especially to countries without well-developed health systems. Any mid- to

long-term governmental support should ideally **steer** the industry to more sustainable practices, backed by efficient policy and economy instruments.

7. Extractive Industries: Oil and Gas and Seabed Mining. Oil and gas. The offshore oil and natural gas industry represents the largest share of today's ocean economy and contributes to many developing economies, particularly in Africa and Latin America, despite important environmental externalities. While Nigeria and Angola are Africa's largest oil and gas producers, an **unprecedented** number of other African countries-among them Ghana, Mauritania, Mozambique, Senegal, Somalia and South Africa-are extending new exploration licenses to offshore companies. The oil and gas sector accounts for about 10% of Nigeria's GDP and around 86% of its exports revenue, which represents 70% of total government revenue. In Angola, oil production and its supporting activities contribute around 50% of the country's GDP and around 89% of exports. The **momentum** for new extraction licenses built on the results of recent oil and gas exploration programmes, particularly from the Atlantic coast. These include the discovery in West Africa of large deposits off the coasts of Senegal in the MSGBC basin (Mauritania, Senegal, Gambia, Guinea-Bissau and Equatorial Guinea), all since 2015. Given the uncertainty surround oil demand and price recoveries in the short to medium term, it is **conceivable** that investments in some offshore oil and gas projects will be delayed or cancelled due to low prices stemming from the reduced demand and oversupply.

8. Seabed Mining. Rising demand for minerals and metals, alongside the **depletion** of land-based resources, is stirring growing commercial interest in exploiting resources on the seabed in national waters and the high seas. To date, ongoing mining projects are targeted for the most part towards extracting diamonds, phosphates and seabed marine sulphide deposits, for example in Namibia and South Africa as well as China, Japan and Korea. On a global scale, these activities are still small and publicly available economic data on the operations are scarce. At this stage, when it moves beyond exclusive economic zones and towards the high seas, mining remains an experimental industry with still-unknown impacts on the marine environment and biodiversity, particularly in areas where knowledge of the ocean floor and deep-water ecosystems is limited. Although deep sea science is progressing, much uncertainty still remains about any future operational seabed mining in the high seas. It could have a wide range of impacts on marine ecosystems that may be potentially widespread and long-lasting, with very slow recovery rates. Some impacts may also be theoretically irreversible for instance **disturbance** of the benthic community where nodules are removed; plumes impacting the near-surface biota and deep ocean; and deposition of suspended sediment on the sea-floor. In consequence, a few countries have called a precautionary approach and even a **moratorium** on seabed mining activities.

9. Shipping and Passenger Transport. Both shipping (i.e. maritime freight transport) and

passenger transport have been growing steadily. International maritime trade has grown almost every year. The top five ship-owning regions—Greece, Japan, China, Singapore, and Hong Kong-account for more than 50% of the world's dead weight tonnage. The main sub-sectors in maritime passenger transport also have been registering growth. Globally, ferries transport approximately two billion passengers a year, on a par with air passenger traffic, and the cruise industry carries some 26 million passengers annually. Overall, the maritime transport landscape has changed markedly in recent years. In parallel, technology and services have been playing an expanding role in logistics and services, and sustainability issues have been looming larger on the maritime transport industry agenda. Additional challenges are likely to arise if the global economic downturn persists.

10. Shipbuilding. Shipbuilding, like other manufacturing activities, is influenced by a **multitude** of factors ranging from global trade, energy consumption and prices to changing cargo types and trade patterns, vessel age profiles, scrapping rates and replacement levels. High-income and upper middle-income Asian countries are the dominant shipbuilding market leaders, with China, Japan and Korea together accounting for about 90% of global new-build deliveries in tonnage in commercial ships. Several European countries (e.g. Denmark, Finland, France, Germany and Italy) are producing highly specialized vessels such as ferries, offshore vessels, and large cruise ships. European countries also account for about a 50% global market share in the marine equipment industry. In Indonesia, as in other developing countries, the shipbuilding industry is mainly focused on supplying the domestic market. While the 2012—25 shipbuilding industry development roadmap aims to increase exports of whole ships, the Indonesian industry is **constrained** by poor access to finance, skilled labour shortages and a tax regime that **incentivizes** the importation of whole ships rather than maritime parts for construction in domestic shipyards. At the end of the value chain, several low-income countries in South Asia (e.g. Bangladesh, India and Pakistan) are involved in the dismantling of the world's ships.

11. The COVID-19 pandemic has generated and is expected to generate many effects on the shipbuilding industry and its broader value chains. Ship demand is driven by global maritime activities that have been severely affected by the pandemic. On the supply side, many shipyards experienced production disruptions in 2020 as governments put in place lockdowns and **quarantine** measures. In addition, ship orders and ship deliveries have been delayed because it has been difficult for ship owners to meet with shipbuilders and conclude deals. The industry will need to adapt to recover.

12. Renewable Energy. Electrification is a major challenge in many developing countries that remain dependent on imported fossil fuel for energy generation. The cost of fossil fuels is a burden on government budgets, business and households, and disproportionally affects people

already struggling with poverty. This is especially the case for SIDS, where on average more than 30% of foreign exchange reserves are allocated each year to cover the cost of fossil fuel imports and where retail energy rates are three to seven times higher than in developed economies. To lower the cost of energy and transition towards greener, low-emission development pathways, several renewable solutions are being tested thanks to recent innovations in offshore wind farms and solar and **geothermal** resources.

13. Offshore wind in particular is a rapidly growing sector. It has expanded at an extraordinary rate over the last 20 years or so in developed and emerging countries. The cost of offshore wind generation has dropped progressively. But the long-term impact of large offshore wind farms on the ocean environment itself is slowly starting to be considered too. The offshore wind sector presents some opportunities but also many specific challenges for low-income countries. Technical difficulties can be vast due to geographic characteristics and remoteness, for SIDS in particular. Offshore wind farms still require rather large upfront investments.

14. Marine renewable energy-wave, current and tidal energy-is also considered an important potential source of power generation for the transition to a low-carbon future. However, ocean energy technologies are for the most part still at the demonstration stage, with only a few prototypes moving towards the commercialization phase. In many cases, the installation of wind turbines on land or as offshore platforms is not possible due to topographical constraints and competition for space with other ocean-based industries, typically coastal tourism. This is particularly the case for many SIDS that are considering these marine renewable energy options.

15. The use of geothermal resources can be explored for some tropical islands, particularly volcanic islands. Most geothermal technologies generate stable and carbon dioxide (CO_2) emissions-free baseload power. The International Energy Agency, in 2018 forecasts for renewable energy to 2023, projected that growth in geothermal capacity as projects in nearly 30 countries come on line, with 70% of the growth in developing countries and emerging economies.

16. Marine Biotechnologies. To date, the potential of marine bio-resources remains largely untapped, although many developing countries have extensive and valuable marine resources such as corals, sponges and fish. As ocean processes become better known, many countries are developing strategies to **foster** marine biotechnology for future pharmaceutical drug development and cosmetic products for health and well-being as well as for food production using algae, biofuel, etc. Marine bio-resources research is already essential in many industries, for instance in the pharmaceutical sector for the development of new generations of antibiotics. Marine genetic resources could be at the core of new solutions to fight pandemics.

As seen earlier, the impacts of the COVID-19 pandemic could have long-lasting effects on ocean-based industries in general. But they could also **accelerate** developments in specific emerging ocean-based sectors, for example marine biotechnologies for medical applications. Developing countries should consider exploring and potentially engaging in a sustainable way in these activities. An important first step to avoiding irreversible damage to fragile ecosystems may be linking with existing knowledge and innovation networks to form partnerships and base any future activities on scientific evidence.

17. To summarize, many of the major trends associated with ocean-based industries will continue in spite of uncertainties. Improving long-term sustainability should remain a core factor in decisions related to ocean-based industries, as policy makers consider strategies to **stimulate** their economies once it is safe to do so. The development of ocean-based industries should go hand in hand with preserving marine natural assets and ecosystem services.

(2,469 words)

https://www.oecd-ilibrary.org/sites/bede6513-en/1/3/2/index.html?itemId=/content/publication/bede6513-en&_csp_=b7545bd11087d48c1bbc2341619b3830&itemIGO=oecd&itemContentType=book#section-d1e3563

New Words

1. aquaculture ['ækwəkʌltʃə(r)]

    n. rearing aquatic animals or cultivating aquatic plants for food 水产养殖；水产业

    Aquaculture will only work, environmentally and economically, with the right sort of fish.

    而水产养殖只有在环境条件、经济条件和鱼的种类适合人工养殖的情况下才能成功。

2. substantially [səb'stænʃəli]

    ad. to a great extent or degree 大量地，可观地；大体上，基本上

    Salaries and associated costs have risen substantially.

    薪金与相关费用大大增加。

3. vulnerable ['vʌlnərəbl]

    a. susceptible to attack; susceptible to criticism or persuasion or temptation; capable of being wounded or hurt 易受攻击的；脆弱的；易受伤害的

    People with high blood pressure are especially vulnerable to diabetes.

    血压高的人尤其容易患糖尿病。

4. enhance [ɪn'hɑːns]

    v. increase 增强，提高，改善

    Large paintings can enhance the feeling of space in small rooms.

    大幅画作能增加小房间的宽敞感。

5. contribute [kən'trɪbjuːt]

v. give one's share of (money, help, advice, etc) to help a joint cause; add to sth.; help to cause sth.(为……)做贡献；促成；导致

Synchronous sounds contribute to the realism of film and also help to create a particular atmosphere.

音画同步有助于增强电影的真实性，也有助于营造一个特定的气氛。

6. trenchant ['trentʃənt]

a. (of comments, arguments, etc) strongly and effectively expressed; characterized by or full of force and vigor(指言论、论据等)有力的，有效的；尖刻的；锐利的；苛刻的

What impresses most in his work is his trenchant pen, his poetic narrative style, and his keen insight into human nature.

他的作品以其犀利的文笔、诗歌般的叙述风格以及对人性的深刻洞察而著称。

7. steer [stɪə(r)]

v. direct the course; be a guiding force, as with directions or advice 引导，指导(某人的行为)；引导，带领(某人去某地)

The new government is seen as one that will steer the country in the right direction.

新政府被认为能将这个国家引向正确的方向。

8. unprecedented [ʌn'presɪdentɪd]

a. having no precedent; novel 前所未有的，史无前例的

Such policies would require unprecedented cooperation between nations.

这样的政策会要求国与国之间前所未有的合作。

9. momentum [mə'mentəm]

n. an impelling force or strength 冲力，推力；动力，势头

The historical tide is surging forward with great momentum.

历史洪流气势磅礴，奔腾向前。

10. conceivable [kən'siːvəb(ə)l]

a. capable of being imagined 可想象的，可相信的

It is quite conceivable that every species might be equally different from every other.

每个物种和其他物种有很大不同，这很好理解。

11. depletion [dɪ'pliːʃn]

n. the act of decreasing something markedly 损耗，耗尽

Overfishing not only causes depletion in individual fish stocks, but also disruption to entire ecosystems and food webs in the ocean.

过度捕捞不仅引起个别鱼种的骤减，而且还引起海洋中整个生态系统和食物网的崩溃。

12. disturbance [dɪ'stɜːbəns]

n. activity that is an intrusion or interruption 干扰，扰乱；骚乱，动乱

He was charged with causing a disturbance after the game.

他被指控在比赛结束后制造骚乱。

13. moratorium [ˌmɒrəˈtɔːriəm]

    n. suspension of an ongoing activity 暂停，中止

    The convention called for a two-year moratorium on commercial whaling.

    会议呼吁两年内暂停商业捕鲸活动。

14. multitude [ˈmʌltɪtjuːd]

    n. a large indefinite number 众多，大量

    These elements can be combined in a multitude of different ways.

    这些因素可以通过无数不同的方式进行组合。

15. constrain [kənˈstreɪn]

    v. restrict, hold back 限制，约束

    It turns out that great artists choose to constrain themselves all the time.

    事实证明，伟大的艺术家总是会选择约束他们自己。

16. incentivise [ɪnˈsentɪvaɪz]

    v. encourage sb. to do sth. 鼓励，激励

    My job is to do all I can to empower, support and incentivize them, to do their best to deliver superior customer service to our customers.

    我的工作就是尽我所能去帮助、支持和鼓励他们，让他们尽自己最大的努力为客户提供卓越的服务。

17. quarantine [ˈkwɒrəntiːn]

    n. isolation to prevent the spread of infectious disease 隔离，检疫

    It is said that all ships coming from abroad should be performed quarantine.

    据说对所有国外来的船舶都要进行检疫。

18. geothermal [ˌdʒiːəʊˈθɜːml]

    a. of or relating to the heat in the interior of the earth [地物] 地热的；地温的

    The contribution of geothermal energy to the world's energy future is difficult to estimate.

    地热能对世界未来能源的贡献是难以估计的。

19. foster [ˈfɒstə(r)]

    v. promote the growth of; help develop, help grow 促进，培养；领养，收养

    Cell phones foster social connections with peers across time and space.

    手机促进了人与人之间跨越时间和空间的社交联系。

20. accelerate [əkˈseləreɪt]

    v. cause to move faster (使)加快，促进

    Exposure to the sun can accelerate the ageing process.

    暴露在日光下会加快老化过程。

21. stimulate [ˈstɪmjuleɪt]

    v. cause to do; cause to act in a specified manner 促进，激发；鼓励；刺激

Likewise, automation should eventually boost productivity, stimulate demand by driving down prices, and free workers from hard, boring work.

同样,自动化最终应该提高生产力,通过降低价格来刺激需求,并把工人从繁重、乏味的工作中解脱出来。

Phrases and Expressions

account for 对……负有责任;对……做出解释;(比例)占……
take into account 考虑;重视;体谅
adjacent to 邻近的,毗连的
albeit with 尽管
take a toll 产生负面影响;造成损失
a case in point 恰当的例子
play out 逐渐发生;结束
on a par with 与……同等;和……一样
for the most part 在极大程度上,多半

Terminology

ice plant 制冰厂
high sea 公海,外海
exclusive economic zone 专属经济区
the benthic community 底栖生物群落
dead weight tonnage 载重吨位
foreign exchange reserves 外汇储备
offshore wind farm 海上风电场
wind turbine 风力涡轮机

Proper Names

Gross Domestic Product (GDP) 国内生产总值
the Organization for Economic Co-operation and Development (OECD) 经济合作与发展组织
the Food and Agriculture Organization (FAO) 联合国粮食及农业组织
Small Island Developing States (SIDS) 小岛屿发展中国家
coronavirus disease 2019 (COVID-19) 2019 新型冠状病毒肺炎

一、翻译策略/方法/技巧:换序译法

　　换序译法也叫结构调整法,是一种常见的翻译方法。冯庆华在其《实用翻译教程》中提出,换序译法即翻译时应根据译文的语言习惯,对原文的词序进行调整,使译文做到最大程度上的

通顺。

不同的文化背景使得英汉两种语言在语法结构上存在着巨大差异。具体来说,英语是形态性语言,重形合,句子结构呈树状分布,句子的主谓宾基本结构是树的主干,定状补成分是树杈,句子中的介词和关系代词等则是各分支连接处。而汉语是语义型语言,重意合,句子结构呈竹节状,按一定顺序层层递进。因此,英语多长难句,句子逻辑思维严谨,主要通过词汇和语法体现出上下文关系;而汉语多短句,句子外形松散,无从句概念和形态变化,主要通过词语内在的含义展示出上下文的语义关系。尽管英汉句法结构里都包含"主语+谓语(+宾语)/表语"基本结构,但是其他句子成分如"定语"和"状语"在顺序和位置上都有很大差别。因此,在翻译的过程中,译者需要时常考虑到两种语言语法结构上的差异,摆脱原文语序和句子形式的约束,按照目的语的行文习惯,重新组合语言,使译文更加流畅、地道。

在实践中,换序译法主要是定语或定语从句的换序以及状语或状语从句的换序,同时也包含其他句子成分换序的情况。

1. 定语或定语从句换序

汉语的定语,无论是单个或是多个连用,通常都习惯放在所修饰的中心词之前。英语则不同,一个单词做定语时,一般放在中心词之前,有时也放在中心词之后,如 the students present（在场的学生）、the material required（所需的材料）;词组、短语和从句做定语时,一般放在中心词之后。多个单词做定语时,汉语排列词序是由大到小、由远到近、由强到弱、由具体到一般,而英语正好相反。例如:

He witnessed ①the sixth ②post-war ③economic crisis ④of serious consequence ⑤that prevailed in various fields ⑥in the USA.

译文:他亲眼看见了⑥美国②战后①第六次④后果严重的⑤波及各领域的③经济危机。

英语原文定语的词序:①次第定语,②时间定语,③本质性定语、中心词,④判断性定语,⑤陈述性定语,⑥国别定语;

汉语译文定语的词序:⑥国别定语,②时间定语,①次第定语,④判断性定语,⑤陈述性定语,③本质性定语、中心词。

2. 状语或状语从句换序

汉语的状语一般位于动词之前,而英语中状语的位置比较灵活。英语中多项状语的词序一般为:条件状语、目的状语、主语、程度状语、谓语、方式状语、频度状语、时间状语、宾语。汉语中多项状语的次序一般为:主语、目的状语、时间状语、条件状语、方式状语、程度状语、谓语、宾语。例如:

①For this reason, our company explained ②solemnly ③to your company ④many times ⑤in February ⑥last year.

译文:我公司①为此⑥于去年⑤二月②郑重地④多次③向贵公司解释。

英语原文状语的词序:①目的状语、主语、谓语,②方式状语,③指涉状语,④频度状语,⑤时间状语中的月份,⑥时间状语中的年份;

汉语译文状语的词序:主语①目的状语、⑥时间状语中的年份,⑤时间状语中的月份,②方式状语,④频度状语,③指涉状语、谓语。

3. 其他换序

根据具体情况,换序译法还可以出现在主语或主语从句换序、宾语或宾语从句换序、同位语或同位语从句换序、插入语或插入语从句换序、倒装句换序、表语或表语从句换序、无生命主语句换序等当中。例如:

Even the wild animals of his homeland, it seemed to Kunta, had more dignity than these creatures.

译文:昆塔觉得,即使是他家乡的野兽也比这群人自尊自重。(插入语换序)

(摘自:冯庆华.实用翻译教程[M].上海:上海外语教育出版社,2010:81,83-84.)

二、译例与练习

Translation

1. Small-scale fisheries remain the backbone of socio-economic well-being for many coastal communities and especially for developing countries in the tropics, where the majority of fish-dependent countries are located. (Para. 2)

小规模渔业仍然是许多沿海社区,特别是热带发展中国家的社会经济福祉的支柱。大多数依赖渔业的国家都位于热带地区。

2. More complex and still at a demonstration stage, open ocean farming projects also have potential for more sustainable fish production. (Para. 3)

开放式海洋养殖项目虽然更为复杂且仍处于示范阶段,但是也具有实现更可持续鱼类生产的潜力。

【英文的表达注重形合,汉语则注重意合,所以英译汉的时候,逻辑关联词的增译尤为重要。比较常见的增译的逻辑关联词像"因为……所以……""不但……而且……""……却……"等起到了顺承、连接、转折等作用。译者在充分尊重原文含义的前提下,通过适当增添原文中无其词但有其意的逻辑关系词"虽然……但是……",显化了句际间逻辑关系,使译文表达更为准确清晰。】

3. Such installations are needed for adding value to the seafood product and obtaining better prices, but also to reduce post-harvest losses that occur in artisanal fisheries. (Para. 4)

这种装置是必需的,既能增加海产品的价值和获得更好的价格,又能减少手工渔业收获后的损失。

4. Small Island Developing States (SIDS) are also particularly dependent on the tourism sector: two out of three SIDS rely on tourism for 20% or more of their GDP. (Para. 5)

小岛屿发展中国家也特别依赖旅游业:三分之二的小岛屿发展中国家国内生产总值中,旅游业占20%或更多。

(小岛屿发展中国家是指一些小型低海岸的国家。这些国家普遍遇到可持续发展的挑战,包括领土面积较小、人口日益增长、资金有限、对自然灾害的抵抗能力较弱和过分依赖国际贸易等。在1992年6月的联合国环境与发展会议上,"小岛屿发展中国家"被定义为一个发展中国家集团。小岛屿发展中国家里只有新加坡被视为发达国家,其他国家都被视为发展中国家或最不发达国家。)

5. An increasingly trenchant issue for many developing countries is balancing the promotion of commercial activities that cater to foreign demand and the need to address environmental concerns. (Para. 6)

对许多发展中国家来说,一个日益尖锐的问题是,如何在促进满足外国需求的商业活动和解决环境问题之间取得平衡。

6. Given the uncertainty surround oil demand and price recoveries in the short to medium term, it is conceivable that investments in some offshore oil and gas projects will be delayed or cancelled due to low prices stemming from the reduced demand and oversupply. (Para. 7)

鉴于中短期内石油需求和价格复苏的不确定性,可以想象,一些海上石油和天然气项目的投资会由于需求减少和供应过剩导致的低价格而被推迟或取消。

7. It could have a wide range of impacts on marine ecosystems that may be potentially widespread and long-lasting, with very slow recovery rates. (Para. 8)

它可能对海洋生态系统产生广泛而持久的影响,恢复速度非常缓慢。

8. To lower the cost of energy and transition towards greener, low-emission development pathways, several renewable solutions are being tested thanks to recent innovations in offshore wind farms and solar and geothermal resources. (Para. 12)

得益于海上风电场、太阳能和地热资源方面的最新创新,相关部门正在对几种可再生能源解决方案进行测试,目的是降低能源成本,向更绿色、低排放的发展道路过渡。

【英语常用非人称主语,让事物以客观的语气呈现。而汉语常用人称主语,注重主体思维,倾向于"什么人做了什么事"。如果对照原文结构将其译为"为了……几种可再生资源解决方案正在被测试中,得益于……"可以看出,机械地对照原文结构得出的译文句子之间逻辑关系不甚明了,有悖于汉语读者"什么人做了什么事"的思维预期。译者将原文的非人称主语"several renewable solutions"转换为增译的人称主语"相关部门",句子结构调整为"得益于……相关部门对……进行测试,目的是……",显化了句际间的逻辑关系,使人物和事件的表述方式更倾向汉语表达习惯,使译文连贯流畅,通顺达意。】

9. An important first step to avoiding irreversible damage to fragile ecosystems may be linking with existing knowledge and innovation networks to form partnerships and base any future activities on scientific evidence. (Para. 16)

要避免对脆弱的生态系统造成不可逆转的破坏,重要的第一步可能是与现有的知识和创新建立伙伴关系,并将未来的任何活动都建立在科学证据的基础上。

10. Improving long-term sustainability should remain a core factor in decisions related to ocean-based industries, as policy makers consider strategies to stimulate their economies once it is safe to do so. (Para. 17)

提高长期可持续性仍应是与海洋产业相关决策的核心因素,因为在确定安全的情况下,政策制定者会考虑刺激经济的战略。

Exercises

1. Fill in the blanks with the proper given words, and then translate the sentences into Chinese.

    innovation   accelerate   take a toll   on a par with   generate   suspend

1) However, it's still unclear if tobacco fumes actually _____ on children's brains, or if something else is at play.

2) If you exceed your credit limit, we have the right to _____ or cancel your account.

3) The government wanted to _____ the reform of the institutions, to find new ways of shaking up the country.

4) The water park will be _____ some of the best public swim facilities around.

5) We intend to _____ among all people a sense of personal responsibility for the environment in which we live.

6) To grow the business, we must promote originality, inspire creativity, encourage _____ and develop management expertise across our team.

2. Translate the following sentences into Chinese.

1) The ocean and its resources are increasingly seen as indispensable to addressing the multiple challenges the planet is set to face in the coming decades.

2) Participants in the conference, which covers topics including maritime law, marine resources assessment and shipping economics, also expressed the importance they attach to the Chinese market.

3) Fish and fishery products increase the economic value of the fish and allow the fishing industry and exporting countries to reap the full benefits of their aquatic resources.

4) In many areas, the winds strength is too low to support a wind turbine or wind farm, and this is where the use of solar power or geothermal power could be great alternatives.

5) Coastal areas are transitional areas between the land and sea characterized by a very high biodiversity and they include some of the richest and most fragile ecosystems on earth, like mangroves and coral reefs.

6) The interdependency of ocean-based industries and marine ecosystems combined with increasingly severe threats to the health of the ocean, have led to a growing recognition of the need for an integrated approach to ocean management.

7) Globally, fish stocks are significantly affected by illegal, unregulated, and unreported (IUU) fishing, though the exact magnitude of the matter is difficult to assess accurately.

8) Long-term interruptions of tourism have significant consequences for the countries whose domestic economy relies on this sector for their domestic economy.

9) Science is crucial to achieving global sustainability and adequate stewardship of the ocean, since it provides the ability to deepen our understanding and monitor the ocean's resources, its health, as well as predict changes in its status.

10) The importance of ocean science will need to remain at the forefront of efforts to face challenges posed by accelerations in the deterioration of ocean health, the changing climate, and ocean economic activity.

3. Translate the following sentences into English.

1) 有效的渔业管理需要改进海洋监视和监测以及渔业科学和数据的可用性。
2) 这些例子有助于对跨海部门的可持续活动的构成达成共识。
3) 除了将更可持续地利用海洋资源纳入海洋产业的主流以外,还需要采取具体的行动来保护和恢复海洋生态系统。
4) 除了跨部门的方法外,由于天然海洋资产的跨界性质,可能还需要采取跨国的方法。
5) 通过控制采矿、过度捕捞等人类活动,海洋保护区可以改善海洋生态系统的总体健康水平,确保海洋生态健康。
6) 在绝望的经济环境下,彼得和无数其他人一样,把大海视为唯一可行的选择。
7) 研究表明,在全球每年捕捞的约7 000万到1亿只鲨鱼中,鱼翅贸易仍占很大比例。
8) 可持续性应该仍然是海洋经济决策中的一个关键因素。
9) 海洋占据了地球表面的三分之二,从这个角度来看,海洋的蓝色海水被认为是这个地球的血液。
10) 产品、价格、地点和促销这四个传统的营销要素也适用于水产品销售。

4. Choose the best paragraph translation. And then answer why you choose the first translation or the second one.

因此,各地建立海洋保护区的目的有所不同。有的是为了保护并恢复海洋生物多样性,保障食品供给;有的是为了补充传统渔业管理疏漏,并通过保护海洋碳储量应对气候变化。事实证明,海洋保护区有助于蓝色海洋经济实现可持续发展,朝着社会公平、环境良好及经济可行的海洋产业发展。蓝色海洋经济是一个经济学术语,与海洋环境的利用和保护有关。如果全球海洋保护区的面积达到71%,那么在不影响目前渔业捕捞量的前提下,生物多样性所能产生的最大效益将会实现91%,海洋碳储所能产生的最大效益将会实现48%。因此,随着越来越多的证据表明MPAs是一种有效的生态保护手段,它们已经被广泛应用于解决过度捕捞等人类活动带来的问题,缓解气候变化带来的压力,确保海洋生物的多样性。

译文一:

Therefore, the purposes of establishing marine reserves are different in different places. Some are to protect and restore marine biodiversity and secure food supplies. Some are to complement traditional fisheries management lapses and to combat climate change by protecting ocean carbon stocks. Practice has proved that marine ecological zones can help blue economy achieve sustainable development. Marine industry helps to achieve social equity, environmental sustainability and economic development. Blue Marine economy is an economic term and is related to the utilization and protection of the marine environment. If 71 per cent of the world's Marine Protected Areas are covered, 91 per cent of the maximum benefits of biodiversity and 48 per cent of the maximum benefits of marine carbon storage would be achieved without affecting current fisheries catches. Thus, more and more successful cases show that MPAs are an effective means of ecological conservation. They have been widely used to solve problems caused by human activities such as overfishing, ease pressures from climate change and ensure the diversity of marine life.

译文二：

Thus, the reasons for creating marine reserves may vary from safeguarding and restoring ocean biodiversity and associated services such as food provisioning to complimenting conventional fisheries management and addressing climate change by protecting marine carbon stocks. Marine reserves have been shown to offer an opportunity to support a sustainable blue ocean economy towards a socially equitable, environmentally sustainable, and economically viable ocean industries. Blue ocean economy is a term in economics that relates to the use and preservation of the marine environment. A recent study has shown that protecting 71 percent of the ocean could generate 91 percent maximum biodiversity benefits and also yield 48 percent carbon benefits without offsetting the current fisheries catches. Hence, as a result of the increasing evidence of MPAs success as effective conservation-based instrument, they have become a widely applied ecosystem-based tool to address the effects of human activities such as overfishing and mitigate climate change impacts and ensure biodiversity protection.

5. **Translate the following passage into Chinese.**

Millions of people and particularly women are involved in artisanal fish processing, making it another important ocean-based industry in developing countries but one that faces some common challenges. Post-harvest facilities such as drying equipment, ice plants and cold storage facilities are often lacking. Such installations are needed for adding value to the seafood product and obtaining better prices, but also to reduce post-harvest losses that occur in artisanal fisheries. When no storage facilities are available in the ports with no ice, the fishers sometimes tend to sell their unsold fish at cheaper price or face spoilage of their catches. The Food and Agriculture Organization (FAO) estimates that approximately 35% of the global harvest is either lost or wasted every year. Economic development across the entire fish production system is therefore highly dependent on enhancing post-harvest processing, as well as exploring further sustainable fishing practices (e.g. certifications and eco-labels). Looking ahead, the impacts of overfishing, climate change, coastal pollution, biodiversity loss and illegal, unreported and unregulated fishing will take a toll on seafood production, as they add to the inherent challenges of artisanal fisheries. Some countries will increasingly need more effective strategies for marine conservation and sustainable fisheries management to rebuild stocks for nutritional security.

● **6. Translate the following passage into English.**

海洋和沿海旅游业高度依赖自然生态系统的质量来吸引游客,因为海滩和清洁水域的娱乐价值对它们极其重要。然而,管理不善的旅游业正在加剧生态系统的退化和脆弱性,危及旅游业自身的经济可持续性。气候变化脆弱性也是依赖旅游业的国家面临的风险,例如珊瑚礁白化事件。小岛屿国家的风险最大,因为旅游业是国民经济的最大部门。影响该部门的其他挑战包括因采砂导致的海滩退化、红树林砍伐以及不断增长的沿海人口对沿海生态系统造成的压力。

三、翻译家论翻译

严复(1854—1921)是我国清末杰出的翻译家与西方哲学思想的传播者。他一生译书甚多,著名的有《天演论》《群己权界说》《穆勒名学》《原富》等一系列介绍西方思想文化制度的典籍。严复在进行翻译实践活动的过程中,提出的翻译理论"信、达、雅"至今仍然在翻译理论与实践中起着举足轻重的作用。

1897年,严复在《天演论》中指出:译事三难:信、达、雅。求其信,已大难矣。顾信矣不达,虽译犹不译也,则达尚焉……此在译者将全文神理融会于心,则下笔抒词,自善互备。至原文词理本深,难于共喻,则当前后引衬,以显其意。凡此经营,皆以为达,为达即所以为信也。易曰:"修辞立诚"。子曰:"辞达而已。"又曰:"言之无文,行之不远。"三者乃文章正轨,亦即为译事楷模。故信达而外,求其尔雅……严复的"信"强调的是原文内容的传递,要忠实于原文,译者要将全文神理,融会于心,然后方可下笔。严复的"达"强调的是行文通顺流畅,不然就会"虽译犹不译也",读者读了译文也不知作者所云。严复的"雅"强调的是文章应该文笔优美,富有文采。"信、达、雅"是一个完整的原则体系。"雅"是对"信"和"达"的深化和补充,是"故信达而外,求其尔雅"。任何一篇文章,都包含思想、语言和风格三个要素,"信、达、雅"可以说是对这三个要素分别提出的不同的翻译标准。严复的"信、达、雅"三字理论的提出,继往开来,言简意赅,意义重大,影响深远。梁启超说:"近人严复,标信、达、雅三义,可谓知言。"郁达夫说:"信、达、雅的三字,是翻译界的金科玉律,尽人皆知。"周作人也说:"信达雅三者为译书不刊的典则,至今悬之国门无人能损一字,其权威是已经确定的了。"

下面是基于严复"信、达、雅"翻译理论的几个例子:

1. "信"

例1:Marine reserves are especially highly protected areas of the oceans and other water bodies such as rivers in which extractive and harmful human activities are banned.

原译:海洋生态区是海洋和其他水体例如河流中尤其高度保护的区域,禁止开采性及其他有害人类的活动。

分析:"信"是忠实,是翻译的基础。"harmful human activities"指的是对海洋生态产生危害的人类活动,译文把它译成了"有害人类的活动",违背了"信"的原则,成为误译,使译文失之毫厘,谬之千里。

改译:海洋生态区是指海洋和其他水体(如河流)中受特殊保护的区域,生态区内禁止开采性及其他有害的人类活动。

2."达"

例2:Audrey Hepburn traveled representing UNICEF, making over 50 emotionally draining and physically dangerous missions into bleak destinations to raise world awareness of wars and droughts.

原译:奥黛丽·赫本代表联合国儿童基金会外出,承担了五十多项情绪上枯竭、身体上危险的任务,到荒凉的地方去唤醒世界人民对战争和旱灾的意识。

分析:"达"是通顺,是翻译的核心。赫本作为联合国儿童基金会的亲善大使,始终牵挂着世界上身处困境的儿童。从句中可知,她频繁外出,四处奔波,承担了大量繁重的任务。所到达的那些地区由于偏远落后,安全问题得不到保证。"emotionally draining and physically dangerous"译为"情绪上枯竭、身体上危险",不符合汉语表达习惯,不够通顺流畅。改译为"劳心劳力、危及生命"更为贴切。

改译:奥黛丽·赫本代表联合国儿童基金会四处奔走,承担了五十多项劳心劳力、危及生命安全的任务,深入荒凉之地,唤起世界人民对战争和旱灾的关注。

3."雅"

例3:Good to the last drop.

译文:滴滴香浓,意犹未尽。

分析:"雅"指用词优雅美好,译文讲求既保留原文意境,又符合读者的认同价值,强调文学价值和艺术价值,做到雅俗共赏。美国前总统罗斯福首次品尝到麦斯威尔(Maxwell)咖啡时,大赞"Good to the last drop"。他的高度赞誉成为麦斯威尔咖啡之后的宣传语。汉语多采用四字成语、四字词语、生活谚语等来表达某人对某事的看法和感受,从而增强说服力与感染力。与直译"好到最后一滴水"相比,译文"滴滴香浓,意犹未尽"一方面更能充分表达咖啡香味浓郁的口感,另一方面两个四字词语能使译文节奏优美,朗朗上口。

(摘自:郭延礼.中国近代翻译文学概论[M].武汉:湖北教育出版社,2005:244-245;
陈福康.中国译学理论史稿[M].上海:上海外语教育出版社,2002:111.)

Text B

Key Pressures on Ocean and Ecosystem Services

By Anonymous Author

Ideological and Political Education：打造海洋的"绿水青山"

海洋是高质量发展的战略要地,保护好海洋环境对于促进沿海地区高质量发展、构建人海和谐关系具有重要意义。目前,我国海洋生态环境状况总体稳中趋好,在保护好海洋生态环境的同时,也为海洋经发展提供有力支撑和保障。

在习近平总书记关于海洋生态文明建设论述的指导下,我国在海洋生态环境修复、污染防治等工作中取得突破,海水水质明显提升,海洋生态环境状况明显好转。目前,我国珊瑚礁、红树林等多个典型海洋生态系统得到有效保护,海洋生物多样性得到显著提高。初步遏制了局部海域红树林盐沼、海草床等典型生态系统退化趋势,区域海洋生态环境明显改善。如今,我国成为全球少数红树林面积净增加的国家之一,记录海洋生物达到2.8万多种,是世界上海洋生物多样性最为丰富的国家之一。每年繁育、迁徙和越冬的水鸟已经达到240多种,全球8条候鸟迁徙路线中有3条经过我国境内。

"十四五"时期,全国海洋生态环境保护工作将以习近平生态文明思想为指导,切实解决海洋生态环境突出问题,持续改善海洋生态环境质量,扎实推进"水清滩净、鱼鸥翔集、人海和谐"的美丽海湾保护与建设,不断提升社会公众临海亲海的获得感、幸福感和安全感,以海洋生态环境高水平保护促进沿海地区经济社会高质量发展。与此同时,我国统筹国际国内,不断提升参与全球海洋环境治理的影响力和话语权。我国以海洋强国战略和"21世纪海上丝绸之路"为引领,深化海洋生态环境保护的双边和多变合作,积极融入全球海洋环境治理体系中,积极探索在应对气候变化、海洋生物多样性保护、海洋塑料垃圾防治等重点领域提供全球公共产品,为推动构建海洋命运共同体提供中国经验、贡献中国智慧。

坚持"创新、协调、绿色、开放、共享"的发展理念,促进海洋经济全面绿色转型,才是海洋可持续发展的长久之策。在海洋经济发展过程中,习近平总书记提出的"绿水青山就是金山银山"理念里"绿水青山"即"海洋生态系统","金山银山"即"优质海洋生物资源(蓝色粮仓)和海洋环境资源(蓝色耕地)"。我们应努力打造海洋的"绿水青山",发展蓝色经济,助力海洋强国建设。

1. The conservation and sustainable use of the ocean is critically important. Ocean and marine ecosystems provide the **intermediate** inputs-ecosystem services-that drive value within sectors by acting as nurseries for fish, providing areas for recreation such as beaches and coral reefs for diving, and providing genetic material for marine biotechnology.

2. Despite the invaluable benefits provided by a healthy ocean, the **cumulative** impacts of **anthropogenic** pressures are pushing the ocean to conditions outside human experience. Additionally, these pressures can **reinforce** each other, **exerting** greater cumulative impacts on marine ecosystems. Global sea-level rise, warming of the ocean, more frequent and severe weather events, and changing ocean currents will **aggravate** the negative impacts of overfishing; illegal, unreported and unregulated fishing (IUU); pollution; and habitat degradation. In marine ecosystems, according to The Intergovernmental Science-Policy Platform on Biodiversity and Ecosystem Service IPBES, direct exploitation of organisms (mainly fishing) has had the largest relative impact, followed by land-/sea-use change, including coastal development for aquaculture and **infrastructure**, and pollution. It is not known for how much longer the ocean can continue to provide its life-sustaining functions and ecosystem services under business-as-usual **scenarios**. The remainder of this article outlines the key pressures on the oceans.

3. Climate change. Anthropogenic carbon dioxide emissions have risen over time, and the ocean has absorbed 20%-30% of the carbon dioxide, leading to ocean acidification. Greenhouse gases in the atmosphere have led to rising sea temperatures and sea levels and shifts in ocean currents. The implications for ocean ecosystems and marine diversity are considerable, already being seen in species and habitat loss, changes in fish stock composition and migration patterns, and higher frequency of severe ocean weather events. The higher frequency of severe ocean weather events particularly affects vulnerable, low-lying coastal communities, including small island developing states. These populations depend on and are vulnerable to the ocean's quality, stability and **accessibility**.

4. Pollution. Most sources of marine pollution are land-based and include industrial, residential and agricultural runoffs and waste such as plastics as well as solid waste. In particular, the runoff of agriculture fertilisers, animal husbandry waste, sewage disposal and industrial effluents **releases** excessive nutrients into the ocean that favour the growth of toxic and harmful species in the ocean (**eutrophication**), altering marine habitats and negatively impacting fisheries. Marine pollution also originates from direct discharge through ship pollution (e.g. ballast water and hot water discharge) and deep-sea mining (e.g. for oil and gas), with the resulting types of pollution consisting of acidification, eutrophication, marine litter, toxins and

underwater noise. Left unchecked, eutrophication can lead to the creation of dead zones, as is occurring in different parts of the world including the Gulf of Mexico, the Black Sea and the Baltic Sea.

5. Marine litter is generated directly or indirectly by very different economic sectors, for example aquaculture and fisheries (e. g. accidental loss, intentional abandonment and **discarding** of fishing gear), shipping and cruise ships (e. g. ship-generated waste), cosmetics and personal care products, textiles and clothing, retail, and increasingly tourism. **Illicit** dumping particularly affects artisanal fisheries and the tourism industry, as the health and safety of persons who use beaches for recreational activities are at risk in areas where litter **accumulates**, with both sectors often representing the primary form of foreign revenue for many developing countries.

6. Plastics are a significant source of ocean pollution. Many of these plastics are extremely long-lived and will remain in the environment for hundreds, if not thousands, of years, meaning the full impacts will only become apparent in the longer term. There is currently **dedicated** research and development into new sustainable petrochemical production routes (from production to use and disposal of products), which may contribute to efforts to **curb** and stop the leakage of plastic pollution and other harmful chemical products into the ocean. New initiatives have also emerged, such as the World Economic Forum's Global Plastic Action Partnership. However, much progress is still required to address the root of chemical pollution, in terms of both the research and development necessary to find potential alternatives that would be less damaging to the environment and changing current production and consumption practices (e. g. building on the circular economy concept).

7. Overfishing, by-catch and IUU fishing and over-exploitation of other natural resources. According to the recent data of the Food and Agriculture Organization, about 35% of the fish stocks in the world's marine fisheries were classified as overfished, with the maximally sustainably fished stocks accounting for about 60% of the total number of assessed stocks. IUU fishing **exacerbates** overfishing and is associated with significant impacts, from both an economic and a food security perspective. Estimating the magnitude of IUU fishing and its many social impacts (e. g. slavery on ships) is complex and depends on many factors, such as the type of fishery, the geographic location and the availability of information. Overexploitation of other natural resources such as shellfish and other organisms is also causing damage to the marine environment.

8. Habitat degradation. Habitat destruction along coastlines and in the ocean results from harmful fishing practices such as dynamite fishing or improper trawling; poor land use practices in

agriculture, coastal development and forestry sectors; other human activities such as mining, dredging and anchoring; and tourism and coastal **encroachment**. For example, logging and vegetation removal can introduce sediments from soil erosion. Harbour development and other land-based activities (such as shrimp aquaculture) can lead to the destruction of mangroves, which serve as nurseries for species of fish and shellfish and provide flood protection. Poor shipping practices and coastal tourist activities such as snorkelling, boating and scuba diving come in direct contact with fragile wetlands and coral reefs, consequently damaging marine habitats and degrading the ecosystem services they provide.

9. **Invasive** alien species. Another serious threat to the marine environment is the introduction of non-native marine species to marine ecosystems to which they do not belong. Most of these alien species had been rapidly introduced to a different habitat through ballast water from commercial shipping operations across the oceans. These foreign organisms are responsible for severe environmental impacts, such as altering native ecosystem by disrupting native habitats, extinction of some marine flora and fauna, decreased water quality, increasing competition and predation among species, and spread of disease. Considering these situations, the International Maritime Organization has made international efforts to address the transfer of invasive aquatic species through shipping, as illustrated by the International Convention for the Control and Management of Ships' Ballast Water which entered into force in September 2017.

10. The degradation of marine ecosystems is extending beyond ecologically and economically sustainable **thresholds**. One of the underlying reasons is that many of the services provided by marine and coastal ecosystems-such as coastal protection, fish nursery, water purification, marine biodiversity and carbon sequestration-are not reflected in the prices of traditional goods and services on the market (and hence referred to as non-market values). While there is often a lack of scientific information to clearly understand the complex links between these marine ecosystem services and their economic value, this undervaluation of marine ecosystem services results in under-investment in their conservation, sustainable use and restoration and lost opportunities for economic growth and poverty reduction, both now and for the future.

(1,197 words)

https://www.oecd-ilibrary.org/

New Words

1. intermediate [ˌɪntəˈmiːdiət]

    a. lying between two extremes in time or space or degree 居中的；中等程度的

    Liquid crystals are considered to be intermediate between liquid and solid.

    液晶被认为介于液态和固态之间。

2. cumulative [ˈkjuːmjələtɪv]

    a. increasing by successive addition 积累的，渐增的；累计的，累积的

    The condition appears to result from the cumulative effect of a number of factors, with atmospheric pollutants the principal culprits.

    这种情况似乎是由许多因素累积的结果，而大气污染物是主要的罪魁祸首。

3. anthropogenic [ˌænθrəpəˈdʒenɪk]

    a. of or relating to the study of the origins and development of human beings 人为的；[人类] 人类起源的

    Many such extinctions are due to natural forces, while others are due to anthropogenic factors.

    有一些动物的绝种是由于自然的限制，有一些是因为人为的因素。

4. reinforce [ˌriːɪnˈfɔːs]

    v. strengthen and support with rewards 加强，加固；增援

    Success in the talks will reinforce his reputation as an international statesman.

    谈判成功将会增强他作为国际政治家的声望。

5. exert [ɪɡˈzɜːt]

    v. bring (a quality, skill, pressure, etc) into use; apply sth.; make an effort 运用，施加（影响）；努力，尽力（exert oneself）

    Rising population combined with improved nutrition standards and shifting dietary preferences will exert pressure for increases in global food supply.

    人口增长、营养标准提高和饮食偏好转变将对全球粮食供应施加压力。

6. aggravate [ˈæɡrəveɪt]

    v. make (a disease, a situation, a situation, an offence, etc) worse or more serious; irritate (sb.) 使加重，使恶化；惹怒，激怒

    If the reports are well-founded, the incident could seriously aggravate relations between the two nations.

    如果这些报道是有根据的，那么该事件将可能导致两国间的关系严重恶化。

7. infrastructure [ˈɪnfrəstrʌktʃə(r)]

    n. the basic structure or features of a system or organization 基础设施，基础建设

    Owing to the level of the damage to factories and infrastructure, it will be weeks or even months before the country's supply chain returns to normal.

    由于工厂和基础设施的损坏程度严重，这个国家的供应链需要数周或甚至数月的时间才能

恢复正常。

8. scenario [sə'nɑːriəʊ]

   n. a postulated sequence of possible events; a setting for a work of art or literature 设想，情节；脚本；情景

   This is a relatively simple and straightforward approach, but it could suffer from a few deficiencies under some scenarios.

   这是一个比较简单且容易操作的方法，但是在有些场景下可能会有一些缺陷。

9. accessibility [ək,sesə'bɪləti]

   n. the quality of being at hand when needed 易使用性，可及性

   Seminar topics are chosen for their accessibility to a general audience.

   专题讨论会的话题是根据普通听众的理解力来选定的。

10. release [rɪ'liːs]

    v. grant freedom to; eliminate (substances); allow (news, etc) to be made known 释放；排放（物质）；发布（新闻等）

    The ability of coal to release a combustible gas has long been known.

    煤能够释放易燃气体，这一点早已为人所知。

11. eutrophication [juːtrəfɪ'keɪʃn]

    n. excessive nutrients in a lake or other body of water, usually caused by runoff of nutrients (animal waste, fertilizers, sewage) from the land, which causes a dense growth of plant life; the decomposition of the plants depletes the supply of oxygen, leading to the death of animal life 富营养化；超营养作用

    Algae are the most frequently used biological monitoring indicators of waters eutrophication.

    藻类是反映水体富营养化最常用的生物监测指标。

12. discard [dɪ'skɑːd]

    v. throw or cast away 扔掉，弃置

    At the same time, workers were required to discard old habits, for industrialism demanded a worker who was alert, dependable, and self-disciplined.

    与此同时，工厂还要求工人们摒弃原来的习惯，因为工业制度需要机敏、可靠和自律的工人。

13. illicit [ɪ'lɪsɪt]

    a. not allowed by law; illegal 非法的；违禁的

    Unfortunately, the cheap alternatives are even more harmful than the illicit drugs they replace.

    不幸的是，这些廉价的替代品比它们所替代的非法药物更有害。

14. accumulate [ə'kjuːmjəleɪt]

    v. gradually get or gather together an increasing number or quality of (sth.) 积累，积攒

    Lead can accumulate in the body until toxic levels are reached.

    铅可以在体内积聚直至到达有毒的程度。

15. dedicated ['dedɪkeɪtɪd]

a. devoted to sth. ;designed for one particular purpose only 专心致志的,献身的;专用的,专门用途的

The Institute's annual award is presented to organizations that are dedicated to democracy and human rights.

该协会的年度奖授予致力于民主与人权的组织。

16. curb [kɜːb]

    v. place restrictions on;hold or keep within limits 控制,抑制

    In my point of view, we are in a prime time to curb the problem from deteriorating.

    在我看来,我们正处在阻止问题恶化的黄金时期。

17. exacerbate [ɪɡˈzæsəbeɪt]

    v. make(pain,disease,a situation) worse;aggravate 使恶化,使加剧

    He also warned that global warming could exacerbate the problem by causing drought.

    他还强调说,全球变暖可能会导致干旱,从而加剧这一问题。

18. encroachment [ɪnˈkrəʊtʃmənt]

    n. entry to another's property without right or permission; any entry into an area not previously occupied 侵入,侵犯;侵蚀

    The survival of the country's famous wildlife is threatened by trophy hunting, climate change and human encroachment.

    该国著名野生动物的生存正受到狩猎、气候变化和人类侵犯的威胁。

19. invasive [ɪnˈveɪsɪv]

    a. tending to spread harmfully 侵略性的,扩散性的

    Next to habitat loss, these invasive species represent the greatest threat to biodiversity worldwide, many ecologists say.

    许多生态学家说,这些入侵物种仅次于栖息地的丧失,是对全球生物多样性的最大威胁。

20. threshold [ˈθreʃhəʊld]

    n. entrance of a house;point of entering or beginning sth. 门槛;开端,起点

    She felt as though she was on the threshold of a new life.

    她觉得好像就要开始新生活了。

Phrases and Expressions

act as 担任,担当,扮演

in particular 尤其,特别

originate from 源于……;来自……

consist of 由……组成;由……构成;包括

at risk 处于危险中

in the long term 从长远的角度来看

in terms of 在……方面；依据；按照
be associated with 和……联系在一起；与……有关
in direct contact with 直接接触；直接联系

Terminology

coral reef 珊瑚礁
ocean acidification 海洋酸化
greenhouse gas 二氧化碳、甲烷等导致温室效应的气体
carbon dioxide 二氧化碳
animal husbandry 畜牧业；畜牧学
ballast water [水运]压载水；压舱水；压舱配重水
dynamite fishing 炸药捕鱼
scuba diving 水肺潜水；轻便（潜水器）潜水
marine flora and fauna 海洋动植物
carbon sequestration 碳封存；碳固定

Proper Names

illegal, unreported and unregulated fishing (IUU) 非法、未报告和无管制捕鱼
The Intergovernmental Science-Policy Platform on Biodiversity and Ecosystem Service (IPBES) 生物多样性和生态系统服务政府间科学政策平台
the World Economic Forum's Global Plastic Action Partnership 世界经济论坛的全球塑料行动伙伴关系
the International Maritime Organization 国际海事组织
the International Convention for the Control and Management of Ships' Ballast Water《船舶压载水控制和管理国际公约》

一、翻译简析

1. Left unchecked, eutrophication can lead to the creation of dead zones, as is occurring in different parts of the world including the Gulf of Mexico, the Black Sea and the Baltic Sea. (Para. 4)

  如果不加以控制，那么富营养化可能导致死亡地带的形成，这在世界各地都有发生，包括墨西哥湾、黑海和波罗的海。

  【许多英语句子并不都是通过逻辑语法词表达因果、转折及连接等关系的，而是通过特定的语法或用词习惯来呈现。因此原文中存在一些隐性逻辑结构，翻译时应该灵活变通，通过增补关联词将原文中隐含的逻辑关系表现在译文中，使译文通顺易懂。】

2. However, much progress is still required to address the root of chemical pollution, in

terms of both the research and development necessary to find potential alternatives that would be less damaging to the environment and changing current production and consumption practices (e.g. building on the circular economy concept). (Para. 6)

然而,要解决化学污染的根源问题,仍需在以下两方面加大力度:研发找到对环境危害较小的潜在替代品,以及改变当前的生产和消费观念(如以循环经济概念为基础)。

【一种语言在另一种语言里找不到对等的语法结构时,可通过结构重组法来解决翻译的难题。在保证词义准确的前提下,将原句进行合理拆分、再组合,可以使译文表达更为流畅。】

3. Harbour development and other land-based activities (such as shrimp aquaculture) can lead to the destruction of mangroves, which serve as nurseries for species of fish and shellfish and provide flood protection. (Para. 8)

港口开发和其他陆地活动(如虾养殖)可能破坏给鱼类和贝类物种提供苗圃并且能够防洪的红树林。

【逆译法是指在充分分析文本内部结构的基础上,按照目标语的语言使用习惯,以逆向的方式调整源语中的语序,转换并重建出符合目标语语言使用习惯的译文。一般来讲,这种逆向的调整主要集中在状语和定语上。逆译法能降低文本阅读难度,帮助读者扫除阅读障碍。】

4. Considering these situations, the International Maritime Organization has made international efforts to address the transfer of invasive aquatic species through shipping, as illustrated by the International Convention for the Control and Management of Ships' Ballast Water which entered into force in September 2017. (Para. 9)

考虑到这些情况,国际海事组织作出了国际努力,旨在解决通过航运带来的入侵水生物种的问题,其中 2017 年 9 月生效的《船舶压载水控制和管理国际公约》就是一个例证。

【英译汉过程中,有时直译某些名词不能向读者表达具体、明确的含义。因此,汉译的时候要在这些名词后面增译一些用来表示行为、现象、属性等概念所属范畴的词,如"现象、问题、作用、效果、影响、情况、……性、……度"等,从而更加准确地补全其背后含义,使译文传达的意思更加具体和完整,同时也符合汉语中词语搭配的习惯。】

5. One of the underlying reasons is that many of the services provided by marine and coastal ecosystems-such as coastal protection, fish nursery, water purification, marine biodiversity and carbon sequestration-are not reflected in the prices of traditional goods and services on the market (and hence referred to as non-market values). (Para. 10)

其中一个根本原因是,海洋和沿海生态系统提供的许多服务——如海岸保护、鱼类苗圃、水净化、海洋生物多样性和碳封存(碳封存是指将捕获、压缩后的 CO_2 运输到指定地点进行长期封存的过程,即以捕获碳并安全存储的方式来取代直接向大气中排放 CO_2)——没有反映在市场上传统商品和服务的价格中(因此被称为非市场价值)。

【术语的翻译是专业性较强文本翻译的首要问题。为减少目标语读者的理解障碍,翻译时可采用释义法,在句子结尾处或词组后对专业词汇加上括号内具体注释。释义法既能保留原语篇的贴切程度,保证文本专业性不变;又能帮助读者填补语义空白,不给读者留下晦

涩难懂之处。】

二、参考译文

海洋和生态系统服务面临的主要压力

1. 海洋的保护与可持续利用至关重要。海洋和海洋生态系统通过充当鱼类的苗圃、提供海滩和珊瑚礁等可供潜水的娱乐区域、为海洋生物技术提供遗传物质等方式，提供中间投入（即生态系统服务），从而推动各部门的价值。

2. 尽管健康的海洋给人类带来了宝贵的好处，但人为压力的累积影响正将海洋推至人类经验之外的境地。而且，这些压力可以相互强化，对海洋生态系统造成更大的累积影响。全球海平面上升、海洋变暖、更频繁更恶劣的天气事件以及不断变化的洋流等因素将加剧过度捕捞、非法和不报告以及不管制捕捞、污染以及栖息地退化等现象所带来的负面影响。根据生物多样性和生态系统服务政府间科学政策平台的相关数据可知，直接开采生物（主要是捕鱼）对海洋生态系统产生的相对影响最大，其次是土地/海洋利用情况发生的变化，包括水产养殖和基础设施的沿海开发，以及污染问题。目前尚不清楚，在一切照旧的情况下，海洋还能继续提供维持生命的功能和生态系统服务多久。本文以下部分将概述海洋面临的主要压力。

3. 气候变化。随着时间的推移，人为的二氧化碳排放量不断增加，海洋吸收了20%～30%的二氧化碳，从而导致海洋酸化。大气中的温室气体已导致海洋温度和海平面的上升以及洋流的变化。海洋生态系统和海洋多样性受到的影响是巨大的，这已经体现在物种和栖息地的丧失、鱼类种群组成和迁徙模式的变化以及严重海洋天气事件发生的更高频率中。更频繁的严重海洋天气事件尤其影响到脆弱的低洼沿海社区，包括小岛屿发展中国家。这些人口依赖海洋的质量、稳定性和可及性，并易受其影响。

4. 污染。大多数海洋污染源来自陆地，包括工业、住宅和农业径流以及如塑料和一些固体的废物。具体而言，当来自农业化肥、畜牧业废物、污水处理和工业废水的径流进入海洋时，会向海洋释放过多的营养物质，而这些营养物质会促进海洋中有毒和有害物种的生长（即富营养化），从而改变海洋栖息地并对渔业产生负面影响。海洋污染还源自船舶污染（如压载水和热水排放）和深海采矿（如石油和天然气开采）直接排放的污染物，由此产生的污染类型包括酸化、富营养化、海洋垃圾、毒素和水下噪音。如果不加以控制，那么富营养化可能导致死亡地带的形成，这在世界各地都有发生，包括墨西哥湾、黑海和波罗的海。

5. 海洋垃圾是由不同的经济部门直接或间接产生的，例如水产养殖和渔业（如意外丢失或故意丢弃的渔具）、航运和游轮（如船舶产生的废物）、化妆品和个人护理产品、纺织和服装、零售业，以及日益增长的旅游业。手工渔业和旅游业往往是许多发展中国家的主要对外收入形式，而非法倾倒垃圾尤其影响到这两个部门，因为在垃圾堆积的海滩进行娱乐活动，人们的健康和安全都受到威胁。

6. 塑料是海洋污染的一个重要来源。许多塑料的寿命极长，将在环境中停留上百数千年,这意味着全面的影响只有在较长时间内才会显现。目前,新的可持续石化生产路线(从产品生产到使用和处置)正处于专门研究和开发中,这也许有助于遏制和阻止塑料污染和其他有害化学产品泄漏到海洋。此外,一些新的倡议如《世界经济论坛的全球塑料行动伙伴关系》也出现在人们的视野里。然而,要解决化学污染的根源问题,仍需在以下两方面加大力度:研发找到对环境危害较小的潜在替代品,以及改变当前的生产和消费观念(如以循环经济概念为基础)。

7. 过度捕捞、副渔获物和非法、不报告和不管制捕捞以及对其他自然资源的过度开发。联合国粮农组织最近的数据表明,世界海洋渔业中约35%的鱼类被归类为过度捕捞,最大可持续捕捞的鱼类约占评估鱼类总数的60%。非法、不报告和不管制捕捞活动加剧了过度捕捞,无论从经济角度还是粮食安全的角度来看,都产生了重大影响。估算非法、不报告和不管制捕捞活动的规模及其造成的诸多社会影响(如船上奴役)这一过程十分复杂,因为这取决于许多因素,如渔业的类型、地理位置和信息的可用性等。对贝类和其他生物等其他自然资源的过度开发,也对海洋环境造成了破坏。

8. 栖息地的退化。沿海和海洋中的栖息地破坏是由以下几个方面造成的:炸药捕鱼或不当拖网捕鱼等有害的捕鱼行为;农业、沿海发展和林业部门的不良土地利用行为;采矿、疏浚和锚泊等其他人类活动以及旅游业和海岸侵蚀等。例如,伐木和植被移除会导致土壤侵蚀产生沉积物;港口开发和其他陆地活动(如虾养殖)可能破坏给鱼类和贝类物种提供苗圃并且能够防洪的红树林;不良的航运行为以及浮潜、划船和水肺潜水等沿海旅游活动由于直接接触于脆弱的湿地和珊瑚礁上而破坏海洋栖息地,降低其提供的生态系统服务。

9. 入侵性外来物种。另一个对海洋环境造成严重威胁的情况是非本地海洋物种被引入不属于它们的海洋生态系统中。大多数外来海洋物种是通过跨洋商业航运船只的压舱水迅速引入不同的栖息地的。这些外来生物对环境造成了严重的影响,如原生生态系统由于原生栖息地被破坏而发生改变、一些海洋动植物的灭绝、水质的下降、物种间日益激烈的竞争和掠夺以及疾病的传播等。考虑到这些情况,国际海事组织作出了国际努力,旨在解决通过航运带来的入侵水生物种的问题,其中2017年9月生效的《船舶压载水控制和管理国际公约》就是一个例证。

10. 海洋生态系统的退化正在超出生态和经济上可持续发展的临界值。其中一个根本原因是,海洋和沿海生态系统提供的许多服务——如海岸保护、鱼类苗圃、水净化、海洋生物多样性和碳封存(碳封存是指将捕获、压缩后的二氧化碳运输到指定地点进行长期封存的过程,即以捕获碳并安全存储的方式来取代直接向大气中排放二氧化碳)——没有反映在市场上传统商品和服务的价格中(因此被称为非市场价值)。虽然经常由于缺乏科学信息而无法清楚地了解这些海洋生态系统服务及其经济价值之间的复杂联系,但正因为低估了海洋生态系统服务的价值,导致了对其保护、可持续利用和恢复的投资不足,因而失去了现在以及未来的经济增长和减贫机会。

Cultural Background Knowledge: Marine Economic Activities
海洋经济活动文化背景知识

汪洋大海占地球表面的71%，是全球经济发展重要的交通和贸易通道。海洋中蕴藏着丰富的生物资源、矿产资源、工业原料、能源和淡水资源等。海洋经济指的是开发利用海洋的各类产业及相关经济活动的总和。

自古以来，人们就意识到海洋经济的重要性。公元前500年，古希腊著名思想家狄米斯托克(Themistocles)就指出"谁控制了海洋，谁就控制了一切"，19世纪美国杰出军事理论家马汉(Mahan)在"海权论"中指出"国家兴衰的决定因素之一在于对海洋的开发和控制"，而当代海洋专家学者们一致认为"谁在海洋科技和海洋经济上强大，谁就占领了新世纪发展的制高点"。

随着海洋经济重要性的日益凸显，国际组织和世界各国对海洋经济活动的关注点从渔盐之利、舟楫之便慢慢扩大到传统海洋产业如海洋渔业、海洋交通运输业、造船业、海洋旅游业、海洋油气业等，再过渡到新兴海洋产业如海洋可再生能源、海洋生物技术、海洋工程装备、深海资源开发等。海洋新兴产业高度依赖高新技术。许多国家纷纷投入大量的人力物力，进行海洋环境监测、海洋生物医药、海洋资源勘探、海洋可再生能源等方向的技术和装备研发，旨在提高海洋经济活动的竞争力。当代世界四大海洋支柱产业为海洋石油工业、滨海旅游业、现代海洋渔业和海洋交通运输业。当代全球海洋经济发达的主要国家和地区有美国、法国、英国、加拿大、中国、日本、澳大利亚、新西兰等。

海洋经济活动在刺激经济增长、创造就业和推动创新等方面发挥着重要的作用，于全人类的未来福祉和繁荣至关重要。然而，海洋资源衰退和环境问题也引发了全球民众的担忧。相对于陆地资源开发，海洋资源开发更具复杂性。国际组织和世界各国在海洋经济发展的过程中，需要秉持绿色发展理念，提升海洋科学技术水平，增强海洋环境治理能力，保护好海洋生物多样性，以实现海洋资源可持续性使用。

Unit Eight
Ocean Art

海洋艺术

Text A

Ocean Artworks
By Ingram Ober and Marisol Rendon

Ideological and Political Education：乘新时代长风 开海洋艺术新局

海洋艺术是以海洋环境、海岛海岸、海洋动物、海洋植物、海洋人物、海洋民俗、海洋历史等海洋元素为基本素材，运用音乐、雕塑、绘画、影视等艺术形式表现海洋生活，再现典型海洋环境中各类海洋人物、塑造海洋艺术形象、诠释人与海洋的关系，海洋与人的艺术本质的艺术作品。具体包括海洋绘画、海洋音乐、海洋雕塑和海洋影视等艺术创作形式。

中国海洋艺术是我国海洋艺术家们自20世纪八九十年代共同探索、创新、发展、开拓出来的中国艺术发展新领域。海洋艺术实践切合了海洋时代发展主题，艺术家们用海洋艺术表达出中华民族海洋强国的梦想，为海洋时代讴歌，为捍卫我国海洋主权做出艺术回应，做出了具有重要时代意义的艺术成就。

21世纪以来，海洋艺术家在多地举办或参加多种艺术交流活动，包括海洋画派开派作品展、海洋雕塑学术研讨会、海洋音乐独奏音乐会、海洋题材影视剧的首映仪式和座谈会，或以海洋艺术作品参加全国和地方展览及座谈，普及海洋艺术，担当起了用海洋艺术讴歌新时代的社会责任。这些以海洋为主题，以绘画、音乐、雕塑和影视剧为艺术形式，以宣传海洋、保护海洋、提升全民海洋意识、配合国家海洋宣传活动为主旨，已成为历年世界海洋日暨全国海洋宣传日活动中富有美誉度和影响力的品牌展览。

目前，海洋艺术界汇聚了一批当代精英艺术家，他们中有从艺六七十年、从事海洋画创作三四十年的老一辈艺术家，也有年富力强、海洋艺术技艺精湛的中年艺术家。近年来，海洋艺术界中青年艺术家新锐突起，他们以扎实的基本功、崭新的艺术面貌，打开了中国海洋艺术的新天地，展示出海洋艺术家的不凡实力。海洋艺术界老中青三代艺术家，为中国海洋画的开创与发展共同努力，做出了历史贡献。

新时代海洋强国建设为中国海洋艺术带来了前所未有的历史发展机遇，风华正茂的海洋艺术家们将从"民族复兴伟业""人民立场""守正创新""讲好中国故事""弘扬正道"出发，增强文化自觉，坚定文化自信，展示中国文艺新气象，铸就中华文化新辉煌。

1. Located less than a mile from the sugar-white sand of Grayton Beach State Park lies an underwater sculpture park, The Underwater Museum of Art (UMA). The museum, which came to life through a partnership between the Cultural Arts Alliance of Walton County (CAA), the South Walton Artificial Reef Association (SWARA) and Visit South Walton, combines art, education and ecosystems-three passions of the South Walton community-in a truly unique way, creating a source of biological **replenishment** and protective marine habitats where one does not exist.

2. 2022 Development: A total of 10 sculpture designs have been selected for inclusion in the 2022 Underwater Museum of Art Deployment.

3. *The Seed and the Sea*. Artist Davide Galbiati's goal at the UMA is to educate the public on the **fragility** of marine ecosystems and the importance of preserving the balance of marine life with all of its members. To succeed in his message, he relies on the metaphor of the Seed in Nature. Nothing is more important in nature than the seed. It represents the matrix that will make it possible to have thousands of trees. For Nature what matters is the seed. It conquered territory, redrew landscapes, transformed **biodiversity**, got involved in fragile interstices, and was reborn after destruction. The information that is contained within it must be transmitted. This is the seed's mission: to transmit. The surface of the statue will allow the development of new plant and animal organisms; the sculpture itself will be transformed into a Seed, into a matrix that will allow a new life and which will have to be protected.

4. *Fibonacci Conchousness*. Artist Anthony Heinz May is known from his recent Roost and Puddle sculpture addition to the *Watersound Monarch Art Trail*. His concrete **conch** shell design for the UMA reflects site-responsive specificity of location of UMA and **existentialism** between museum-goers, natural/human-built environments and **precarious** human-nature relationships. The conch will lay on its side with **flanges** extending from a **welded** frame substrate of steel rod/wire mesh underneath layered concrete. This tested true **prototype** holds the highest structural integrity and best suitable for the natural underwater environment as well transport/install **methodologies**. Conch shells can be found along Florida Panhandle beaches while combing sands near the water's edge, however in small sizes and typically commandeered by rogue hermit crabs. The increasing **scarcity** of conches housing sea snails and **mollusks** from years of harvesting Florida waters has made them illegal for anyone to remove. Several narratives of the conch include sacred Native American histories, musical instrumentation, used in cultural recipes, as well exemplified in mathematical formula established by Leonardo Fibonacci in the 13th century. Architecture uses ratios in designs **elucidated** by the conch as a form of pure aesthetic. In **reclamation** by **algal** plant life and for organisms to

anchor, the intentions of his proposal continue expansion of his public art portfolio which includes concepts involving nature, humans and technology. Reinvestment of the organic existence of large conch shells once **omnipresent** in these tropical waters paid homage to nature, natural cycles and patterns. Remnants of conch shells wash ashore along the Northwestern Panhandle of Florida as archeological fragments depicting severity of history in travel to where it lay in the sand. The perilous trip of conch shells, affected by storms, laws of **entropy** and human intervention in natural environments, is reversed in his sculpture which depicts the conch shell as a complete and unbroken whole.

5. Hawaii-based artist Janetta Napp is creating an abstract cement sculpture, New Homes that alludes to a row of cone snail egg casings reimagined as three vertical ovule panels. In total, the three panels together will weigh approximately 2,090 lbs and will be 36″ long. This piece is titled *New Homes* because each panel will have identical 6″ diameter holes and randomly scattered .5″ diameter **indentations** approximately .5″ deep. One hole will line up across all three panels so that if a diver is facing the front of the sculpture, they could see through to the other side. These holes and indentations will create resting places and encourage marine life to settle. Each panel will be set approximately 1′ apart and will alternate front and back to provide an **asymmetrical** appearance like a row of cone snail egg **casings**. To create this artwork, Napp will use clean concrete cement reinforced with rebar and stainless-steel mesh connected with stainless steel ties to create a rough grid within, reinforcing each panel. Her fascination with the aquatic world has led her to volunteer for marine research projects with the University of Hawaii sparking her interest in the combination of science and art. By creating an artificial reef structure, she can contribute to the conservation of coral reefs.

6. From the depths of our reefs, to the soft tissue in our heads controlling our every move, the reaction-diffusion pattern expressed in *Arc of Nexus* from artist Tina Piracci exemplifies the synergy and wonder of the macrocosm we live in today. Enchanted by the uncanny echo of these patterns across various scales, the artist aims to illuminate similar **algorithmic** arrangements through the intersection of science and art. Inspired by Vitruvius and DaVinci, the divine connections found in nature influence Piracci to create and research within the context of the natural world. This imaginary portal acts as a passage between realms inviting the viewer to investigate and understand the world around them. The process of this work included drawing this diffusion pattern from personal photos gathered on diving trips around various coasts in Florida, some of which were restoration trips with the Coral Restoration Foundation. With a sister sculpture located in St. Petersburg, this doorway acts as the underwater portal to its counterpart. Doors and portals are often a theme in Piracci's work as they allude to "another realm."

Through dreams and **weird** coincidences, the artist finds this notion of a portal intriguing as a threshold between worlds. Inspired by her passed brother who visits her in dreams through misplaced mysterious doorways, these works provide the artist with the hope of another world. Through exploring the patterns found in nature, Piracci emphasizes the magical nature of the world as we can find the same structures in our eye's irises out in the cosmos. Connections like these bring life to the artist as she knows she must protect nature as it is the one thing she holds sacred.

7. *We All Live Here*. Artist Marisol Rendón believes being underwater changes our experience of gravity and time. It makes us aware of our breath. We are acutely aware we are visitors to another world bound by very different rules. This change of perspective and the mindfulness it helps to generate is a main ingredient in experiencing art within UMA, and is the foundation for fantasy, interspecies empathy, and activism. *We All Live Here* proposes a further change of perspective as we peer in through the open portals of a submerged submarine and the fish that will find refuge within its form peer back out at us. *We All Live Here* will echo the playful and ever recognizable **silhouette** of the Beatles Yellow Submarine. Its round volumetric form constructed of stainless steel and clean concrete mortar invites us to let our imaginations wander into a fantasy realm where ocean animals come to visit us in their own "submarine," or, where unlikely heroes battle the injustices of uncaring Blue Meanies. Physically the form of the submarine will be hollow with special attention paid to proper turtle ingress and egress points by strategically "removing" panels from the hull. As to keep the submarine playful and not feel as though it has been wrecked at the bottom of the ocean the piece will be elevated above the mounting **plinth** on a series of organic forms that mimic large bubbles. Further interactive possibilities will be explored through some of the **faux** mechanical details of the vessel, like the 4 periscopes, propeller, portholes and such. It is Rendón's hope that as that catchy refrain "We all live in a yellow submarine" plays in visitors' heads they remember the creatures they saw that day sharing space within that vessel.

8. The Gulf of Mexico and live music are two common chords that bring people together on 30A according to artist Vince Tatum. His sculpture, *Common Chord*, combines these two local loves by joining music with nature in perfect harmony. The sculpture is a celebration of the natural beauty that surrounds us all and brings us together. Whether it's gathering on the beach with Osprey soaring overhead, playing in the Gulf while **stingrays** glide below, or dancing like nobody's watching while the band plays into the night. The natural beauty of it all, brings us together. It's the *Common Chord*. The sculpture will be a beneficial addition to UMA as it is designed to be a thriving marine habitat that will add visual interest for divers. The hollow stingrays and sound hole

features of the guitar will make **cozy** coral-nooks for creatures to take up residence. The body of the sculpture will encourage coral growth with an ample clean cement surface and quickly become its own marine ecosystem.

9. 2021 Deployment. A total of seven sculptures have been selected for inclusion in the 2021 Underwater Museum of Art Deployment.

10. *Building Blocks* is the realization of a concept artist Zachary Long had about a year ago. He wanted to build a metal sculpture that would become the building blocks for new life to take place. He imagined a beautiful stainless structure that was bold, strong, and growing yet delicately balanced and struggling to cling to life. Zachary could see many changing angles and spaces allowing colorful sea life to be displayed and housed against the large blocks (which seem very small on an oceanic scale). These delicately balanced blocks are a reminder that life is fragile but that with some attention and time some of the most fragile and important organisms on our planet can thrive. To create any reason whatsoever to get people to care, become interested, and engaged in our incredibly diverse and amazing underwater neighbors. This piece of art will be coming from Oklahoma City in the middle of our country. Even those who do not have an ocean in their backyard can make changes and spread awareness. Zachary hopes the selection of this piece will bring up discussions in Middle America where people feel more disconnected from the problems facing our oceans. He wants to show others you can be part of a solution if you get creative with what you have, no matter where you live.

11. *Dawn Dancers*. Designer Shohini Gosh is a Denver-based artist originally from New Delhi, India. *Dawn Dancers* is a sculpture of two seahorses doing a dance. Seahorses are a flagship species, **charismatic** symbols of the coral reefs, estuaries and seaweed coastlines. The presence of Seahorses indicates the health of a reef system. *Dawn Dancers* is a silhouette of two seahorses doing the **hypnotically** romantic mating dance, looking to creating a home at the Walton beach reef forever. My **stenciled** silhouette sculpture allows the underwater tides and sea life to move through the design and gives ample space for the corals and seagrass to grow on it without hiding the shape. This design will evolve into a fascinating sculpture of seahorses with a living and growing surface of coral on them.

12. *Eco-Bug* by Florida-based artist Priscilla D'Brito allows her to introduce the Eco-Bug, the beginning of a new series of aquatic insects that will venture the underwater world. UMA would be the first to have the Eco-Bug as this concept design will be spread throughout the world. The Eco-Bug can be accompanied by creative exotic plant

sculptures as they journey the bottom of the ocean creating colonies. These insects will be **magnified** and **accentuated** to overtake the underwater world as it will contribute to be the home to diverse marine life. Their many limbs and robust segmented bodies will provide a **sturdy** base for proper installation and for the extensive function to foster marine life and coral growth throughout their bodies.

13. *From the Depths* by artist Kirk Seese evokes a childhood wonder about the mythical creatures that live in the depths of the sea. The concrete sculpture portrays a large stylized fish, something you might see as an illustration on a map to warn sailors about the **treacherous** waters ahead. With its mouth open, it offers a wide cave for smaller fish to hide in and has a 36″ diameter turtle escape hole towards the back. The artist poses these questions: Will it seem too lifelike for the real fish to trust it? Will they swim in its mouth once they realize it's not a threat? Will the sight of it scare the medium and large fish away, leaving the smaller ones in its mouth protected? Only time will tell.

14. Artist Jonathan Burger will construct an eight-foot-tall mask form looking upwards towards the light filtering down through the water as the form for his sculpture, *Hope*. The piece will only depict the front of the face, with a rough edge along the sides, leading down into a round neck form. Inspired by the broken forms of Greek and Roman sculptures, and by the work of Igor Mitoraj, the concept for this work deals with climate change, rising sea levels and the need for humanity to work together to solve these issues. As climate change affects our planet and causes sea levels to rise, many people who have previously lived on dry land above the water will find themselves flooded, much like the face of the sculpture. But this outcome is not entirely ensured, and can be slowed and hopefully prevented by the actions of our governments, corporations, and personal behaviors. The face looks up towards the light of the sun filtering down through the water in a symbol of this hope, which symbolizes we will realize the scope of our actions and work to prevent such outcomes.

15. Husband and wife team, design duo, and dive buddies Ingram Ober and Marisol Rendón will co-create *Three Wishes*. When we dive we are experiencing magic, the magic of being weightless, of traveling in a foreign environment, of shedding all but the most essential of concerns. For us, every beam of light, every stone, animals large and small seem imbibed with magic and we are lucky to experience it. *Three Wishes* is about that magic and the search for it. Bringing together the desert-like environment at the UMA site, a sublime sense of wonder derived from a change of perspective and scale, and the underlying search for magic and treasure wrapped up in each **foray** under the waves, we propose the construction of a giant scale genie's lamp. Geometrically constructed from stainless

steel rod in a 3D wire form format the surface of the lamp will be clean concrete. The imposing form of this lamp will strike a strong silhouette, at once at home within the shifting sands of Grayton Beach seafloor and strangely out of place surrounded by ocean life. The surface of the lamp will feature high relief geometric textures and indentions adding surface area and "nooks" for sea creatures to reside, the natural overhang of the lamp's form provides structure and shelter for marine life. This artwork is not however intended simply for marine life to interact with, it comes to life with the addition of a foreign element, divers and the air we carry with us. Low on the belly of the lamp will be a few small ports below which a diver, posing for a picture pretending to rub the lamp, can purge air from their octopus regulator, and that air will enter the lamp and be carried to the spout of the lamp where it emerges like a genie to grant us wishes and fill our lives with magic.

(2,580 words)

https://www.visitsouthwalton.com/underwater-museum-art/

New Words

1. replenishment [rɪˈplenɪʃmənt]

    n. the process by which something is made full or complete again 补充,充满

    Natural replenishment of this vast supply of underground water occurs very slowly.
    靠自然补充大量地下水是十分缓慢的。

2. fragility [frəˈdʒɪləti]

    n. the situation being fragile 脆弱性,易碎性

    Older drivers are more likely to be badly injured because of the fragility of their bones.
    年纪较大的司机因骨骼易碎而更可能会受重伤。

3. biodiversity [ˌbaɪəʊdaɪˈvɜːsəti]

    n. the existence of a wide variety of plant and animal species 生物多样性

    When a species goes extinct, it dramatically changes the landscape of biodiversity.
    一个物种的灭绝会极大地改变生物多样性的状况。

4. conch [kɒntʃ]

    n. the shell of a sea creature which is also called a conch 海螺壳;海螺

    His ordinary voice sounded like a whisper after the harsh note of the conch.
    听过海螺刺耳的声音后,他那平常的讲话再听起来就像是悄声细雨。

5. existentialism [ˌegzɪˈstenʃəlɪzəm]

    n. the theory that humans are free and responsible for their own actions in a world without meaning 存在主义

    Marx's new concept of nature is the natural view of existentialism of history.

马克思的新自然观是历史存在论的自然观。

6. precarious [prɪˈkeərɪəs]

   a. not safe or certain; dangerous 不稳的，不确定的

   They maintain a precarious balance only by careening wildly back and forth.

   他们仅靠胡乱向后倾或向前倾来保持不稳定的平衡。

7. plinth [plɪnθ]

   n. a block of stone on which a column or statue stands (雕像或柱子的)底座，柱基

   They are, as is typical for villages, placed on a stone plinth that sits in the water.

   端立于水中的石基上，这是典型的村庄样式。

8. weld [weld]

   v. to join pieces of metal together by heating their edges and pressing them together 焊接；熔接；锻接

   In extreme instances, the contacts may weld together.

   极端情况下，触点会焊接在一起。

9. prototype [ˈprəʊtətaɪp]

   n. the first design of sth. from which other forms are copied or developed 原型；雏形；最初形态

   Even as it's the prototype for originality, it's also something very disturbing and harmful.

   即使是独创性的原型，它也是非常令人不安和有害的东西。

10. methodology [ˌmeθəˈdɒlədʒi]

    n. a set of methods and principles used to perform a particular activity 方法；原则

    The author comments on the entire methodology on plan design and three attention points during this process.

    作者全面论述了计划制订的整套方法以及规划设计过程中的三个注意事项。

11. charismatic [ˌkærɪzˈmætɪk]

    a. having charisma 有超凡魅力的；有号召力(或感召力)的

    With her striking looks and charismatic personality, she was noticed far and wide.

    她以出众的相貌和富有魅力的个性闻名遐迩。

12. scarcity [ˈskeəsəti]

    n. there is not enough of it and it is difficult to obtain it 缺乏；不足；稀少

    In the wild, these birds store food for retrieval later during periods of food scarcity.

    在野外，这些鸟会储存食物以备日后食物短缺时再取用。

13. elucidate [ɪˈluːsɪdeɪt]

    v. to make sth. clearer by explaining it more fully 阐明；解释；说明

    Research is used to elucidate and shape the final product, price, place, promotion and related decisions.

营销研究一般用来解释和制定最终的产品、价格、分布、推广和相关决策。

14. reclamation [ˌrekləˈmeɪʃ(ə)n]

n. the process of changing land that is unsuitable for farming or building into land that can be used 开垦

Spartina has been transplanted to England and to New Zealand for land reclamation and shoreline stabilization.

大米草已经被移植到英格兰和新西兰进行土地复垦和海岸线加固。

15. algal [ˈælgəl]

a. relating to algae 海藻的

These cause the algal cells to expel the oil almost while they have generated it.

这使得海藻细胞可以在制造出油料的同时迅速排出。

16. omnipotent [ɒmˈnɪpətənt]

a. having total power;able to do anything 万能的;全能的;无所不能的

And so he imagines that God, as omnipotent, could make two plus two equal five.

所以他猜想,万能的上帝可以让二加二等于五。

17. entropy [ˈentrəpi]

n. a measurement of the energy 熵

The present study is to investigate the application of sample entropy measures.

现在的研究采用了复杂性分析中的样品熵算法。

18. indentation [ˌɪndenˈteɪʃn]

n. a cut or mark on the edge or surface of sth. 缺口;凹陷;凹痕

Using a knife, make slight indentations around the edges of the pastry.

用刀绕馅饼边沿刻出小缺口。

19. asymmetrical [ˌeɪsɪˈmetrɪk(ə)]

a. having two sides or halves that are different in shape, size, or style 不对称的

But when he held them side by side, he was troubled to see that they were slightly asymmetrical.

但是,当他把它们并排放在一起时,看上去并不十分对称,他有点困惑。

20. casing [ˈkeɪsɪŋ]

n. a covering that protects sth. 箱;盒;套;罩

Many of them have dropped their Blackberry over and over with little more than some scratches on the casing.

他们中的许多人一次又一次地把他们的黑莓手机掉在地上,而手机外壳上只有几处划痕。

21. algorithmic [ˌælgəˈrɪðəmɪk]

a. of or relating to or having the characteristics of an algorithm 算法的

How might algorithmic composition technology change the face of music?

算法作曲的出现将会如何改变音乐的面貌呢?

22. weird [wɪəd]

a. very strange or unusual and difficult to explain 奇异的;不寻常的;怪诞的

Nobody wants to break the silence, so the atmosphere in the room is pretty weird.

没有人愿意打破沉默,所以屋子里的氛围十分怪异。

23. silhouette [ˌsɪluˈet]

n. the dark outline of an object that you see against a light background 暗色轮廓

The dark silhouette of the castle ruins stood out boldly against the fading light.

城堡遗迹的黑暗轮廓在暗淡光线下显得格外突出。

24. faux [fəʊ]

a. artificial, but intended to look or seem real 人造的;仿制的

Some jeggings have faux zip-flies and pockets, while others just have an elastic waistband and no pockets.

有些牛仔打底裤会有假的拉链和裤兜设计,有些则带松紧腰带,没有裤兜。

25. stingray [ˈstɪŋreɪ]

n. a type of large flat fish with a long tail which it can use as a weapon 黄貂鱼

If you step on a stingray and get the spine inside your leg, it hurts very badly.

如果你踩在黄貂鱼的身上被刺扎到腿,后果不堪设想。

26. cozy [ˈkəʊzi]

a. of a house or room that is cozy is comfortable and warm 温暖舒适的

But a cozy-armchair sits in a corner and a bright curtain screens off their bed.

角落里摆着一张舒适的扶手椅,床前拉着鲜艳的帘子。

27. stencil [ˈstensl]

n. a thin piece of metal, plastic or card with a design cut out of it, that you put onto a surface and paint over so that the design is left on the surface; the pattern or design that is produced in this way (印文字或图案用的)模板;(用模板印的)文字或图案

Then place a stencil over the drink and sprinkle with cocoa powder.

然后在饮料上放一个模具,撒上可可粉。

28. hypnotically [hɪpˈnɒtɪkli]

ad. by means of hypnotism 催眠地

This being looks at her with a hypnotically deep and steady gaze, as if penetrating her with its eyes.

天使如催眠般深深凝望着她,仿佛要用目光将她穿透。

29. magnify [ˈmæɡnɪfaɪ]

v. to make sth. look bigger than it really is, often by using a lens or microscope 放大

If you really magnify the spectrum of the sunlight, you could identify more than 100,000 of them.

如果你真的放大太阳光的光谱,你可以分辨出多达10万条的光谱线。

30. accentuate [əkˈsentʃueɪt]

   v. to emphasize sth. or make it more noticeable 着重;强调;使突出
   Each element needs to be properly lit to accentuate the design and function artistically.
   每种元素都需要适当照亮,以便艺术地突出设计和功能。

31. sturdy [ˈstɜːdi]

   a. strong and not easily damaged 结实的;坚固的
   They would peel large sheets of bark from the tree to form lightweight yet sturdy canoes.
   他们会从树上剥下大片的树皮,做成轻便而结实的独木舟。

32. treacherous [ˈtretʃərəs]

   a. dangerous, especially when seeming safe 有潜在危险的
   But for you and your teen, the traps of adolescence are all too real and treacherous.
   但是对你们和你们的孩子来说,所有青春期的陷阱都是切切实实、危险无比的。

33. foray [ˈfɒreɪ]

   n. a short journey to find a particular thing or to visit a new place 短途(寻物);短暂访问(新地方)
   Your life is a brief foray on Earth that started one day for no reason and will inevitably end.
   你的人生对于无因而生且必将完结的地球来说,只是一位不速之客的短暂打扰而已。

Phrases and Expressions

get involved in 涉及,参与,卷入
pay homage to 向……表示敬意
line up across 一字排开
be composed of 由……组成
act as 起到……作用
ingress and egress 进进出出
take up 占据(时间或空间)
cling to 紧紧抓住
bring up 提出,引发
wrapped up 全神贯注于

Terminology

ovule 胚珠
flange 法兰

synergy 协同作用,协同增效作用
irises 鸢尾花
periscope 潜望镜
propeller (船等的)螺旋桨
porthole 舷窗

Proper Names

Grayton Beach State Park 格雷顿海滩州立公园
Underwater Museum of Art (UMA) 海洋艺术博物馆
Cultural Arts Alliance of Walton County (CAA) 沃尔顿县文化艺术联盟
South Walton Artificial Reef Association (SWARA) 南沃尔顿人工礁协会
Underwater Museum of Art Deployment 海洋艺术展列馆
Coral Restoration Foundation 珊瑚修复基金会
Beatles Yellow Submarine 甲壳虫型黄色潜水艇
Blue Meanies 蓝色坏心族

一、翻译策略/方法/技巧:转态译法

转态译法就是在翻译过程中把原文中的被动语态转换成译文中的主动语态,或把原文中的主动语态转换成译文中的被动语态。一般来说,无论是书面语中,还是口头语中,英语比汉语用更多的被动语态。英语中那些用了被动语态的句子在翻译成汉语时,我们可以根据汉语的语言习惯把被动语态转换成主动语态;汉语中用主动语态的一些句子在翻译成英语时,我们也可以根据具体情况把主动语态转换成被动语态。

例1:

原文:She was buffeted by the wind, brushed with ghostly hands by the low-reaching shrubs; she was half sobbing with terror, but nevertheless she ran onward.

译文:狂风不断拍打着她,低矮的灌木像鬼手一样拉扯着她,她吓得抽泣起来,但还是继续往前跑。

分析:译文中,按照汉语的语言习惯,将原文的主语译为译文的宾语,原文"by"后的宾语译为译文的主语,被动的"被拍打"和"被拉扯"变成主动的"拍打"和"拉扯",能够渲染恐怖急迫的氛围。

例2:

原文:Conducted by a master of ceremonies, these seventy youths rode slowly across the hippodrome, covered in gold and silver and dressed as little girls.

译文:在司仪的指挥下,这七十个年轻人慢慢骑过竞技场,他们穿金戴银,打扮得像小女孩一样。

分析:原文中有三个表示被动的动词,"conducted""covered""dressed",如果翻译成被

动句,不可避免会出现翻译腔,所以根据原文语序直接顺译成主动句,原文主语在译文中仍作主语,原文的动词直接由被动变成主动,"被金银覆盖"翻译成四字短语"穿金戴银",既符合汉语语言习惯,又真实还原了原文情境。

关于转态译法,我们要注意两点:一是英译汉时汉语译文可以采用主动语态,但不一定要有主语,而汉译英时汉语原文中的无主句译成英语后必须添加主语。另一点要注意的是有些英语被动句在翻译成汉语时仍以被动语态出现,但避开用"被"字,而根据具体的上下文用"受""为""由""获"等表示被动概念的词语代替。

例3:

原文:For alternative management strategies and regulations, potential impacts must be carefully evaluated prior to their implementation, and ineffective or risk-prone management procedures should be excluded before they cause ecological harm.

译文:对于替代管理策略和法规,实施之前必须仔细评估其潜在影响,造成生态危害之前应排除无效或容易产生风险的管理程序。

分析:无主语的用法在科技英语中最为常用,凡是说不出动作执行者(施事者),或无从说出施事者时,都可以用无主语的形式来处理。原文中,"evaluated"和"excluded"这两个被动态的动词没有施事者,所以把它翻译成主动语态的无主句,符合科技英语客观简练的特点。

总之,翻译时经常进行语态转换是十分必要的。但故意强调被动意义或主动意义时,或者转换语态后译文显得生硬时,切不可强行转换。

(摘自:冯庆华.实用翻译教程[M].上海:上海外语教育出版社,2010:98.)

二、译例与练习

Translation

1. The museum, which came to life through a partnership between the Cultural Arts Alliance of Walton County (CAA), the South Walton Artificial Reef Association (SWARA) and Visit South Walton, combines art, education and ecosystems-three passions of the South Walton community-in a truly unique way, creating a source of biological replenishment and protective marine habitats where one does not exist. (Para.1)

该博物馆在沃尔顿县文化艺术联盟、南沃尔顿人工礁石协会和南沃尔顿旅游网站的通力合作之下建立,它以一种极为独特的方式将艺术、教育和生态系统融为一体(南沃尔顿社区推重的三大价值观),从而在稀缺之地为生物补给和保护性海洋栖息创建来源地。

【这个句子的主干被一个非限定性定语从句和介词短语分隔成两个部分,最后又以一个现在分词短语来说明主干谓语动词的最终目的和效果为何。该句子在结构上还有一个特点就是并列成分和插入语较多。在翻译并列词语或词组时,除了要保证选词的准确和表达的通顺之外,还要特别注意不要引起表达上的累赘感。例句在翻译三个并列的机构名称时,用了直接翻译的方法。如果在表述上略显冗长和累赘的话,可以采取先总说后分列的译法,比如"在三家机构(沃尔顿县文化艺术联盟、南沃尔顿人工礁石协会和南沃尔顿旅游网站)的通力合作之下建立"。该句中还有一处插入语,相较于中文,英文中的插入语使用更为频繁,往往用两个逗号或破折号隔开。翻译英文中的插入语时,如果直接翻译会影响全句的表达节奏的话,可以如参考译文那样将该插入语的中文译文放到一个括号中。这样处理,既可以保证译文的准确,还可以保证译文表达的顺畅。】

2. Several narratives of the conch include sacred Native American histories, musical instrumentation, used in cultural recipes, as well exemplified in mathematical formula established by Leonardo Fibonacci in the 13th century. (Para. 4)

不论是美洲原住民的神话历史及其文化传统中演奏的种种乐器,还是13世纪列奥纳多·斐波那契建立的数学公式中的例证中,都有关于海螺的记述。

【这个句子在结构上属于典型的演绎式思维方式,也就是先给出一个相对抽象和总括的表述,原文中就是关于海螺的记述,然后再分别加以事例型列举,原文中就是这些关于海螺的记述分别出现在何种载体之上。相较于英文,中文在语言思维的使用上还是多偏重于归纳型,也就是先列举相关事例,最后再得出一个相对抽象和总括性的表述。这样看来,在翻译该句时,需要先译原文中的诸种载体,最后再译出这些载体在记载内容上的共同之处。也就是说,要从后往前翻,这样更加符合中文的语言表达思维。】

3. The perilous trip of conch shells, affected by storms, laws of entropy and human intervention in natural environments, is reversed in his sculpture which depicts the conch shell as a complete and unbroken whole. (Para. 4)

暴风骤雨,热熵定律还有人类对自然环境的干扰都会让海螺壳的旅程险象环生。然而,在这座雕塑中,这场旅行却以完全相反的方式表达出来——这枚海螺壳虽有破损但依旧保持完整。

【考虑到这个句子的主语部分包含的信息比较多,如果硬将其译成一个中文词组的话,会导致表达十分累赘,这种情况下,可以考虑使用"转句译法",将主语部分转译成一个独立的小句。然后再利用一些连词将其与后半部分连接起来。在选用连词时,需要考量前后两部分之间的逻辑关联。原文中的reversed一词表明前后部分之间乃是转折关系,因此选用"然而……却"来连接前后部分。】

4. To create this artwork, Napp will use clean concrete cement reinforced with rebar and stainless-steel mesh connected with stainless steel ties to create a rough grid within, reinforcing each panel. (Para. 5)

为了创作这件艺术品,纳普将使用以钢筋加固的干净混凝土水泥和不锈钢扎带连接的不锈钢网,在内部形成一个粗糙的网格,加固每个面板。

5. From the depths of our reefs, to the soft tissue in our heads controlling our every move, the reaction-diffusion pattern expressed in *Arc of Nexus* from artist Tina Piracci exemplifies the synergy and wonder of the macrocosm we live in today. (Para. 6)

从珊瑚礁的深处,到我们大脑中控制我们一举一动的软组织,艺术家蒂娜·皮拉齐在

《关系之弧》这件作品中表达的反应-扩散模式体现了我们今天生活的宏观世界中的协同作用和奇特之处。

6. The process of this work included drawing this diffusion pattern from personal photos gathered on diving trips around various coasts in Florida, some of which were restoration trips with the Coral Restoration Foundation. (Para. 7)

这项工作的过程包括以下步骤：根据在佛罗里达州各处海岸潜水旅行时收集的个人照片，绘制出某种扩散模式，其中一些旅行是"珊瑚恢复基金会"为了恢复珊瑚植被而组织的。

【这个句子在结构上比较齐整，中间只有一个逗号分隔。另外，句子中的宾语成分前后的修饰成分比较多。在处理这样的句子时，不宜将其译成一个同样齐整的中文句子。这样的话，语言表达效果上显得冗长繁杂，也不符合中文以小句为主的表达习惯。此种状况下，可以考虑使用"转句译法"将原文句子进行拆分。为了更加明确清晰地表达原文意思，可以将其拆分成三个部分：The process of this work included, drawing this diffusion pattern from personal photos gathered on diving trips around various coasts in Florida 以及 some of which were restoration trips with the Coral Restoration Foundation。考虑到句子表述上的节奏，译文在原文原意基础上加上一个冒号，稍作停顿后提示下文内容。在翻译最后一个非限制性定语从句时，最好重复一下所修饰的 trip 的中文对应词，这样可以增强句子之间的衔接性和关联性。此技法也是翻译非限定性定语从句时常用的。】

7. Its round volumetric form constructed of stainless steel and clean concrete mortar invites us to let our imaginations wander into a fantasy realm where ocean animals come to visit us in their own "submarine" or, where unlikely heroes battle the injustices of uncaring Blue Meanies. (Para. 7)

这件作品的圆形空间乃是由不锈钢和干净的混凝土砂浆支撑起来的。这种外形让观者兴趣盎然，引导着他们的想象力漫步到一个幻想的境界之中。在那里海洋动物乘坐着他们自己的"潜水艇"登门造访，在那里传说中的英雄为了维护正义与冷漠的蓝色坏心族英勇作战。

【这个句子结构上十分齐整，仍然需要利用"转句译法"将这个结构齐整地按照内部的逻辑关系拆分成几个相对独立的部分，然后分别进行翻译，最后按照逻辑关联和中文的表达习惯重新进行组合。需要特别注意的是，原文中的两个词语——invite 和 unlikely 在翻译过程中需要根据原文语境进行适当引申，切不可照搬词典中的解释，这样极易造成译文的表达不通顺，违反了"达"的原则。第一个词 invite 如果翻译成"邀请"的话，就会让读者十分费解，一件冰冷无生命的艺术作品如何向观者发出邀请呢？原文作者其实是想说，这件作品的艺术魅力很强，让观者不由自主地产生兴趣，进而浮想联翩。故此，译文中将其引申翻译成"兴趣盎然"。第二个词 unlikely 如果照搬"不可能的"释义的话，显然是不通顺的。这里作者其实是想说这样的"英雄人物"仅仅存在于传说中而非现实世界里。此种意义上说，是"不可能的"。这样的话，就可以将其引申译成"传说中的"。】

8. Inspired by the broken forms of Greek and Roman sculptures, and by the work of Igor Mitoraj, the concept for this work deals with climate change, rising sea levels and the need

for humanity to work together to solve these issues. (Para. 14)

受到残留的古希腊和古罗马雕塑作品的影响，还受到依格·米拉托吉作品的启发，该作品传达的创作理念关涉到气候变化、海平面上升以及人类协力解决这些问题的必要性等方面。

9. Bringing together the desert-like environment at the UMA site, a sublime sense of wonder derived from a change of perspective and scale, and the underlying search for magic and treasure wrapped up in each foray under the waves, we propose the construction of a giant scale genie's lamp. (Para. 15)

鉴于海洋艺术博物馆遗址的准沙漠环境，视角和规模的变化所带来的颇具崇高意味的惊奇感，还有波涛之下对魔法和宝藏的每一次全神贯注的水下探索，我们提议建造一个巨型的带有鳞片的神灯。

10. Low on the belly of the lamp will be a few small ports below which a diver, posing for a picture pretending to rub the lamp, can purge air from their octopus regulator, and that air will enter the lamp and be carried to the spout of the lamp where it emerges like a genie to grant us wishes and fill our lives with magic. (Para. 15)

在灯的底部下方有几个小端口，潜水员可在这些小端口下摆姿势拍照留念，看上去就像是在摩擦神灯。这些小端口主要用来排放章鱼调节器排出的空气，然后这些空气又重新进入神灯内部并被带到神灯的出口。当空气从出口冒出时，活像一位神仙实现了我们的愿望，让我们的生活发生了翻天覆地的变化。

【这个句子的翻译重点和难点在于语义的理解。该句的大意是描写某件艺术品的内部设计机理，因此在文字描述上比较细致、抽象，尤其是没有示意图的辅助下，很难马上领会作者设计上的匠心之处。从原文来看，这件艺术品的独特之处在于它利用空气流动的原理，模拟了阿拉伯神话故事中阿拉丁神灯施展魔力的场景，将神话传说中的虚拟场景实境化。只要理解了作者此番独特的创作初衷，文字组织上就得心应手了。因此，翻译的基础是准确理解原文原意，也就是严复先生所讲的翻译三原则中的"信"。最后，原文最后的 genie 一词，还是翻译成"神仙"更为符合中文读者的文化认知习惯，而不要翻译成西方文化色彩浓重的"精灵"。再就是 magic 一词，不要照搬词典中的"魔力"的释义，这样会让译文比较生硬，属于表面化翻译。既然是神仙帮助实现愿望，那定会使原来的生活发生彻底转变，这样就可以引申译为"翻天覆地的变化"。】

Exercises

1. Fill in the blanks with the proper given words, and then translate the sentences into Chinese.

　　omnipotent　　prototype　　get involved in　　precarious　　treacherous　　accentuate

1) Energy conservation has tended to _____ the situation in some cases.

2) Troops were displaying an obvious reluctance to _____ quashing demonstrations.

3) But Google's assault comes at a time when the once-_____ software giant looks vulnerable.

4) Even as it's the _____ for originality, it's also something very disturbing and harmful.
5) Washington struck me as a _____ place from which to publish such a cerebral newspaper.
6) Although a man of blood and violence, Richard was too impetuous to be either _____ or habitually cruel.

2. Translate the following sentences into Chinese.
1) The sculpture is inspired by the beautiful homes of Diatoms, which often use circle packing to generate an ornate organic geometry from silicate.
2) Doing a simple segment of the body like the rib cage is a great standalone piece but could also grow with additions to the skeleton frame over time.
3) With this project he not only wants to create an aesthetically pleasing sculpture, but he also looks to create a sustainable habitat for sea life and corals.
4) The roots, in the style of a banyan tree, would have deep grooves that provide a perfect breeding ground/habitat for fish, algae, coral, and other marine life.
5) The combination of beauty, brilliance and resilience is an enviable trait and the octopus has it all.
6) The artist believes having a sculpture of a saguaro cactus on the bottom of the ocean floor will offer a unique juxtaposition carrying multiple layers of interpretation.
7) Aspiration is a silhouette of a young girl's face looking up in the longing at the wonders of the underwater world.
8) Ascending from the mouthpiece will be a trail of bubbles, many of which will serve as a framework for the students' individual designs and allow the sculpture to function as fish habitats.
9) As a self-taught ocean artist, Gaffrey's work challenges the norms of art, pushing boundaries and constantly changing.
10) Over time the interstitial spaces will be filled in with living oysters and other marine species.

3. Translate the following sentences into English.
1) 海洋艺术或海事艺术是描绘或吸取海洋主要灵感的形式多样的艺术门类。
2) 随着文艺复兴时期的山水艺术的出现，所谓的海洋景观成为作品中的一个重要元素，但直到后来才是纯粹的海景。
3) 有了浪漫的艺术，大海和海岸被许多风景画家当作创作的题材，没有船的作品第一次变得普遍。
4) 从史前时代开始，船只和船只的表现就出现在艺术中，但海军直到中世纪后期才开始成为一种特殊的艺术类型。

5) 对于画家佛朗哥·萨拉斯·博尔克斯(Franco Salas Borquez)而言,沉默的海洋,作为一个观念,乃是永恒景观的代表。
6) 海洋绘画与其他绘画题材不同,是以海洋为主要创作动机,而且很难跟其他领域隔别开来。
7) 有一副真正的海景画,《圣朱利安和圣玛莎之旅》,但其中两页在 1904 年被烧毁,只能在黑白照片中得见真容。
8) 海洋艺术科学馆将海洋绘画、海洋地图、海洋模型等海洋文化艺术资源融于一身。
9) 除了神话和统治者的称赞之外,地图绘画也是本世纪海洋绘画的重要部分。
10) 有了浪漫的艺术,大海和海岸成为许多风景画家的题材。自此没有船的作品第一次变得普遍。

4. Choose the best paragraph translation. And then answer why you choose the first translation or the second one.

海洋艺术,尤其是海洋绘画作为一种与景观分离的艺术出现要从 17 世纪的荷兰黄金时代绘画开始。海洋绘画是荷兰黄金时代绘画中的一个重要产物,反映和记录了海外贸易和海军力量对荷兰的重要性。我们可以把这些画看成是历史的进程或是艺术家表达个人思想的途径,总之走近观看这些作品很难不被这惊涛骇浪打动。有趣的是东方绘画的场面与西方绘画略有不同,早在 12 世纪我们就可以看到中国卷轴中的海景画。而日本艺术家在借鉴中国艺术的同时利用他们精细的画笔描绘了具有东方特色的海浪翻腾。

译文一:

The emergence of marine art, especially marine painting as an art separated from the landscape, began with the Dutch Golden Age painting in the 17th century, of which marine painting is an important product, reflecting and recording overseas trade and naval power importance to the Netherlands. We can regard these paintings as the process of history or a way for artists to express their personal thoughts. In short, it is hard not to be moved by the turbulent waves when looking at these works closely. In a contrast, it is interesting that the scenes in Eastern paintings are slightly different from those in Western paintings, in which we can see seascape paintings in Chinese scrolls as early as the 12th century, while borrowing from Chinese art, Japanese artists used their delicate brushes to depict the churning waves with oriental characteristics.

译文二:

The emergence of marine art, especially marine painting as an art separated from the landscape, began with the Dutch Golden Age painting in the 17th century. Marine painting is an important product of Dutch Golden Age painting, reflecting and recording overseas trade and naval power importance to the Netherlands. We can regard these paintings as the process of history or a way for artists to express their personal thoughts. In short, it is hard not to be moved by the turbulent waves when looking at these works closely. It is interesting that the scenes in Eastern paintings are slightly different from those in Western paintings. We can see seascape paintings in Chinese scrolls as early as the 12th century. While borrowing from Chinese art, Japanese artists used their delicate brushes to depict the churning waves with oriental characteristics.

5. Translate the following passage into Chinese.

Building Blocks is the realization of a concept artist Zachary Long had about a year ago. He wanted to build a metal sculpture that would become the building blocks for new life to take place. He imagined a beautiful stainless structure that was bold, strong, and growing yet delicately balanced and struggling to cling to life. Zachary could see many changing angles and spaces allowing colorful sea life to be displayed and housed against the large blocks (which seem very small on an oceanic scale). These delicately balanced blocks are a reminder that life is fragile but that with some attention and time some of the most fragile and important organisms on our planet can thrive. To create any reason whatsoever to get people to care, become interested, and invested in our incredibly diverse and amazing underwater neighbors. This piece of art will be coming from Oklahoma City in the middle of our country. Even those who do not have an ocean in their backyard can make changes and spread awareness. Zachary hopes the selection of this piece will bring up discussions in Middle America where people feel more disconnected from the problems facing our oceans. He wants to show others you can be part of a solution if you get creative with what you have, no matter where you live.

6. Translate the following passage into English.

俄罗斯浪漫主义画家伊万·康斯坦季诺维奇·艾瓦佐夫斯基(Ivan Konstantinovich Aivazovsky, 1817—1900),世界公认的最伟大的海洋艺术大师。艾瓦佐夫斯基1817年出生在克里米亚黑海港口费奥多西亚的一个亚美尼亚家庭。艾瓦佐夫斯基是一位以浪漫主义为主的画家,作品也带有现实主义的风格,通常表现的都是规模宏大的戏剧性的场景,主要包括人类与海洋之间的浪漫斗争,以及所谓的"蓝色海洋"和城市景观。在他的早期创作中,色彩表现极为丰富,直到他艺术生涯的最后二十年,色彩表现才开始变得细腻沉稳,创作了一系列银色调的海景作品。

三、翻译家论翻译

陈定安在谈到翻译之难时认为翻译之难难在文化的不同。他进一步指出,有些东西在一种文化中是不言而喻的,而在另一种文化里却是很难理解的;同一个词或成语在不同国家中含义却往往不同。翻译绝不能只着眼于语言转换,而是透过语言表层,了解其深层内涵和文化含义。因此,译者必须深谙所要交流的民族语言与文化。至于文化差异的可译性多高,却取决于译者的文化素养和语言的功底,取决于译者的智慧和主观能动性。是移植还是替代,是意译还是注译,这都取决于总的艺术效果,使人们通过上下文猜词悟意,把握住文章的真正含义。

而翻译文化差异的关键,在于透彻的理解。只有彻底了解,才能"全文神理,融会于心","下笔抒词,自善其备",才能译出既保持异国情调,又为读者所接受的,最自然的,最接近原文的译文。现举几个较为成功的译例来加以说明。

例1:两块肩胛骨高高凸出,印成一个阳文的"八"字。

译文:And his shoulders blades stuck out so sharply, an inverted V seemed stamped there.

原文出自鲁迅先生的小说名篇《药》,描写的是久患痨病的华小栓瘦骨嶙峋的模样。杨宪益和戴乃迭两位先生在翻译时,尤其是在处理"阳文的'八'字"时,并未望文生义地翻译成 the character of "eight" cut in relief 之类的文字,而是译成 an inverted V。表面上看,似乎是"篡改"了原文之意,其实却是更为准确、地道地传达了原文之意。所谓"阳文",指的是印章或某些器物上所刻或所铸的凸出的文字或花纹。若直接译成 eight,对于不识中文的外国读者来说,简直是一头雾水。因此,两位先生便巧妙地利用阳文书写的"八"和英文字母 V 字形上的相似,进行文化移植对译,从而直观而准确地传达了原文之意。这里展现出来的不仅是两位翻译大师高超的语言技能,更有两个人学贯中西的深厚文化底蕴。

例2:Every family is said to have at least one skeleton in the cupboard.

译文:据说家家户户多多少少都有家丑。

英语语言文化中,skeleton 这个词极具宗教文化色彩,有"邪恶恐怖"以及"不可告人之秘"的意思,一般用来表示某种罪恶或者对某人或是外界发出要远离某事物的警告等。短语 skeleton in the cupboard 字面意思是"橱柜中的骷髅",其实是以形象的方式表示有意将某种罪恶或丑闻藏起来,不让外人发现。然而,对于中国读者来说,这些文化信息都十分生疏。只能借用中文中表达类似含义的短语来翻译。根据上下文,家庭中怕外人知晓的事情常被称作"家丑"。

上述两则译例启发我们,不管是在学翻译还是在教翻译的过程中,不但要学语言,还要学语言文化,包括交际模式、习俗、价值观、思维方式、处世态度,才能真正掌握交际工具——语言,并使语言成为真正交际工具。翻译也是如此。译者不能追求词句的等值,拘泥于字面意义,而要力求把字里行间的深层含义与文化的真正含义传达出来。

(摘自:陈定安.英汉比较与翻译[M].北京:中国对外翻译出版公司,2002:279,281.)

Text B

Four Underwater Art Museums

By Anonymous Author

Ideological and Political Education：用海洋艺术讲好中国故事

习近平总书记在党的二十大报告中指出："坚守中华文化立场，提炼展示中华文明的精神标识和文化精髓，加快构建中国话语和中国叙事体系，讲好中国故事、传播好中国声音，展现可信、可爱、可敬的中国形象。"作为备受关注的艺术形式，海洋艺术在讲故事、塑形象、展风貌上具有独特优势。新时代新征程，中国在世界上扮演越来越重要的角色，这呼唤着影视海洋艺术创作推出与之相匹配的精品力作，向世界讲述中国人的生活变迁和心灵世界。

用海洋艺术讲中国故事，须立足中国大地。一切艺术都来源于生活。当代中国的发展进步，为海洋艺术创作提供了最丰富的题材库，从中择取最能代表中国变革和中国精神的题材进行艺术表现，必能创造出深入人心的精品佳作。

用海洋艺术讲中国故事，要坚守中华文化立场。文艺的民族特性体现了一个民族的文化辨识度。中国传统文化中关于海洋的绘画、音乐、雕塑、戏曲和书法都直观展示出中国文艺的民族形式与样式，而蕴藏其中的，更有无比厚重的中华文化传统。直至今天依然深植于中国人的内心，不仅潜移默化地影响着中国人的思想方式和行为方式，也通过丰富多彩的艺术作品感染着其他国家和地区的人民。海洋艺术要讲好中国故事，必须从传统艺术中汲取养分。

用海洋艺术讲中国故事，应打开全球视野。中国人民历来具有深厚的天下情怀，当代海洋艺术要把目光投向世界、投向人类。各国人民的处境和命运千差万别，但对美好生活的不懈追求、为改变命运的不屈奋斗是一致的，也是最容易引起共鸣的。人类生活在同一个地球村，面对许多共同的海洋问题，关于海洋有着许多共同的价值追求，对这些共同问题和追求的艺术呈现，对人类共同价值的艺术表达，体现了中国当代文艺的胸怀与抱负。

中国共产党团结带领中国人民，百年奋斗、砥砺前行，开辟了中国特色社会主义道路，创造了经济快速发展和社会长期稳定两大奇迹，创造了人类文明新形态，使国际社会更加希望解码中国的发展道路和成功秘诀，了解中国人民的生活变迁和心灵世界。可以说，新时代为文艺繁荣发展、面向世界提供了前所未有的广阔舞台，讲好中国故事，海洋艺术大有可为。

1. Underwater **Pavilions**: Underwater Pavilions is artist Doug Aitken's large-scale **installation** produced by Parley for the Oceans and presented in partnership with The Museum of Contemporary Art, Los Angeles (MOCA). The work consists of three temporary underwater sculptures, floating beneath the ocean's surface that swimmers, **snorkelers**, and scuba divers swim through and experience.

2. **Geometric** in design, the sculptures create underwater spaces **synthesizing** art and science as they are constructed with carefully researched materials and will be moored to the ocean floor. Part of each structure is mirrored to reflect the underwater **seascape** and create a kaleidoscopic observatory for the viewer, while other surfaces are rough and rock-like. The environments created by the sculptures will constantly change with the currents and the time of day, focusing the attention of the viewer on the rhythm of the ocean and its life cycles.

3. Underwater Pavilions engages the living ocean ecosystem as the viewer swims into and through the sculptures, which create reflective abstractions. The work operates as an observatory for ocean life, creating a variety of **converging** perceptual encounters. The sculptures will continuously change due to the natural and manmade conditions of the ocean, creating a living presence and unique relationship with the viewer. Both aesthetic and scientific, Underwater Pavilions puts the local marine environment and the global challenges around ocean conversation in dialogue with the history of art, inviting the viewer to write a contemporary narrative of the ocean and to participate in its protection.

4. Doug Aitken (b. 1968) is an American artist and filmmaker whose work explores every medium, from sculpture, film, and installation to architectural intervention. His work has been featured in exhibitions around the world at institutions including the Whitney Museum of American Art, The Museum of Modern Art, Vienna Secession, the Serpentine Gallery in London, and the Centre Georges Pompidou in Paris. Aitken earned the International Prize at the Venice Biennale in 1999 for the installation electric earth. He also received the 2012 Nam June Paik Art Center Prize and the 2013 Smithsonian American Ingenuity Award: Visual Arts. He lives and works in Los Angeles, California.

5. Parley for the Oceans: Parley for the Oceans is the New York-based organization and global network where creators, thinkers, and leaders raise awareness for the beauty and fragility of our oceans and **collaborate** to end their destruction. Founded in 2012 by Cyrill Gutsch, Parley believes the power for change lies in the hands of the consumer, and the duty to **empower** them lies in the hands of the creative industries. Artists, musicians, actors, filmmakers, designers, journalists, architects, inventors, and scientists have the tools to mold reality and reshape our

future on this planet. With a focus on ocean plastic pollution, overfishing and climate change, and deep sea exploration, Parley **implements** comprehensive strategies to ensure we are fast enough to meet the ultimate deadline, before we lose a treasure we have only just started to explore and still don't fully understand: the fantastic blue universe beneath us—the oceans.

6. The Museum of Contemporary Art, Los Angles (MOCA): Founded in 1979, MOCA's vision is to be the defining museum of contemporary art. In a relatively short period of time, MOCA has achieved astonishing growth with three Los Angeles locations of architectural renown; a world-class permanent collection of more than 6,800 objects, international in scope and among the finest in the world; **hallmark** education programs that are widely **emulated**; award-winning publications that present original scholarship; **groundbreaking monographic**, touring, and thematic exhibitions of international **repute** that survey the art of our time; and cutting-edge engagement with modes of new media production. MOCA is a not-for-profit institution that relies on a variety of funding sources for its activities.

7. Museum of Underwater Art: Townsville, Queensland Australia: The Museum of Underwater Art is not your traditional art gallery—it's a museum located in the Great Barrier Reef Marine Park in Townsville, Queensland, Australia, and features globally known artists and their sculptures. The focal point of the art installations is the conservation, education, and restoration of the Great Barrier Reef.

8. The unusual underwater gallery was created by Jason de Caires Taylor, who is the first to utilize the ocean as his exhibit area. Taylor first began by displaying his pieces underwater in Mexico and Spain at the MUSO and Museo Atlantico, but quickly shifted his focus to the Great Barrier Reef due to its need for conservation and the beauty it beholds underwater. Taylor also created the Museum of Underwater Art because he wants people to know that the reef is still in excellent condition, and has some of the most refined corals on the planet.

9. Spotting an **illuminated** sculpture rising from the ocean is a rare occurrence, but the Museum of Underwater Art flawlessly created this unique sight. In 2019, the Museum of Underwater Art opened its first art installation, Ocean **Siren**. The sculpture appears as if a woman is emerging out of the ocean, standing tall with a shell **clutched** in her hand. The statue was modeled and based on Takoda Johnson, a 12-year-old girl a part of the native Wulgurukaba people. The Wulgurukaba are the landowners of the Strand, Townsville, where the statue resides. At dusk, the Ocean Siren's true colors come about.

10. The art piece is fully illuminated with color-changing LED lights that change based on the

temperature of the sea. The data used to track the temperature comes directly from the weather station on the Great Barrier Reef. The **captivating** statue is an indication warning that our sea temperatures are continuing to dangerously rise. The mission of the *Ocean Siren* is to inspire positive change for our planet's environment and to begin conservation action on the reef. The most promising view of the Ocean Siren can be seen from the beachfront **promenade** at the Strand.

11. The *Ocean Siren* was created by Jason deCaires Taylor, the creator of the museum. Also cultivated by Taylor are the sculptures of 20 marine biologist students located underwater at John Brewer Reef, in Townsville. They were installed in 2020 and are part of the largest art piece at the Museum of Underwater Art. A massive stainless steel structure houses the sculptures of the underwater scholars, who have been given the name "Reef Guardians." The sculptures have been given this title because their purpose is to be a suitable environment for new marine life to inhabit. It also encourages species to populate the Great Barrier Reef, which will help restore it. Scuba diving is encouraged to capture the entirety of the Coral Greenhouse. The sculptures can also be spotted by snorkeling, giving a distant view of the art piece and the fish who now freely occupy the reef guardians.

(1,109 words)

https://www.underwaterpavilions.com/#underwaterpavilions

New Words

1. pavilion [pəˈvɪliən]

    n. a building that is meant to be more beautiful than useful, built as a shelter in a park or used for concerts and dances（公园中的）亭，阁；（音乐会、舞会的）华美建筑

    A roof-top observation desk will provide stunning views of the lake and near-by Italian pavilion.

    站在屋顶观景台可欣赏到湖泊和附近的意大利凉亭的壮丽景色。

2. installation [ˌɪnstəˈleɪʃ(ə)n]

    n. the act of fixing equipment or furniture in position so that it can be used 安装；设置

    The major components of our energy systems, such as fuel production, refining, electrical generation and distribution, are costly installation that have lengthy life span.

    我们能源系统的主要组成部分，如燃料生产、提炼、发电和分配，用的是昂贵的、使用寿命较长的装置。

3. snorkeler [ˈsnɔːklə]

    n. the practitioner of swimming on or through a body of water while equipped with a diving mask, a shaped breathing tube called a snorkel, and usually swimfins 浮潜者

Smack in the middle of the blazing Chihuahuan Desert, a snorkeler scans the bottom of a spring-fed pool for aquatic life.

在灼热的奇瓦瓦沙漠中心，一名潜水员在生机盎然的池塘下搜索水生物。

4. geometric [ˌdʒiːəˈmetrɪk]

   a. of or like the lines, shapes, etc. used in geometry, especially because of having regular shapes or lines 几何(学)的；(似)几何图形的

   He began constructing objects that have circles, squares and other geometric shapes.

   他开始建造具有圆形、方形和其他几何形状的物体。

5. synthetize [ˈsɪnθɪsaɪz]

   v. combine (a number of things) into a coherent whole 综合，融合

   It is the comprehensive art that synthetize music, drama, verse, dance, the stage art, etc.

   它是融音乐、戏剧、诗歌、舞蹈、舞台美术等为一体的综合性艺术。

6. seascape [ˈsiːskeɪp]

   n. a picture or view of the sea 海景；海景画

   The pale blue walls of this room hang the black lacquer framed seascape.

   这个房间淡蓝色的墙壁上挂着玄色漆框的海景图。

7. converge [kənˈvɜːdʒ]

   v. to move towards a place from different directions and meet 汇集；聚集；集中

   The adult ones converge on remaining portions of healthy coral and feed hungrily.

   成年珊瑚聚集在健康珊瑚的剩余部分上，狼吞虎咽地进食。

8. collaborate [kəˈlæbəreɪt]

   v. to work together with sb. in order to produce or achieve sth. 合作；协作

   The ability to work well with others and collaborate on projects is a sought-after ability in nearly every position.

   几乎在每个职位上，与他人良好的合作和项目协作能力都是很受欢迎的。

9. empower [ɪmˈpaʊə(r)]

   v. to give sb. the power or authority to do sth. 授权；给(某人)……权力

   You must delegate effectively and empower people to carry out their roles with your full support.

   你们必须有效地下放权力，使人们能够在你们的全力支持下履行他们的职责。

10. implement [ˈɪmplɪment]

    v. to make sth. that has been officially decided start to happen or be used 使生效；贯彻；执行；实施

    The government promised to implement a new system to control financial loan institutions.

    政府许诺要实施新的制度来控制金融贷款机构。

11. hallmark [ˈhɔːlmɑːk]

n. a feature or quality that is typical of sb.or sth. 特征；特点

Chopsticks are highly praised by Westerners as a hallmark of ancient oriental civilization.

筷子被西方人高度赞扬为古代东方文明的标志。

12. emulate ['emjuleɪt]

    v. to try to do sth.as well as sb.else because you admire them 努力赶上；同……竞争

    Too many companies try to emulate a leader or compete head on in a market.

    太多的企业试图赶超行业的龙头或与市场上的行业龙头相竞争。

13. groundbreaking ['graʊndbreɪkɪŋ]

    a. breaking new ground, innovative, pioneering 开辟新领域的，具有革新意义的

    The research, published in the journal *Nature*, was hailed as groundbreaking by fellow researchers.

    这份发表在《自然》杂志上的研究成果被其他同领域研究者赞誉为具有突破性意义。

14. monographic [ˌmɒnəˈɡræfɪk]

    a. concerned with a single subject or aspect of a subject 专题的

    This monographic film is a real reflection of the living conditions of people in the western region.

    这部专题片真实地反映了西部人民的生存情况。

15. repute [rɪˈpjuːt]

    n. the opinion that people have of sb.or sth. 名誉；名声

    In those days the lion was much admired in heraldry, and more than one king sought to link himself with its repute.

    在那些日子里，狮子在纹章学中受到极大的尊敬，不止一个国王试图把自己和它的名声联系起来。

16. illuminate [ɪˈluːmɪneɪt]

    v. to shine light on sth. 照明；照亮；照射

    Thanks to fiber optics, it is now possible to illuminate many of the body's remotest organs and darkest orifices.

    多亏有了光纤，如今才能够照见人体中许多最微小的器官和最暗的腔体。

17. siren [ˈsaɪrən]

    n. any of a group of sea creatures that were part woman and part bird, or part woman and part fish, whose beautiful singing made sailors sail towards them into rocks or dangerous waters 塞壬（古希腊神话中半人半鸟或半人半鱼的女海妖，以美妙歌声诱使航海者驶向礁石或进入危险水域）

    The television stayed on day and night, singing like a Siren in the crowded house.

    在拥挤的餐厅中，电视机昼夜不断地播放如海中女妖般的歌声。

18. clutch [klʌtʃ]

    v. to hold sb.or sth.tightly 紧握；抱紧；抓紧

Mistress Mary did not mean to put out her hand and clutch his sleeve but she did it.

玛丽小姐并不想伸出手来抓住他的衣袖，但她还是这么做了。

19. captivate [ˈkæptɪveɪt]

    v. to keep sb.'s attention by being interesting, attractive, etc. 迷住；使着迷

    I would captivate them with money before telling them that we needed to deforest their land.

    我会先用钱迷惑住他们，然后再告诉他们我们要采伐森林。

20. promenade [ˌprɒməˈnɑːd]

    n. a public place for walking, usually a wide path beside the sea 公共散步场所；（常指）滨海步行大道

    It is reached via a covered promenade whose wooden walls are perforated by 89 flower-shaped Windows.

    通过隐蔽的通道可到达，通道的木墙上开了 89 个花型窗。

Phrases and Expressions

in partnership with 与……建立合作关系
due to 因为，由于
in dialogue with 与……建立关联
in excellent condition 处于良好的状况之下
at dusk 傍晚，黄昏时分
come about 现身，出现

Terminology

architectural intervention 建筑学干预
Wulgurukaba 乌尔古鲁卡巴人
LED (Light Emitting Diode) 发光二极管

Proper Names

Underwater Pavilions 海洋展馆
Parley for the Oceans 海洋谈判，一家创立于 2012 年的海洋环境保护的非营利性公益组织
The Museum of Contemporary Art, Los Angeles (MOCA) 洛杉矶现代艺术博物馆
Whitney Museum of American Art 惠特尼美国艺术博物馆
The Museum of Modern Art 现代艺术博物馆
The Vienna Secession 维也纳现代艺术先锋运动
The Serpentine Gallery in London 伦敦蛇形画廊
The Centre Georges Pompidou in Paris 巴黎乔治·蓬皮杜中心
The Venice Biennale 威尼斯双年展

The 2012 Nam June Paik Art Center Prize 2012年度白南准艺术中心奖
The 2013 Smithsonian American Ingenuity Award 2013年度史密森尼美国匠心奖
Museum of Underwater Art: Townsville, Queensland Australia 澳大利亚昆士兰汤森威尔海洋艺术博物馆
Great Barrier Reef Marine Park in Townsville, Queensland, Australia 澳大利亚昆士兰汤森威尔大堡礁海洋公园
Museo Atlantico 大西洋博物馆
John Brewer Reef, in Townsville, Queensland Australia 澳大利亚昆士兰汤森威尔约翰·布鲁尔礁
The Coral Greenhouse 珊瑚温室

一、翻译简析

1. Geometric in design, the sculptures create underwater spaces synthesizing art and science as they are constructed with carefully researched materials and will be moored to the ocean floor. (Para.2)

　　这些雕塑在设计上采用了几何图形,因为是用经过仔细研究的材料建造而成的,从而创造了一个集艺术和科学于一身的水下空间,它们也将被安置在海底。

　　【这个句子的结构并不复杂,整体上来看,就是由一个连词 as 连缀成的复合句,表示的是因果关系。在翻译因果关系的英文句子时,需要注意的是中英文在表述因果关系时的差别。中文比较多的是先因后果,而英文相对比较灵活。因此在翻译这个句子时,可以考虑将后面的 as 引导的原因状语置前,这样在表述上更加符合中文的习惯。】

2. The environments created by the sculptures will constantly change with the currents and the time of day, focusing the attention of the viewer on the rhythm of the ocean and its life cycles. (Para.2)

　　雕塑创造的环境会随着洋流和时间的变化而不断变化,从而将观众的注意力集中在海洋的节奏及其生命周期的变化上。

　　【这是个带有分词做伴随状语的句子。翻译时采用直译法进行翻译。先翻译由过去分词修饰的主语,译为"……创造的……",伴随状语 focusing......on......译为"将观众的注意力集中在……"。】

3. Both aesthetic and scientific, Underwater Pavilions puts the local marine environment and the global challenges around ocean conversation in dialogue with the history of art, inviting the viewer to write a contemporary narrative of the ocean and to participate in its protection. (Para.3)

　　海洋展馆将当地海洋环境,还有围绕着海洋问题产生的全球性挑战,与艺术史关联了起来,既顾及了美学,又顾及了科学,从而引发了观众书写当代海洋故事并参与海洋保护的兴致。

　　【这个句子在结构上十分紧密,翻译上的难点主要集中在结构的重构上。其实,从句子的语义上看,both aesthetic and scientific 是一个结论性的叙述,是在为海洋展馆的特殊举动做一个赞赏性评价。中文在句子逻辑顺序的安排上,比较倾向于归纳型,也就是说先列举事实和依

据,然后做出结论。相比之下,英文则是倾向于演绎型,也就是先给出结论,再列举事实或依据。这种语言逻辑顺序安排上的基本差异要求译者在很多翻译实境中需要进行句子的重构。像该句将"既顾及了美学,又顾及了科学"置于其后,就更符合中文表达的习惯和心理。】

4. With a focus on ocean plastic pollution, overfishing and climate change, and deep sea exploration, Parley implements comprehensive strategies to ensure we are fast enough to meet the ultimate deadline, before we lose a treasure we have only just started to explore and still don't fully understand: the fantastic blue universe beneath us—the oceans. (Para.5)

该组织专注于海洋塑料污染、过度捕捞和气候变化以及深海探索,并不断实施综合战略以确保我们不至于错过解决问题的最后时机,不至于失去我们脚下这片奇幻的蓝色世界。当前,我们才刚刚迈出探索的步伐,对这片世界依旧所知甚少。

【这个句子结构上比较复杂一些,其中既有常见的从句和介词短语,也有比较少用的同位语。内部组成成分如此繁杂的英语句子,在翻译时的主要难点在于诸成分间的相互调整。首先,可以将 with 介词短语升格成一个独立句子,与主干中的 implement 并列做主语 Parley 的谓语动词,这样翻译可以避免句首无主语而直接陈述导致的突兀之感。另一个较大的调整就是将冒号后的同位语调整到 lose 后面,与 treasure 合二为一。这样处理可以避免先翻译 treasure 而造成的语义理解上的困难。另外,该句子的翻译过程中,还需注意两处语词上的处理。首先是 fast enough to meet the ultimate deadline,直译乃是"加快行动以免错过最后期限"。如使用"最后期限"之类的表达,会让读者不知所云。这里作者实际上说的是解决海洋问题的"最后时机"。再就是后文中的 before。表面上看,这是个肯定意义的连词,但实际上也隐含着否定的含义。因此,译文中采用了否定的译法,一是与前文保持叙述上的一致,再就是突出后果之严重性,以引起读者的注意。】

5. The Museum of Underwater Art is not your traditional art gallery—it's a museum located in the Great Barrier Reef Marine Park in Townsville, Queensland, Australia, and features globally known artists and their sculptures. (Para.7)

海洋艺术博物馆不是一间传统艺术画廊,它位于澳大利亚昆士兰州汤斯维尔大堡礁海洋公园内,拥有全球知名艺术家及其雕塑作品。

二、参考译文

四家海洋艺术博物馆

1. 海洋展馆:海洋展馆是艺术家道格·艾特肯的大型作品,由"海洋谈判"组织制作,并与洛杉矶当代艺术博物馆(MOCA)合作展出。该作品由三个水下临时雕塑组成,浮于海面之下,游泳者、浮潜者和水肺潜水员可以穿梭其间并亲身体验。

2. 这些雕塑在设计上采用了几何图形,因为是用经过仔细研究的材料建造而成的,从而创造了一个集艺术和科学于一身的水下空间,它们也将被安置在海底。每个结构的某个部分都安装有镜面,以反映水下海景,并为观众创造一个万花筒般的观景台,其他部分则如岩石般粗糙不平。不仅如此,雕塑创造的环境会随着洋流和时间的变化而不断变化,从而将观众的注意力集中在海洋的节奏及其生命周期的变化上。

3. 海洋展馆将鲜活的海洋生态系统融为一体,当观众游入并穿过雕塑时,会创造出一种反射性的镜像。该作品作为观察海洋生物的地点,创造了多种多样的视觉碰撞。由于海洋的自然和人为条件的不同,雕塑将不断随之变化,创造出生动的景象并与观众建立独特的关系。海洋展馆将当地海洋环境,还有围绕着海洋问题产生的全球性挑战,与艺术史关联了起来,既顾及了美学,又顾及了科学,从而引发了观众书写当代海洋故事并参与海洋保护的兴致。

4. 道格·艾特肯(生于1968年)是一位美国艺术家和电影制作人,他的作品几乎涉猎到了所有艺术媒介,从雕塑、电影、装置艺术到建筑学干预,无所不包。他的作品曾在世界各地的机构中展出,包括惠特尼美国艺术博物馆、现代艺术博物馆、维也纳现代艺术先锋运动、伦敦蛇形画廊和巴黎乔治·蓬皮杜中心。艾特肯在1999年的威尼斯双年展上凭借装置艺术作品《带电的地球》获得国际大奖。他还获得了2012年白南准艺术中心奖和2013年史密森尼美国视觉艺术创意奖。目前他在加利福尼亚州洛杉矶生活和工作。

5. "海洋谈判":它是一个总部位于纽约的全球性组织,在全世界很多地方都有分支机构。在这里那些创新者、思想家和领导者们致力于唤醒人们对美丽但脆弱的海洋的保护意识,并协力阻止人们对海洋的破坏活动。该组织由西里尔·古切于2012年创立,他坚信变革的力量掌握在消费者手中,而那些创意产业则肩负着赋予消费者权力的责任。艺术家、音乐家、演员、电影制作人、设计师、记者、建筑师、发明家和科学家能塑造现实和重塑人类在这个星球上的未来。该组织专注于海洋塑料污染、过度捕捞和气候变化以及深海探索,并不断实施综合战略以确保我们不至于错过解决问题的最后时机,不至于失去我们脚下这片奇幻的蓝色世界。当前,我们才刚刚迈出探索的步伐,对这片世界依旧所知甚少。

6. 洛杉矶当代艺术博物馆(MOCA):洛杉矶当代艺术博物馆成立于1979年,立志成为当代艺术的标志性博物馆。不出几年,洛杉矶当代艺术博物馆取得了惊人的发展——在洛杉矶拥有三栋享有盛誉的建筑,拥有超过6 800件世界级永久收藏品,视野国际化,业内名列前茅,开设广受效仿的标志性教育项目,设立展现原创作品的出版奖项,组织开创性的国际知名专题展览、巡回展览和专题展览,以检视当下的艺术发展,还时刻关注新媒体制作模式的最新动态。洛杉矶当代艺术博物馆是一家非营利机构,其活动经费主要依赖各种捐助。

7. 海洋艺术博物馆——澳大利亚昆士兰州汤斯维尔:海洋艺术博物馆不是一间传统艺术画廊,它位于澳大利亚昆士兰州汤斯维尔大堡礁海洋公园内,拥有全球知名艺术家及其雕塑作品。其艺术作品关注的是大堡礁的保护、知识普及和生态恢复。

8. 这间独特的海洋画廊由詹森·德·卡里尔·泰勒创建,他是利用海洋作为展览区域的第一人。泰勒首先在墨西哥的墨索和西班牙的大西洋博物馆展示他的海洋艺术作品,但出于生态保护和维护自然景致的考量,他很快将注意力转移到大堡礁上。泰勒还创建了海洋艺术博物馆,因为他想让人们知道珊瑚礁仍然处于良好状态,这座博物馆里还拥有地球上几株最精致的珊瑚。

9. 发现从海中升起的发光雕塑是罕见的,但海洋艺术博物馆完美地创造了这种独特的景象。2019年,海洋艺术博物馆开放了其首个艺术作品《海妖》。这座雕塑看起来就像一个女人从海里浮出水面,高高地站着,手里拿着一个贝壳。它是以高田·约翰逊为原型制作的,高田·约翰逊是一名12岁的女孩,是当地的乌尔古鲁卡巴人。乌尔古鲁卡巴人是汤斯维尔斯特兰德的土著,而该地也是这座雕塑的安置地。只有到了黄昏时分,海妖才展露出本真面貌。

10. 这件艺术品完全沐浴在变色的LED灯光之下,LED灯会随着海水温度而变化。用于追踪温度的数据均由大堡礁的气象站提供。这座看似迷人的雕塑其实是一个警示牌,时刻警示我们海水温度正在持续危险地上升。《海妖》肩负的使命是为地球环境带来积极的变化,并启发人们为保护珊瑚礁而行动起来。从斯特兰德的海滨长廊可以看到《海妖》给人类带来的希望之光。

11. 《海妖》是由博物馆的创建人詹森·德·卡里尔·泰勒创作的。他还指导了汤斯维尔约翰·布鲁尔礁水下的20名海洋生物学学生的雕塑作品。它们于2020年正式下水,是该馆最大的艺术作品群中的一部分。一个巨大的不锈钢结构中安放着海洋学者们的雕塑,他们被称为"珊瑚礁的守护者"。之所以给这些雕塑起这个名字,是因为它们的目的是为新的海洋生物提供适宜的栖息环境。它还为物种在大堡礁的繁衍生息提供便利,这将有助于大堡礁的生态恢复。潜水是将整个珊瑚温室一览无余的最佳方式。另外,浮潜也可以发现这座雕塑,从远处可以看到艺术品的全貌,还可以看到在雕塑间自由穿梭的鱼,它们时刻守护着珊瑚礁。

Cultural Background Knowledge: Ocean Art
海洋艺术文化背景知识

海景画(Seascape)一词和海景画的形式起源于风景画(Landscape),是海洋、海滩、海岸线、海上船只、航海图像等这些元素构成的艺术作品,海景画也是海洋艺术的一种。在英国,海景画最早使用于环境规划和土地使用,区域包括毗邻的陆地、海岸线和一些地区的海域,通过海陆间的能见度和沿海的地形来进行评估。长期以来,海洋一直是历史上许多著名画家关注的焦点和灵感的来源,就像其他的生物一样,海洋似乎有它自己的"个性",海面时而平静神秘,时而又波涛汹涌且充满诱惑。许多伟大的艺术家都曾尝试描绘海洋,以及海洋是如何与人类的行为相互影响的。对于许多画家来说,用画笔捕捉大海的独特魅力并非易事,但也有技术熟练的画家摸索出了描绘大海的方式。

"水乃万物之本原,是一切元素的起源"。古希腊哲人泰勒斯在寻找世界起源的真相时得出如此结论。希腊人与海的关系一直是决定性的,因为他们的存在、身份和历史都与这环拥着大陆和岛屿的液态元素紧密相关。希腊艺术家从崇高的海洋中汲取灵感,将海洋作为人类潜意识的物理模拟物和希腊民族心理不可分割的一部分,并将深切感悟到人类社会的基本真理转化为令人敬畏的视觉图像。

光对画面的影响是至关重要的，特别是在海景画中。大约在 1813 年，欧洲的艺术家们离开四面是墙的室内工作室，开始在户外绘画和创作风景画，这种绘画风格被称作"外光派绘画"。外光派绘画方法最早是在英国艺术家约翰·康斯特·布尔的带领下流行于欧洲，从 1860 年开始为印象画派出现打下了基础。在 19 世纪 70 年代，随着便携画架和管装颜料的生产与普及，"外光派绘画"方式越来越流行。

　　有"光之子"之称的英国画家约瑟夫·马洛德·威廉·透纳对风景和海景画的热爱来源于色彩的巧妙利用和绘画技巧上的不断尝试，并且将色彩作为其作品的重要组成部分。他于 1839 年创作的油画作品《被拖去解体的战舰无畏号》与康斯坦丁诺斯·沃拉纳基斯的《海景》略有相似。画面中描绘了"无畏号"被解体前的最后一次航行，战舰在画面左侧小小汽船身后，向朦胧的远方驶去。远处的天空、夕阳和云间的光辉渲染在海平面上，与战舰形成鲜明的对比，仿佛在对"无畏号"的谢幕致敬。作为印象主义画派先驱，透纳的绘画创作脱离了对具象事物的描绘，更关注于对光线和色调呈现出的氛围意境。

Unit Nine
Fishery Culture

渔业文化

Text A

Eliminating Government Support to Illegal, Unreported and Unregulated Fishing

By Claire Delpeuch, Emanuela Migliaccio and Will Symes

Ideological and Political Education：传承渔业文化 建设现代化渔业强国

实施乡村振兴和文化强国战略时代背景下,讲好休闲渔业的"中国故事"。党的十九大作出了中国特色社会主义进入新时代的重大政治论断,确立了新时代的指导思想和基本方略,进一步指明了党和国家事业的前进方向。中央农村工作会议对实施乡村振兴战略做了系统部署,对渔业渔政工作提出了新的要求。在决胜全面建成小康社会和建设现代化国家的新征程中,要系统谋划新时代渔业发展思路,制定时间表、路线图、施工单,锲而不舍,接续奋斗,只争朝夕,扎扎实实走好新时代的长征路,加快建设现代化渔业强国。

十九大明确提出实施乡村振兴战略,并将其作为七大战略之一写入党章,促进农业农村发展提到了前所未有的高度。乡村振兴战略的实施对渔业发展带来了重大战略机遇。渔业要按照20字总要求,全面对接、系统谋划、统筹推进。要抓住产业兴旺这个着力点,不断发展渔业生产力,做大做强渔业产业。按照生态宜居的新要求,渔业不仅要为人民群众提供丰富的水产品,还要提供优美的水生态环境。促进乡风文明,要求注重传承渔文化,不断满足城乡居民日益增加的休闲、度假和健康需求,提供更好的休闲文化生活。推进治理有效,就要创新渔业渔区管理,依法治渔,维护渔区稳定。实现生活富裕,就要不断提高渔业发展质量效益,实现渔民增收致富,提高渔民群众获得感、幸福感。坚持人与自然和谐共生,是新时代的基本方略之一。

十九大要求,坚持节约资源和保护环境的基本国策,建设美丽中国。要在长江流域水生生物保护区实施全面禁捕,加大近海滩涂养殖污染治理力度,继续实施重要生态系统保护和修复工程,健全耕地、草原、森林、河流、湖泊休养生息治理,建立山水、林田、湖草系统治理的思维,要切实践行"两山"理念,保护绿水青山,还渔业美丽本色,再塑"日暮紫鳞跃,圆波处处生"的景象,为人民提供更多清新美丽的渔业生态环境产品,使渔业成为渲染美丽中国的一道亮丽风景线,给党中央、给习总书记交上满意答卷。

1. Illegal, unreported and unregulated (IUU) fishing and fishing-related activities in support of IUU fishing continue to seriously **undermine** and threaten the sustainability of fisheries, the livelihoods of coastal communities and the ocean economy. Largely unseen, IUU fishing complicates the stock **assessments** that underpin evidence-based fisheries management, while causing law-abiding fishers to face unfair competition over resources and in markets. Furthermore, IUU fishing results in losses of important tax revenue. IUU fishing can also threaten food security, for example by diverting fish away from local markets in regions and communities that depend on local seafood and may pose food safety risks due to the mislabeling of illegal products. It is also sometimes associated with conflicts over scarce resources and **disputed** waters; transnational criminal activities; and the exploitation of forced labour. Whilst the negative effects of IUU fishing are well understood, it is nevertheless the case that IUU fishing can benefit from government support for fisheries. This report assesses how to avoid this.

2. Improving management as well as monitoring, control, and **surveillance** (MCS), **sanctioning** IUU fishing activities and reducing the expected net benefits of IUU fishing more generally are key to its eradication. Preventing support from government budgets **inadvertently** benefitting IUU fishing is another important lever to reduce its profitability. For this reason, eliminating support to IUU fishing has been the subject of a number of international commitments on domestic and international reforms, including in Sustainable Goal (SDG) 14, which includes a call for "eliminating **subsidies** that contribute to IUU fishing and refraining from introducing new such subsidies". The goal for achieving SDG Target 14.6 was 2020. Negotiations at the World Trade Organization (WTO) to phase out subsidies that contribute to IUU fishing, as well as subsidies to unsustainable fishing more generally, are ongoing. Nevertheless, countries are addressing such subsidies through reforms of their domestic law and regulations. Disciplines have also been included in recent trade agreements.

3. This report considers how to avoid supporting IUU fishing in a broad sense, without limiting the concept of IUU fishing to one particular definition. The report reviews how OECD Members and Partner economies engaging in the OECD Fisheries Committee (COFI) can ensure that government support that they may provide to fisheries does not contribute to IUU fishing, on the basis of a survey conducted in 2021, and suggests avenues to more effectively close public budgets to IUU fishing.

4. Legal and regulatory systems, as well as fisheries support, vary significantly globally. As a result, ways of excluding IUU fishing from government support also vary. Three main approaches were seen from the review. Some economies use specific mechanisms to deny or

withdraw support in relation to IUU fishing, which are set in **overarching** legislation and regulation. Others use specific mechanisms that are set in individual support programmes' agreements or contracts. And others rely on the withdrawal of fishing authorizations, which, combined with the need for an authorization to be **eligible** for support, may, **implicitly**, suspend support **eligibility**.

5. Several challenges are common to the three approaches. First, by nature, IUU fishing is hard to observe and document. Establishing links between IUU fishing activities-most often identified in relation to a vessel-and the individuals and companies that benefit from these activities and from public support can be even more difficult.

6. Second, delineating what actions should trigger the denial, **withholding**, or withdrawal of support is complicated. IUU fishing, as described in the International Plan of Action against IUU Fishing (IPOA-IUU) of the Food and Agriculture Organization of the United Nations (FAO), covers a range of different fishing activities and contexts, including industrial vessels fishing illegally in the waters of a foreign country, or fishing in the high seas by a vessel without nationality; but also small-scale fishers failing to diligently report their catch; use of prohibited gear; or fishing in excess of a quota in the coastal areas of the fisher's own country. The sustainability and socio-economic **implications** of excluding such different types of activities from support, and the opportunity cost of doing so, vary accordingly.

7. At the same time, a range of fishing activities that a common-sense interpretation would consider unregulated, unreported or insufficiently regulated may not be covered by the IPOA-IUU description. This can be the case, for example, of fishing on the high seas concerning species or areas outside the competence of any Regional Fisheries Management Organization or Arrangement (RFMO/A), which is not co-operatively regulated in a way that would allow for the sustainable management of the resources.

8. Furthermore, fishing-related activities, such as transshipment-whereby fish are transferred from fishing boats onto larger refrigerated vessels, which then carry the fish to port-and the **provisioning** of personnel, fuel and other supplies at sea, are generally less subject to legislation and regulation and harder to monitor and sanction than the fishing activity itself, and thus can play a central role in IUU fishing.

9. Finally, some types of government support to fisheries are made available to the sector as a whole or to all in the sector. This is often the case of public investment in infrastructure or of tax exemptions for example. Excluding particular individuals, companies and vessels from

associated benefits may prove more challenging in such cases.

10. **Notwithstanding** these challenges, this report finds that a range of actions are being and can be taken by governments to cut support to IUU fishing, through both specific and implicit mechanisms. Based on these findings, the report recommends actions that can be undertaken by countries to not only maximize the chances that individuals and companies with links to IUU fishing are excluded from government support.

11. Firstly, make all support conditional on the vessel being flagged to the supporting country and having fishing authorization. In addition to conditions that are typically included in authorizations processes, such as position transmission through vessel monitoring systems or reporting of catch, where appropriate, the authorizations should require unique vessel identifiers, such as an International Maritime Organization (IMO) number, where appropriate; and detailed information on vessel beneficial owners.

12. Secondly, make use of appropriate processes to effectively exclude from all types of support all potential recipients linked to IUU fishing (understood in a broad sense) by being transparent about the consequences of IUU fishing before support is given and the use of support-related enforcement actions. Ensuring proportionality of government action by giving due consideration to the nature of the IUU fishing activity, and the context in which it happened. **Delineating** who is concerned; for how long; and whether past support needs to be recovered. Without necessarily tying action on support to other IUU-enforcement action (such as processes related to IUU vessel-listing).

13. Thirdly, adopt a definition of IUU fishing and fishing-related activities, under national legislation, with the objective of cutting support to those engaging in these activities, including when they happen outside the jurisdiction of the supporting country. The IPOA-IUU description is the most commonly used reference for defining IUU fishing in national legislation, and the Agreement on Port State Measures to Prevent, Deter and Eliminate Illegal, Unreported and Unregulated Fishing (PSMA) for defining fishing-related activities. A national definition may be helpful when co-operating internationally on IUU fishing.

14. Fourthly, better regulate and monitor the transshipment of fish and other fisheries-related activities such as at-sea vessel supplying, including with authorization and reporting obligations. Enhance monitoring of fishing and fishing-related activities to better detect and enforce against IUU fishing; implement fully the key provisions of the PSMA and, where possible, become a party to the Agreement. Denying use of ports to those suspected of or

involved in IUU fishing and IUU fishing-related activities (except for purposes of inspection or in situations of force majeure) will also directly cut access to associated infrastructure and services supported by governments. Improve information-sharing within and between government agencies, countries and increase **transparency** on the processes in place to cut support to IUU fishing and their implementation and, where compatible with privacy legislation, on the recipients of fisheries support, including non-specific support and support to fishing-related activities.

15. Given the **inherent** difficulty of excluding individuals and companies with links to IUU fishing from government support, the report also recommends actions that can minimize the risk of government support benefitting IUU fishing. Reduce or redirect support away from policies that have the most potential to increase fishing effort and capacity and consequently drive higher levels of IUU fishing. This is notably the case of support that reduces the costs of vessels and fuel.

16. The multi-faceted nature of IUU fishing poses a number of challenges to governments looking to exclude it from support. Eliminating support to IUU fishing and fishing-related activities presents genuine challenges for governments as the sustainability and socio-economic **implications** of sanctioning (and excluding from support) different types of IUU fishing activities, and the opportunity cost of doing so, vary and sometimes remain unclear.

17. The International Plan of Action against IUU Fishing (IPOA-IUU) of the Food and Agriculture Organization of the United Nations (FAO) describes IUU fishing as covering a wide range of different fishing activities and contexts, including industrial vessels fishing illegally in the waters of a foreign country, or fishing in the high seas by a vessel without nationality; but also small-scale fishers failing to diligently report their catch; use of prohibited gear; or fishing in excess of a quota in the coastal areas of the fisher's own country. IUU fishing is thus a broad concept that includes activities which vary in terms of their impacts on the sustainability of fisheries and the resources they rely upon. At the same time, a range of fishing activities that a common-sense interpretation of IUU fishing would consider "unregulated", "unreported" or "insufficiently regulated" may not be covered by the IPOA-IUU description.

18. This can be the case, for example, of fishing on the high seas that concerns species or areas outside the competence of any RFMO/A that is not co-operatively regulated in a way that would allow for evidence-based sustainable management of the resources. It is

also the case of fishing-related activities which are central to IUU fishing. Taking the perspective that all these types of IUU fishing and fishing-related activities can be unsustainable, and with the objective of directing government support away from unsustainable activities, this paper considers how to avoid supporting IUU fishing understood in a broad sense, which may go beyond the IPOA-IUU description and the definitions of IUU fishing adopted in the domestic **legislation** of some economies. Doing so, this paper considers and recommends a range of possible actions to exclude IUU fishing from support, with options that are compatible with different contexts.

19. The international community has recognized the need to eliminate support to IUU fishing and made it a priority for action for over two decades. The FAO voluntary IPOA-IUU called on countries to avoid support to IUU fishing in 2001. This objective was then included in the 2002 Report of the World Summit on Sustainable Development and, later, became Target 14.6 of the Sustainable Development Goal (SDG) 14 of the United Nations' (UN) 2030 Agenda for Sustainable Development, which also charges the World Trade Organization (WTO) with establishing **multilateral** disciplines on inter alia support that benefits IUU fishing. This is an area of focus for on-going WTO negotiations on fisheries subsidies. Progress on cutting support to IUU fishing has been slow. At the end of 2020, the deadline for reaching SDG Target 14.6, over 40% of the economies covered in the OECD Review of Fisheries reported not having or not implementing fully domestic restrictions on support for operators determined to have engaged in IUU fishing.

20. However, there has been some progress. The Comprehensive and Progressive Agreement for TransPacific Partnership (CPTPP); the United States-Mexico-Canada Agreement (USMCA); and the recently concluded trade agreements between the United Kingdom of Great Britain and Northern Ireland and New Zealand (UK-NZ TA) and Australia (UK-AUS TA) contain **provisions** to prohibit subsidies to IUU fishing. According to some parties, these disciplines, and the associated reporting requirement, have already facilitated the design and adoption of domestic legislation and regulation to prevent supporting IUU fishing.

(2,015 words)

https://www.oecd-ilibrary.org

New Words

1. undermine [ˌʌndəˈmaɪn]

    v. to make sth. especially sb's confidence or authority, gradually weaker or less effective 逐渐削弱(信心、权威等);使逐步减少效力

    The central state, though often very rich and very populous, was intrinsically fragile, since the development of new international trade routes could undermine the monetary base and erode state power.

    尽管中央帝国通常非常富有和人口稠密,但它本质上是很脆弱的,因为新的国际贸易路线的发展可能会破坏货币基础和侵蚀国家权力。

2. assessment [əˈsesmənt]

    n. an opinion or a judgement about sb./sth. that has been thought about very carefully 看法;评估

    A small step has been taken in the direction of a national agency with the creation of the Canadian Coordinating Office for Health Technology Assessment, funded by Ottawa and the provinces.

    在渥太华和各省的资助下,加拿大成立了卫生技术评估协调办公室,朝建立一个全国性机构的方向迈出了一小步。

3. underpin [ˌʌndərˈpɪn]

    v. to support or form the basis of an argument, a claim, etc. 加强,巩固,构成(……的基础等)

    What we seem to be sacrificing in all our surfing and searching is our capacity to engage in the quieter, attentive modes of thought that underpin contemplation, reflection and introspection.

    在我们所有的浏览和搜索中,似乎最神圣的是我们从事于更安静、专注的思维模式的能力,而这种思维模式是沉思、反思和内省的基础。

4. dispute [dɪˈspjuːt]

    n. an argument or a disagreement between two people, groups or countries; discussion about a subject where there is disagreement 争论;辩论;争端;纠纷

    Air traffic controllers have begun a three-day strike in a dispute over pay.

    空中交通管制员在一场薪酬纠纷中开始了为期3天的罢工。

5. surveillance [sɜːˈveɪləns]

    n. the act of carefully watching a person suspected of a crime or a place where a crime may be committed(对犯罪嫌疑人或可能发生犯罪的地方的)监视

    On June 7 Google pledged not to "design or deploy AI" that would cause "overall harm," or to develop AI-directed weapons or use AI for surveillance that would violate international norms.

谷歌在6月7日承诺不会"设计或部署"可能会造成"全面伤害"的人工智能,也不会开发人工智能制导武器,或将人工智能用于违反国际准则的监控。

6. sanction ['sæŋkʃn]

n. an official order that limits trade, contact, etc. with a particular country, in order to make it do sth. such as obeying international law 制裁

Another sanction in their accession treaties is that other EU members may refuse to recognize court judgments.

他们的加入条约中还有另外一项附加条款,即其他欧盟成员国可以拒绝认可该国的法庭判决书。

7. eradication [ɪˌrædɪ'keɪʃ(ə)n]

n. the complete destruction of every trace of something 根除,消灭

The unintentional consequence has been to halt the natural eradication of underbrush, now the primary fuel for mega fires.

意外的后果是阻止了对灌木丛的天然清理,而它们现在成了超级大火的最主要的燃料。

8. inadvertently [ˌɪnəd'vɜːt(ə)ntli]

ad. by accident; without intending to 无意地;不经意地

Whether the public bus ride, or riding, are often inadvertently seen some people holding a Rubik's Cube in turn, or couples, or single.

无论是坐公交,还是坐车,常常看见一些人(他们或者情侣,或者单身)拿着魔方在转。

9. subsidy ['sʌbsədi]

n. money that is paid by a government or an organization to reduce the costs of services or of producing goods so that their prices can be kept low 补贴;补助金;津贴

The townsfolk don't see it this way and the local council does not contribute directly to the subsidy of the Royal Shakespeare Company.

市民们并不这么看,地方议会也不直接资助皇家莎士比亚剧团。

10. refrain [rɪ'freɪn]

v. (formal) to stop yourself from doing sth. especially sth. that you want to do 克制;节制;避免

Following the Harvard scandal, Mary Miller, the former dean of students at Yale, made an impassioned appeal to her school's professors to refrain from take-home exams.

哈佛丑闻发生之后,耶鲁大学前学生事务主任玛丽·米勒慷慨激昂地呼吁教授避免学生在家考试。

11. overarch [ˌəʊvər'ɑːtʃ]

v. to form an arch over 在……上方成拱形

It was proved that the centre of the drop didn't overarch water fall.

早已证实出雨滴的中心物质并不构成降雨的真正原因。

12. eligible ['elɪdʒəb(ə)l]

a. qualified for or allowed or worthy of being chosen 符合条件的,合格的;(婚姻)合适的,合意的

Although women first served on state juries in Utah in 1898, it was not until the 1940s that a majority of states made women eligible for jury duty.

虽然早在1898年,已有妇女在犹他州担任州陪审团,但直到20世纪40年代,大多数州才允许妇女担任陪审员。

13. implicitly [ɪmˈplɪsɪtli]

ad. without doubting or questioning 含蓄地;暗示地;无疑问地;无保留地

Instead of casting a wistful glance backward at all the species we've left in the dust I.Q.-wise, it implicitly asks what the real costs of our own intelligence might be.

这一问题不是惆怅地回望那些我们人类在智力上已远远超越的物种,而是含蓄地询问我们的智力的真正代价可能是什么。

14. eligibility [ˌelɪdʒəˈbɪləti]

n. the state of being eligible 资格,合格

The care was substantially provided by voluntary services which worked together with local authorities as they long had with eligibility based on income.

这些关怀主要是由志愿服务机构提供的,这些机构长期与地方当局合作,根据收入来确定服务资格。

15. withhold [wɪðˈhəʊld]

v. to refuse to give sth. to sb. 拒绝给;不给

Or they withhold information to avoid responsibility, wanting someone else to make a decision even if it is wrong.

为了躲避责任,他们总是隐瞒一些信息;他们总是希望别人看拍板做决定,哪怕这个决定明显错误。

16. implication [ˌɪmplɪˈkeɪʃn]

n. a possible effect or result of an action or a decision 可能的影响(或作用、结果)

The implication is a sharp decline in global trade, which has plunged partly because oil-producing nations can't afford to import as much as they used to.

这意味着,全球贸易大幅下滑的部分原因,是产油国无法像过去那样进口那么多资源。

17. provision [prəˈvɪʒ(ə)n]

n. the act of supplying sb. with sth. that they need or want; sth. that is supplied 提供;供给;给养;供应品

I am of the opinion when designing and building new towns, planning officials should give more consideration to the provision of public parks and sports facilities rather than ubiquitous malls.

我的观点是,在设计和建造新城时,负责规划的官员应该更多地考虑提供公共公园和体育设施,而不是无处不在的购物中心。

18. notwithstanding [ˌnɒtwɪθ'stændɪŋ]

   prep./conj. without being affected by sth.; despite sth. 虽然，尽管

   Notwithstanding important recent progress in developing renewable fuel sources, low fossil fuel prices could discourage further innovation in, and adoption of, cleaner energy technologies.

   尽管最近在开发可再生燃料方面取得了重要进展，但化石燃料价格较低这一因素可能会阻碍清洁能源技术的进一步创新和应用。

19. delineate [dɪ'lɪnieɪt]

   v. to describe, draw or explain sth. in detail（详细地）描述，描画，解释

   Across time and space, food has always been used to delineate social distinctions, whether in Roman dining rooms or modern gourmet supermarkets.

   穿越时间和空间，饮食一直以来都被用来描绘社会的区别，不论是在古罗马的饭厅里还是在现代的美食超级市场上。

20. transparency [træns'pærənsi]

   v. the quality of sth. such as glass, that allows you to see through it 透明；透明性

   This type of integrity requires well-enforced laws in government transparency, such as records of official meetings, rules on lobbying, and information about each elected leader's source of wealth.

   这种诚信要求政府透明度方面的法律得到很好的执行，比如官方会议记录、游说规则，以及每位当选领导人的财富来源信息。

21. inherent [ɪn'herənt]

   a. that is a basic or permanent part of sb./sth. and that cannot be removed 固有的；内在的

   My task is to build fluency while providing the opportunity inherent in any writing activity to enhance the moral and emotional development of my students.

   我的任务就是在任何写作活动中提供一些内在的机会来培养他们语言的流畅程度，以此来增强学生道德和情感上的发展。

22. legislation [ˌledʒɪs'leɪʃn]

   n. a law or a set of laws passed by a parliament 法规；法律

   The government has in the past given qualified support to the idea of tightening the legislation.

   政府过去对加强立法的观点给予了有保留的支持。

23. multilateral [ˌmʌlti'lætərəl]

   a. in which three or more groups, nations, etc. take part 多边的；多国的

   The International Finance Corporation (IFC) and the Multilateral Investment Guarantee Association (MIGA) established the Compliance Advisor/Ombudsman Office (CAO) in 2000.

● 2000年,国际金融公司和多边投资担保机构成立了合规顾问/投诉办公室。

Phrases and Expressions

Proper Names
Terminology

in support of 支持;以便支持;为了支持;拥护
be associated with 和……联系在一起;与……有关,与……有关系
combined with 加上;连同;联合
in relation to 关于;涉及
in excess of 超过;较……为多
in addition to 除……之外(还有,也)
allow for 考虑到,虑及;使……成为可能
be compatible with 一致;适合;与……相配
be engaged in 参与;从事于;忙于

Terminology

tax revenue 税收;赋税收入
high seas 公海;远海
tax exemption 免税

Proper Names

Illegal, unreported and unregulated (IUU) 非法、不报告和不管制捕鱼
OECD (Organization for Economic Co-operation and Development) 经济合作与发展组织
Food and Agriculture Organization (FAO) 联合国粮食农业组织
Regional Fisheries Management Organization or Arrangement (RFMO/A) 区域渔业管理组织
International Maritime Organization 国际海事组织
Unreported and Unregulated Fishing (PSMA) 关于港口国预防、制止和消除非法、不报告、不管制捕鱼的措施协定

一、翻译策略/方法/技巧:正反译法

任何反说转正说和正说转反说以及任何相反概念的转译都属于正反译法。

由于国家、历史、地理、社会文化背景和生活习性的不同,汉英两种语言在表达正说和反说时有很大差异,尤其英语在否定意义的表达上更为复杂,有时形式否定而实质肯定,或形式肯定而实质否定。在两种语言互译时,原文中正说的句子可能不得不处理成反说,或是用反说表达更为合适;反之亦然。翻译中,这种把正说处理成反说、把反说处理成正说的译法,

就称为正反译法。这样才能确切表达原意并符合语言的规范。这种正说和反说的相互转换是翻译技巧中的一个重要方法，它属于引申和修辞范围。那么，什么是正说和反说呢？英语词句中含有"never""no""not""non-""un-""im-""in-""ir-""-less"等成分以及汉语词句中含有"不""没""无""未""甭""别""休""莫""非""毋""勿"等成分的为反说，不含有这些成分的为正说。正说和反说包括的词类范围很广，不仅包括动词、形容词、副词、名词、介词和连词，而且包括各种词组、短语和从句。

例1：I lay awake almost the whole night.

译文一：我整晚醒着躺在那儿。

译文二：我躺在那里，几乎一夜没合眼。

例2：Where is your brother? He's still in bed.

译文一：你哥哥在哪儿？他还在床上。

译文二：你哥哥在哪儿？他还没起床。

英译汉，正转反：英语句中隐含有否定概念的词或是短语，译文则从反面表达，beyond、absent、stop、bad、avoid、exclude、except、doubt、resistant、refuse、few、little 等。

例3：Children were excluded from getting in the building.

孩子不许进入这幢楼房。

例4：Such a chance denied me.

我没有得到这个机会。

例5：The explanation is pretty thin.

这个解释站不住脚。

英译汉，反转正：英语从反面表达，译文从正面表达，使译文合乎汉语的表达习惯，通顺易懂。

例6：No smoking!

严禁吸烟

例7：No deposit will be refunded unless ticket produced.

凭票退还押金。

例8：Don't lose time in posting this letter.

赶快把这封信寄出去。

适当的使用正反译法可以使句子更加通顺流畅，符合汉语的表达习惯。但正反译法又是一个比较灵活的译法，我们在翻译实践中，对具体情况还要进行具体分析，不要生硬地套用这些规则，而要根据具体的语言环境采用合乎汉语习惯的译法。

二、译例与练习

Translation

1. Whilst the negative effects of IUU fishing are well understood, it is nevertheless the case that IUU fishing can benefit from government support for fisheries. This report assesses how to avoid this. (Para.1)

虽然非法、不报告和不管制捕鱼的负面影响是众所周知的,但非法、不报告和不管制捕捞可以从政府对渔业的支持中获益。本报告评估了如何避免这种情况。

2. Improving management as well as monitoring, control, and surveillance (MCS), sanctioning IUU fishing activities and reducing the expected net benefits of IUU fishing more generally are key to its eradication.(Para.2)

改善管理以及监测、控制和监视,制裁非法、不报告和不管制捕鱼活动,并更广泛地减少非法、不报告和不管制捕鱼的预期净效益,是根除非法、不报告和不管制捕鱼的关键。

3. The report reviews how OECD Members and Partner economies engaging in the OECD Fisheries Committee (COFI) can ensure that government support that they may provide to fisheries does not contribute to IUU fishing, on the basis of a survey conducted in 2021, and suggests avenues to more effectively close public budgets to IUU fishing.(Para.3)

报告审查了经合组织成员和参与经合组织渔业委员会的伙伴经济体如何根据2021年进行的一项调查,确保它们可能向渔业提供的政府支持不会促进非法、不报告和不管制捕捞,并建议了更有效地将公共预算关闭非法、不报告和不管制捕鱼的途径。

4. Three main approaches were seen from the review. Some economies use specific mechanisms to deny or withdraw support in relation to IUU fishing, which are set in overarching legislation and regulation.(Para.4)

从审查中可以看到三个主要方法。一些经济体利用特定机制拒绝或撤回对非法、不报告和不管制捕鱼的支持,这些机制是在总体立法和法规中规定的。

【本句中which引导的是一个非限定性定语从句,翻译此类句子,一般可采用以下方法:一般来说,非限制性定语从句较少译成带"的"的定语词组,在翻译成汉语时可以将从句与主句分译,独立成句。】

5. At the same time, a range of fishing activities that a common-sense interpretation would consider unregulated, unreported or insufficiently regulated may not be covered by the IPOA-IUU description.(Para.7)

与此同时,在常识解释中被认为不受管制、未报告或监管不足的一系列捕鱼活动可能不在《关于预防、制止和消除非法、不报告和不管制捕鱼的国际行动计划》说明的范围内。

6. Furthermore, fishing-related activities, such as transshipment-whereby fish are transferred from fishing boats onto larger refrigerated vessels, which then carry the fish to port-and the provisioning of personnel, fuel and other supplies at sea, are generally less subject to legislation and regulation and harder to monitor and sanction than the fishing activity itself, and thus can play a central role in IUU fishing.(Para.8)

此外,与渔业有关的活动,例如转运——即将鱼从渔船转移到更大的冷藏船上,然后将鱼运至港口——以及海上人员、燃料和其他用品的供应,一般不像渔业活动本身那样受立法和管制,也更难监测和制裁,因此可以在非法、不报告和不管制捕鱼中发挥核心作用。

【本句中出现了两个破折号,作用是解释性的插入语,相当于一个括号,如果前后两个破折号引出的部分与第一个破折号之前的成分是修饰关系,就要用汉语的括号取代英语的破

折号,本句中前后两个破折号引出的部分与第一个破折号之前的成分是并列关系,那么在翻译的时候破折号仍可保留。】

7. In addition to conditions that are typically included in authorizations processes, such as position transmission through vessel monitoring systems or reporting of catch, where appropriate, the authorizations should require unique vessel identifiers, such as an International Maritime Organization (IMO) number, where appropriate; and detailed information on vessel beneficial owners. (Para. 11)

除了通常包括在授权程序中的条件,如通过船舶监测系统传送位置或报告渔获物,授权还应酌情要求唯一的船舶标识,如国际海事组织(海事组织)编号;及船舶实际拥有人的详细资料。

8. Improve information-sharing within and between government agencies, countries and increase transparency on the processes in place to cut support to IUU fishing and their implementation and, where compatible with privacy legislation, on the recipients of fisheries support, including non-specific support and support to fishing-related activities. (Para. 14)

改善政府机构、国家内部和之间的信息共享,提高减少对非法、不报告和不管制捕捞支持的现行程序及其实施的透明度,并在符合隐私立法的情况下,提高渔业支持接受国的透明度,包括非具体支持和对渔业相关活动的支持。

9. Reduce or redirect support away from policies that have the most potential to increase fishing effort and capacity and consequently drive higher levels of IUU fishing. (Para. 15)

减少或改变对最有可能增加捕捞量和能力的政策支持,从而推动非法、不报告和不管制捕鱼达到更高水平。

10. The international community has recognized the need to eliminate support to IUU fishing and made it a priority for action for over two decades. (Para. 19)

国际社会认识到有必要取消对非法、不报告和不管制捕鱼的支持,并在20多年来将其作为优先行动事项。

Exercises

1. Fill in the blanks with the proper given words, and then translate the sentences into Chinese.

  withdraw legislation infrastructure implicit jurisdiction interpretation

1) Yet for the first eight decades of the amendment's existence, the Supreme Court's _____ of the amendment betrayed this ideal of equality.

2) So when the premiers gather in Niagara Falls to assemble their usual complaint list, they should also get cracking about something in their _____ that would help their budgets and patients.

3) Large-scale enterprises tend to operate more comfortably in stable and secure

circumstances, and their managerial bureaucracies tend to promote the status quo and resist the threat implicit _____ in change.

4) Although population, industrial output and economic productivity have continued to soar in developed nations, the rate at which people _____ water from aquifers, rivers and lakes has slowed.

5) Religion and politics, they believed, should be kept clearly separate, and they generally opposed humanitarian _____.

6) Governments of rapidly developing countries incorporate waste minimization thinking into the transport _____ and storage facilities currently being planned, engineered and built.

2. Translate the following sentences into Chinese.

1) For thousands of years fishermen of Zhoushan Archipelago have lived on fishery and created a rich ocean culture of which the proverbs related to fishery are a representative part.

2) Tilapia and silver carp are the most common fishes in tropical reservoirs; their mixed culture is a typical mode for fishery tropical reservoirs.

3) Fishery activity was one of the majority culture activities of the inhabitants in southwestern region during the Han-Jin periods.

4) It consists of direct consumptive coefficient of fishery production, construct of value, analysis of the national economic results and a comparison of the economic targets of various types of culture.

5) Both sides have conducted fruitful cooperation in the areas such as finance, trade, terrestrial heat, fishery and culture, and kept sound contacts and coordination in global and regional affairs.

6) Therefore, artificial inland fishery culture had wider prospect than artificial Marine fishery culture.

7) But since it has been largely covered with year-round ice throughout modern history, it is the least mapped ocean.

8) The Atlantic played a particularly formative role in U.S. history through immigration—for the ancestors of most Americans, the ocean was a treacherous passageway to a better life.

9) This is a lively rattle bag of a history of the Pacific slope and of how the Pacific Ocean came to be an American lake.

10) Where he is famed for his history of the whaler that inspired Herman Melville's "Moby Dick" and for arguing plausibly in his "Sea of Glory" that the ocean, not the West, was America's first frontier.

3. Translate the following sentences into English.
1）东北渔猎民族皮服文化的形成源于当时极其低下的生产力水平和其特殊的渔猎生产方式。
2）人们认为农耕文化和渔业文化是有联系的，它们不是孤立存在的。
3）然而，阿拉斯加渔业真正的文化核心和灵魂是鲑鱼。
4）作为一个岛国，渔业是当地经济的重要支撑，当地文化的一部分。
5）岛上最流行的文化名胜是卡洛科力鱼塘，古代的夏威夷人曾经在此形成了成熟的水产业形式。
6）洪湖湿地在水产、航运和旅游等方面具有较高的经济价值，在美学、文化教育、科研等方面拥有较大的社会价值。
7）池塘养殖是我国最重要的水产养殖生产方式，2004年其产量占到淡水养殖产量的70.36%。
8）本发明涉及水产养殖饲料技术领域，特别是一种低盐度养殖凡纳对虾添加剂预混料。
9）本发明涉及一种养殖池塘的水质在线监测方法和系统，属于池塘养殖领域。
10）水产养殖专业的研究集中如何对淡水和海水的水产品（包括鱼类、贝类及水生植物等）进行育种、养殖、捕捞及销售。

4. Choose the best paragraph translation. And then answer why you choose the first translation or the second one.

我国幅员辽阔、河湖众多、海岸线绵长，拥有地域特色鲜明的渔业文化资源。作为优秀传统文化的重要组成部分，摸清现阶段渔业文化资源家底对履行与促进文化保护、科学推进渔业文化保护与发展具有重要意义。但目前有关渔业文化资源的调查评价和分类归档工作尚未得到普及或仅局限于个别区域开展，资源内涵和外延尚未得到充分认识，资源体系架构不够明晰，资源利用也较为粗浅。

译文一：
Our country has a vast territory, many rivers and lakes, a long coastline, has fishery cultural resources with distinct regional characteristics. As an important part of excellent traditional culture, it is significant to find out the background of fishery culture resources at the present stage for the implementation and promotion of cultural protection and scientific promotion of fishery culture protection and development. But, at present, the investigation, evaluation, classification and archiving of fishery cultural resources have not been popularized or are only carried out in individual regions. The connotation and extension of resources are not fully understood, the framework of resource system are not clear enough, and the utilization of resources is shallow.

译文二：
China has a vast territory, numerous rivers and lakes, and a long coastline. It has fishery cultural resources with distinctive regional characteristics. As an important part of

excellent traditional culture, it is of great significance to find out the status of fishery cultural resources at the present stage for fulfilling and promoting cultural protection and scientifically promoting the protection and development of fishery culture. However, at present, the survey, evaluation, classification and filing of fishery cultural resources have not been popularized or are only limited to individual regions, the connotation and extension of resources have not been fully understood, the resource system structure is not clear enough, and the utilization of resources is relatively shallow.

5. Translate the following passage into Chinese.

Illegal, unreported and unregulated (IUU) fishing and fishing-related activities in support of IUU fishing continue to seriously undermine and threaten the sustainability of fisheries, the livelihoods of coastal communities and the ocean economy. Largely unseen, IUU fishing complicates the stock assessments that underpin evidence-based fisheries management, while causing law-abiding fishers to face unfair competition over resources and in markets. Furthermore, IUU fishing results in losses of important tax revenue. IUU fishing can also threaten food security, for example by diverting fish away from local markets in regions and communities that depend on local seafood and may pose food safety risks due to the mislabeling of illegal products. It is also sometimes associated with conflicts over scarce resources and disputed waters; transnational criminal activities; and the exploitation of forced labour. Whilst the negative effects of IUU fishing are well understood, it is nevertheless the case that IUU fishing can benefit from government support for fisheries. This report assesses how to avoid this.

6. Translate the following passage into English.

关于文化的概念，有广义与狭义之分。狭义的文化指社会意识形态方面的精神财富，广义的文化则指人类在社会历史实践中所创造的物质财富和精神财富的总和。渔业是人类早期直接向大自然索取食物的生产方式，是人类最早的生产行为。那时，靠山吃山指的是狩猎，靠水吃水指的是捕鱼。或者说，先人以水域为依托、利用水生生物的自然繁衍和生命力，通过劳动获取水产品，我们称之为渔业。勤劳朴实的先祖，不仅创造了丰富的物质文明，而且还创造了灿烂的渔文化。

三、翻译家论翻译

尤金·A.奈达(Eugene A. Nida)，语言学家，翻译家，翻译理论家，被誉为"当代翻译理论之父"。1914年11月11日，出生于美国俄克拉何马州，这位在学术界赫赫有名的人物，主要学术活动都围绕《圣经》翻译展开。在《圣经》翻译的过程中，奈达从实际出发，发展出了一套自己的翻译理论，最终成为翻译研究的经典之一。奈达理论的核心概念是"功能对等"。

所谓"功能对等"，就是说翻译时不求文字表面的死板对应，而要在两种语言间达成功能上的对等。为使源语和目的语之间的转换有一个标准,减少差异,奈达从语言学的角度出发,根据翻译的本质,提出了著名的"动态对等"翻译理论,即"功能对等"。他指出,"翻译是用最恰当、自然和对等的语言从语义到文体再现源语的信息"。奈达有关翻译的定义指明,翻译不仅是词汇意义上的对等,还包括语义、风格和文体的对等,翻译传达的信息既有表层词汇信息,也有深层的文化信息。"动态对等"中的对等包括词汇对等、句法对等、篇章对等和文体对等四个方面。在这四个方面中,奈达认为,"意义是最重要的,形式其次"。形式很可能掩藏源语的文化意义,并阻碍文化交流。因此,在文学翻译中,根据奈达的理论,译者应以动态对等的四个方面作为翻译的原则,准确地在目的语中再现源语的文化内涵。为了准确地再现源语文化和消除文化差异,译者可以遵循以下的三个步骤。第一,努力创造出既符合原文语义又体现原文文化特色的译作。然而,两种语言代表着两种完全不同的文化,文化可能有类似的因素,但不可能完全相同。因此,完全展现原文文化内涵的完美的翻译作品是不可能存在的,译者只能最大限度地再现源语文化。第二,如果意义和文化不能同时兼顾,译者只有舍弃形式对等,通过在译文中改变原文的形式达到再现原文语义和文化的目的。例如,英语谚语"white as snow"翻译成汉语可以是字面意义上的"白如雪"。但是,中国南方几乎全年无雪,在他们的文化背景知识中,没有"雪"的概念,如何理解雪的内涵?在译文中,译者可以通过改变词汇的形式来消除文化上的差异。因此,这个谚语在汉语中可以译作"白如蘑菇""白如白鹭毛"。再如,英语成语"spring up like mushroom"中"mushroom"原意为"蘑菇",但译为汉语多为"雨后春笋",而不是"雨后蘑菇",因为在中国文化中,人们更为熟悉的成语和理解的意象是"雨后春笋"。第三,如果形式的改变仍然不足以表达原文的语义和文化,可以采用"重创"这一翻译技巧来解决文化差异,使源语和目的语达到意义上的对等。"重创"是指将源语的深层结构转换成目的语的表层结构,也就是将原语文章的文化内涵用译语的词汇来阐述和说明。例如"He thinks by infection, catching an opinion like a cold","人家怎么想他就怎么想,就像人家得了伤风,他就染上感冒。"在此句的英文原文中,原文的内涵并不是靠词汇的表面意义表达出来的,而是隐藏在字里行间里。因此,如按照英汉两种语言字面上的对等来翻译,原句译为"他靠传染来思维,像感冒一样获得思想",这样,原文的真正意义就无法清楚地表达。

事实上,在汉语中很难找到一个完全与英文对等的句型来表达同样的内涵。于是,译者将源语的深层结构转换成目的语的表层结构,即用目的语中相应的词汇直接说明、解释原文的内涵,以使译文读者更易接受译作。根据奈达的翻译理论,文化差异的处理是与从语义到文体将源语再现于目的语紧密相联的。只有当译文从语言形式到文化内涵都再现了源语的风格和精神时,译作才能被称作是优秀的作品。

(摘自:https://www.sohu.com/a/196926867_99937634)

Text B

Pacific Monuments to Fishing

By Laura Parker

Ideological and Political Education:转型升级 助力渔业绿色发展

党的十八大以来,按照"绿水青山就是金山银山"的理念,全国渔业系统践行创新、协调、绿色、开放、共享的新发展理念,大力推进渔业绿色发展。

2017年,我国主动提出海洋渔业资源总量管理目标,启动实施海洋渔业资源总量管理制度。截至目前,我国近海实际捕捞量控制在1 000万t以内,沿海11个省(区、市)已全部开展限额捕捞管理试点工作。为进一步保护和合理利用海洋生物资源,农业农村部多次调整完善海洋伏季休渔制度。

2017—2018年,对海洋伏季休渔的时间和作业类型作出更加严格的规定,同时科学设定了休渔期间特许捕捞品种;2021年再次优化海洋伏季休渔时间,将休渔分界线由三条减为两条,科学稳妥有序扩大伏休期间特许捕捞品种,初步实现海洋渔业资源合理利用、渔民增收、休渔秩序平稳有机统一。根据党中央、国务院决策部署,2021年1月1日起,长江流域重点水域实行为期10年的常年禁捕,11.1万艘渔船、23.1万名渔民如期退捕上岸。这是以习近平同志为核心的党中央从战略全局高度和长远发展角度作出的重大决策,是落实长江经济带共抓大保护措施、扭转长江生态环境恶化趋势的关键之举。据持续监测显示,长江禁捕1年多来,江豚群体在长江中下游江段出现的频率明显增加,长江刀鱼已上溯至长江中游和鄱阳湖。长江水生生物资源状况逐步好转。

2017年以来,农业农村部先后印发通知,对珠江、闽江、海南省内陆水域、海河、辽河、松花江的禁渔管理作出规定并不断调整。2018年黄河流域正式实施休禁渔制度,2022年3月农业农村部又印发《关于进一步加强黄河流域水生生物资源养护工作的通知》,对黄河禁渔期制度进行了调整和完善,延长了禁渔时间、扩大了禁渔范围。目前黄河上游已实现常年禁渔,黄河下游休渔时间达4个月,我国内陆七大重点流域禁渔期制度和我国主要江河湖海休禁渔制度实现了全覆盖。

1. Managers of the American fisheries operating in the Pacific Ocean have asked President Donald Trump to open four national marine monuments to commercial fishing—a request that could **inhibit** protections in areas set aside, in part, for dwindling fish populations to renew themselves.

2. The Western Pacific Regional Fishery Management Council is a quasi-governmental body that sets the fishing seasons and annual catch limits. The council, also known as WESPAC, told the president that "quick action is urgently needed" to meet "**exceptionally** high retail demand" for canned tuna as a result of the global corona virus **pandemic**.

3. "We note that the fishing restrictions in the Pacific marine national monuments are **impeding** America's three main tuna fisheries in the Pacific and the StarKist tuna cannery in American Samoa from operating at optimal levels and that these fishing restrictions are unnecessary as they have no proven conservation benefit," Kitty Simonds, the council's executive director, wrote to Trump in a letter dated May 8. The request from the council arrived in Washington a day after Trump signed an executive order unveiling a broad initiative aimed at promoting economic growth and competitiveness of the U. S. seafood industry. The order included an invitation to the eight regional fishery management councils in the United States to recommend ways to reduce regulatory burdens on domestic fishing.

4. The WESPAC request to undo restrictions in the marine national monuments, if granted by Trump, could have far-ranging consequences. It will almost surely set off a political battle in Congress and a legal battle in court over the limits of presidential powers. Two lawsuits seeking clarity on presidential authority regarding national monuments are already in federal court before separate judges and some experts watching monument politics wonder if the Trump administration has the appetite to launch a third. "So far, we're not expecting that much is going to come of this request. But if it does, we're prepared to go to court," says David Henkin, an attorney in the Honolulu office of Earth justice, an environmental organization that campaigned to establish the largest Pacific marine monument. "President Trump is without executive authority to open up monuments to fishing, and can't **dismantle** them in a **piecemeal** fashion." WESPAC's claim that the restrictions on fishing in the monuments have brought economic hardship to the tuna fleet is the subject of a long-running debate. In an effort to clarify the issue, a team of economists and scientists published a study in Nature earlier this year that found the monuments' expansion had "little if any negative impacts" on the catch. In fact, the team found that Hawaii's longline fleet caught more fish after the monuments were expanded.

5. "Opening up the marine national monuments to fishing would be like opening up Yellowstone National Park to industrial-scale hunting," says Enric Sala, a marine biologist and the founder of National Geographic's Pristine Seas program, who helped with both the creation and expansion of the Pacific Remote Islands monument. Henkin adds that the pandemic does not provide a **justification** for expanding fishing because the issue created by COVID is not one of lack of supply, but lack of demand. "There is a glut of fish rotting in the fish markets in Honolulu," he says. "Demand for Ahi Tuna has plummeted as all the hotels and restaurants have closed. The boats aren't out fishing not because the monuments are close to them, but because they're not making any money".

6. Presidents have held the power to protect federal lands and water from development since 1906, when President Theodore Roosevelt signed the **Antiquities** Act into law. Since then, every president except Nixon, Reagan, and George H. W. Bush has created national monuments. The total now stands at 158 monuments, including the most recent addition, a 373-acre Civil War camp in Kentucky that President Obama designated as a National Historic Landmark in 2016 and Trump upgraded to monument status in 2018. National monuments have been resized, **abolished**, and turned into national parks. But the presidential ability to alter monuments created by their **predecessors**, including altering their boundaries or prescriptive protections, has never been tested in court until now.

7. In 2017, Trump was sued by several environmental groups and Native American tribes after he shrank two monuments situated in southern Utah's red rock country—the Grand Staircase Escalante, created by President Clinton in 1996, and Bears Ears, created by Obama in 2016. The two monuments comprised a combined total of 4.2 million acres. Trump made them nearly 4 million acres smaller in the largest reduction of federal land protection in the nation's history. A federal judge rejected the government's effort to dismiss the suit, as well as its request to transfer the case from Washington, D. C., to a federal court in Utah. The latest round of briefs is due in June.

8. Meanwhile, in a lawsuit brought by several commercial fishing associations, the D. C. Circuit Court of Appeals last December rejected their claim that the Northeast Canyons and Seamounts Marine National Monument had been illegally created by President Obama in 2016. The court concluded that Obama acted within his authority as president to create the monument, which **sprawls** over nearly 5,000 square miles 130 miles off the coast of Cape Cod. It protects a collection of underwater canyons and mountains that date back 100 million years and is home to sea turtles, several species of whales, and deep-sea, cold-water corals.

9. All four Pacific marine monuments were created by President George W. Bush. In 2006, he **designated** the Northwestern Hawaiian Islands marine monument to protect ocean around the northwestern Hawaiian Islands out to the 200-mile limit of the exclusive economic zone (EEA). It was later renamed the Papahanaumokuakea.

10. Marine National Monument. In 2009, Bush created the other three: Rose Atoll marine monument, near American Samoa; Marianas Trench marine monument; and Pacific Remote Islands marine monument, a string of islands in the central Pacific off Hawaii. Obama enlarged the Pacific Remote Islands marine monument in 2014 and Papahānaumokuākea in 2016, which he **quadrupled** in size, making them the fifth and third largest, respectively, marine protected areas in the world. The Western Pacific fishery council has opposed both the monuments' creation and expansion, calling them an **infringement** on its authority to regulate fishing in American Pacific waters. Jointly, the expanded monuments protect more than one million square miles—areas larger than Alaska, Texas, and California combined, and their sheer size served to **bolster** the fishing council's claim that the monuments cause economic hardship to the tuna fishing fleet.

11. When Trump launched a study early in his term to hunt for land or marine monuments he could shrink, the council recommended scaling back the monuments, although only two—Rose Atoll and the Pacific Remote Islands marine monuments—were included in the final recommendations then-Interior Secretary Ryan Zinke presented to Trump. Trump did not act on recommendations. (Read more about the Western Pacific fishery council here.) U.S. Rep. Ed Case, D-Ha., a long-time critic of the Western Pacific council's fisheries management, says the request is unsurprising, given the council's history, but nevertheless should be regarded as yet another threat to the protected areas. "Some of the fishery councils do a very, very good job to manage fishing on a **sustainable** basis," he says, "WESPAC is far more slanted on the **extraction** side of the equation, and I have long been concerned. The desire to open up marine monuments and manage fisheries on borderline sustainability is not the solution." The president's executive order asked the fishery councils to submit recommendations within 180 days. They are due November 8, Election Day.

(1,267 words)

(https://www.nationalgeographic.com/science/article/fishery-managers-seek-to-open-pacific-monuments-to-fishing)

New Words

1. inhibit [ɪn'hɪbɪt]

    n. to prevent sth. from happening or make it happen more slowly or less frequently than normal 阻止；阻碍；抑制

    There are many types of fear, but the two that inhibit iconoclastic thinking and people generally find difficult to deal with are fear of uncertainty and fear of public ridicule.

    恐惧有很多种，但其中会抑制打破常规的思维且人们通常很难应对的两种恐惧，是对不确定性的恐惧和对公众嘲笑的恐惧。

2. exceptionally [ɪk'sepʃənəli]

    ad. used before an adjective or adverb to emphasize how strong or unusual the quality is（用于形容词和副词之前表示强调）罕见，特别，非常

    I know you've all been working very hard recently and we've been exceptionally busy, especially with the wedding last weekend and the trade fair straight after that.

    我知道你们最近工作都很努力，我们也特别忙，尤其是忙着上周末的婚礼和紧接着的交易会。

3. pandemic [pæn'demɪk]

    a. a disease that spreads over a whole country or the whole world（全国或全球性）流行病；大流行病

    A new data shows that the global AIDS pandemic will cause a sharp drop in life expectancy in dozens of countries, in some cases declines of almost three decades.

    一项新的数据显示，全球艾滋病疫情将导致数十个国家的预期寿命大幅下降，在有些国家寿命减幅近 30 年。

4. impede [ɪm'piːd]

    v. to delay or stop the progress of sth. 阻碍；阻止

    As Knapp and Michaels might say what's at stake in calling attention to the way in which language does impede communication?

    耐普和迈克尔斯也许会说，当语言阻碍交流的情况引起关注时候，又带来了什么问题呢？

5. initiative [ɪ'nɪʃətɪv]

    n. a new plan for dealing with a particular problem or for achieving a particular purpose 倡议；新方案

    This universal consistency among education experts indisputably demonstrates an immutable principle of learning: initiative and correct methods are fundamental to academic success.

    教育专家的这种普遍共识无可争辩地证明了一个不可改变的学习原则：主动性和正确的方法是学术上成功的基础。

6. attorney [ə'tɜːni]

    n. a professional person authorized to practice law; conducts lawsuits or gives legal advice 律师

(尤指代表当事人出庭者);代理人

His attorney presented testimony that he had indeed applied for jobs and was listed with several employment agencies, including the State Employment Agency, but there weren't any jobs.

他的律师提供的证词显示,他确实申请了工作,并在几家职业介绍所登记过,其中包括州职业介绍所,但那里没有任何职位。

7. dismantle [dɪsˈmænt(ə)l]

v. to end an organization or system gradually in an organized way(逐渐)废除,取消

It is far easier to promise a bit more aid than to dismantle food subsidies that favour Western farmers but can devastate African ones.

承诺给予更多的补助要比取消食物补贴来得容易。取消食物补贴会使西方农场主受益,但非洲农民会因此而遭受灭顶之灾。

8. piecemeal [ˈpiːsmiːl]

a. done or happening gradually at different times and often in different ways, rather than carefully planned at the beginning 逐渐做成(或发生)的;零敲碎打的;零散的

If there are many conceptual issues, you can bring them to office hours, but please realize that there are always piecemeal hours.

如果有很多概念方面的问题,你完全可以把它们带到办公时间来,但是请认识到,总是有零零碎碎的时间的。

9. justification [ˌdʒʌstɪfɪˈkeɪʃ(ə)n]

n. a good reason why sth. exists or is done 正当理由

The most plausible justification for higher taxes on automobile fuel is that fuel consumption harms the environment and thus adds to the costs of traffic congestion.

汽车燃油税最合理的理由是,燃料消耗会损害环境,从而增加交通拥堵的成本。

10. antiquity [ænˈtɪkwəti]

n. an object from ancient times 文物;古物;古董;古迹

Although several ancient cultures practiced mummification, mummies from ancient Egypt are generally more well-preserved than mummies of similar antiquity from other cultures.

尽管有几个古代文化有木乃伊化的习俗,但古埃及的木乃伊通常比其他文化中类似古代的木乃伊保存得更好。

11. abolish [əˈbɒlɪʃ]

v. to officially end a law, a system or an institution 废除,废止(法律、制度、习俗等)

Do not think that I have come to abolish the Law or the Prophets; I have not come to abolish them but to fulfill them.

莫想我来要废掉律法和先知;我来不是要废掉,乃是要成全。

12. predecessor [ˈpriːdəsesə(r)]

n. a thing, such as a machine, that has been followed or replaced by sth. else 原先的东西;被替

代的事物

Though I'm still not keen on the design of the Kindle, it is a vast improvement on its predecessor and certainly tolerable.

虽然我仍然不喜欢电子阅读器(Kindle)的设计,但它和其前身相比已经有了很大的改进,当然也可以接受。

13. sprawl [sprɔːl]

n. to spread in an untidy way; to cover a large area 蔓延;杂乱无序地拓展

However, public infrastructure did not keep pace with urban sprawl, causing massive congestion problems which now make commuting times far higher.

然而,公共基础设施并没有跟上城市扩张的步伐,这造成了大规模的交通拥堵,也大大拉长了通勤时间。

14. canyon ['kænjən]

n. a deep valley with steep sides of rock(周围有悬崖峭壁的)峡谷

For the more adventurous, we offer rewards beyond mere sightseeing—from a three-day hike across the Grand Canyon to a ride along China's Yangtze River.

对于更喜欢冒险的人,我们提供了除纯粹的观光以外的奖励——从大峡谷三天徒步旅行到沿长江骑行。

15. designate ['dezɪgneɪt]

n. to say officially that sth. has a particular character or name; to describe sth. in a particular way 命名;指定

They had pushed the agency to designate the bird as "endangered", a status that gives federal officials greater regulatory power to crack down on threats.

他们敦促该机构将这种鸟列为"濒危物种",这个地位可以赋予联邦官员更大的监管权力来消除对该物种的威胁。

16. quadruple ['kwɒdrʊp(ə)l]

v. to become four times bigger; to make sth. four times bigger(使)变为4倍

Now, Japan's quadruple disaster, earthquake, tsunami, nuclear alert and power shortages, has put the supply chain under far greater stress.

现在,日本的四重灾难,地震、海啸、核泄漏和电力供应不足,使供应链处于巨大的压力之下。

17. infringement [ɪnˈfrɪndʒmənt]

n. an action or situation that interferes with your rights and the freedom you are entitled to.(对他人权利或自由等的)侵犯

Patents were filed more for strategic purposes, to be used as bargaining chips to ward off infringement suites or as a means to block competitors' products.

专利申请更多的是出于战略目的,被用作抵御侵权案件的讨价还价筹码,或者作为阻止竞争对手产品发展的手段。

18. bolster ['bəʊlstə(r)]

v. to improve sth.or make it stronger 改善；加强

Since the former is most common among those inclined towards indifferent relationships, their predominance can bolster individuals' sense of self-worth.

由于前者在那些倾向于淡漠关系的人群中最为常见，它们的优势可以增强个体的自我价值感。

19. sustainable [səˈsteɪnəb(ə)l]

a. involving the use of natural products and energy in a way that does not harm the environment（对自然资源和能源的利用）不破坏生态平衡的，合理利用的

With production anticipated to increase by 25% between now and 2030, sustainable energy sourcing will become an increasingly major issue.

从现在到2030年这段时间，石油产量预计将增加25%，而可持续能源的来源将成为一个日益重要的话题。

20. extraction [ɪkˈstrækʃn]

n. the act or process of removing or obtaining sth.from sth.else 提取；提炼；拔出；开采

Community members may reside temporarily in one of the lower zones to manage the extraction of products unavailable in the homeland.

社区成员可以暂时居住在一个较低的区域，以管理在本土无法获得的产品的提取。

Phrases and Expressions

in part 部分地；在某种程度上
set off 出发；引起；动身；使爆炸；抵销；分开
go to court 起诉；朝见君主
be home to 是……所在地；是……的所在地
act on recommendation 根据建议行事

Terminology

annual catch limits 年度捕捞限额
exclusive economic zone 专属经济区
federal court 联邦法庭

Proper Names

Western Pacific Regional Fishery Management Council 西太平洋区域渔业管理委员会
National Geographic's Pristine Seas 国家地理杂志原始海洋
D.C.Circuit Court of Appeals 华盛顿上诉法院
Cape Cod 科德角（美国地名）

一、翻译简析

1. The WESPAC request to undo restrictions in the marine national monuments, if granted by Trump, could have far-ranging consequences. It will almost surely set off a political battle in Congress and a legal battle in court over the limits of presidential powers. (Para.4)

  西太平洋区域渔业管理委员会要求取消对海洋国家纪念碑的限制,如果特朗普批准,可能会产生深远的影响。它几乎肯定会在国会引发一场政治斗争,并在法庭上引发一场关于总统权力限制的法律斗争。

  【第一句话的主语是 request,翻译成汉语的时候转译成动词词性更加自然;原句的主语由后面的不定式 to undo……和后面的过去分词结构 granted by……修饰,翻译的时候要转译为句子更合适。】

2. Presidents have held the power to protect federal lands and water from development since 1906, when President Theodore Roosevelt signed the Antiquities Act into law. (Para.6)

  自 1906 年西奥多·罗斯福总统签署《古物法》成为法律以来,总统一直拥有保护联邦土地和水资源不受开发的权力。

  【本句 when 引导的是一个非限定性的定语从句,翻译的时候采用分译的方法解释前面的先行词,由 when 引导的定语从句再转译为时间状语放在句首更符合汉语的表达习惯。】

3. In 2017, Trump was sued by several environmental groups and Native American tribes after he shrank two monuments situated in southern Utah's red rock country—the Grand Staircase Escalante, created by President Clinton in 1996, and Bears Ears, created by Obama in 2016. (Para.7)

  2017 年,特朗普在缩小了位于犹他州南部红石县的两座纪念碑——克林顿总统 1996 年创建的大楼梯和奥巴马 2016 年创建的熊耳后,被几个环保组织和美国原住民部落起诉。

  【英语中状语从句的位置可以放在主句的前面,也可以放在主句的后面,而汉语的习惯是把状语放在前面,所以本句翻译的时候把 after 引导的时间状语从句提到主句之前。】

4. Meanwhile, in a lawsuit brought by several commercial fishing associations, the D.C. Circuit Court of Appeals last December rejected their claim that the Northeast Canyons and Seamounts Marine National Monument had been illegally created by President Obama in 2016. (Para.8)

  与此同时,在几家商业渔业协会提起的诉讼中,华盛顿上诉法院去年 12 月驳回了他们的指控,即东北峡谷和海山海洋国家纪念碑是奥巴马总统在 2016 年非法创建的。

  【本句中 that 引导的是一个同位语从句,对其前面的抽象名词 claim 进行解释说明,翻译的时候先翻译主句,然后用"就是……"或者"即……"引导出同位语从句,或者把同位语从句译成独立的句子,由冒号或破折号引出。】

5. When Trump launched a study early in his term to hunt for land or marine monuments he could shrink, the council recommended scaling back the monuments, although only

two—Rose Atoll and the Pacific Remote Islands marine monuments—were included in the final recommendations then-Interior Secretary Ryan Zinke presented to Trump. (Para.11)

当特朗普在任期早期启动一项研究,寻找他可以缩小的陆地或海洋纪念碑时,该委员会建议缩小这些纪念碑,尽管当时的内政部长莱恩·辛克向特朗普提出的最终建议中只包括两个玫瑰环礁和太平洋偏远岛屿海洋纪念碑。

【本句中出现两个破折号,用在一个解释性的插入语的前面和后面(相当于一个括号),作用是解释说明第一个破折号之前的 two。因此,在翻译的时候把破折号之间的部分按定语翻译即可。】

二、参考译文

太平洋渔业纪念碑

1. 在太平洋海域作业的美国渔业管理者已要求唐纳德·特朗普总统开放四个国家海洋保护区,以供商业捕捞,这一要求可能会阻碍对一些区域的保护,这些区域的预留部分原因是鱼类种群不断减少,以进行自我更新。

2. 西太平洋区域渔业管理委员会是一个准政府机构,负责规定捕鱼季节和年度捕捞限额。该委员会(也被称为 WESPAC)告诉总统,由于全球冠状病毒大流行,迫切需要"快速行动",以满足对金枪鱼罐头的"异常高的零售需求"。

3. 该委员会执行董事凯蒂·西蒙兹在 5 月 8 日给特朗普的一封信中写道:"我们注意到,太平洋海洋国家保护区的捕鱼限制正在阻碍美国在太平洋的三个主要金枪鱼渔业和美属萨摩亚的星科瑞斯特金枪鱼罐厂以最佳水平运营,这些捕鱼限制是不必要的,因为它们没有得到证明的保护效益。"该委员会的要求在特朗普签署一项行政命令的第二天抵达华盛顿,该行政命令公布了一项旨在促进美国海产品行业的经济增长和竞争力的广泛倡议。该命令包括邀请美国 8 个区域渔业管理委员会就减少国内渔业监管负担的方法提出建议。

4. 西太平洋区域渔业管理委员会要求取消对海洋国家纪念碑的限制,如果特朗普批准,可能会产生深远的影响。它几乎肯定会在国会引发一场政治斗争,并在法庭上引发一场关于总统权力限制的法律斗争。两起寻求澄清总统在国家纪念碑方面权力的诉讼已经在联邦法院提起,分别由法官和一些关注纪念碑政治的专家提出,他们想知道特朗普政府是否有兴趣发起第三起诉讼。"到目前为止,我们预计这一要求不会带来太多结果。但如果真的发生了,我们准备上法庭",地球正义檀香山办公室的律师大卫·亨金说。地球正义是一个环境组织,致力于建立最大的太平洋海洋纪念碑。"特朗普总统没有向捕鱼开放纪念碑的行政权力,也不能零敲碎打地拆除它们"。西太平洋区域渔业管理委员会声称,对保护区捕鱼的限制给金枪鱼船队带来了经济困难,这是一个长期辩论的主题。为了澄清这个问题,一个由经济学家和科学家组成的团队今年早些时候在《自然》杂志上发表了一项研究,发现古迹的扩张对捕捞"几乎没有负面影响"。事实上,研究小组发现,在纪念碑扩建后,夏威夷的延绳钓船队捕获了更多的鱼。

5. 海洋生物学家、《国家地理》"原始海洋"项目创始人安里克·萨拉说:"向渔业开放海洋国家纪念碑,就像向工业规模的狩猎开放黄石国家公园一样。"萨拉帮助创建和扩建了太平洋偏远岛屿纪念碑。汉基恩补充说,大流行并没有为扩大渔业提供理由,因为新冠造成的问题不是供应不足,而是需求不足。"火奴鲁鲁的鱼市上有大量腐烂的鱼,"他说。"随着所有酒店和餐馆的关闭,对金枪鱼的需求急剧下降。这些船不出去捕鱼,不是因为纪念碑离他们很近,而是因为他们赚不到钱。"

6. 自1906年西奥多·罗斯福总统签署《古物法》成为法律以来,总统一直拥有保护联邦土地和水资源不受开发的权力。从那时起,除了尼克松、里根和乔治·布什,所有总统都建立了国家纪念碑。现在总共有158个纪念碑,包括最近增加的一个,肯塔基州一个373英亩的内战营地,奥巴马总统在2016年将其指定为国家历史地标,特朗普在2018年将其升级为纪念碑。国家纪念碑被重新调整、废除,变成国家公园。但直到现在,总统改变前任创建的纪念碑的能力,包括改变它们的边界或规定性保护,从未在法庭上受到考验。

7. 2017年,特朗普在缩小了位于犹他州南部红石县的两座纪念碑——克林顿总统1996年创建的大楼梯和奥巴马2016年创建的熊耳后,被几个环保组织和美国原住民部落起诉。这两个纪念碑总共占地420万英亩。特朗普将它们缩小了近400万英亩,这是美国历史上联邦土地保护的最大降幅。一位联邦法官驳回了政府驳回该诉讼的努力,也驳回了政府将此案从华盛顿特区转移到犹他州一家联邦法院的请求。最新一轮案情摘要将于6月提交。

8. 与此同时,在几家商业渔业协会提起的诉讼中,哥伦比亚特区巡回上诉法院去年12月驳回了他们的指控,即东北峡谷和海山海洋国家纪念碑是奥巴马总统在2016年非法创建的。法院的结论是,奥巴马在他作为总统的权力范围内创建了这座纪念碑,它位于科德角海岸附近,占地近5 000平方英里。它保护着一批可以追溯到1亿年前的水下峡谷和山脉,是海龟、几种鲸鱼和深海冷水珊瑚的家园。

9. 这四个太平洋海洋纪念碑都是由乔治·布什总统创建的。2006年,他指定了西北夏威夷群岛海洋纪念碑,以保护夏威夷群岛西北部200英里专属经济区范围内的海洋。后来更名为帕帕哈瑙莫夸基亚。

10. 2009年,布什创建了另外三个海洋国家纪念碑:美属萨摩亚附近的玫瑰环礁海洋纪念碑、马里亚纳海沟海洋纪念碑以及太平洋偏远岛屿海洋纪念碑。这是太平洋中部夏威夷附近的一系列岛屿。奥巴马在2014年和2016年分别扩建了太平洋偏远岛屿海洋纪念碑和帕帕哈瑙莫夸基亚国家海洋保护区,将其扩大了4倍,分别成为世界第五和第三大海洋保护区。西太平洋渔业委员会反对建立和扩大这些纪念碑,称它们侵犯了该委员会在美国太平洋水域管理渔业的权力。加在一起,这些扩大后的纪念碑保护了超过100万平方英里的土地——比阿拉斯加、得克萨斯州和加利福尼亚州加起来还要大。它们的巨大规模支持了渔业委员会的说法,即这些纪念碑给金枪鱼捕鱼船队带来了经济困难。

11. 当特朗普在任期早期启动一项研究,寻找他可以缩小的陆地或海洋纪念碑时,该委员会建议缩小这些纪念碑,尽管当时的内政部长莱恩·辛克向特朗普提出的最终建议中只包括两个玫瑰环礁和太平洋偏远岛屿海洋纪念碑。特朗普没有按照建议采取行动。美国众议员艾德说,考虑到该委员会的历史,这一要求并不奇怪,但应该被视为对保护区的另一个威胁。"一些渔业委员会在可持续的基础上管理渔业方面做得非常好",他说:"西太平洋区域渔业管理委员在开采方面要倾斜得多,我长期以来一直担心。开放海洋纪念碑和在边缘可持续性上管理渔业的愿望并不是解决方案。"总统的行政命令要求渔业理事会在180天内提交建议。他们的期限是11月8日,选举日。

Cultural Background Knowledge: Fishery Cultural
渔业文化背景知识

当一种文化的内容变化引起其结构性、全局性、整体性变化之时,便形成文化变迁;文化只有在不断地变迁中,才能获得发展和进步。

旧石器、新石器时代,我国渔文化曾具有超越一切的主导地位:由鱼骨的穿凿、涂饰,发展到彩绘、刻划、雕凿、陶塑、研磨,一切新技术几乎都投向了渔文化的创造;黄河流域和长江流域出现了不期而遇的多点交映局面,留下了大量的鱼图、鱼物和捕鱼、食鱼、信鱼、拜鱼的社会信息。在原始社会,渔文化以图腾崇拜、生殖信仰和物阜祈盼为主,围绕物质和人"两种生产",发挥组织、教化与改造的功能作用。进入阶级社会以后,渔文化经历了宗教化、制度化、哲学化与艺术化的重建,其内涵日趋复杂,在社会的物质生活、精神生活与仪式礼制中,显示认识、整合、选择、向心、满足等功能作用。随着农耕的发展,特别是龙的冲击,"龙尊鱼卑"的人为划分导致人们信仰重心的位移,造成鱼龙混杂、尊卑互映的文化形态。随着社会实践与认知范围的扩大、民族融和与文化交流的拓展、人为宗教的兴盛,渔文化的神秘性在俗信化的趋势中淡化,不断地在适应与整合中变迁,在民间文化领域得到长久承传,在现实生活中求得了生存与发展。

相传盘古开天地,混沌初开、天地相连,鳌鱼献出四腿、顶天立地于东南西北四角。这是一种舍己为人的牺牲精神。而大禹治水时,见洛水有一神龟游来,它的背上负有一至九之数,谓之"洛书"。伏羲氏见龙马负图出河,又根据图文画成八卦。"河图洛书"成为时代太平之象征,龙马神龟的贡献也表现出一种企盼吉祥的奉献精神。传说大禹治水时劈开黄河中游的一座山,形成了两岸峭壁对峙、河水湍急的"龙门",河鱼若能跃过就化为龙,否则头额触破败退而回。

黄河鲤鱼逆流而上、纷纷跳跃悬水数十仞的龙门,这也成为我们常说的"鲤鱼跃龙门",是一种奋发向上的进取精神,寓意打开美好前程的大门。鱼类昼夜不合双眼,时刻保持警觉、清醒,因而其成了道、佛两教的法器。这是一种难能可贵的自律精神。然而,所有这些,都构成一个共同的协调相处的生态图景——鱼水情谊。"鱼水和谐"既是大自然生态协调的方式,更是渔文化的精神实质。在建设和谐社会中,鱼水和谐给予我们特别的启示和引导,将显示出越来越顽强与执着的时代生命力。

Unit Ten
Marine Mythology

海洋神话

Text A

Women in the Marine Mythology of Ancient Mediterranean

By George Pararas-Carayannis and Amanda Laoupi

Ideological and Political Education：追溯神话传承密码 构筑共有精神家园

海洋神话产生于蒙昧时期或人类早期，隐藏远古时期生存环境极其恶劣、生产条件极端落后条件下的环境条件要素，刻录人类的生产生活痕迹，隐射人类的思想观念及其发展变化轨迹，具有其他任何载体都无法替代的功用，是不可多得的"活化石"。这种"活化石"是万物之源，是文化的核心，是国家的珍贵财富，是国家软实力构筑的根源。

海洋神话是人类早期的产物，具有"原始思维"根性。这种"原始思维"蕴含着早期的宇宙观、世界观、自然观、社会观和人生观。"原始思维"的存在，其背后深层根源是人类生存的压力。人类农业时代到来之前处于采集时代、渔猎时代，游动式采集、迁徙式渔猎为人类生存的根本方式。追逐渔猎成为人类早期的本能。由于史前时期人类口语的简单性、无文字性，人类族群的活动无从考证，海洋文化更是如此。远古时期的人类活动遗存只能从神话、原始宗教、图腾、血缘要素、特殊符号、特殊烙印，以及极少数考古遗址、器具、岩画去考究。中华远古先民在成千上万年的渔猎、采集、战争、贸易、漂移、迁徙过程中，产生出无数次族群性小迁徙及民族历史大迁徙，形成了分布于世界各地或世界某区域的神话线路、洪水线路、渔猎线路、血脉线路、图腾线路、文化碎片线路等，这些线路或轨迹用其他方法难以破解，但如果以"原始思维"去透视其发展轨迹及内涵，便可揭示出另一个世界图景。

功能主义认为，神话的功能是弥补现实差距。因而，与社会事实反差越大，神话的弥补功能就越强烈。要理解远古海洋神话密码，绝不能单从故事的表面去理解内容，而是要理解背后的结构，以及结构背后的另一个结构。要学会理解"原始思维"，理解这种思维的结构、特点、性质及方向，并借助于"原始思维"还原过去图景，增强还原力、透视力，才可能真正揭示出海洋神话背后的秘密，才有可能真正透视世界。

1. Introduction. Throughout human history and in all ancient societies women constituted a significant and active force in sustaining the development of communities, safeguarding resources, educating youth and ensuring continuity of social, cultural and historical heritage values. Although this role is not **explicitly** stated in ancient texts, the impact and influence of women is evident by implied symbolisms in mythology. For example, the circum-Mediterranean area, a cradle of civilization, embodied a rich variety of feminine symbolic expressions that echoes the socio-economic structures of past societies, as well as the impact of the sea upon their fate. Noteworthy is also the fact that water was initially part of feminine symbolism.

2. The connection between the feminine element and water involving Mediterranean coastal communities dates back to Prehistoric Times, as aquatic features, marine disasters and natural phenomena (**tsunami**, flooding, stormy winds & rainfalls, **submergence** of islands and coastal areas, coastal erosion and **transgression/regression** of the seashore, sea currents, **isthmuses** and straits, tides and **whirlpools**) were strongly interrelated with human life and the progress of civilization (i. e. navigation, **archaeoastronomy**, socio-economic contacts via a sea communication network, wars and geopolitical conflicts). Women helped increase the awareness of youth through communication and education and contributed to the **remediation** of damage caused by environmental or man-induced hazards. Youths, upon reaching maturity, were better prepared to assume roles of leadership in **alleviating** the impact of environmental hazards threatening communities and their resources.

3. The aim of this paper is to: (a) illustrate the presence and importance of the **aforementioned** feminine elements in the marine mythology of ancient Mediterranean through philological and archaeological evidence, along with other social and religious testimonies; (b) determine their **spatio-temporal** distribution within the process of symbols' migration, and (c) group them into coherent cycles (thematic, phyletic, and other) in order to **elucidate** their diffusion and importance in the ancient world.

4. In brief, the analysis of the mythological symbolisms illustrates the important and continuous role that women have always played in protecting marine resources and in helping conserve the heritage of mankind—a role that must be properly acknowledged, appreciated and encouraged, now and in the future.

5. Grouping the Mediterranean Symbolisms. People of the circum-Mediterranean area of the ancient world embraced a rich variety of feminine symbolisms. Survival of their coastal communities, productivity of their aquatic ecosystems and omens of the priesthood were all dependent on female deities. Even the splendor and glory of the

accomplishments of these seafaring nations were dependent upon divine feminine interventions (i. e. the goddess Athena helping Odysseus). And conversely, so were their hardships and losses (i. e. Poseidon's wrath—Scylla & Charybdis). Water was considered a feminine realm by these ancient communities and was represented by a variety of rich and diverse feminine, marine symbolisms.

6. Indicative examples are the following six categories:

   1) The **primordial** forces of waters (the Sumerian Nammu—sea goddess and creator of heaven, earth, the Egyptian watery chaos out of which Nun emerged and Isis as protector of seamen, the Phoenician Astarte called Asherar-yam 'our lady of the sea', the Greek Tethysand Eurynome);

   2) Sea creatures and monsters (Scylla & Charybdis, Keto, Sirens, Circe & Kalypso);

   3) Nymphs and other aquatic deities (the Minoan Diktynna & Britomartys, Thetis and Amphitrite, other Nereids & Oceaneids, Aphrodite);

   4) Heroines with a "suffering" connection to the sea (Andromeda, Danae, Alcyone, Ino-Leucothea and her child Melikertes—Palaimon);

   5) Other sea figures whose names were associated with the seas (Myrto, Gorge and Hyrie); and

   6) Some special cases with multi-layered symbolism such as Ariadne and her watery symbol—the labyrinth, Leto/Asteria & Helle—who fell onto the sea (Hellespont).

7. The Archetypal Symbolism of Creation: Waters: A Women's Realm. The world was once thought to be composed of the four basic elements of water, fire, earth and air. This concept is of little use to modern science, which has defined many more elements than these original basic four. However, the four elements still maintain a powerful symbolism within the overall realm of imaginative experience, possessing a strong correspondence to internal states and emotions.

8. In human spirituality, the element of water was always connected to the female nature. The cold, moist properties of water symbolize the enclosing, generating forces of the womb, intuition and the unconscious mind. And by way of the color blue—the color of light, electricity and the oceans—consciousness awareness spirals and cultivates the feminine intuitive, creative part of the brain.

9. Historical symbolic representations of water—such as the alchemical/magical symbol of an inverted triangle that symbolizes the downward, gravitational flow of water—and the Cup or **Chalice**—symbolic of the water triangle—parallel the ancient feminine symbolisms of a downward pointing triangle (the representation of female genitalia) and the feminine elements of intuition, **gestation**, psychic ability, and the subconscious,

respectively. Moreover, the Cup also stands as a symbol of the Goddess, the womb and the female generative organs.

10. One of the earliest symbols in human history is the zigzag, which was used by Neanderthals around 40,000 B. C. which according to Marija Gimbutas, represents water. The 'M' symbol is interpreted as shorthand for the zigzag, and it is found on water containers. The chevron (repetitive form of the 'V') is often found along with the meander, which is also a water symbol. On the other hand, the **meander** was a symbol of the Great Mother Goddess and her life-giving and nourishing aspects. It was from the divine waters of the Mother's womb that life came into existence. Some of the earliest depictions of the Goddess showed her as a hybrid woman/water bird. Without water, life cannot be sustained, and for ancient people waterfowl were an important source of food. Consequently, as water is an archetype from where all life flows, Gimbutas claims it to be representative of the Mother Goddess as the Life-Giver. And, although evidence for the Goddess culture is still disputable, the new feminist views in archaeology have made many archaeologists re-evaluate their concepts of civilization.

11. Later on, in Ancient Egypt, the **hieroglyphic** sign for water was a horizontal zigzag line. The small sharp crests on this sign appear to represent wavelets or ripples on the water's surface. Egyptian artists indicated bodies of water, such as a lake, or a pool or the **primeval** ocean, by placing the zigzag line in a vertical position and then multiplying it in an equally spaced pattern. Of significance is the ancient Egyptian name for water, "uat", which also meant the color green and, for ancient Egyptians, characterized the hard green stone, the emerald, and the green **feldspar**. Of these stones, the emerald is associated with romantic love and the sensual side of nature; and sacred to the goddess Aphrodite/Venus, a sea-divinity who was born from the sea. Some early Greek thinkers conceived the sea-divinities as feminine primordial powers, since the Goddess of the Sea took a part in the creation of the world, due to an old belief, in which life began in the water.

12. The oceans and other large bodies of water in the world are a type of middle ground between the activity of rivers and the passiveness and reflection of lakes. In the ancient world, the protectors of these water systems, the Sea-deities, the Limnades (of the lakes, marshes and swamps), and the Naiades (of the springs, fountains and rivers), maintained the balance. The symbol of **agitated** 'troubled waters' has traditionally related to the illusions and vanities of life. Agitated waters are more subject to climatic conditions involving wind, while deep waters such as seas, lakes and wells have a symbolism related to the dead and the supernatural. Conversely, water plays a major

part in various weather phenomena. Rainstorms and snowstorms involve the free-fall of water from above to below. Floods occur when containment of water fails. Tsunamis, hurricanes and tornados involve the movement of water and turbulence in the seas. Clouds, fog, humidity and mist symbolize in-between states where water is mixed with air. Like a time of twilight between night and day, fog and mist are the 'twilight' states of water and air.

13. The Dual Nature of Sea Creatures. Since ancient times, washing with water has signified, in both a literal and metaphoric sense, the process of cleansing, purification, transformation and **metamorphosis**. Although water itself may contain the power to bring about change, it also serves as the medium through which a god, goddess, or priest exercises change. Change in the form of physical transformation or metamorphosis is characteristic of the female nature through **menstruation** and birth. Water, identified as female and associated with women, symbolizes seduction and transformation, a powerful and often feared aspect of women by men. Death and destruction were usually the fate of mortal men that were seduced by divine females, as depicted by the paradigms of Aphrodite, Artemis, Circe and Kalypso.

14. Moreover, many cultures believed that life sprang from the primordial waters that symbolize life and eternal youth. On the other hand, too much water could be harmful and life threatening. Water in large quantities, such as in the sea, contains a power, which can sustain life but can also destroy life and good order. Numerous traditions around the world have stories of sea monsters symbolizing the violent threat of the sea. In Babylonian mythology, the sea monster Tiamat threatened to overthrow the gods. In other stories, human sacrifice was necessary to appease these sea monsters. The sacrificial victim was often a young virgin who, as in the cases of Andromeda, may be lucky enough to be rescued at the last moment by a male hero. Some sea monsters, which remained located in a single place, were represented as female. The most notorious of those were Scylla and Charybdis, symbols of marine disasters and doom in Greek mythology. Equally, the mysterious depths of the ocean and the potentially destructive aspects of the sea were also personified as female in form, such as Mermaids and Sirens, the latter memorably encountered by Odysseus.

15. On the contrary, compassion, salvage and initiation were also attributed to the female nature. Female goddesses, like Isis and Athena, were protectors of mortals undertaking long open-sea journeys (such as the Argonautics and the voyage of Odysseus); and Aphrodite and Ino-Leucothea were the super-natural protectors of fishermen, sailors, navigation and the entire sea world.

16. Gendered Landscapes of the Past and Their Implications for the Present. The basic masculine and feminine symbolism of the elements finds a correspondence in place symbolism and the gendered landscapes. The most distinctive characteristics of world ecosystems relate to climatic conditions and physical landscape. Climate directly relates to the amount of water contained in ecosystems and the major aspect of physical landscapes is verticality. In this sense, the major natural areas of the world can be divided between those that are dry, wet, low or high. The quality of dryness and height is related to the elements of air and fire and that of wetness and lowness to water and earth. Thus, using these criteria we interpret the division of the natural world into masculine and feminine places.

17. Although the sea world has always been a feminine realm since early prehistory, gender dichotomies have existed. Anthropological linguistics of the Mediterranean illustrates the 'taboos' of the aquatic psychology and marine folklore, expressed through sexual dimorphism. For example, the ship (a male symbol), which has a feminine name in English (ship) and Latin (navis), 'penetrates' into the female sea. But nevertheless, women were usually not allowed to go on board. Gender dichotomies such as these have served to reinforce gender differences between activities over time, and further define feminine and masculine roles in society. The majority of **ethnographic** evidence has suggested that, during ancient times, gender differences between activities in feminine realms existed, such as open-sea fishing, a mainly male task, and shellfish gathering, a mainly female task.

18. The degree to which there are innate differences between male and female behaviors is one of the most challenging issues in the study of gender behavior today. Understanding gender preferences for particular types of tasks assists both men and women, and ultimately youths, to expand their awareness through communication and education into new fields of work and/or learn how to utilize, to a better degree, the resources around them.

19. Women's productive roles in the value of water systems are crucial as they relate to using and managing these resources. Water is essential for all forms of life and crucial for human development. Water systems, including oceans, wetlands, coastal zones, surface waters and acquifers provide a vast majority of environmental goods and services, including drinking water, transport and food.

20. Today, women should continue to be positive agents of change, in both developmental and environmental causes, because the female nature has a more holistic, symbiotic relationship with the surroundings. Although often limited by preconceived

assumptions regarding their gender's role in society, women, by nature, would make better and more effective nurturers of the sea-world.

21. It is suggested therefore, during the process of environmental impact assessments for developmental projects, the identification of gender dichotomies, as well as comprehensive gender analyses should be included as an essential ingredient of the formula. This approach empowers women and may result in more efficient, sustainable development of local areas and the overall welfare of communities.

22. Throughout the circum-Mediterranean area, great matriarchal civilizations flourished (e. g. Anatolian, Minoan & Cycladic, Etruscan), embedded with a rich variety of feminine symbolic expressions. The implied symbolisms in mythology strongly suggest that women constituted a significant and active force in sustaining the development of ancient communities, safeguarding their resources, nurturing youth and ensuring continuity of social and cultural values. Because of the more holistic and symbiotic relationship with their surroundings, women have always been positive agents of change in protecting aquatic resources—a role that must be properly acknowledged, appreciated and encouraged, now and in the future.

(2,305 words)

https://www.researchgate.net

New Words

1. explicitly [ɪkˈsplɪsɪtli]
   ad. in a way that is clear and exact 明确地，明白地
   It should be explicitly stated exactly what the grant covers.
   应清楚说明拨款的确切适用范围。

2. tsunami [tsuːˈnɑːmi]
   n. an extremely large wave caused by a violent movement of the earth under the sea 海啸
   More than 600,000 people were displaced by the tsunami.
   超过60万人被海啸搞得背井离乡。

3. submerge [səbˈmɜːdʒ]
   v. to plunge, sink, or dive or cause to plunge, sink, or dive below the surface of water （使）潜入水中，（使）没入水中；浸没
   She was taken to hospital after being submerged in an icy river for 45 minutes.
   她在冰冷的河水中浸了45分钟后被送进了医院。

4. transgression [trænzˈɡreʃ(ə)n]
   n. the act or process of breaking a law or moral rule, or an example of this 违反，违法；侵犯；过失，错误

Adolescence is a period marked by fascination with the transgression of rules.

青春期是一个以迷恋违反规则为特征的时期。

5. regression [rɪˈgreʃn]

   n. the fact of an illness or its symptoms (= effects) becoming less severe(病症)消退

   The treatment was more effective than tamoxifen, which causes complete regression in only about 30% of cases.

   这种治疗比三苯氧胺更为有效,只有约 30% 的患者在接受三苯氧胺治疗后病症完全消退。

6. isthmus [ˈɪsməs]

   n. a narrow piece of land with water on each side that joins two larger areas of land 地峡

   North and South American species migrating across the Isthmus now came into competition with each other.

   北美和南美的物种迁徙穿过地峡,现在开始互相竞争。

7. whirlpool [ˈwɜːlpuːl]

   n. a small area of the sea or other water in which there is a powerful, circular current of water that can pull objects down into its center 漩涡

   The recent rain had churned up the waterfall into a muddy whirlpool.

   近来的雨水把那条瀑布搅成了泥水漩涡。

8. archaeoastronomy [ˌɑːkiəʊəˈstrɒnəmi]

   n. the scientific study of the beliefs and practices concerning astronomy that existed in ancient and prehistoric civilizations 考古天文学(指对古代天文学进行研究的考古学科)

   Archaeoastronomy depends for its discoveries on experts in many disciplines.

   考古天文学上的发现有赖于许多学科的专家。

9. remediation [rɪˌmiːdiˈeɪʃn]

   n. the process of improving or correcting a situation 补救;矫正,纠正

   After several years of cultivation and harvest, the site would be restored at a cost much lower than the price of excavation and reburial, the standard practice for remediation of contaminated soils.

   经过几年的耕种和收获,该遗址将以远低于挖掘和重新掩埋费用的成本进行修复,这也是修复污染土壤的标准做法。

10. alleviate [əˈliːvieɪt]

    v. to make something bad such as pain or problems less severe 减轻;缓和,缓解

    The drugs did nothing to alleviate her pain/suffering.

    这些药物对减轻她的病痛/痛苦没有丝毫作用。

11. afore-mentioned [əˈfɔːˌmenʃənd]

    a. mentioned earlier 前面提到的;上述的

    The aforementioned Mr. Parkes then entered the cinema.

    前面提到的这位帕克斯先生接着进入了电影院。

12. spatio-temporal [ˌspeɪʃiəʊˈtempərəl]

a. relating to both space and time, or to space-time (= the part of Einstein's Theory of Relativity that adds the idea of time to those of height, depth, and length)时空的

Where scientists think about viruses, electrons or stars, philosophers think about spatiotemporal continuants, universals and identity.

在科学家研究病毒、电子或恒星时,哲学家思考的是时空连续性、普遍性和同一性。

13. elucidate [ɪˈluːsɪdeɪt]

v. to explain something or make something clear 阐明;解释

The reasons for the change in weather conditions have been elucidated by several scientists.

已经有数位科学家对天气状况变化的原因做了解释。

14. primordial [praɪˈmɔːdɪəl]

a. existing at or since the beginning of the world or the universe(从)太古时代存在的,原始的

The planet Jupiter contains large amounts of the primordial gas and dust out of which the solar system was formed.

木星含有大量的原始气体和尘埃,太阳系就是由这些气体和尘埃形成的。

15. archetypal [ˌɑːkɪˈtaɪpl]

a. typical of an original thing from which others are copied 原型的;典型的

The archetypal extrovert prefers action to contemplation, risk-taking to heed-taking, certainty to doubt.

典型的外向者更喜欢行动而不是沉思,喜欢冒险而不是谨慎,喜欢确定性而不是疑虑。

16. chalice [ˈtʃælɪs]

n. in Christian ceremonies, a large, decorative gold or silver cup from which wine is drunk(基督教仪式中用金或银装饰的)圣餐杯,高脚酒杯

Some churches have also begun to allow "intinction" or dipping bread in communion wine rather than sharing the chalice.

一些教堂也开始允许"面包蘸酒",即在圣餐酒中蘸面包,而不再共用圣餐杯。

17. gestation [dʒeˈsteɪʃn]

n. (the period of) the development of a child or young animal while it is still inside its mother's body 怀孕(期),妊娠(期)

The baby was born prematurely at 28 weeks gestation.

胎儿在怀孕的第28周早产。

18. meander [miˈændə(r)]

n. a bend or curve, as in a stream or river; an aimless amble on a winding course 河流(或道路)弯道,河曲

Major changes in a river course result in both lateral shifts of the old meander belt and in the development of new belts.

河道中大的变动,造成老河曲带的侧向迁移和新河曲带的产生。

19. hieroglyphic [ˌhaɪərəˈɡlɪfɪk]

a. written in, constituting, or belonging to a system of writing mainly in pictorial characters 象形文字的；(尤指艺术)风格化的；似象形文字的

It contains the same exact text written in three different alphabets: Greek, demotic, hieroglyphic.

它包含了完全相同的文本,用三种不同的字母书写:希腊语,通俗语,象形文字。

20. primeval [praɪˈmiːv(ə)l]

    a. ancient; existing at or from a very early time 原始的；早期的

    The railway cuts through a primeval forest.

    铁路穿越原始森林。

21. feldspar [ˈfeldspɑː(r)]

    n. a common type of mineral found especially in igneous rocks such as granite 长石(一种常见矿石,尤见于火成岩,如花岗岩中)

    Color is an important indicator to identify potash feldspar.

    颜色是鉴别钾长石的一个重要标志。

22. agitate [ˈædʒɪteɪt]

    v. to make someone feel worried or angry 使焦虑；使躁动不安

    I didn't want to agitate her by telling her.

    我不想告诉她,免得她焦虑。

23. metamorphosis [ˌmetəˈmɔːfəsɪs]

    n. a complete change 彻底的变化

    Under the new editor, the magazine has undergone a metamorphosis.

    在新主编的领导下,这份杂志彻底改头换面了。

24. menstruation [ˌmenstruˈeɪʃn]

    n. an occasion when a woman menstruates 行经,月经来潮

    Menstruation may cease when a woman is anywhere between forty-five and fifty years of age.

    月经可能在女性45至50岁之间停止。

25. ethnographic [ˌeθnəˈɡræfɪk]

    a. relating to ethnography 人种志的,人种论的；人种志著作的

    Fortunately, he had largely completed assembling the ethnographic material.

    幸运的是,他大体完成了收集民族志材料的工作。

Phrases and Expressions

date back to 追溯到；始于；可追溯到
assume the role of 担任……的角色,承担……的任务
be associated with 与……有关
by way of 通过……方式；取道；经过
come into existence 成立,建立
conceive……as 认为……是

be subject to 受支配,从属于;可以……的;常遭受……
in a sense 从某种意义上说;就某种意义来说;就某种意义而言
be characteristic of 具有……的特性,为……所特有
identify……as 认定为;确定为

Terminology

spatio-temporal distribution 时空分布
circum-Mediterranean area 环地中海地区
gender dichotomies 性别二分法
sexual dimorphism 性别二态现象;两性异型;雌雄二型;雌雄异型
matriarchal civilizations 母系文明

Proper Names

Prehistoric Times 史前时代;史前时期

一、翻译策略/方法/技巧:抽象译法

　　为了译文的忠实与通顺,我们往往有必要把原文中带有具体意义或具体形象的抽象单词、词组、成语或句子进行抽象化处理。这种译法,我们称之为抽象译法。这种译法一方面把比喻形象上较为具体的单词词组或成语进行抽象概括,如:开门见山 come straight,狗急跳墙 do something desperate。另一方面,抽象译法也包括带有范畴词的具体化名词在语言形式上的抽象化译法,主要是在翻译过程中去掉后面的范畴词,如:谦虚态度 modesty,发展过程 development,无知的表现 innocence。西方思维传统注重科学、理性,重视分析、实证,在论证、推演中认识事物,然后对其本质进行总结、归纳和抽象思维,挖掘其规律。中国传统思维注重实践经验、整体思考,常常借助直觉体悟从总体上模糊而直接地把握认识对象的内在本质和规律。另外,英美人的抽象思维占据主流趋势,这也离不开英语语言所倚赖的哲学传统和拼音语言系统,倾向于使用表达同类事物的整体词语来表达具体事物或客观现象,尤其是喜欢用抽象性的名词、不定式、动名词乃至主语从句来充当话题的主体;相对而言,中国人的形象思维较发达,这既与中国传统哲学有关,也与汉语的象形文字有关。抽象表达法在英语里使用得相当普遍,尤其常用于社会科学论著、官方文章、报刊评论、法律文书和商业信件等文体中。G.M.Young 曾指出,"an excessive reliance on the noun at the expense of the verb will, in the end, detach the mind of the writer from the realities of here and now, from when and how and in what mood the thing was done, and insensibly induce a habit of abstraction, generalization"。英语的抽象表达法主要见于大量使用抽象名词。

　　例如:

　　1. Most people in America, minority people in particular, are convinced that injustices exist in the economic system.

　　译文:大多数美国人,尤其是少数民族,都确信美国经济制度中存在着不公正现象。

2. Preparations for the summit meeting continued.

译文:最高级会议的准备工作继续进行。

然而,汉语用词倾向于具体,常常以实的形式表达虚的概念,以具体的形象表达抽象的内容。汉语缺乏像英语那样的词缀虚化手段。汉语没有形态变化,形式相同的词可以是名词,也可以是动词,还可以是形容词或其他词。如元代马致远的《天净沙·秋思》:

　　枯藤老树昏鸦,
　　小桥流水人家,
　　古道西风瘦马,
　　夕阳西下,
　　断肠人在天涯。

整个散曲基本上由事物的名词构成,几乎没有虚词,全诗借用具体的形象表达了抽象的内容,语言简洁、逻辑清晰、意义明确、主题鲜明、节奏感强,形散神聚,构成一幅羁旅漂泊者的画面。由于英语的形态特征,尽管译者 Cyril Birch 尽可能再现原文的风格,其译文还是必须使用形态或形式词语,如 at、the、and、with 和分词后缀-ing,以及单复数和第三人称等。请欣赏这首散曲的译文:

　　Autumn Thoughts
　　Dry vine, old tree, crows at dusk,
　　Low bridge, stream running, cottage,
　　Ancient road, west wind, lean nag,
　　The sun westering,
　　And one with breaking heart at the sky's edge.

总的来说,英语民族擅长用抽象概念表达具体的事物,比较重视抽象能力的运用;而汉民族更习惯于运用形象的方法表达抽象的概念,不太重视纯粹意义的抽象思维。但是,抽象与具体的分类只是相对而言,英语存在着大量的非常具体的描述,汉语中也不乏抽象的表达。为了表达的需要,在翻译过程中,我们也常将英语中的形象表达做抽象化处理来翻译出其深层的含义。如 keep body and soul together(使身心在一起——维持生计),wake a sleeping dog(弄醒熟睡的狗——惹是生非)。因此,在英汉翻译的过程中,抽象与具体是要视特定的情况进行转换的。

(摘自:冯庆华.实用翻译教程[M].上海:上海外语教育出版社,2010:53.)

二、译例与练习

Translation

1. In brief, the analysis of the mythological symbolisms illustrates the important and continuous role that women have always played in protecting marine resources and in helping conserve the heritage of mankind—a role that must be properly acknowledged, appreciated and encouraged, now and in the future. (Para. 4)

简而言之，对神话象征主义的分析展示了妇女在保护海洋资源以及帮助保护人类遗产方面一直发挥着重要的作用，妇女所扮演的角色以及发挥的作用必须得到适当认可，赞赏和鼓励，无论是在现在还是将来。

2. In human spirituality, the element of water was always connected to the female nature. The cold, moist properties of water symbolize the enclosing, generating forces of the womb, intuition and the unconscious mind. (Para.8)

在人的灵性世界（神话信仰）中，水元素总是与女性的本性相连。水具有冷、湿的特性，是封闭的、能够孕育生命的女性子宫的象征，除此之外，水也象征女性所特有的直觉和潜意识。

【这个句子由两个句子组成，第一句是概括，第二句话是具体说明，翻译的关键是理清句子间的逻辑。第二句句子结构比较简单，主谓宾结构，修饰语有点多，如果直接按照句子顺序翻译成汉语，则不符合汉语多短句的习惯，可以在翻译时适当地对句子进行拆分，把句子的主语单独成句，然后再翻译谓语和第一个宾语部分，原文中symbolize是谓语动词，翻译成汉语时处理成"是……的象征"，这样就和后面的"象征……"形成呼应，又避免了重复，然后把第二个和第三个宾语单提出来，为了呼应前面总括句中"跟女性的本性相连"，在处理第二个和第三个宾语时虽然原句中没有女性相关的字样，翻译时还是进行了增译，增加了"女性所特有的"，体现与总括句的逻辑衔接，并加了一个"除此之外"，凸显前后的句子逻辑关系。】

3. And by way of the color blue—the color of light, electricity and the oceans—consciousness awareness spirals and cultivates the feminine intuitive, creative part of the brain. (Para.8)

蓝色，即光、电和海洋的颜色，能够唤醒人的意识知觉，蓝色还有助于提升女性直觉，也就是能够唤醒、激发人的大脑中产生创造性的那个部分。

【这个句子结构不复杂，有两个并列谓语，然后一个介词短语引导的方式状语，翻译的关键是理清句子间的逻辑。句子的后半部分有两个名词 the feminine intuitive 和 creative part of the brain，这两个名词之间是解释与说明的关系，因此，在翻译时就没有按照原文的形式翻译，而是按照汉语的习惯加上了"也就是"，起到了解释说明的作用。另外，本句的开头就用了一个介词短语作方式状语，原句的主语 consciousness awareness 翻译时也没遵循原文的形式，而是从方式状语中提炼出新主语，即名词蓝色，然后把蓝色为信息起点来提供信息，这样就和上文形成了有机整体，语言连贯性更好。】

4. Historical symbolic representations of water—such as the alchemical/magical symbol of an inverted triangle that symbolizes the downward, gravitational flow of water—and the Cup or Chalice—symbolic of the water triangle—parallel the ancient feminine symbolisms of a downward pointing triangle (the representation of female genitalia) and the feminine elements of intuition, gestation, psychic ability, and the subconscious, respectively. (Para.9)

历史上曾经使用过多种代表水的符号——比如呈倒三角的炼金术符号（魔法符号），该符号象征由于重力，水会向下流动——又比如水杯或圣杯——水三角的象征——跟古代社会所采用的女性象征,即向下三角形(代表女性生殖器)对应,也分别对应各种女性元素,包括女性直觉、妊娠、心理能力和潜意识。

5. Consequently, as water is an archetype from where all life flows, Gimbutas claims it to be representative of the Mother Goddess as the Life-Giver. And, although evidence for the Goddess culture is still disputable, the new feminist views in archaeology have made many archaeologists re-evaluate their concepts of civilization. (Para. 10)

水是所有生命之源,鉴于此,金布塔斯提出水是母亲女神,给予万物生命。尽管人们对能证明女神文化的证据仍存在争议,考古学界还是出现了新的女权主义观点,该观点的出现迫使考古学家们重新评估他们对文明的定义。

【这个句子虽然很长,但是并不算难,翻译的关键是理清句子间的逻辑。这里,as 引导一个原因状语从句,解释金布塔斯持有观点的原因,claim......as 是一个词组,阐述了金布塔斯的观点,在翻译成汉语时,结合汉语多短句子的特征,把这个句子进行了拆分,处理成两个短句,即他"提出水是母亲女神,给予万物生命"。后面 although 引导让步状语从句,引导虽然这种观点还没有被广泛接受,引出后来的考古学界新思想倾向,同上所述,这里也把一句话拆成了两句进行翻译,更符合汉语思维习惯。】

6. Of significance is the ancient Egyptian name for water, "uat", which also meant the color green and, for ancient Egyptians, characterized the hard green stone, the emerald, and the green feldspar. (Para. 11)

重要的一点是:古埃及人给水起名"uat",这个词在埃及语中意味着绿色,对于古埃及人来说,水具有坚硬的绿色石头、祖母绿和绿色长石的典型特征。

7. Some early Greek thinkers conceived the sea-divinities as feminine primordial powers, since the Goddess of the Sea took a part in the creation of the world, due to an old belief, in which life began in the water. (Para. 11)

一些早期的希腊思想家认为女性的原始力量来自海神,这点在古希腊的信仰中可以找到佐证,他们认为:海洋女神参与了创造世界。他们还认为:生命起源于水。

【这个句子虽然很长,但是句子并不算难,翻译的关键是理清句子间的逻辑。这里,since 引导一个原因状语从句,解释希腊思想家持有观点的原因,后面 due to 即这种观点的出处,in which 表明早期希腊信仰中的核心思想还有哪些,只要理顺了思路和逻辑关系,然后按照汉语的前后顺序组织文字即可。】

8. Agitated waters are more subject to climatic conditions involving wind, while deep waters such as seas, lakes and wells have a symbolism related to the dead and the supernatural. (Para. 12)

搅动的水更容易受到比如风等气候条件的影响,而海洋、湖泊和水井等深水则具有与死者和超自然有关的象征意义。

9. Water, identified as female and associated with women, symbolizes seduction and transformation, a powerful and often feared aspect of women by men. (Para. 13)

水经常与女性形象联系在一起,被人格化为女性的水,往往象征着诱惑和转变,这是女性力量强大的一面,也是经常让男性对女性产生恐惧的一面。

10. Equally, the mysterious depths of the ocean and the potentially destructive aspects of the sea were also personified as female in form, such as Mermaids and Sirens, the latter memorably encountered by Odysseus. (Para. 14)

同样,海洋深处极其神秘,另外,海洋还具有潜在的破坏性,所有这些特点通常都被拟人化成为文学作品中的女性形象,例如美人鱼和海妖,奥德修斯在海上遭遇后者,其形象让人印象深刻,令人久久难忘。

Exercises

1. Fill in the blanks with the proper given words, and then translate the sentences into Chinese.

  innate assume feminine assessment encounter sacrifice

1) It's saying everybody has to be counted as an equal even though at the end of the day, one can be ____ for the general welfare.

2) We are born with very different genetic tendencies which society encourages as either "masculine" or "_____".

3) Depression is a common thymogenic mental disease with prominent and persistent low mood as its major clinical feature. It becomes a common disease and frequently _____ disease in modern society.

4) Bodily excretions, death, and rotten smells can be signs of danger or disease, triggering our _____ sense of disgust.

5) The Greeks _____ that the structure of language had some connection with the process of thought, which took root in Europe long before people realized how diverse languages could be.

6) Using caregiver _____ and the children's self-observations, she rated each child's overall sympathy level and his or her tendency to feel negative emotions after moral transgressions.

2. Translate the following sentences into Chinese.

1) You need sound financial advice and a strong plan if you're going to start your own business—it can't be all castles in Spain.

2) Most of the time, playing classical music for high schoolers is like casting pearls before swine. But every so often a few kids appreciate it.

3) He joined the party and turned out to be a wolf in sheep's clothing because he was sent to infiltrate our group for the extreme right wingers.

4) Ms. Anderson always separates the sheep from the goats in her classes. She puts naughty boys in front so as to keep an eye on them.

5) Science deals with things in a practical way. Science means honest, solid knowledge, allowing not an iota of falsehood, and it involves Herculean efforts and grueling toil.

6) Iran's nuclear program is likened to the Sword of Damocles, ever perilously hanging over the head of America, giving it sleepless nights.

7) Stop tantalizing me by having braised pork in front of the screen! It's such a torture for us who can't celebrate the new year at home.

8) With the advent of the Internet boom, he graduated to become an astute investor and had the Midas touch in whatever company he invested in.

9) She won the court case, but it was a Pyrrhic victory because she had to pay so much in legal fees.

10) The old stock had deteriorated and was all mixed up so the manager decided to cut the Gordian knot, sell it all as scrap, and re-stock with new materials.

3. Translate the following sentences into English.
1) 现在电视上掀起了一股舞蹈比赛的热潮，主角是一些并不以舞蹈天赋著称的名人。
2) 可悲的是，他的改革开启了潘多拉魔盒，引发了一系列的国内问题。
3) 历史学家总把一个王国的沦落归咎为红颜祸水是不公平的。
4) 琼斯先生在会上做了长篇发言，大家都认为这是在故意拖延时间。
5) 那个项目看起来好像很有希望，结果招来许多灾祸。
6) 我好奇他们怎么会在夜里吵起来。那辆坏掉的车似乎是争执之端。
7) 只有经历过艰辛和灾难，才可以彰显出真正的友谊。
8) 俄国人对把阿拉斯加廉价割让给美国人一直感到懊悔。
9) 这位歌唱家本星期在巴黎演唱的全部门票都销售一空——显然，公众认为这将是她最后一次告别演出。
10) 虽然这个城市改革任务异常艰巨，他们还是不懈坚持着。

4. Choose the best paragraph translation. And then answer why you choose the first translation or the second one.

大荒之中，有山名曰日月山，天枢也。吴姖天门，日月所入。有神，人面无臂，两足反属于头山，名曰嘘。颛顼生老童，老童生重及黎，帝令重献上天，令黎邛下地。下地是生噎，处于西极，以行日月星辰之行次。

有系昆之山者，有共工之台，射者不敢北乡。有人衣青衣，名曰黄帝女魃。蚩尤作兵伐黄帝，黄帝乃令应龙攻之冀州之野。应龙蓄水，蚩尤请风伯雨师，纵大风雨。黄帝乃下天女曰魃，雨止，遂杀蚩尤。魃不得复上，所居不雨。

译文一：
Inside the Great Wildness, there is a mountain called Mount of the Sun and the Moon which is the pivot of the sky. Wujutianmen is where the sun and the moon set. There is a

god who has a human face and no arms. With two feet bent reversely onto his head, he is called Xu. King Zhuanxu gave birth to Laotong. Laotong gave birth to Chong and Li. The God of heaven ordered Chong to hold up the sky and Li to press down the earth. After finishing his job, Li gave birth to Ye. Then Li lives at the West Pole, presiding over the movement of the sun, the moon and the stars.

There is Mount Xikun and there is the Terrace of Gonggong. Bowmen dare not shoot arrows in the direction of the north. There is a woman who wears green clothes and is called Nuba of Emperor Huangdi. Chiyou once raised an army and attacked Emperor Huangdi. The emperor then ordered Yinglong to attack the army of Chiyou in the wilderness of Jizhou. As Yinglong hoarded up all the water, Chiyou then invited the God of Wind and the God of Rain to make a big rainstorm. Seeing this, Emperor Huangdi sent down Nuba from the sky to stop the rain. Nuba killed Chiyou, but she was unable to go back to heaven and the place she stayed always suffered from a drought.

译文二：

In the middle of the Great Wildness, there is a mountain. Its name is Mount Sunmoon. It is the pivot of the sky. Mount Crygiantess-skygate is here and it is where the sun and the moon set. There is a goddess-human here who has a human face and no arms, and her two feet are doubled black and joined to the top of her head. Her name is Breathout. The great god Fond Care gave birth to Old Child. Old Child gave birth to Layers and Jetblack. The great god commanded Layers to bear the sky up, and he commanded Jetblack to press the earth down. Jetblack came down on earth and gave birth to Choke, and Jetblak stays at the West Pole in order to move the sun, the moon and the stars in their due motion and towards their stations.

Here is Mount Constant offspring. This is where the Terrace of Common Work is situated. Bowmen do not dare to face in its direction. There is someone on this mountain wearing green clothes. Her name is Droughtghoul, daughter of the great god Yellow. The god Jest Much invented weapons. He attacked the great god Yellow. The great god Yellow then ordered Responding Dragon to do battle with Jest Much in the Wildness of Hopeisland. Responding Dragon hoarded up all the water. But the god Jest Much asked the Lord of the Winds and the Leader of the Rains to let loose the strong winds and heavy rain. So the great god Yellow sent down his sky daughter called Droughtghoul and the rain stopped. Then she killed Jest Much. Droughtghoul could not get back to the sky. The place where she lives on never has rain.

5. Translate the following passage into Chinese.

Throughout human history and in all ancient societies women constituted a significant and active force in sustaining the development of communities, safeguarding resources, educating youth and ensuring continuity of social, cultural and historical heritage values. Although this role is not explicitly stated in ancient texts, the impact and influence of women is evident by implied symbolisms in mythology. For example, the circum-Mediterranean area, a cradle of civilization, embodied a rich variety of feminine symbolic expressions that echoes the socio-economic structures of past societies, as well as the impact of the sea upon their fate. Noteworthy is also the fact that water was initially part of feminine symbolism. The connection between the feminine element and water involving Mediterranean coastal communities dates back to Prehistoric Times, as aquatic features, marine disasters and natural phenomena (tsunami, flooding, stormy winds & rainfalls, submergence of islands and coastal areas, coastal erosion and transgression/regression of the seashore, sea currents, isthmuses and straits, tides and whirlpools) were strongly interrelated with human life and the progress of civilization (i.e. navigation, archaeoastronomy, socio-economic contacts via a sea communication network, wars and geopolitical conflicts). Women helped increase the awareness of youth through communication and education and contributed to the remediation of damage caused by environmental or man-induced hazards. Youths, upon reaching maturity, were better prepared to assume roles of leadership in alleviating the impact of environmental hazards threatening communities and their resources.

6．Translate the following passage into English．

尽管中国古代神话没有十分完整的情节，神话人物也没有系统的家谱，但它们却有着鲜明的东方文化特色，其中尤为显著的是它的尚德精神。这种尚德精神在与西方神话，特别是希腊神话进行比较时，显得更加突出。在西方神话，尤其是希腊神话中，对神的褒贬标准多以智慧、力量为准则，而中国古代神话对神的褒贬则多以道德为准绳。这种思维方式深植于中国的文化之中。几千年来，这种尚德精神影响着人们对历史人物的品评与现实人物的期望。

三、翻译家论翻译

英译汉：理解是关键

庄绎传(1933—)北京外国语大学高级翻译学院教授。庄教授认为，英译汉的关键就是透彻理解原文。只有真正理解原文，才能准确译出。若勉强去译，便会采取机械的办法，逐字翻译，许多误译就是这样产生的。

例1：We want to get all the parties back to the negotiating table.

例2：Their differences have been thrown into sharp relief by the present crisis.

虽然 party 一词可以指"政党"，但此处与 negotiating table 相联系，便指"谈判的一方"了。所以，例1的意思是：我们想把有关各方拉回到谈判桌上来。differences 一词本身是有

"差别"的意思,但在这个上下文里,它却指"意见分歧"。例 2 的意思是:目前的危机使得他们的分歧更加引人注目。

例 3:He was found guilty of murder.

例 4:There is no right of appeal against the decision.

涉及法律时,find 不一定表示"发现",而可以指"裁决""判决"。appeal 也不一定表示"呼吁",而可以指"上诉"。因此,例 3 的意思是:经裁决,他犯有谋杀罪。例 4 的意思是:关于这项判决,没有上诉权。

例 5:The end result of her hard work was a place at medical school.

例 6:To graduate with honors from college.

在学校教育方面,work 就指学习,a place 就是 an opportunity to study at a university,也就是一个入学名额,而不是一个工作职位。with honors 指的是"以优异的成绩",而不是"感到荣幸"。因此,例 5 的意思是:她勤劳学习,终于进了医学院。例 6 的意思是:以优异的成绩从大学毕业。

例 7:This new production radically reinterprets the play.

例 8:The doorway is a 19th century reconstruction of Norman work.

在文化方面,production 和戏剧相联系,就指"一次演出"。因此,例 7 的意思是:这次演出体现了对这部戏的全部理解。例 8 是什么意思呢?能不能译作:门廊是 19 世纪罗马建筑的翻版?不行。首先,Norman 不是罗马,而是指 11 世纪欧洲大陆的诺曼人征服英国后在英国流行的诺曼式建筑风格;其次用"19 世纪"修饰"罗马建筑"也是不行的。例 8 的意思是:门廊是 19 世纪时模仿诺曼式建筑修建的。

例 9:You'll be expected to replace any broken glasses.

例 10:Around here, you leave school at sixteen and next thing you know, you're married with three kids.

在生活方面,所谈内容往往与当地的风俗习惯相联系。例 9 就是店主对顾客说的一句话,意思是:玻璃杯如有损坏,你要负责赔偿。例 10 的用词很简单,能不能译作"这儿,你十六岁时离开了学校,接着,你带着三个孩子结了婚。"或译作"……你和有三个孩子的人结了婚。"从原文的时态看,这里说的不是一次性的已经完成的动作,而是一种反复出现的现象,句中的 you 也不是指具体的某人,而是泛指。这样就可以看出这句话说的是当地一种普遍的生活方式。因此,例 10 的意思是:这一带的人十六岁中学毕业,接着就结婚,生三个孩子。

例 11:I hate to say I told you so.

例 12:Ed couldn't make it so they sent me instead.

例 13:Go on—read it to us.

英语有许多习语(idioms),其含义往往不是从字面上可以看出的。以上三例中的 I told you so,make it 和 go on 都是习语,翻译时,不能取其字面上的含义,而要把它看作一个整体来处理。如果不知道它的意思,那就要到词典里去查一查。如果你手边有一本 *Oxford Advanced Learner's Dictionary*,在词条 tell 里就可以查到 I told you (so),解释为:used when sth. bad has happened, to remind sb. that you warned them about it and they did not listen to you. 得到这个解释之后,就能看出例 11 不是"我真不想说是我告诉你的",而是"我不愿意显得自己有先见之明。"用同样的办法查 make it,可以查到 4 条解释,第 3 条解释为:to be able to be present at a place. 因此例 12 的意思就不是"埃德做不出来……",而是"埃德去不了,所以他们就派我去了。"Go on 共有 8 条解释,最后一条是:used to encourage sb. to do sth. 因此,例 13 的意思就不是"继续——给我们读下去",而是"念吧——念给我们听听。"综上所述,一个词用在不同的场合会有不同的含义,译者不能只想到自己最熟悉的那个含义,而要充分利用上下文,依靠能够获得的相关信息,判断出词的确切含义。遇到习语,更要勤查词典,切忌望文生义。

(摘自:庄绎传. 翻译漫谈[M]. 北京:商务印书馆,2015:112.)

Text B

Why Noah's Ark Will Never Be Found
By Erin Blakemore

Ideological and Political Education：从神话到现实 书写文化认同

海洋文化的特点是流动性。海洋跟陆地不同，陆地之间相通，需要修路，需要马路、公路或铁路；而海洋完全不同，世界四大洋及诸多海域自然连为一片，天生为一体。只要条件许可，只需一叶扁舟，就可以从海洋的某个角落通向其他任何一个角落。海洋文明的过程是相互流通、相互开放的过程。因而，海洋文明的移动、交流、扩散和传播自远古时期就发生，神话的扩散传播就是重要体现。正如洪水兄妹婚神话，不仅在国内汉族和诸多少数民族根深蒂固，在日本、东南亚、印度、欧美尤其是印第安人群体里也有较广泛的传播。这种现象很大程度上有远古海洋文明的扩散交流传播的原因。

海洋神话具有一定的哲理性，在远古时代生产力极度低下、人类认识能力极度受限制的情况下，神话在很大程度上承担认识宇宙、认识自然、辨析人类自身的辨析功能和全民教化功能，在即时语境下，具有某种"真理性"或类似"真理性"功能。然而，真理并不是一成不变的，而是随着环境条件的变化而不断发展变化。随着社会生产力不断进步，人类的认识能力极限不断被突破，真理本身也在不断突破。正是这种"活化石"，构成中华民族文化的血脉基因和不可见的深厚底蕴，构成民族的文化根系和灵魂，构成整个民族乃至中华民族的"潜意识"或"无意识"。正是通过对这种"活化石""原始思维"的对比鉴别、互动反思，更能使不断发展的人类社会真正审视历史、反思自身、透视世界，进而透视原理，透视出真理，找到人类发展的内在原始动力，从而找到巨大历史动能，推动人类文明不断发展飞跃。

1. Noah's Ark is among the best known and most **captivating** of all Old Testament stories: After creating humans, God became so displeased with them that he struck Earth with an all-encompassing flood to wipe them out—with one noteworthy (and seaworthy) exception: the biblical patriarch and his family, accompanied by pairs of each of the planet's animals, who rode out the **deluge** in an enormous wooden vessel. For people who accept the religious text as a historically accurate account of actual events, the hunt for archaeological evidence of the Ark is equally captivating, inspiring some **intrepid** faithful to comb the slopes of Armenia's Mt. Ararat and beyond for traces of the wooden vessel.

2. In 1876, for example, British attorney and politician James Bryce climbed Mount Ararat, where Biblical accounts say the Ark came to rest, and claimed a piece of wood that "suits all the requirements of the case" was in fact a piece of the vessel. More modern Ark "discoveries" take place on a regular basis, from an **optometrist's** report he'd seen it in a rock formation above the mountain in the 1940s to a claim **Evangelical** pastors had found **petrified** wood on the peak in the early 2000s.

3. But searches for the Ark draw everything from **exasperation** to disdain from academic archaeologists and biblical scholars. "No legitimate archaeologist does this," says National Geographic Explorer Jodi Magness, an archaeologist at the University of North Carolina at Chapel Hill, of modern searches for evidence of Noah. "Archaeology is not treasure hunting," she adds. "It's not about finding a specific object. It's a science where we come up with research questions that we hope to answer by excavation."

4. Flood or Fiction? Stories of destructive floods and those who survive them **predate** the *Hebrew Bible*, the oldest parts of which are thought to have been written in the 8th century B.C. Legends about a deluge that destroys civilization at the **behest** of a supernatural deity can be found in multiple Mesopotamian texts, from the *Epic of Gilgamesh*, which was written around the early second millennium B.C., to a recently **deciphered** Babylonian **cuneiform** tablet from about 1750 B.C. describing how the ark was built.

5. Could these flood myths be based in fact? "There does seem to be geological evidence that there was a major flood in the Black Sea region about 7,500 years ago," says National Geographic Explorer Eric Cline, an archaeologist at George Washington University. But scientists disagree on the extent of that event, just as historians of the era differ on whether writings about a deluge were inspired by real life. It seems likelier that floods were simply experienced in different places and at different times—and that those events naturally made their way into the world's oral and written **lore**.

6. Complicating the issue even further, scholars differ on the precise location of Noah's Ark according to the Hebrew Bible. In the *Book of Genesis*, the ark came to rest "upon the mountains of Ararat" located in the ancient kingdom of Urartu, an area that now includes Armenia and parts of eastern Turkey and Iran—not the single, iconic peak that bears its name today. "There's no way we can determine where exactly in the ancient Near East it occurred," says Magness. And both Cline and Magness say that even if artifacts from the Ark have been or will be found, they could never be conclusively connected to historical events.

7. "We have no way of placing Noah, if he really existed, and the flood, if there really was one, in time and space," says Magness. "The only way you could determine that would be if you had an authentic ancient inscription"—and even then, she points out, such an inscription could refer to another Noah, or another flood. That hasn't stopped the **proliferation** of **pseudoarchaeology** that upholds the Bible as literal truth. The fruitless searches are often aligned with adherents of "young-earth creationism," the belief that, despite evidence to the contrary, Earth is only thousands of years old.

8. Same Evidence, Very Different Conclusions. Such groups use secular archaeological evidence to **bolster** their literal interpretation of Scripture—and simply disregard or attempt to disprove evidence to the contrary. But they don't all share the same tactics. Answers In Genesis, a self-described **apologetics** ministry that focuses on scientific issues and even runs a Noah's Ark-themed amusement park in Kentucky, acknowledges the **ubiquity** of flood-related myths beyond the *Old Testament* story of Noah, and even **concedes** that the Ark could never be found.

9. "We do not expect the Ark to have survived and been available to find after 4,350 years," says Andrew A. Snelling, a geologist and Director of Research for Answers in Genesis who has spent decades attempting to prove Earth's youth. Snelling differs from archaeologists, however, about why the vessel's remains will never be found. "With no mature trees available for Noah and his family to build shelters after they got off the Ark, there is every reason to expect they **dismantled** the Ark (which they didn't need anymore) to salvage timber from it," he says. While the ministry does not rule out the potential of one day finding the Ark, Snelling rues what he calls "questionable claims" by Ark-seekers that "blunt the potential impact of a true discovery."

10. For Magness, who currently leads excavations at a late-Roman **synagogue** in Galilee, the search for Noah's Ark not only confuses the public, but diminishes excitement about actual archaeological finds, even ones that offer support for parts of the Bible such as the existence

of the House of David. "We know a lot about the biblical world, and it's very interesting," she says.

11. Setting the Record Straight. Part of the problem, says Cline, is that the public has unrealistic expectations of the discipline of archaeology—and popular media highlights the thrill of the chase instead of the slow **accretion** of archaeological knowledge. "We're not like Indiana Jones," he says. "It's a scientific procedure. It's painstaking. But what excites us does not necessarily excite other people." In his younger years, says Cline, he spent significant time and energy attempting to **rebut** the **purported** biblical evidence that **enchants** the public year after year. Eventually, though, he quits—and now focuses his time on both his expeditions and translating his research for those willing to accept the results of the scientific process. "People are gonna believe what they want to believe," he sighs. That won't change any time soon—so in the meantime, he's focused on unearthing an 18th-century B.C. Canaanite palace at Tel Kabri in what is now northern Israel. Following a **pandemic**-related pause on fieldwork, he anticipates returning next summer to continue excavating a painted plaster floor at the Old Testament-era site. "For us, [the floor] is incredibly important, because it shows international relations and contacts from almost 4,000 years ago," he says. "It's not Noah's Ark, but it's a painted floor," the archaeologist says, "which is good enough for me."

(1,190 words)

https://www.nationalgeographic.com

New Words

1. captivating ['kæptɪveɪtɪŋ]

a. holding your attention by being extremely interesting, exciting, pleasant, or attractive 非常有趣的,迷人的,吸引人的

It's a place where desert dust mingles with briny breezes to create a special kind of magic which lures travelers with year-round sunshine, cool waters, and captivating views.

这里沙漠尘埃与微风吹拂,营造了一份特殊的魅力,而且全年阳光普照,又有清凉海水和迷人美景,吸引了很多游客。

2. deluge ['deljuːdʒ]

n. a very large amount of rain or water 暴雨;洪水

This little stream can become a deluge when it rains heavily.

遇到暴雨的时候,这条小溪的水势就会猛涨。

3. intrepid [ɪn'trepɪd]

a. extremely brave and showing no fear of dangerous situations 勇猛的,无畏的

Some intrepid individuals were still prepared to make the journey.

某些无畏的人仍然准备作此旅行。

4. optometrist [ɒpˈtɒmətrɪst]

　　n. someone whose job is examining people's eyes and selling glasses or contact lenses to correct sight problems 验光师；配镜师

　　I know of a big optometrist store nearby.

　　我知道这附近就有一家配眼镜的店。

5. evangelical [ˌiːvænˈdʒelɪkl]

　　a. belonging to one of the Protestant Churches or Christian groups that believe the teaching of the Bible and persuading other people to join them to be extremely important 基督教福音派的

　　Evangelical Christians form a core of voters who want to restrict or even outlaw abortion in the United States.

　　以福音派基督徒为核心的选民，他们希望在美国限制甚至禁止堕胎行为。

6. petrify [ˈpetrɪfaɪ]

　　v. to frighten someone a lot, especially so that they are unable to move or speak 把……吓呆，使惊呆

　　I think you petrified poor Frazer—he never said a word the whole time you were here.

　　我想你把可怜的弗雷泽吓呆了，你在的时候他一句话也没说。

7. exasperation [ɪɡˌzæspəˈreɪʃ(ə)n]

　　n. the feeling of being annoyed, especially because you can do nothing to solve a problem 恼怒，愤怒；激怒

　　There is growing exasperation within the government at the failure of these policies to reduce unemployment.

　　由于这些减少失业的政策不能奏效，政府内部越来越多的人感到懊恼。

8. predate [ˌpriːˈdeɪt]

　　v. to have existed or happened before another thing 早于……存在（或发生）

　　These cave paintings predate any others which are known.

　　这些洞穴绘画早于其他任何已知的同类绘画。

9. behest [bɪˈhest]

　　n. order 命令；邀请；请求

　　The budget proposal was adopted at the mayor's behest.

　　在市长的要求下这项预算计划获得采纳。

10. decipher [dɪˈsaɪfə(r)]

　　v. to discover the meaning of something written badly or in a difficult or hidden way 辨认；破解，破译

　　Can you decipher the writing on this envelope?

　　你能辨认出信封上的字吗？

11. cuneiform [ˈkjuːnɪfɔːm]

a. of a form of writing used for over 3,000 years until the 1st century BC in the ancient countries of Western Asia 楔形文字的

She wrote hymns to the gods in cuneiform.

她用楔形文字写下了献给神祇的颂歌。

12. lore [lɔː(r)]

n. traditional knowledge and stories about a subject(祖辈流传下来的)知识,传说

According to local lore, the water has healing properties.

根据当地的传说,这种水有治病的功效。

13. proliferation [prəˌlɪfəˈreɪʃn]

n. the fact of something increasing a lot and suddenly in number or amount 激增

The past two years have seen the proliferation of TV channels.

过去的两年见证了电视频道(数量的)激增。

14. pseudoarchaeology [ˈsuːdəʊɑːkiˈɒlədʒi]

n. fake archaeology 伪考古学

Psuedo-archaeology refers to the nonscientific misapplication, misinterpretation, and/or misrepresentation of the archaeological record.

伪考古学是指对考古记录的非科学误用、误解和/或歪曲。

15. bolster [ˈbəʊlstə(r)]

v. to support or improve something or make it stronger 支撑;加固;提高;改善

She tried to bolster my confidence/morale (= encourage me and make me feel stronger) by telling me that I had a special talent.

她说我有特别的才能,想要增强我的信心。

16. apologetics [əˌpɒləˈdʒetɪks]

n. a branch of theology devoted to the defense of the divine origin and authority of Christianity 护教学(基督教神学的一部分,研究教条的辩证)

Apologetics is thus primarily a theoretical discipline, though it has a practical application.

因此护教学首先是一种理论上的(训练),尽管它也有实际的应用。

17. ubiquity [juːˈbɪkwəti]

n. the fact that something or someone seems to be everywhere(人或物)无处不在,普遍存在

We overcome the constraints and push past the boundaries and we then forget the boundaries existed as we accelerate breakthroughs on the path to ubiquity.

我们打破了限制,扩大了边界,然后在加速计算机推广的过程中,就忘了边界的存在。

18. concede [kənˈsiːd]

v. to admit, often unwillingly, that something is true(常指不情愿地)承认

The government has conceded (that) the new tax policy has been a disaster.

政府承认新的税收政策是彻底失败的。

19. dismantle [dɪsˈmænt(ə)l]

v. to take a machine apart or to come apart into separate pieces 拆开，拆卸

She dismantled the washing machine to see what the problem was, but couldn't put it back together again.

她拆开洗衣机想看看出了什么问题，但是却装不起来了。

20. synagogue [ˈsɪnəɡɒɡ]

n. a building in which Jewish people worship and study their religion 犹太教堂，犹太会堂

They had their synagogue and the five books of Moses.

他们有会堂和摩西的五卷书。

21. accretion [əˈkriːʃ(ə)n]

n. a gradual increase or growth by the addition of new layers or parts 堆积，积聚；逐渐增加，增大

The room hadn't been cleaned for years and showed several accretions of dirt and dust.

房间多年没有打扫，积了厚厚的灰尘。

22. rebut [rɪˈbʌt]

v. to argue that a statement or claim is not true 反驳；驳斥；驳回

She has rebutted charges that she has been involved in any financial malpractice.

面对指控，她反驳说她没有参与任何金融营私舞弊案。

23. purport [pəˈpɔːt]

v. to pretend to be or to do something, especially in a way that is not easy to believe 声称，标榜

They purport to represent the wishes of the majority of parents at the school.

他们声称自己代表了该校大多数学生家长的愿望。

24. enchant [ɪnˈtʃɑːnt]

v. to attract or please someone very much 使陶醉；使入迷

The audience was clearly enchanted by her performance.

观众显然陶醉于她的演出中。

25. pandemic [pænˈdemɪk]

a. (of a disease) existing in almost all of an area or in almost all of a group of people, animals, or plants（疾病）大规模流行的，广泛蔓延的

In some parts of the world malaria is still pandemic.

在世界上一些地区疟疾仍在大规模流行。

Phrases and Expressions

wipe……out 完全摧毁

come to rest 停止移动

on a regular basis 定期地；经常；经常地

at the behest of 按照……的要求；在……的命令下
make their way into 打进
bear sb's name 以……的名义
refer to 提到，谈及，谈起
be aligned with 与……一致
to the contrary 相反地
rule out the potential 排除……的可能性
set……straight 确保（某人）了解实情，使（某人）了解真相
in the meantime 与此同时；在此期间；其间

Proper Names

Old Testament（基督教《圣经》的）《旧约》
Noah's Ark 诺亚方舟
Mt. Ararat 亚拉腊山；亚拉拉特山；亚拉赫山
National Geographic 国家地理杂志

一、翻译简析

1. *Noah's Ark is among the best known and most captivating of all Old Testament stories: After creating humans, God became so displeased with them that he struck Earth with an all-encompassing flood to wipe them out—with one noteworthy (and seaworthy) exception: the biblical patriarch and his family, accompanied by pairs of each of the planet's animals, who rode out the deluge in an enormous wooden vessel.* (Para.1)

诺亚方舟是《旧约》所有故事中最著名、最吸引人的一个；在创造了人类后，上帝对人类感到不满，于是用滔天洪水席卷地球，毁灭人类——只有一个值得注意的（与航海有关的）例外：圣经中的族长诺亚及其家人，带着地球上每种动物各一对，乘着一艘巨大的木船，在大洪水中幸存下来。

【这个句子虽然很长，但实际不难，由冒号和破折号组成，冒号和破折号主要是起到解释和说明的作用，稍微复杂点的句子就是第二句，由 so……that 和后面的不定式 to wipe them out 作目的状语，过去分词 accompanied 充当伴随状语以及一个 who 引导的非限制性定语从句构成，翻译时如果把 so……that 结构直接转换成汉语"如此……以至于"，译文英语腔就会很明显，于是处理成"上帝对人类感到不满，于是用滔天洪水席卷地球"。这时，我们需要考虑把该句中的谓语动词 struck Earth 和目的状语 to wipe them out 结合在一起翻译更合适，于是处理成汉语的两个四字结构"席卷地球，毁灭人类"，不用特别突出目的状语成分，符合汉语意合的特点，把句子的逻辑关系隐含在内。另外，英语多用分词作伴随状语，如本句中的"accompanied by pairs of each of the planet's animals"，而汉语则多用短句，且很少使用被动句。因此，英译汉时，把被

动的翻译成主动的,因此译成"带着地球上每种动物各一对",这样更符合汉语习惯。最后,英语是讲究语法的语言,本句中最后一句是非限制性定语从句,有一个关系代词 who,汉语翻译时处理成无主句,和前面的带着顺承下来,更符合汉语习惯。】

2. For people who accept the religious text as a historically accurate account of actual events, the hunt for archaeological evidence of the Ark is equally captivating, inspiring some intrepid faithful to comb the slopes of Armenia's Mt. Ararat and beyond for traces of the wooden vessel. (Para. 1)

对于那些认为宗教文本就是对历史真实事件的准确描述的人而言,寻找方舟的考古证据同样令人着迷。这些忠实的信徒搜索了亚美尼亚的亚拉拉特山以及更远的地方,搜寻木船的痕迹。

【这个句子是由一个主句和一个现在分词引导的伴随状语构成,翻译时如果按照原句结构直接转换成汉语,译文会显得臃肿不堪。英语是葡萄型结构,汉语是竹竿型结构,因此在英译汉时,需要对原文的逻辑关系进行梳理,把握主干句和枝蔓句,然后按照汉语的逻辑重新组合,这样就把原句的一个句子处理成汉语的两个句子,现在分词引导的伴随状语就按照汉语的思维方式另成一句,翻译时,把后面 inspire 的宾语 some intrepid faithful 单独提炼出来,变成句子主语,这样就以"这些忠实的信徒"作为句子叙述的出发点,重新来组织句子,这样更符合汉语思维习惯。】

3. More modern Ark "discoveries" take place on a regular basis, from an optometrist's report he'd seen it in a rock formation above the mountain in the 1940s to a claim Evangelical pastors had found petrified wood on the peak in the early 2000s. (Para. 2)

步入现代,经常有人称"发现"了方舟:20 世纪 40 年代,一位验光师说自己在山上的岩层中看到了方舟;21 世纪初,福音派牧师宣称在山顶发现了木头化石。

【这个句子由一个 from……to 引导的几个例子构成,这个 from……to 结构起到列举的作用,对前面的 discoveries on a regular basis 进行呼应,翻译时如果把 from……to 这种结构直接转换成汉语,译文会带有英语腔。英语多用动词转换的名词,汉语多用动词,因此在翻译此句时,把 discoveries 处理成动词,然后再列举各个声称取得新发现的例子,就更符合汉语习惯,另外,英语原句中时间状语放在句尾,翻译成汉语时,把时间状语提前,一来符合汉语习惯,二来可以以时间为线索列举各种例子。最后,英语原句中各种列举的例子均是以名词形式出现,用 report 和 claim 作先行词,后面一个同位语从句,原句名词结构比较长,在英译汉处理时,把名词结构处理成汉语短句,把发现人的身份提前,作为句子主语,然后把 report 和 claim 两个名词翻译时处理成动词,变成句子的谓语,这样更符合汉语习惯。】

4. Legends about a deluge that destroys civilization at the behest of a supernatural deity can be found in multiple Mesopotamian texts, from *the Epic of Gilgamesh*, which was written around the early second millennium B. C., to a recently deciphered Babylonian cuneiform tablet from about 1750 B. C. describing how the ark was built. (Para. 4)

在美索不达米亚的多个文献中,都可以找到在超自然神灵的命令下,洪水摧毁文明的传说,从公元前 2000 年早期的《吉尔伽美什史诗》,到最近破译的公元前 1750 年前后的巴比伦楔形文字碑,其中都有建造方舟的描述。

【这个句子由一个that引导的限制性定语从句和which引导的非限制性定语从句,以及一个现在分词作后置定语的句子构成,翻译时如果把这种结构直接转换成汉语,译文会显得主次不分,逻辑不清。这时,我们需要把句子的逻辑进行重新捋顺。英语句子用了被动句,翻译成汉语转换成主动句,"在美索不达米亚的多个文献中,都可以找到",这样就规避了被动句,另外was written也做了类似处理,直接翻译成"从公元前2000年早期的《吉尔伽美什史诗》",这样就规避了"被写于"这样的字样,更符合汉语习惯。最后,describing现在分词作后置定语,翻译时译者没有把修饰语前移,修饰语如果前置,文字过于臃肿,翻译时保留了原文的顺序,翻译时增加了"其中"字样,这样就和前面的名词建立起了联系,逻辑同样完整。】

5. Part of the problem, says Cline, is that the public has unrealistic expectations of the discipline of archaeology—and popular media highlights the thrill of the chase instead of the slow accretion of archaeological knowledge. (Para. 11)

克莱恩说,部分问题在于公众对考古学科的期望不切实际,而且大众媒体强调追逐的快感,而非慢慢积累考古知识。

【这个句子是由一个that引导的表语从句组成,表语从句中含有两个并列句,句子很长,但是结构并不复杂。翻译时,只要捋顺句子逻辑关系,然后按照汉语思维习惯组织语言即可。本句中,says Cline是插入语,汉语习惯于把这是谁说的话提前,表明某一说法的出处。另外,英文原句中has unrealistic expectations of,在翻译时也进行了处理,把unrealistic从定语翻译成了谓语,翻译时把原句的形容词转换成汉语的动词,符合汉语多使用动词的特征,后面对accretion翻译时,也做了类似词性转换处理,翻译时把名词处理成动词。】

二、参考译文

为什么一直找不到诺亚方舟?

一个多世纪以来,人们一直在搜寻《旧约》里从大洪水中幸存下来的诺亚方舟。然而考古学家表示,这是一件徒劳无功的事。

1. 诺亚方舟是《旧约》所有故事中最著名、最吸引人的一个;在创造了人类后,上帝对人类感到不满,于是用滔天洪水席卷地球,毁灭人类——只有一个值得注意的(与航海有关的)例外:圣经中的族长诺亚及其家人,带着地球上每种动物各一对,乘着一艘巨大的木船,在大洪水中幸存下来。对于那些认为宗教文本就是对历史真实事件的准确描述的人而言,寻找方舟的考古证据同样令人着迷。这些忠实的信徒搜索了亚美尼亚的亚拉拉特山以及更远的地方,搜寻木船的痕迹。

2. 例如1876年,英国律师、政治家詹姆斯·布莱斯攀上了亚拉拉特山,宣称一块木头"符合所有要求",是木船的一部分;根据《圣经》记载,方舟在此停靠。步入现代,经常有人称"发现"了方舟;20世纪40年代,一位验光师说自己在山上的岩层中看到了方舟;21世纪初,福音派牧师宣称在山顶发现了木头化石。

3. 但对于寻方舟这件事,考古学家和圣经学者表达了从恼怒到鄙视的各种情绪。"真正的考古学家不会做这种事",国家地理探险家、北卡罗来纳大学教堂山分校的考古学家乔迪·

马格内斯在谈到寻找诺亚方舟的证据时说。"考古学不是寻宝游戏",她补充道:"不是找到特定的物体。这是一门科学,我们提出研究问题,并通过挖掘来寻找答案。"

4. 真假难辨。毁灭性大洪水和幸存者的故事早于《希伯来圣经》,一般认为,《希伯来圣经》最早的部分成书于公元前8世纪。在美索不达米亚的多个文献中,都可以找到在超自然神灵的命令下,洪水摧毁文明的传说,从公元前2000年早期的《吉尔伽美什史诗》,到最近破译的公元前1750年前后的巴比伦楔形文字碑,其中都有建造方舟的描述。

5. 这些洪水神话是否有事实依据?"确实有地质证据表明,大约7500年前,黑海地区发生过大洪水",国家地理探险家、乔治华盛顿大学的考古学家埃里克·克莱恩说。但科学家在那次事件的规模方面莫衷一是,就像历史学家对于洪水的相关描述是否受到现实生活启发,也存在分歧。更有可能不同地方、不同时间的人们都经历了洪水——这些事件自然而然成为全世界的口头和书面传说。

6. 更复杂的是,根据《希伯来圣经》,学者们对诺亚方舟的具体位置持不同意见。在《创世记》中,方舟停靠在古代乌拉尔图王国的"亚拉拉特山上",这片地区包含现在的亚美尼亚、土耳其东部和伊朗,而非今天以这个名字命名的标志性山峰。"我们无法确定究竟这件事发生在古代近东的具体地点",马格内斯说。克莱恩和马格内斯都表示,即使已经或即将找到方舟上的文物,也无法确定它们与历史事件有关。

7. "我们无法在时间和空间上确定诺亚方舟和洪水的位置,如果二者真的存在的话",马格内斯说:"唯一确定的方法是找到真正的古代铭文",即便如此,她指出,这样的铭文可能指向另一个诺亚方舟,或者另一场洪水。这并没有让伪考古停下脚步,他们坚信《圣经》上的每个字都是真的。一无所获的搜寻往往得到"年轻地球创造论"追随者的支持;虽然有相反的证据,但这些人认为地球只有几千年历史。

8. 证据相同,结论不同。这些团体利用世俗考古证据支持他们对《圣经》的字面解释,并试图反驳相反的证据,或者视而不见。他们采取了不同策略。"创世记中的答案"组织(Answers In Genesis)自称为护教团体,关注科学问题,甚至在肯塔基州经营着诺亚方舟主题游乐园;他们承认,除了《旧约》中的诺亚方舟故事外,有关洪水的神话无处不在,甚至承认可能永远找不到方舟。

9. "我们不指望方舟经历了4 350年还能保存下来",地质学家、创世记中的答案组织的研究主管安德鲁·斯内林说。他用几十年时间试图证明地球还很年轻。然而,关于为什么找不到方舟的残骸,斯内林不同意考古学家的看法。"诺亚和家人离开方舟后,没有成熟的树木供他们搭建避难所,我们有理由相信他们因此拆除了方舟(他们不再需要方舟了),回收木材。"虽然该组织没有排除有朝一日找到方舟的可能性,但斯内林对方舟寻找者的"可疑论述"感到懊恼,因为这会"削弱真正发现的潜在影响"。

10. 马格内斯目前正在加利利负责罗马晚期一座犹太教会堂的挖掘工作。在她看来,寻找诺亚方舟不仅令公众困惑,也降低了人们对实际考古发现的兴趣,甚至是支持《圣经》部分内容的那些发现,比如大卫之家。"我们很了解圣经世界,这一点很有意思",她说。

11. 澄清真相。克莱恩说,部分问题在于公众对考古学科的期望不切实际,而且大众媒体强调追逐的快感,而非慢慢积累考古知识。"我们不是印第安纳·琼斯",他说:"这是一个科学过程,很艰苦。在我们看来激动人心的事物,不一定会让其他人兴奋。"克莱恩说,年轻时,他花了大量时间精力,试图反驳所谓的圣经证据,每年都有公众被这些证据所吸引。然而最终他放弃了,现在他把时间集中在探险上,并向那些愿意接受科学研究结果的人们介绍自己的研究。"人们总是相信自己愿意相信的事情",他叹息道。这一点短期内不会改变,因此目前他专注于在今天以色列北部的特拉卡布里挖掘一座公元前18世纪的迦南宫殿。现场作业因为疫情暂停,他预计明年夏天会回到这座旧约时代的遗址,继续挖掘石膏地板。"对于我们来说,(这块地板)非常重要,因为它展现了近4 000年前的国际关系和联系。""这不是诺亚方舟,而是粉刷地板",这位考古学家说:"我认为这已经足够好了。"

Cultural Background Knowledge: Marine Mythology
海洋神话文化背景知识

　　神话是一种文化积淀,也是一种民族意识的积淀,直至今天,它还在以潜意识的方式影响着现代人的生活。对于原始神话的认识,有助于对民族文化、民族意识的寻根溯源。每个民族的起源都与海洋有着千丝万缕的联系,海洋文明主要以两种方式被传承,"海洋神话是远古人类的精神记忆,考古遗存则是远古人类的物质形态记忆。"而根据奥地利心理学家荣格的"原型理论",一个民族的文化印记就存在于神话中,这也是识别每个民族的性格符码。因此,"海洋神话作为远古人类的精神记忆,各个民族呈现出来的也就各异"。除中国外,几大古老文明也都曾孕育出生动的、极具民族特色的海洋神话。在古埃及人的想象中,世间的一切来源于神的创造,而这些神又来自混沌的海洋。这片混沌的海洋最古老的主宰者叫努恩。努恩被描绘成腰部浸在水里的人形或蛙首人身。努恩在某种程度上意味着"原始水域"。努恩之妻为蛇首人身。他们创造了其他诸神,故努恩又称为"众神之父"。

　　同时,对航海女神伊西丝的崇拜遍布整个埃及,埃及人认为,伊西丝是天狼星。天狼星与尼罗河水的泛滥有着密切的关系,并把天狼星偕日同升与尼罗河水泛滥同时发生的那一天定为一年的第一天,即新年的开端。伊西丝具有使他们的"生命之源"——尼罗河水准时泛滥的力量以及使生命再生的超能力。同样,苏美尔神话中创造和确立宇宙万物及其秩序的是水神(又是智慧之神)恩基。印度雅利安人的《吠陀本集》中记载的一个重要的神叫伐楼拿,他以一条如鳄鱼、鲨鱼、海豚的怪兽为坐骑,周游于陆海之间。平时,他深居海底宫殿,掌管着天下的海洋江河。古希腊人则把海洋神话演绎得最为完整,最为著名的是海神波塞东系列。

Unit Eleven
Harbour Culture

海港文化

Text A

Reconstruction and Reuse of Port Heritage

By Jing Wang

Ideological and Political Education：从"中国制造"到"中国智造"

中国港口凭借着领先的吞吐量规模、集约高效的岸线利用水平、辐射全球的航线网络等优势占据了世界一流港口的较多席位。2020年,世界排名前十的港口中,中国占了七席,包括上海港(排名第一)、深圳港、宁波港、维多利亚港(香港)、广州港、青岛港、天津港。中国港口货物吞吐量长期稳居世界第一。2022年的交通部数据显示,宁波舟山港连续14年位居全球第一,紧随其后的是唐山港和上海港。尽管上海港的吞吐量有所下滑,仍以突破4 730万标准箱的集装箱吞吐量,连续13年蝉联全球第一。同时,根据《新华·波罗的海国际航运中心发展指数报告(2022)》显示,上海超过第四、五名的香港和迪拜,成为中国第一、世界第三的航运中心,仅次于新加坡和伦敦。

随着科技水平的提升,自动化码头近年来也逐步成为中国沿海港口的"标配"。中国经济信息社与交通运输部水运科学研究院发布的《世界一流港口综合评价报告(2022)》显示,中国已建和在建自动化码头数量居世界首位。上海洋山四期自动化码头是目前全球规模最大的自动化集装箱码头,装卸运输均采用自主研发的智能系统。它是国内唯一一个"中国芯"的自动化码头,也是国内唯一一个软件系统纯粹由"中国智造"的自动化码头,标志着中国港口行业在运营模式和技术应用上实现里程碑式跨越升级与重大变革的关键技术。2021年9月,洋山四期自动化码头被"复制"到以色列海法新港,这是以色列60年来的第一个新码头,也是中国企业首次向发达国家输出"智慧港口"先进科技和管理经验。

港口建设是一项门槛极高的技术活,中国已经实现了全产业链、全要素的比较优势,技术水平已是名副其实的世界第一,也成为继"中国高铁""中国核电"后走向世界的又一张"中国名片"。如今,世界各地分布着近百个中国承建、参建、运营及合营的港口。它们承载着中国人的智慧与汗水,书写着"中国制造"向"中国智造"转变的传奇,更肩负着"一带一路"、人类命运共同体和疫后时代全球经济复苏的历史使命。

1. Throughout the history of humankind, ports have been the hub of coastal towns, and the changes **undergone** there have reflected the town's historical, social and economic evolution. The port is a unique landscape edge where the city and water meet, where water is a significant mean of transport, along with navigable ports, lakes and waterways. Urban **waterfronts** are often **vibrant** trading and industrial sites, as well as the "front door" of the city. The port heritage in the port city waterfront has highly visible regional characteristics, as recording the history of urban development, which is a unique cultural landscape in most cities. These landscapes are rich in history and culture and should be considered assets, as historically significant classics that can enhance preservation, design and possibilities for innovative practice in planning. While acknowledging the complexity and potential for conflict, it is argued that a transformed form of port heritage can contribute to community vitality and local presence. Nevertheless, with the deepening of urbanization and the adjustment of industrial structure, many cities' waterfront space is no longer used as the main industrial transportation channel, storage and transportation of goods, but rather a **gorgeous** turnaround of a once-abandoned waterfront through organic renewal.

2. Nowadays, waterfronts are the most important urban areas which reflect the competitive development strategies of 21st century cities, and they are the locations hosting large investments targeting the **regeneration** of abandoned port areas and their economic **revitalization** and the sustainability of the urban economies. Some experts pointed before the **advent** of the industrial age, and coastal areas were heavily used also accompanied by the **vigorous** development of human activities, which was gradually replaced by industry. With the current pressure of the demand for urban upgrading, the public's awareness of civic society is gradually increasing, and the waterfront edge of the city has been rediscovered. Meanwhile, the natural water elements of cities play a vital role in establishing a balance between nature and society. Therefore, since the 1980s, urban waterfront regeneration projects have become useful, sustainable development tools for urban planning and political internationalization. The literatures and policies on sustainable urban development have increased significantly in the last decade. From the perspective of the sociology of sustainable development, there is possibility to discover the potential of waterfront heritage, with practicing the design concept of adaptive reuse on it, as well as analyzing the industrial heritage of the coastal area during the urban renewal process, to eventually realize the social sustainability. Urban waterfront as historical gateway is always essential for the economic development and socio-cultural identity of many port cities. In connection with the changes in transport modes and industrial operation in the 1960s and 1970s, most port cities around the world and their waterfronts have undergone **unprecedented** transformations in both physical form and economic function. The role of the waterfront, to which these cities could be said to owe their existence, has changed with the passage of time. Transitions in three aspects led to the decline of the waterfront, namely the expansion of city size, the reforms in transportation technologies and changes in

industry.

3. Within the global scale, waterfront regeneration projects shape the relationships between places, uses and visions in the interurban competition, and offer extraordinary opportunities by investigating the harbor, city functions from their economic, environmental, social aspects. Along with the globalization of the economy, the current competition between the countries left its place to interurban and the sea, also port cities have become the laboratories for urban regeneration processes which develop residential, commercial, tourism and recreational function. Of port cities in rapid decline in the 20th century, the waterfront became an important conversion target for urban regeneration in the UK. And later small and medium-sized urban waterfronts such as Paddington Port, Liverpool Albert Dock and others have also begun to regenerate. Today, more than 200 waterfront regeneration projects have been established in the UK, demonstrating the important meaning of waterfront regeneration to the UK's urban regeneration. In the world-wide era of waterfront revitalization, the **prominent** achievements accomplished by cities in the UK in their waterfront development drew the attention of the world planners and researchers. They have not only been documented as successful cases, but also have been carefully analyzed in order to find the principles behind their achievements. The following parts are case studies provided by Paddington port regeneration project and Liverpool waters regeneration project.

4. Port of Paddington and Liverpool History Background. The development of the Paddington Port area is one of the largest and most successful examples of waterfront regeneration in London at the last decade. Once a distribution center for coal mines and construction materials destined for the city, the area takes advantage of its waterfront location around Paddington Station. This regeneration into a major regional center contains large office and residential developments of the area, which began in the 1980s and is now largely completed, has been **hailed** as one of the best examples of urban regeneration in the UK, with its high-quality waterfront public space construction, advanced waterfront design experience and construction mode, provides a reference for many cities which are currently in full swing. The Paddington Port pool area is located on the western edge of London's central activity area, between Hyde Park and Regent's Park, and is part of the Westminster borough. Paddington port is a branch of London's main canal, the Union Canal, in the heart of the city, and since the industrialization of the 1840s period, it has become an important transport hub on London's network of canals, with goods from central and northern England traveling along the canal was carried to London and the Paddington Liverpool is located in the North West of England and is an important port city along the River Mersey. The waterfront is the main area of Liverpool's urban regeneration, with regeneration concentrated on the west bank of the River Mersey, the city center. Liverpool's waterfront regeneration journey has been broadly divided into three phases: the first two phases are more culture and tourism driven, with government investment or government-led multi-investment is the main focus; the latter phase began to intensify purely

commercial and residential development, with mainly private investment. A total of six main projects have been completed so far, covering a total area of over 1.8 million square meters and spanning more than 30 years.

5. Paddington Port Regeneration Project. Paddington area became a rundown and outdated problem area in central London as a result of the advent of the de-industrialization period in the 1990s. The area's canals were surrounded by low-density warehouses, car parks and waste disposal sites at that time, and the waterfront space was completely closed off to **pedestrians**. The buildings are backed up to the canal water, the surrounding community cannot feel the presence of the canal, and the canal has even become a community barrier between. There was an inaccessible abandoned area in public opinions. In terms of regional human qualities, Paddington has a rich cultural identity and is a critical area that reflects the urban qualities of London. St. Mary's Hospital, which is 170 years old, is the birthplace of Prince George; the laboratory where penicillin was invented is also located here. Meanwhile, the famous Paddington Bear was named by the author based on the region's Paddington railway station; those rich cultural elements provided the foundation for the region's Revival design (the story of Paddington).

6. Establishment of paths and boundaries, stitching together the urban fabric. The pathways in the area consist of a series of existing and new streets and waterways that reflect the character of the region as well as **stitching** together the urban fabric of the region and its surroundings. The main goal of the **spatial** path organization is to promote accessibility, **permeability**, and spatial identity of the area. The design succeeds in defining the 'primary' and 'secondary', the 'positive' and 'negative' of these paths. The pathways in the region are divided into three levels. The first level is the main carriageway in the area and its borders. For instance, the "main gate" on the south side, Reid Road, has been upgraded in its entirety in terms of building facade, street furniture and environmental greenery. Linking the area to the upmarket settlements to the south, it also creates entrance streets that reflect the character of the area. The second level is the public right-of-way roads in the area, including North and South Pier Street. They are located on the backside of the development buildings and are used to ease internal **vehicular** traffic and provide space for related municipal services such as trash pickup, ensure the safety, comfort and continuity of the main pedestrian paths in the area on both sides of the canal. The third level is a complete and coherent pedestrian path around the canal harbor pond. The design uses the canal as the centerpiece of the public space for the entire area, creating a continuous path around the canal from the end of the harbor pool on the east side to the Little Venice complete on the west side. The masterplan for the area established six functional areas, including: the Hilton Metropolitan Hotel, Paddington Harbour Pool Mix, West Quay Residential, Paddington Health Management University, Paddington Railway Station and Paddington Central Office, which are skillfully linked by the canal and surrounding

public space.

7. Liverpool Waters Regeneration Project. Top-down physical transformation of the waterfront (1980—1997). In the early 1980s, the British Government launched "Act 1980", which was designed to tackle the problem of the Mersey River and involved the physical regeneration of the 3.5 million m² docklands, including the Liverpool Waterfront. In order to deliver the regeneration scheme more efficiently, the Government established the Merseyside Development Corporation (MDC) in 1981 as a single decision-making body to organize the regeneration scheme from top to bottom. The Agency has four main tasks: reuse of **derelict** land, development of new types of trade, creation of an attractive environment, security of residential housing, etc. MDC does not implement project construction, but rather leads the investment on behalf of the Government. Since then, Merseyside Maritime Museum 1986, TATE Liverpool 1986 and the Beatles Story Museum 1990 have been introduced gradually in the city. These projects have continued to improve the physical image of the pier, while at the same time introducing 'cultural tourism' to the site as an external attraction. The regeneration program, which ended in 2003, has seen an investment of over £25 million and the museums on the pier attract visitors that generate 2 million trips per year, which proved the success of this renewal program.

8. Liverpool Urban competition-driven image building of the waterfront (1997—2012). When the Labour Government came to power in 1997, it focused on raising the status of cities and encouraging urban competition. The physical development of the city shifted to image creation and branding. Liverpool responded quickly to this wave by setting up a renewed organization and issued an updated Strategic Regeneration Framework (SRF) in 2002 to update the work after its introduction, in which the preservation of the city's historic sites and the building of the city's image and brand were two key areas of focus. Based on SRF's guidance, Liverpool has had two notable successes. The first was the inscription of parts of the city on the World Heritage Site (WHS), in 2004. The WHS has established a Character Area, which is a total of six feature parcels, primarily along the waterfront; the Buffer Zone contains most of the waterfront. The directory also regulates the height of new buildings in different locations. The second achievement was Liverpool gained the recognition as European Capital of Culture 2008 (ECoC 2008). This award aims to promote the regeneration of the city through festivals and events that connect with European culture externally and internally. The Liverpool Government has combined the delivery of ECoC events with the physical regeneration of the waterfront to significantly enhance the city's brand.

9. Private investment-led construction of waterfront business districts (post-2012). Liverpool Waters was the main waterfront regeneration project undertaken during this phase, which was **initiated** in 2007 by the developer PEEL and in 2010 the details of the development were identified as well as a development application was **submitted**. Government approval was

received to **commence** in 2012. The project is located in the north dock area of Liverpool, 42% of which is within the WHS Conservation Area. The development site begins on the north side of the dock top area and extends northwards to Bramley Moore Dock, covering an area of about 60 hm^2, with an estimated investment of more than £5.5 billion over a time span of about 20-40 years.

10. In summary, reconstructing and reusing port heritage in the waterfront are not only a saving of resources, but also the best choice to record the waterfront architectural culture and witness of the revival history of the waterfront. The original form of the port heritage can be transformed into a public space to support social activities and revitalize the **barren** waterfront space. Converting port heritage into public space can establish site boundaries, which also helps to stimulate the vitality of the heritage space. As a precious non- renewable resource, heritage is a medium of life for generation and **transmission**. Therefore, it is **imperative** that the regeneration of port heritage be changed, which can improve the sustainable development of waterfronts in many cities.

(2,289 words)

https://www.researchgate.net/publication/365979345_Ports_heritage_as_public_space_-the_waterfront_regeneration_in_Penarth

New Words

1. undergo [ˌʌndəˈgəʊ]
    v. go through (mental or physical states or experiences) 经历,经受
    An additional reason for getting tech companies to give a higher priority to security is that cyberspace is about to undergo another massive change.
    让科技公司更加重视安全问题的另一个原因是,网络空间即将经历另一个巨大的变化。

2. waterfront [ˈwɔːtəfrʌnt]
    n. the area of a city (such as a harbor or dockyard) alongside a body of water 滩,海滨;水边
    The resort has waterfront lodging, golf, a spa, marina, and cruises on a historic yacht.
    这个度假胜地拥有海滨住宿、高尔夫、温泉浴场、码头和乘坐历史悠久游艇巡游活动。

3. vibrant [ˈvaɪbrənt]
    a. vigorous and animated 充满活力的,充满生机的,生气勃勃的
    Learning memory systems, mental sports and juggling are all wonderful ways to maintain a vibrant and lively mind.
    学习记忆系统、脑力运动和杂耍都是保持大脑活力的好方法。

4. gorgeous [ˈgɔːdʒəs]
    a. dazzlingly beautiful 美丽动人的;令人愉快的;绚丽的,华丽的
    Absolutely, what you were describing is her romantic ideal of nature of wilderness as being the

beautiful sunset, the gorgeous view, the beautiful day.

当然，你所描述的就是她对于自然和荒野的浪漫想法：美丽的日落，壮丽的景象，美好的天色。

5. regeneration [rɪˌdʒenəˈreɪʃn]

   n. the activity of spiritual or physical renewal 再生，重生；重建

   Regeneration is vast in its scope and its timescales but almost unimaginably delicate in the undertaking.

   从时间和空间上讲，城市重建都是浩大的工程，是一项难以置信的巧夺天工的事业。

6. revitalization [ˌriːˌvaɪtəlaɪˈzeɪʃn]

   n. bringing again into activity and prominence 复兴，复苏

   Education is a fundamental revitalization of the country and nation, promoting quality education is placed in front of our country and nation a very urgent task.

   教育是振兴国家和民族的根本，推进素质教育是摆在我们国家和民族面前一项非常紧迫的任务。

7. advent [ˈædvent]

   n. arrival that has been awaited (especially of something momentous) 出现，到来，问世

   Historians suggest that the advent of mass tourism began in England during the industrial revolution with the rise of the middle class and the availability of relatively inexpensive transportation.

   历史学家认为，大众旅游的出现始于工业革命时期的英国，当时中产阶级崛起，交通费用也变得相对便宜。

8. vigorous [ˈvɪɡərəs]

   a. characterized by forceful and energetic action or activity 精力充沛的，充满活力的；强壮的，强健的

   In many of these cities such as Denver, a diverse and vigorous economy has offered stable employment for residents.

   在丹佛等许多城市，多样化和充满活力的经济为居民提供了稳定的就业机会。

9. unprecedented [ʌnˈpresɪdentɪd]

   a. having no precedent; novel 前所未有的，史无前例的

   Recent years have brought middle and small-sized enterprises businesses unprecedented opportunities—as well as new and significant risks.

   近年来中小企业遇到了前所未有的机遇，同时也伴随着新的重大风险。

10. prominent [ˈprɒmɪnənt]

    a. conspicuous in position or importance 重要的，著名的；显眼的，突出的

    It's particularly striking centuries later when these issues are still prominent in public discussions about social justice and women's rights.

    几世纪后，在关于社会公正和妇女权利的公开讨论中，这些仍然是突出的问题，这一点尤其

令人震惊。

11. hail [heɪl]

    v. praise vociferously; greet enthusiastically or joyfully 赞扬,欢呼;呼喊,招呼

    We hail as masterpieces the cave paintings in southern France and elsewhere, dating back some 20,000 years.

    我们把法国南部和其他地方的洞穴壁画赞誉为杰作,它们的历史可以追溯到大概2万年前。

12. pedestrian [pəˈdestrɪən]

    n. a person who travels by foot 步行者,行人

    One way they've come up with some ways to attract more people, to shop downtown was by creating pedestrian malls.

    他们想出了一个办法来吸引更多的人来市中心购物,那就是建造步行街。

13. stitch [stɪtʃ]

    v. fasten by sewing; do needlework 缝,缝补;缝合(伤口)

    The surgeon would pick up his instruments, probe, repair, and stitch up again.

    外科医生会拿起器械进行探查、修复,然后再缝合。

14. spatial [ˈspeɪʃ(ə)l]

    a. pertaining to or involving or having the nature of space 空间的,与空间有关的

    At the moment, urban landscapes are highly managed and limited in their spatial extent.

    目前,城市景观被高度管理,空间范围受到限制。

15. permeability [ˌpɜːmiəˈbɪləti]

    n. the property of something that can be pervaded by a liquid (as by osmosis or diffusion) 渗透性;弥漫

    Extraction of water depends on two properties of the aquifer: porosity and permeability.

    水的开采取决于含水层的两个性质:孔隙度和渗透性。

16. vehicular [vəˈhɪkjələ(r)]

    a. of or relating to or intended for (motor) vehicles 车辆的;用车辆运载的

    Care is also required to ensure that the vehicular traffic diverted to other streets does not cause a worse problem.

    我们也必须小心,确保在实施改道后,其他街道不会出现更严重的交通问题。

17. derelict [ˈderəlɪkt]

    a. worn and broken down by hard use (建筑物或土地)破旧的,弃置的

    This stoked interest in the derelict housing, and units were increasingly rehabilitated and preserved.

    这激起了人们对废弃房屋的兴趣,这些单元逐渐被恢复和保存。

18. initiate [ɪˈnɪʃieɪt]

    v. bring into being, take the lead 开始实施,发起

    People who are talked to have equally positive experiences as those who initiate a conversation.

被交谈的人与主动发起谈话的人有着同样积极的体验。

19. submit [səbˈmɪt]

   v. hand over formally 呈递，提交

   It is laid down that all candidates must submit three copies of their dissertation.

   根据规定所有的学位答辩人均须提交论文一式三份。

20. commence [kəˈmens]

   v. take the first step or steps in carrying out an action 开始，着手

   We are working within the law and we will commence work only when all environmental clearances have been granted.

   我们现在从事的商业活动是合法的，并且只有在获得了环境许可之后我们才会展开工作。

21. barren [ˈbærən]

   a. providing no shelter or sustenance; not bearing offspring 贫瘠的；不结果实的

   The formerly once barren land has been changed into stretches of fertile fields.

   昔日的荒地今天成了良田。

22. transmission [trænzˈmɪʃ(ə)n]

   n. the act of sending a message from one person or place to another 播送，发送；传递，传播

   Some of the data is missing because of various failures in the thermometer, the transmission channel or the recording computer.

   由于温度计、传输通路或者进行记录的计算机的各种失误，有一些数据会丢失。

23. imperative [ɪmˈperətɪv]

   a. requiring attention or action 极重要的，必要的

   It is imperative that governments around the world enact laws to protect the environment at home and abroad.

   世界各国政府必须制定法律来保护国内外的环境。

Phrases and Expressions

along with 连同……一起；与……一道
play a......role in 扮演……角色，起……作用
from the perspective of 从……的角度来看
in connection with 与……有关；与……相连
draw the attention of 引起……的注意
takes advantage of 利用
in full swing 活跃；蓬勃高涨
in terms of 依据……；按照……；在……方面
on behalf of 代表……；为了……

Proper Names

the Merseyside Development Corporation (MDC) 默西塞德郡发展公司
the Labour Government 工党政府
Strategic Regeneration Framework (SRF) 战略再生框架
the World Heritage Site (WHS) 世界遗产
European Capital of Culture 2008 (ECoC 2008) 2008年欧洲文化之都

一、翻译策略/方法/技巧：转句译法

转句译法又叫作拆分法(division)，这是因为这种翻译技法本质上乃是一种拆分。在日常翻译实践中，经常会发现这样一种情况，无论是在英语还是汉语中都有这样一类词语、词组或短语，尽管十分短小，却语义内涵丰富。在翻译成译语时，如果将其保持原来位置不动，整个译文在表达效果上就会很不通顺，让人感到十分怪异。此种情况下，就需要译者将原文中这样的词语、词组或短语单独拆分出来，分列成一个相对独立的表达单位，以达到准确通顺地表达原文语义的目的。这种处理方法就是所谓的"转句译法"。这种翻译技法因为凸显了英汉两种语言在构句上的主要差异，在英汉互译中经常使用。下面结合一些典型的译例进一步分析说明。

Characteristically, Mr. Smith concealed his feelings and watched and learned.

译文：史密斯先生并未表露自己内心的情绪，只是察言观色，心中有数。这也是他为人处世的一贯做派。

英语原文中的 characteristically 一词，作为插入语，是用来修饰整个句子的，用于总结说明基于史密斯的日常表现而做出的关于其性格特征的总结。如果照搬词典中的"具有某种特征地"作为译文，显然在通顺度上欠佳，也会引起读者阅读上的困惑。这种情况下，就可以尝试使用转句译法，将该词从原句中拆分出来，译成一个相对独立的小句，置于译文之末。这样处理，既能保证表达上的通顺，又能完整再现该词在语义表达上总结的功能。

A movie of me leaving that foxhole would look like a shell leaving a rifle.

译文：我飞身离开那个掩体的场景，如若拍成电影，极像一发出膛的炮弹。

这个句子在语义理解上有一定难度。原文从语义上实际上可以分成三个部分：A movie of, me leaving that foxhole 和 would look like a shell leaving a rifle。后两部分表达上实际上运用了明喻的修辞格，前者是本体，后者是喻体。这两部分在翻译上难度不大。稍费周章的是 A movie of 的翻译。从语义上来说，它实际上是将明喻所描述的场景定格成一个电影片段，以突出句子表达的视觉形象性。这样的构句安排在语义表达和语言想象力上都很有跳跃性。这也意味着需将其从原文中拆分出来译成一个独立小句才能充分展现和发挥其功能。无疑转句译法是最为恰当的一种处理方法。

他们在人来人往的候车室里坐下来看点书。

译文：They sat down in the waiting-room to do some reading. People came to and from there.

原文中的"候车室"有一个定语"人来人往"。因为是主谓短语做定语，不太容易译成某个英文词语或短语，故此需拆分出来单独成句，也就是需要借助转句译法来处理。译文中这种处理方法是将其彻底拆分成一个独立的简单句，这样是可以的。与此同时，考虑到英文在构句上注重形合的特性，也可以将"人来人往"拆分成修饰"候车室"的定语从句。因此，也可以译成：They sat down in the waiting-room, where people came to and from, to do some reading. 由此而见，转句译法在拆分上也是有程度差异的。究竟以何种程度为宜，还需视具体翻译环境而定。

二、译例与练习

Translation

1. The port heritage in the port city waterfront has highly visible regional characteristics, as recording the history of urban development, which is a unique cultural landscape in most cities. (Para.1)

　　港口城市滨水区的港口遗产具有非常明显的地域特征，它记录了城市发展的历史，是大多数港口城市独有的文化景观。

2. Nowadays, waterfronts are the most important urban areas which reflect the competitive development strategies of 21st century cities, and they are the locations hosting large investments targeting the regeneration of abandoned port areas and their economic revitalization and the sustainability of the urban economies. (Para.2)

　　如今，滨水区是反映21世纪城市竞争发展战略的最重要的城市区域，也是进行大规模投资的地点，旨在恢复废弃港口区及其经济振兴和城市经济的可持续性。

3. From the perspective of the sociology of sustainable development, there is possibility to discover the potential of waterfront heritage, with practicing the design concept of adaptive reuse on it, as well as analyzing the industrial heritage of the coastal area during the urban renewal process, to eventually realize the social sustainability. (Para.2)

　　从可持续发展社会学的角度来看，通过实践滨水遗产适应性再利用的设计理念，以及在城市更新过程中对滨海地区的工业遗产进行分析，可能会挖掘出滨水遗产的潜力，最终实现社会的可持续性。

4. Within the global scale, waterfront regeneration projects shape the relationships between places, uses and visions in the interurban competition, and offer extraordinary opportunities by investigating the harbor, city functions from their economic, environmental, social aspects. (Para.3)

　　在全球范围内，滨水重建项目促进了城市竞争中地点、用途和愿景之间的关系，并通过调查港口城市在经济、环境、社会等方面的功能情况，为城市发展提供了绝佳的机会。

5. This regeneration into a major regional center contains large office and residential developments of the area, which began in the 1980s and is now largely completed, has been hailed as one of the best examples of urban regeneration in the UK, with its high-quality

waterfront public space construction, advanced waterfront design experience and construction mode, provides a reference for many cities which are currently in full swing. (Para. 4)

这个改造项目主要包含大型办公和住宅开发区域,始于20世纪80年代,目前已基本完工。凭借其高质量的滨水公共空间建设、先进的滨水设计经验和建设模式,该项目被誉为英国城市改造的最佳典范之一,为许多城市目前如火如荼的发展提供了借鉴。

6. The pathways in the area consist of a series of existing and new streets and waterways that reflect the character of the region as well as stitching together the urban fabric of the region and its surroundings. (Para. 6)

该地区的道路由一系列现有的和新的街道和水道组成,这些街道和水道既能反映该地区的特点,又能将该地区及其周围的城市结构拼接在一起。

【这个英文句子是一个含有定语从句的复杂句。为了让目的语读者能够更好地理解原文,可以采用断句译法。对于原文中的长句,可以划分语法层次,再按照译文的语言习惯调整搭配,最后依语意断句翻译。原文定语从句由 that 引导,修饰先行词 streets and waterways,由于中文中并不存在从句结构,所以翻译时需要在 that 处断句,将原文的长句切分开,让译文更加符合汉语的表达习惯,从而达到良好的交际目的。】

7. The regeneration program, which ended in 2003, has seen an investment of over £25 million and the museums on the pier attract visitors that generate 2 million trips per year, which proved the success of this renewal program. (Para. 7)

复兴计划于2003年结束,投资超过2 500万英镑。码头上的博物馆每年吸引的游客多达200万人次,证明了这一复兴计划是成功的。

8. This award aims to promote the regeneration of the city through festivals and events that connect with European culture externally and internally. (Para. 8)

该奖项旨在通过与欧洲文化内外相联系的节日和活动促进城市的复兴。

【这里"奖项"是指"欧洲文化之都"称号。1985年,时任希腊文化部长的前女演员梅丽娜·梅尔库丽和法国文化部长杰克·朗提出了设立年度文化之都的想法,通过突出欧洲文化的丰富性和多样性,提高人们对欧洲共同历史和价值观的认识,使欧洲人更加紧密地联系在一起。欧盟委员会负责管理这一称号,每年欧盟部长理事会正式指定一个城市,举办一系列具有浓厚欧洲特色的文化活动。筹备欧洲文化之都可以为城市带来巨大的文化、社会和经济效益,有助于促进城市更新,改变城市形象,提高其在国际范围内的知名度和形象。迄今为止已有40多个城市被指定为"欧洲文化之都"。】

9. The development site begins on the north side of the dock top area and extends northwards to Bramley Moore Dock, covering an area of about 60 hm^2, with an estimated investment of more than £5.5 billion over a time span of about 20-40 years. (Para. 9)

开发场地始于码头顶部区域的北侧,向北延伸至布拉姆利摩尔码头(Bramley Moore Dock),占地面积约60公顷,预计投资超过55亿英镑,时间跨度约20~40年。

10. Therefore, it is imperative that the regeneration of port heritage be changed, which can

improve the sustainable development of waterfronts in many cities. (Para. 10)

因此,改变港口遗产的再生方式,促进许多城市滨水区的可持续发展势在必行。

【英语和汉语在语法结构上存在许多差别。翻译时如果对原文的词序照抄照搬,这样的机械性译文会显得滑稽可笑。利用换序译法,根据目的语的语言表达习惯对原文的词序进行调整,重新组合语言,能够使译文通顺地道。主语从句换序是换序译法的其中一种情况,常见于"It is/was/has been /had been……that"的句型中,其中 it 做从句的形式主语,也就是说从句才是真正的逻辑主语,翻译时应将主语从句前置。】

Exercises

1. Fill in the blanks with the proper given words, and then translate the sentences into Chinese.

　　issue　　undertake　　namely　　stimulate　　coherent　　turnaround

1) In the past, electric vehicles encountered their problems, _____, a limited driving range and very few recharging points, which limited their use.

2) The department already _____ a preliminary environmental impact assessment, which seems generally supportive of the project.

3) Innovative science produces new propositions in terms of which diverse phenomena can be related to one another in more _____ ways.

4) The government now backs up the domestic car industry, lending it money and overseeing its _____ plans.

5) Although as yet unproven, physical exercises specifically designed to _____ the cerebellum may become mainstream interventions.

6) In a gesture to help preserve history, the Greek government established a committee to _____ the professional restoration of the Acropolis.

2. Translate the following sentences into Chinese.

1) Although the city's material and immaterial values are linked in a very complex and interactive way, the concept of heritage and current conservation practices have inevitably attracted the public's attention.

2) With the rise and fall of industrialization, the function of the waterfront has undergone historic changes, with much of the industry now moving away from the waterfront district, leaving behind a large number of abandoned industrial sites, most of which are adjacent to fragile waterfront ecosystems.

3) Preservation on the historical identity and green space management are the major contributing approaches in urban regeneration strategies to achieve social sustainability in historical public space of cities.

4) As one of the active factors to promote urban vitality and cultural character, port

heritage in urban waterfront areas is essential in urban preservation and urban regeneration, providing opportunities for cities to reconnect with their water's edge.

5) Building and adapting these sites implies a constant battle between technology and the forces of nature, requiring the application and development of the most suitable method in the construction process.

6) Historical port space is one of the main pillars in social life that has effects on economic, social, architectural and symbolic dimensions.

7) It seems paradoxical that, despite port heritage indisputable cultural, historical and technical value, in cities, port constructions that compose the waterfront of our cities are forgotten heritage space.

8) In recent years, there has been a rediscovery of the waterfront, led by urban sustainability, and a realization that these areas have the potential to create unparalleled opportunities for regeneration-creating new uses and re-invigorating places vitality and the sense of belonging with a port heritage that was once abandoned or marginalized.

9) Therefore, ports are indispensable as a gateway to promote the development of maritime trade in cities, and the origin and prosperity of many coastal settlements are attributed to waterline transportation and trade, while such cities and ports are usually closely related in function and space.

10) The unique location of the junction of waterside and the city has been the birthplace of many stories in the early stages of urban development from ancient times.

3. Translate the following sentences into English.

1) 由于港口相关活动的迁移，港口城市的郊区往往成为城市重组时被废弃的地区。

2) 历史上，第一批滨水重建项目始于20世纪60年代，当时波士顿、巴尔的摩和旧金山等城市引领了这一潮流。

3) 要让居民相信滨水重建的好处及其与港口活动的可能联系，往往不是一件容易的事情。

4) 大多数城市都明白，港口的工业遗产应该得到加强而不是根除。

5) 港口与城市之间牢固而密切的关系往往是历史环境的结果，但由于港口对城市环境造成的负面影响，这种关系已经被破坏。

6) 如果开发得当，历史悠久的港口城市不仅可以巩固城市的特色，还可以带来经济效益，有助于整个城市的可持续发展。

7) 该项研究的结论提醒人们，重建不仅仅是恢复结构，还包括捕捉历史港口城市的精神。

8) 根据联合国教科文组织的定义，文化遗产包括有形的纪念碑、遗址和物品，以及世代传承的非物质传统和生活方式。

9) 人们普遍认识到，过去的遗产应该影响未来的发展，但如何实现这一点以及它带来什么好处因地而异。

10) 应利用港口城市的历史文化、事件、地标、现有建筑和自然，赋予海滨重建的特色和意义。

4. **Choose the best paragraph translation. And then answer why you choose the first translation or the second one.**

无论规模大小或位居何处，港口城市都受到了工业衰退的强烈影响，被迫进行自我改造。我们曾经熟悉的、有着特定身份和文化印记的港口城市，现在已经被以高科技生产、科研机构、旅游、休闲等服务为一体的新型海洋城市取而代之。由此可见，西方国家港口城市在其历史发展道路上发生了重大变化。在这个转变的过程中，最重要的挑战之一是改变思维和心态。越是特色鲜明的港口城市，其形象、文化和特征就越独特和专业化，应对这些挑战的难度就越大。

译文一：

No matter big or small and where they are located, port cities have been strongly affected by the industrial recession and are forced to have self-improvements. The port cities that we used to be familiar with, with specific identities and cultural imprints, have now been replaced by new marine cities integrating high-tech production, scientific research institutions, tourism, leisure and other services. It can be seen that port cities in western countries have had big changes in their historical development. One of the most important challenges in this transformation process is to change the mindset. The more special the port city is, the more unique and specialized its image, culture and identity are, the more difficult it will be to deal with these challenges.

译文二：

Regardless of size or location, port cities were strongly affected by industrial decline and forced to reinvent themselves. The port city we once knew was coined by a particular identity and culture, but now a new type of maritime city emerges with high-tech production, research institutions, tourism, leisure and other services. So, the port city in western countries underwent serious changes in its historic development path. In this process of transformation, one of the most important challenges was and is to change thinking and mentalities. The more distinctive the port city, the more exclusive and specialized its profile, culture and character, the greater the difficulty of responding to these challenges.

5. **Translate the following passage into Chinese.**

The pathways in the region are divided into three levels. The first level is the main carriageway in the area and its borders. For instance, the "main gate" on the south side, Reid Road, has been upgraded in its entirety in terms of building facade, street furniture and environmental greenery. Linking the area to the upmarket settlements to the south, it also creates entrance streets that reflect the character of the area. The second level is the public right-of-way roads in the area, including North and South Pier Street. They are located on the backside of the development buildings and are used to ease internal vehicular traffic and provide space for related municipal services such as trash pickup, ensure the safety, comfort and continuity of the main pedestrian paths in the area on both sides of the canal. The third level is a complete and coherent pedestrian path around the canal harbor pond. The design uses the canal as the centerpiece of the public space for the entire area, creating a continuous path around the canal from the end of the harbor pool on the east side to the Little Venice complete on the west side.

6. Translate the following passage into English.

许多城市正在对废弃的旧滨水区进行大规模重建,打算将其变成商业、文化、旅游或高档住宅区。这些城市利用滨水重建,给人们提供新的就业机会、更好的住房条件、改善的社区设施和更多的旅游机会,从而振兴城市地区、促进当地经济。滨水区的重建也成了重新点燃城市与港口之间关系的机会。在某些情况下,滨水区改造主要是一项美学工程,尽管项目在多大程度上可以振兴城市地区仍未得到证实。然而,即便如此,它们也代表着土地使用从工业用地向后工业用地的转变。

三、翻译家论翻译

林语堂(1895—1976),是一位蜚声海内外的现代英汉双语作家、文学家和翻译家,是国内外罕有的一位学贯中西的学者。其文学作品代表作有以英文讲述中国人故事的《京华烟云》、着力于介绍中国文化的《吾国吾民》等,翻译作品代表作有《浮生六记》《道德经》《孔子的智慧》《老子的智慧》《英译庄子》等。1975年,林语堂成为第一个获得诺贝尔文学奖提名的中国作家。

林语堂于1933年为吴曙天编著的《翻译论》作序,发表了近万言的《论翻译》,首次提出将"艺术""心理学""语言学""美学"等概念与翻译结合。林语堂认为"翻译是一门艺术",并提出"忠实、通顺、美"三大翻译标准。在《论翻译》开篇,林语堂就明确提出:"谈翻译的人首先要觉悟的事件,就是翻译是一种艺术……翻译的艺术所依赖的是'第一是译者对原文文学上及内容上透彻的了解;第二是译者有相当的国文程度,能写清顺畅达的中文;第三是译事上的训练,译者对于翻译标准及手段的问题有正当的见解'。"对于翻译标准,林语堂的"忠实、通顺、美"是对严复的"信、达、雅"思想的进一步发展。正如林语堂指出:"这翻译的三层标准,与严氏的'译事三难',大体上是正相比符的。忠实就是'信',通顺就是'达',至于翻译与艺术文(诗文戏曲)的关系,当然不是'雅'字所能包括。"对于"忠实""通顺"标准,林语堂认为译作需做到:非字译、须传神,非绝对、须通顺。即翻译过程中不能逐字翻译,要以句为意象着笔,在语境下整体考虑语义;译文不但要达意,还要传神,忠实原文的言外之意;由于中西文法的差异,翻译不可能做到绝对忠实原文;译文要符合译入语的表达习惯,地道流畅。不同于严复的"雅",林语堂认为翻译标准的"美"要求译者要懂得基本的翻译方法和技巧,要有较高的个人文学艺术修养,能明白原文之美,也能在译文中体现"美"——"声音之美,意义之美,传神之美,文气文体形式之美"。简而言之,林语堂的翻译理论就是"美译"理论,即以句译为形式基础,内容上传情,求美达意。"忠实"和"通顺"标准的最终目标是达到"美"的标准。

以下为林语堂的翻译作品《浮生六记·闲情记趣》中的一个译例：

原文：若夫园亭楼阁，套室回廊，叠石成山栽花取势，又在<u>大中见小，小中见大，虚中有实，实中有虚，或藏或露，或浅或深</u>，不仅在"周回曲折"四字，又不在地广石多，徒烦工费。

译文：As to the planning of garden pavilions, winding corridors and out-houses, the designing of rockery and the training of flower-trees, one should try to show <u>the small in the big, and the big in the small, and provide for the real in the unreal and for the unreal in the real. One reveals and conceals alternately, making it something apparent and sometimes hidden</u>. This is not just rhythmic irregularity, nor does it depend on having a wide space and great expenditure of labour and material.

对原文画线部分的翻译既达意又传神，与原文使用四字结构产生的韵律感和节奏感效果一致，具有形式上和音律上的美，看上去对仗工整，读起来抑扬顿挫、朗朗上口，是"忠实、通顺、美"翻译标准的综合体现。林语堂深厚的英文功底和渊博的中国传统文化修养使得他翻译的作品不仅"忠实""通顺"，更是一件"美"的艺术品。

（摘自：林语堂.语言学论丛：林语堂名著全集第 19 卷[M].长春：东北师范大学出版社，1994:35-36；沈复.浮生六记：英汉对照[M].林语堂，译.北京：外语教学与研究出版社，1999:96.）

Text B

The Port of Rotterdam

By Anonymous Author

Ideological and Political Education：传承中华优秀文化 增强文化自信

港口作为国与国、城与城、人与人在经贸和文化相互交流的特定平台，以其不可复制的包容性、接纳性和亲和力，衍生出了世界海港风格各异的港口文化。

自古以来，中国港口文化内涵丰富，如中原文化孕育了自强不息的环渤海港口文化、吴越文化孕育了海纳百川的长江三角洲港口文化、八闽文化孕育了开拓创新的东南沿海港口文化、岭南文化孕育了务实创新的珠江三角洲港口文化、粤西文化孕育了开拓进取的西南沿海港口文化以及万古长江同舟共济的内河港口文化等。2 000多年前，一条以中国徐闻港、合浦港等港口为起点的海上丝绸之路成就了世界性的贸易网络，中国在与世界展开商贸活动的同时，也带动了不同文化的交流碰撞，推动了世界的进步和发展。当下，中国正在启动与东盟及世界各国共建21世纪海上丝绸之路的重大战略，历史上曾创下的海洋经济观念、和谐共荣意识、多元共生意愿，将为国家发展战略再次提供丰厚的历史基础。"友善、包容、互惠、共生、坚韧"的海上丝绸之路的文化内涵，对于建设21世纪海上丝绸之路，对于中国与世界更深层次的互动，无疑具有深刻的启迪和极其重要的当代意义。

中国港口在加快了专业化、大型化、现代化建设的同时，也加速了对老港区的功能调整和技术改造，将历史文化积淀与当代新锐文化特质兼收并蓄，将"海港文化、创意科技、人文关怀"融为一体，促进海港文化遗产的深层次开发。优秀的滨水区重建案例有变废弃滩涂为生态湿地的遂宁五彩缤纷路、营造出丰富多样"湾景观"和"湾生态"的深圳湾滨水景观等。海滨重建和再利用港口遗产，不仅是节约资源，也是记录海滨建筑文化和见证海滨复兴历史的最佳选择，实现了港口与自然、港口与城市、港口与市民等方面的和谐相处。

中国港口以其丰富的文化内涵、非凡的发展速度、辉煌的成就获得世人瞩目。当前，中国已经成为世界港口大国、航运大国、集装箱运输大国，正在向港口强国迈进。青年一代深刻地感受到中国智慧、中国方案、中国力量和中国特色社会主义市场经济的优势，大大增强了中国特色社会主义道路自信、理论自信、制度自信、文化自信。

1. The port of Rotterdam is by far the largest deep-water port in Europe and also ranks among the top ten worldwide. One of the reasons is certainly its location in the delta of the Rhine and Meuse rivers and thus directly on the North Sea. One difference to the other North Sea ports is the possible **draught** of 24 metres, which makes the port of Rotterdam **accessible** even to the giants of the seas. The port has a surface area of around 100 square kilometres, with an extension of just under 40 kilometres.

2. Development of the port of Rotterdam. In the beginning there was a fishing village on the Rotte. The small river gave Rotterdam its name and flows into the Meuse in the city area. Those who are today in the city can have a view to the Hoogstraat that with its today 820 metres marked the division between land city and water city and **therewith** represented a kind of harbour boundary. In the first years since its origin in the 14th century, the harbour of Rotterdam was rather of regional significance and only in the course of the industrialization, there was an increasing growth.

3. In the 19th century, the southern side of the Meuse, facing away from the city centre, was particularly relevant for port development. The Feijenoord district, with its world-famous football club, was where most of the dockworkers lived, while the port itself continued to grow from the north bank. A real **milestone** was the opening of the Nieuwe Waterweg, a 20-kilometre stretch of waterway linking the **estuaries** of the Rhine and Meuse with the North Sea, which was opened in 1872. Only ten years later, the city of Rotterdam took over the port and ensured rapid expansion. Port **facilities** were offered for rent, as were loading cranes and storage areas, and could also be used on an hourly basis. This model proved to be immensely successful, so that the port was able to grow further.

4. In the 1920s, the Rijnhaven, Waalhaven and Maashave followed in the Port of Rotterdam and at the end of the decade the first settlements of the **petrochemical** industry followed, which would continue to shape the face of the port in the years to come. After the Europoort was built in the 1950s, Rotterdam became the largest port in the world in 1962. Especially the transport and direct processing of crude oil were and still are relevant.

5. In addition, capacities for bulk goods were also created in the area of Europoort, and the largest ships of this type can almost only **anchor** in the Dutch metropolis, which means that they mainly handle ore or grain. Especially in the post-war period, the port of Rotterdam benefited from the fact that the Rhine access road remained **intact**, which was continuously expanded for shipping. A direct comparison with the port of Hamburg shows that, due to the division of Germany, the port soon lost the Elbe as its "**hinterland**" and thus suffered a locational disadvantage.

6. The Port of Rotterdam in the Past and Present. The recent past is also characterized by container handling at the port of Rotterdam. The first containers were unloaded in 1966 and as early as 1967 the ECT (European Container Terminal) was put into operation in the Eemshaven. Since the 1980s, container ships have also landed on the Maasvlakte, an **artificial** island not far from the mouth of the Maas. However, it became apparent that the constant expansion was no longer **compatible** with nature conservation and the interests of the local population, so that the port of Rotterdam disappeared further and further from the city centre. This was made more attractive by various development programmes, while the port is mainly located on the Maasvlakte and the second Maasvlakte, which is currently under construction. **Statistically** speaking, the Port of Rotterdam remained the world's number one port until into the 21st century and was first **overtaken** by Shanghai and Singapore a few years ago.

7. Port of Rotterdam Figures. The figures that make up the port of Rotterdam are impressive. In 2018, the handling of sea freight **amounted** to almost 469 million tonnes, which represents a renewed slight growth compared to previous years. The number of ships calling at the Port of Rotterdam each year is 30,000 in the maritime sector alone and 105,000 inland waterway vessels. Around 320,000 people live in the Port of Rotterdam, which generates a proud seven percent of the Dutch gross domestic product. Above all oil is relevant for the port of Rotterdam, which is also due to the pipelines with direct access to the Ruhr area and to Antwerp. In addition, the port of Rotterdam is home to four large **refineries** and various companies from the oil and chemical industries are responsible for processing. The same applies to the gas companies and companies that trade directly in oil.

8. The Port of Rotterdam is also relevant for coal, fruit and vegetables and containers and is a European leader in all the above-mentioned areas. Remarkable is the perfect **logistics** with the possibility of transporting goods quickly to the hinterland via the motorways as well as using the shipping routes across Europe via the Rhine, Main, Main-Danube Canal and Danube. Furthermore, it goes without saying that the port of Rotterdam also has a large freight terminal and a **correspondingly** large amount is transported by rail. The Betuwe route from 2007 is particularly relevant here, linking Rotterdam to Zevenaar in the Gelderland region, from where it runs towards Emmerich and the Rhine-Ruhr region.

9. How is the port of Rotterdam managed? The port of Rotterdam is managed by Havenbedrijf Rotterdam or the Rotterdam Port Authority. We are talking about a company with 1,200 employees that has a **turnover** of 710 million euros (2019). The public limited company (PLC) is not listed on the stock exchange and is owned 70 percent by the **municipality** of Rotterdam and 30 percent by the Dutch state. The company is divided into various departments, including financial and information management or even the port authority, health authorities, tourism, etc.

10. Innovations at the Port of Rotterdam. One of the plus points of the port of Rotterdam is its high innovative power. Since 2018, for example, hybrid ships have been increasingly used, resulting in significantly lower emissions during regular **patrols**. Furthermore, Rotterdam has a so-called innovation ecosystem and promotes companies in the port area that want to position themselves as pioneers, especially in ecological terms. In Rotterdam, digitization was pushed ahead decisively at an early stage, which is also demonstrated by the numerous "Port Forward" solutions. The goal of CO_2 neutrality was also **issued** and has since been pursued both in the area of **infrastructure** and energy generation. According to its own statements, Rotterdam wants to develop into a "waste-to-value hub".

(1,142 words)

https://www.marvest.de/en/magazine/ships/the-port-of-rotterdam/

New Words

1. draught [drɑːft]

    n. the depth of a vessel's keel below the surface (especially when loaded)（船的）吃水深度
    What is the maximum draught of the giant ship now?
    这艘巨轮现在的最大吃水是多少？

2. accessible [əkˈsesəb(ə)l]

    a. capable of being reached 可进入的；可使用的
    We've been talking about the printing press, how it changed people's lives, making books more accessible to everyone.
    我们一直在讨论印刷机，讨论它如何改变了人们的生活，让每个人都能读到书。

3. therewith [ðeəˈwɪð]

    ad. with that or this or it 于是；以其；与此
    This should produce a greater sense of cultural teaching and therewith learning in intercultural conflicts.
    这将能够营造一种更好的文化教学感，并随之能够在跨国界的文化冲突中学习。

4. milestone [ˈmaɪlstəʊn]

    n. a significant event in one's life (or in a project) 转折点；里程碑
    Lewis and Brooks argued that an important developmental milestone is reached when children become able to recognize themselves visually without the support of seeing contingent movement.
    路易斯和布鲁克斯认为，当儿童不借助于观察偶然行为，也能通过视觉认出自己时，他们就达到了发育的一个重要里程碑。

5. estuary [ˈestʃuəri]

    n. the wide part of a river where it nears the sea 河口；江口
    In the 1970s Britain toyed with the idea of building a big new airport in the Thames estuary to the east of London.
    1970年代，英国提出在泰晤士河口到伦敦东部地区建设一个新的大机场。

6. facility [fəˈsɪləti]

   n. a service that an organization or a piece of equipment offers you 设施，设备

   This facility was then in its third decade of production and was beginning to show signs of decline.

   当时该设施已经快有三十个年头了，并开始出现衰退的迹象。

7. petrochemical [ˌpetrəʊˈkemɪkl]

   n. any compound obtained from petroleum or natural gas 石油化学产品

   The group said it will run its refineries at full capacity and cut petrochemical production to boost output of gasoline and diesel for domestic use.

   该集团称将通过发挥其所属炼油厂的最高生产能力、减少汽柴油等石化产品的对外出口等措施来保障国内供给。

8. anchor [ˈæŋkə(r)]

   v. secure a vessel with a mechanical device that prevents a vessel from moving 抛锚，泊（船）

   Large quantities of giant ships with ore or grain anchor in this famous port every day.

   每天都有大量载有矿石或谷物的巨轮停泊在这个著名的港口。

9. intact [ɪnˈtækt]

   a. not impaired or diminished in any way; undamaged in any way 完好无损的

   Well over half of those ships were carrying cargo stored in large ceramic jars, many of which were preserved largely intact on the ocean floor.

   这些船只中有一半以上装载着储存在大型陶瓷罐子里的货物，其中许多罐子被完好无损地保存在海底。

10. hinterland [ˈhɪntəlænd]

    n. a remote and undeveloped area 内地；穷乡僻壤；靠港口供应的内地贸易区

    China's urbanization beginning started from the coastal cities and then extended to the hinterland along the artery of traffic.

    中国城市化首先开始于沿海港口城市，再沿着交通路线向内地延伸。

11. artificial [ˌɑːtɪˈfɪʃ(ə)l]

    a. not arising from natural growth or characterized by vital processes 人造的，人工的

    Managers have learned to grapple with networking, artificial intelligence, computer-aided engineering and manufacturing.

    管理人员已经学会如何去应付网络、人工智能、计算机辅助工程和制造等问题。

12. compatible [kəmˈpætəb(ə)l]

    a. able to exist and perform in harmonious or agreeable combination 兼容的；可共存的；可和睦相处的

    Fujitsu took over another American firm, Amdal, to help it to make and sell machines compatible with IBM in the United States.

    富士通接管了另一家美国公司安岛，以帮助其在美国生产并销售与IBM兼容的机器。

13. statistically [stəˈtɪstɪkli]

    ad. with respect to quantitative data 统计地；统计学上

The conclusion was that social isolation is statistically as dangerous as high blood pressure, smoking and obesity.

结论是,从统计数据上看,社交孤立与高血压、吸烟和肥胖一样危险。

14. overtake [ˌəʊvəˈteɪk]

    v. catch up with and possibly pass 追上;(发展或增长)超越,超过

    Coming out of the final bend, the runner stepped up a gear to overtake the rest of the pack.

    赛跑者绕过最后一个弯道后开始加速,以图超越同组的其他选手。

15. amount [əˈmaʊnt]

    v. add up in number or quantity 达到,总计

    Ben Jumper, the owner of the storage business, said he expected his losses would amount to tens of millions of dollars.

    这家仓储公司的老板本·斯特沃表示,他预计自己的损失将达到数千万美元。

16. refinery [rɪˈfaɪnəri]

    n. an industrial plant for purifying a crude substance 提炼厂,精炼厂

    At a refinery, the crude oil from underground is separated into natural gas, gasoline, kerosene, and various oils.

    在炼油厂,从地下提取的原油被分离成天然气、汽油、煤油和各种其他类型的油。

17. logistics [ləˈdʒɪstɪks]

    n. handling an operation that involves providing labor and materials be supplied as needed 后勤,组织工作;物流

    The skills and logistics of getting such a big show on the road pose enormous practical problems.

    这样的一个大型节目进行巡演在技术和后勤方面都会面临大量实际问题。

18. correspondingly [ˌkɒrəˈspɒndɪŋli]

    ad. in a manner agreeing in amount, magnitude, or degree 相应地,相对地

    As his political stature has shrunk, he has grown correspondingly more dependent on the army.

    随着政治声望不断降低,他相应地变得更依赖于军队了。

19. turnover [ˈtɜːnəʊvə(r)]

    n. amount of business done by a company within a certain period of time (一定时期内的)营业额,成交量

    The dimensions of the market collapse, in terms of turnover and price, were certainly not anticipated.

    股市崩盘的规模,无论从成交量还是价格来看,都是出乎意料的。

20. municipality [mjuːˌnɪsɪˈpæləti]

    n. town, city or district with its own local government; governing body of such a town, etc. 自治市;市政当局

    Every province, autonomous region and municipality has set up a census office and will report to a main center staffed by officials from government departments.

每个省、自治区、直辖市都设立了人口普查办公室,并向由政府部门官员组成的主要中心报告。

21. patrol [pəˈtrəʊl]

   n. the activity of going around or through an area at regular intervals for security purposes 巡逻,巡查

   He began to waver in his resolution when a patrol wagon rolled up and more officers dismounted.
   当一辆巡逻车开过来并且从车上下来更多的警察时,他的决心开始动摇了。

22. issue [ˈɪʃuː]

   v. publish (books, articles, etc) or put unto circulation (stamps, banknotes, shares, etc); make sth. known 出版或发表;发行;颁布

   The minister issued a terse statement to the press denying the charges.
   部长向新闻界发表了一份简短的声明,否认那些指控。

23. infrastructure [ˈɪnfrəstrʌktʃə(r)]

   n. the basic structure or features of a system or organization 基础设施,基础建设

   The project finances small scale irrigation systems, feeder roads and other types of community infrastructure.
   该项目资助小型灌溉系统、支线公路和其他类型的社区基础设施。

Phrases and Expressions

by far 到目前为止
in the course of 在……过程中;在……期间
take over 接管;接收
on a/an……basis 在……的基础上
in addition 另外;此外
benefit from 得益于;得利于
put……into operation 使生效;使运转,使开动
make up 组成;补足;化妆;编造
call at 拜访,访问;停靠(车站)
be home to 是某人/某事的家园
apply to 适用于;应用于
push ahead 向前推进

Terminology

crude oil 原油
bulk goods 散装货物
container handling 集装箱装卸
sea freight 海运

freight terminal 货运站，货运码头
the stock exchange 证券交易所
hybrid ship 混合动力船
CO_2（carbon dioxide）二氧化碳

Proper Names

the port of Hamburg（德国）汉堡港
ECT（European Container Terminal）欧洲集装箱码头
gross domestic product(GDP) 国内生产总值
the Rotterdam Port Authority 鹿特丹港务局
public limited company (PLC)股份有限公司

一、翻译简析

1. Those who are today in the city can have a view to the Hoogstraat that with its today 820 metres marked the division between land city and water city and therewith represented a kind of harbour boundary.（Para. 2）

  如今，城市里的人们可以到呼格斯特拉特街区观光游玩，它长达 820 米，把陆地城市和水上城市分隔开来，因此也被视为港口边界。

  【这是一个包含了主语从句和宾语从句的复杂句，翻译时如果原封不动地直译过来，译文会显得臃肿不堪，且结构不明确、语义模糊。这时，我们可以采用断句译法，根据原文的语法结构和句子含义把长句断开，然后再根据汉语的表达习惯，采用增译法，添加必要的单词、词组、分句或者完整句，以保证译文的准确性和完整性。】

2. A real milestone was the opening of the Nieuwe Waterweg, a 20-kilometre stretch of waterway linking the estuaries of the Rhine and Meuse with the North Sea, which was opened in 1872. (Para. 3)

  1872 年开通的运河新水道具有里程碑的意义。这是一条 20 千米长的水道，把莱茵河和默兹河的河口与北海连接起来。

  【这个句子含有一个同位语和一个非限制性定语从句，要把原文的内容连同其形式一起翻译成地道的汉语几乎不可能。这时，我们可以把非限制性定语从句转换成前置定语，然后利用转句译法，把包含丰富语义的同位语转译成汉语的一个句子，从而使得译文更加清晰通顺，符合汉语的表达习惯。】

3. In addition, capacities for bulk goods were also created in the area of Europoort, and the largest ships of this type can almost only anchor in the Dutch metropolis, which means that they mainly handle ore or grain. (Para. 5)

  此外，欧罗港地区也提升了散装货物的运输能力，这种类型的超大船舶几乎只能在荷兰大都市停泊，主要装载矿石或谷物。

  【无论是书面语还是口头语，英语中的被动语态都要比汉语的要多。但是，汉语中并不会用特定语态的改变来表示被动的意味，而是通过"被、受、给"等词语来表示被动，所以英汉互译中，两者的语态不能很好地对应。这时，我们可以采用转态译法，把原句的被动语态转为汉语

的主动语态进行翻译。这样译文就变得更加通顺,避免翻译腔较重的问题。】

4. However, it became apparent that the constant expansion was no longer compatible with nature conservation and the interests of the local population, so that the port of Rotterdam disappeared further and further from the city centre. (Para. 6)

然而,显而易见的是,由于不断扩张已不再符合自然保护的理念以及当地居民的利益,因此鹿特丹港离市中心越来越远。

【由于英汉两种语言表达习惯存在巨大差异,在翻译的过程中,我们常常需要根据语义或者修辞的需要,采用增译法适当增添原文中无其形但有其意的词,将原文隐含的信息增补出来。翻译这个句子时,通过增译了逻辑关联词"由于……因此……"和行为概念所属范畴词"……的理念",能更加准确地补全其背后含义,使译文传达的意思更加具体和完整,同时也符合汉语中词语搭配的习惯。】

5. Remarkable is the perfect logistics with the possibility of transporting goods quickly to the hinterland via the motorways as well as using the shipping routes across Europe via the Rhine, Main, Main-Danube Canal and Danube. (Para. 8)

货物可以通过高速公路快速到达内陆地区,也可以通过莱茵河、美因河、美因河-多瑙河运河和多瑙河等航运路线穿越欧洲,这种高效便捷的物流系统令人惊叹。

【汉语属于主题显著型语言,句子多采用主题-述题结构,即句子前面是一个话题,后面则是对话题的评述或描述。因此,翻译此句时应把原文句子拆成两部分,先翻译主题部分 with the possibility......Danube,再翻译述题部分 Remarkable is the perfect logistics。另外,这个句子中 perfect 属于比较抽象的一个单词,翻译的时候我们可以采用具体译法,对上下文和语境进行仔细的分析,并且结合词语本身的意思,进行具体的翻译,把词语的意思与整个句子结合在一起,使译文产生与原文同样的效果。】

二、参考译文

鹿 特 丹 港

1. 鹿特丹港是迄今为止欧洲最大的深水港,也是世界十大港口之一。其中一个原因当然是由于它位于北海沿岸,地处莱茵河和默兹河的三角洲。与北海其他港口的一个不同之处是,鹿特丹港吃水最深处可达 24 米,即使是特大轮船也能在此停靠。该港口的表面积约为 100 平方千米,码头总长约 40 千米。

2. 鹿特丹港的发展。起初,鹿特河上有一个渔村。鹿特丹因这条小河而得名,它一直绵延至市区的默兹河。如今,城市里的人们可以到呼格斯特拉特街区观光游玩,它长达 820 米,把陆地城市和水上城市分隔开来,因此也被视为港口边界。鹿特丹港诞生于 14 世纪,在最初的几年里,鹿特丹港仅仅具有相当大的区域意义。到了工业化时期,它才得到了越来越大的发展。

3. 19 世纪,默兹河南侧背对着市中心,对港口发展尤为重要。费耶诺德区拥有世界闻名的足球俱乐部,是大多数码头工人居住的地方,而港口本身则继续从北岸发展壮大。1872 年开通的运河新水道具有里程碑的意义。这是一条 20 千米长的水道,把莱茵河和默兹河的河口与北海连接起来。仅仅十年后,鹿特丹市接管了港口并确保了港口的快速扩张。港口设施以及装载起重机和储存区可供租用,也可按小时计费。事实证明,这种模式非常成功,因此港口得以进一步发展。

4. 20世纪20年代,莱茵港、威尔哈文和马萨韦相继在鹿特丹港出现。20世纪20年代末,第一批石化工业园随之出现,这在未来几年继续塑造着港口的面貌。20世纪50年代欧罗港建成后,鹿特丹港于1962年成为世界上最大的港口。原油的运输和直接加工尤其发达,过去是,现在仍然是。

5. 此外,欧罗港地区也提升了散装货物的运输能力,这种类型的超大船舶几乎只能在荷兰大都市停泊,主要装载矿石或谷物。特别是在战后时期,鹿特丹港得益于完好无损的莱茵河通道,并不断将其扩大以用于航运。与鹿特丹港不同的是,由于德国的分裂,汉堡港很快失去了易北河这个"腹地",因此处于区位劣势。

6. 鹿特丹港的过去和现在。近期以来,鹿特丹港的集装箱装卸也是一大特色。第一批集装箱于1966年卸载。早在1967年,欧洲集装箱码头就在埃姆斯港投入运营。自20世纪80年代以来,集装箱船也在马斯弗拉克特码头登陆,这是一个距离马斯河河口不远的人工岛。然而,显而易见的是,由于不断扩张已不再符合自然保护的理念以及当地居民的利益,因此鹿特丹港离市中心越来越远。各种开发计划使其更具吸引力,而港口主要位于马斯弗拉克特码头和目前正在建设中的马斯弗拉克特二号码头。从统计数据来看,鹿特丹港一直到21世纪都仍然保持着世界第一大港口的地位,几年前才首次被上海港和新加坡港超越。

7. 鹿特丹港数字。有关鹿特丹港的数字令人印象深刻。2018年,海运吞吐量达到近4.69亿吨,与前几年相比再次小幅增长。仅在海运部门,每年停靠鹿特丹港的船舶数量就达3万艘,内河船舶数量为10.5万艘。鹿特丹港大约有32万人口,却创造了占荷兰国内生产总值7%的价值,这真令人感到自豪。首先,石油与鹿特丹港的经济创收密不可分,这也得益于直接通往鲁尔地区和安特卫普的管道。此外,鹿特丹港有四家大型炼油厂,石油和化工行业的多家公司负责加工。这同样适用于天然气公司和直接从事石油贸易的公司。

8. 鹿特丹港的运输还包含煤炭、水果、蔬菜和集装箱等方面,均处于欧洲领先地位。货物可以通过高速公路快速到达内陆地区,也可以通过莱茵河、美因河、美因河-多瑙河运河和多瑙河等航运线路穿越欧洲,这种高效便捷的物流系统令人惊叹。当然,鹿特丹港还有一个大型货运码头,那里的大量货物可以通过铁路运输。2007年开通的贝蒂沃路线尤为重要,它将鹿特丹连接到盖尔德兰地区的泽弗纳尔,再从那里通往埃默里奇和莱茵-鲁尔地区。

9. 鹿特丹港是如何管理的?鹿特丹港由鹿特丹港管理局或鹿特丹港务局管理。我们所谈论的是一家拥有1 200名员工的公司,其营业额为7.1亿欧元(2019年)。这家公共有限公司并未在证券交易所上市,70%的股份由鹿特丹市政府拥有,30%的股份由荷兰政府拥有。公司分为多个部门,包括财务和信息管理,甚至还有港务局、卫生局、旅游局等。

10. 鹿特丹港的创新。鹿特丹港经久不衰的秘诀之一是其强大的创新能力。例如,自2018年以来,混合动力船舶的使用越来越多,这大大降低了定期巡逻时的排放量。此外,鹿特丹有一个所谓的创新生态系统,并促进港口地区的公司将他们自己定位为各方面(特别是在生态方面)的先锋。在鹿特丹,数字化在早期阶段就得到了果断的推进,这也体现在众多的"端口转发"解决方案上。鹿特丹港还发布了二氧化碳中和目标,并在基础设施和能源生产领域一直在追求这一目标。根据其自身的声明,鹿特丹希望发展成为一个"变废为宝中心"。

Cultural Background Knowledge: Harbour Culture
海港文化背景知识

　　海港文化是海洋文化的重要构成部分，是国与国、城与城、人与人在港口进行经济、文化、商贸往来中创造的物质和精神财富的总和。任何一个港口的背后总有一种固有的文化背景在支撑、伴随和影响着这个港口的发展。而不同的地域和时代则沉淀出风格各异的海港文化风貌。

　　欧洲港口是底蕴深厚的，所建立的港口文化框架对世界其他港口文化产生了深远的影响。典型的欧洲港代表有英国的伦敦港、荷兰的鹿特丹港、德国的汉堡港、比利时的安特卫普港等。伦敦港通过广泛传播自身的文化价值理念与游戏规则，打造出了航运金融、保险、司法等方面的规则文化。鹿特丹港通过充分发挥港口的潜能，使当地人们形成了港口文化认同感，从而对港口经济活动产生了巨大的文化推动力量。汉堡港通过妥善协调港口规划与城市规划之间的矛盾，使得港区与城区得以和融共生。安特卫普港则从地域优势的独特视角出发，将港口定位为内外兼顾的、服务型的管理者，成为欧洲繁荣的商业港口。

　　亚洲港口在短时期内实现了对于欧美港口的经济赶超，令世人瞩目。具有代表性的亚洲港口有中国的上海港、新加坡港、韩国的釜山港、日本的横滨港等。截止到2022年1月，上海港连续十二年坐稳世界集装箱第一大港的位置。新加坡港是亚太地区第二大的港口，是港城联动的亚洲模式的代表，比较成功地解决了港口与城市之间的空间与功能冲突协调问题。釜山港是韩国最大的商港，是在夹缝中求得生存的典型代表，"一元化管理"是它管理与经营策略方面的重要特点。横滨港是日本第一大港，也是世界十大集装箱港口之一，它十分重视城市、港口和海洋的互相协调。

　　美洲港口普遍有着多元文化融合、积极进取精神的独特文化。旧金山港和纽约-新泽西港是美洲港口中两个典型的代表。旧金山的文化是多样性的，旧金山港在经济发展过程中也始终贯彻这样的理念，十分重视港口与社区以及自然环境的关系。纽约-新泽西港以纽约自由女神像为中心，半径为25英里的区域，共约1500平方千米。该港在发展过程中，坚定不移地维护当地港口区域的生态资源，走出了一条绿色港口的可持续发展之路。

Unit Twelve
Sea Fiction

海洋小说

Text A

Mysteries on the Island
By Lauren Wolk

Ideological and Political Education：建设海洋生态文明 完成五大重点任务

党的十八大把生态文明建设纳入中国特色社会主义事业"五位一体"的总体布局，提出建设美丽中国的目标。党的十八届三中全会首次确立了建立生态文明制度体系。学习贯彻党的十八届三中全会精神，以建设海洋强国为目标，加快海洋生态文明建设，是当前和今后一个时期内海洋领域的一项重要工作。今后，海洋生态文明建设需要转方式、重保护、保底线、抓示范、建制度、形成全民参与共同推进的大格局，全力完成五大重点任务。

1. 维护海洋生态系统健康和安全

针对中国近海海洋环境的现状，采取有效措施，全力遏制海洋生态环境持续恶化的趋势，维护海洋生态系统健康，构建海洋安全屏障。一是严格控制陆源污染物入海。二是维护海洋自然再生产能力。三是选择典型生态受损海岛开展生态整治修复，推进无居民海岛开发利用示范基地建设。四是提高海洋灾害风险管理与应急处置能力。五是深入开展海洋生态文明示范区建设，探索海洋生态文明建设推进模式。

2. 优化海洋空间开发布局

按照整体、协调、优化和循环的思路，促进海域资源的合理开发与可持续利用。

3. 实施海洋科技创新驱动战略

深入实施全国"科技兴海"规划纲要，大力推进海洋科技与海洋经济的深度融合，加快促进海洋产业转型升级。

4. 建立海洋生态文明制度体系

建立系统完整的海洋生态文明制度体系，实行最严格的源头保护制度、损害赔偿制度、责任追究制度，完善环境治理和生态修复制度，形成有利于生态文明建设的激励约束机制，用制度保护生态环境。

5. 大力增强海洋生态文明意识

生态文明建设是需要人民群众共同参与、共同建设、共同享有的社会主义宏伟事业。畅通公众参与海洋环境保护的渠道，建立完善公众参与机制，在全社会树立和弘扬海洋生态文明理念，培育公民的海洋生态意识、生态道德和生态行为，形成广泛持久保护海洋、爱护海洋、共筑美好蓝色家园的氛围。开展海洋生态文明建设全民行动，大力增强海洋生态文明意识。

1. On the crossing from Penikese, Mr. Sloan **huddled** in the **stern**, shivering and ducking at the mere suggestion of spray coming over the bow, and I felt guilty at the pleasure of a warm June afternoon while he suffered. "Miss Maggie makes the best soup in the world," I said. "She'll fix you up in no time." When I said the word soup, Mr. Sloan finally smiled. "I would be obliged," he said, "for even an old ear of corn. Soup would be heaven." Miss Maggie gave him a wry look. "Far from that," she said, "but it should help put you right." "Here," Osh said, pulling a twist of oil cloth from his pocket. Inside was a **stash** of **jerky** he carried, just in case. Mr. Sloan unwrapped it and immediately tore off a bite, chewing hard, and closed his eyes, holding the jerky in both hands near his mouth, like a squirrel with the last of the winter nuts.

2. When we reached Cuttyhunk, I ran ahead to fetch Cinders so Mr. Sloan wouldn't have to walk, while Miss Maggie went to the post office where Mr. Johnson kept the telegraph machine. We all knew that he'd send for the police in Falmouth and then, with his next breath, begin to spread the news that Mr. Sloan had been held **captive** in the **leprosarium** by a mysterious southerner. The islanders were used to **calamity**. When a **hurricane** blew through, as one had just a year before—tearing off roofs and casting boats up on the shore—everyone suffered at least a little. And when a ship wrecked in the Graveyard, people sometimes died—mariners or lifesavers or both—and everyone on the Elizabeths felt the stab of such disaster. This was different. When word got out about Mr. Sloan, excitement would spread like heat lightning across the islands. There would be talk of nothing else.

3. "What was he digging for?" I asked Mr. Sloan as we all sat at Miss Maggie's table, eating soup and bread. There wasn't much in the garden yet beyond spring onions and baby **kale**, but she still had some potatoes and carrots in the root cellar, and cooking had brought them back to life. So had butter. And I believe that Mr. Sloan would have eaten an old shoe if she'd served it to him hot. He sounded a little like Mouse while he ate, **purring** and blinking as he bent low over the steam rising from his bowl. "Let the poor man eat," Miss Maggie said. "You can ask him questions later." But Mr. Sloan shook his head. "No, please, it's quite all right. I'm happy to tell you what I know." He sat up straight. "What was he digging for?" he repeated thoughtfully. "I didn't know that he was digging. I saw him only when he came to leave food every morning and take away my ······ chamber pot." He turned pink and looked away from Miss Maggie. "I'm sorry, ma'am." "Nothing to be sorry about," she said with a **frown**. "He dug holes all over the island and even in the cottages," I said. Which seemed to surprise Mr. Sloan, but then he nodded and said, "Which must be why he stayed on Penikese for so long. He was

digging up the island. But what for?" "Treasure, maybe?" I said. Which drew something like a laugh from Mr. Sloan. "What treasure? Lots of geese on Penikese but none of them laying golden eggs." "So you can't say if he found anything before he left?"

4. Mr. Sloan shook his head. "All I can tell you is that he was cruel. And more than cruel, for leaving me to die when he might have set me free before he sailed off in my boat. I couldn't have chased him. And he **busted** up my radio so I couldn't call for help. No one would have come to Penikese for days, maybe weeks. He could have gotten clean away." He swallowed hard and I was afraid he might start to cry again. "That's enough now," Miss Maggie said. But Mr. Sloan held up his hand and said, "Oh, I'm all right. A little talk won't do me any harm." He nodded to me and spooned up some more soup.

5. "You don't know anything more about him?" I asked. He paused, looking back. "I can tell you what he asked me," he said. "When we first landed on Penikese, he pulled me ashore and to the top of the bluff there by the **pier** and said, 'Where did she live?'" "Where did who live?" I asked. "That's exactly what I said! Those very words. And he replied, 'The leper,' as if there were only one. As if I knew anything at all about those people." He raised his eyebrows and shook his head. "It was long after they left before I went over to live."

6. "And then what happened?" I asked. "I told him I didn't know a thing about the lepers except where they'd lived and died. I showed him both places. The hospital and the cottages. The graveyard. And then he looked around and back at me and said, 'What about the nurse?'" Mr. Sloan shook his head again. "I think he was a bit **daft**. Or he just didn't understand that I hadn't been there when the others were. I told him that. I told him again that I didn't know anything about a nurse or the lepers or anything else about any other humans who had ever lived on Penikese. Period."

7. Mr. Sloan ate the last of his soup and bread and leaned back with a sigh. "But that wasn't good enough for him." At this point, Miss Maggie fetched some **bandages** and told Mr. Sloan to roll up his pant legs so she could have a look at his **scrapes**. Osh and I grinned at each other as he tried to put her off and she bullied him into it. "If you insist," he finally said, baring his knees.

8. I washed out my bowl and stood in the doorway waiting for Osh, but he tipped his chin at me and said, "Go on home. I'll be staying here." Miss Maggie looked at him curiously. "In the barn," he said, turning his hat in his hands. "I'll be fine," Miss

Maggie said. "Just the same," Osh said.

9. Mr. Sloan closed his eyes and began, very softly, to **snore**. "Help me get him to bed and I can promise you he'll sleep till morning," she whispered. "Just the same," Osh said again. "I'll be nearby if you need anything." He came to stand with me. "Will you be all right by yourself?"

10. To be honest, I didn't much like the idea of being alone like that and I nearly said so. But I did like how Osh wanted to look after Miss Maggie. And I did like how he knew I could look after myself. "I'll be fine," I said. I touched his sleeve. "Will you come home in the morning or will I come to you?" He put his hand on my head. "I'll be home at first light," he said. I had never said such a good-bye to Osh, but I said it now and went out through the door without him.

11. It was odd to spend the night alone on our island. Mouse tried to take Osh's place by saying very little and sitting quietly among his painting things. I lit a lantern when it got dark and carried it with me wherever I went, which wasn't very far in such a small place. My bed was in one corner, Osh's in another. The rest of the space was for cooking and sitting, a table and an easel. Our washtub, where we occasionally did a proper laundry, followed by a proper bath, was outside. So was our fishing gear and everything else.

12. I sat by the cold fireplace and considered reading a book, but I'd read everything I had. I thought about practicing my sums, but that seemed a poor way to spend this unusual **solitude**. What, I wondered, was done best alone? That, I decided, was what I should do. But there was nothing I could do that was not better with Osh there. And so I started to think, not about anything in particular. I didn't try to solve a problem, or chase an idea, or invent a notion. I just let my mind drift. And I let my body follow it, first, in the chair by the fireplace, then at the window, looking at the sea, then out through the door and onto the beach, to lie in the cool sand, my head pillowed on my laced hands. Mouse followed me, though not in a straight line, and then sat next to my head and licked her paws until something **scuttled** through the grasses nearby, and she gave chase.

13. The stars were enormous. They pulsed as if they were breathing. I heard something splash offshore. A bass, perhaps, or a diving bird. A huge sand flea, heavy as a pebble, jumped first onto my chest and then onto my cheek, and I **lurched** upward, batting it

away before it could bite me. I rubbed the spot where it had landed. Beneath my fingers, my skin was soft and cool. Smooth. Even the little birthmark on my cheek.

The little feather.

The little feather I'd always had.

Even before I'd drifted to this shore.

14. I stood up, very still in the darkness. The sea was louder than it had been a moment before, the stars bigger. I went carefully back inside the cottage, **retrieved** the lantern, and set it carefully on my small table next to my bed. The **cinnamon** box was on my windowsill. I opened it slowly, took out the letter, unfolded it, held it in the lantern light and read again: if I could…… for now…… hope you…… bright sea…… better off…… lambs…… little feather…… I left something…… day it might help…… I remembered the carvings on the wall of the cottage on Penikese. A small lamb. Alongside it, a little feather. I put the letter back in the box, carried the lantern to our one small mirror, turned my face and examined the mark on my cheek. A little feather. Dr. Eastman had written that there was no third baby, just one sent to New Bedford and one buried. But the letter in my hand, as ruined as it was, told me a different story. I read it again and again. Then I took its mysteries to bed with me, and found them waiting in the morning, right where I'd left them.

15. I was up and washed and dressed and out the door before daylight was more than a pale idea. I met Osh beyond the bass stands, coming home. He stopped at the sight of me. "Miss me, did you?" he said, almost smiling. I nodded. It was true. I had. But that was not all. "I think I know who my real parents are," I said. And the almost smile was gone, as if it had never been. He walked past me, past the bass stands, down over the bluff and across the sand bridge that was nearly dry for once, the tide dead low. "You'd best be along to Miss Maggie's," he said **brusquely** as I hurried to keep up with him. "The police will be there soon. You and Mr. Sloan can tell your tales together."

16. Outside the house, he stopped only to grab a bait net and a pail before heading for the tidal pools at the base of the rocks where **minnows** and sand eels often found themselves **stranded**. He began to scoop them up and into the pail. They flashed silver in the early light. "Osh," I said. "What's the matter?" But he didn't answer me. "Are you angry?" Though I couldn't see why he would be, he looked at me and away again. "I'm not angry," he said. "Do what you have to do." "I will," I replied. "Like you said I should. You said I should pay attention to things. Remember?" "I do," he said. "Then why are you upset with me?" He looked at me, full on, and I could see how sad

he was. "Am I not real?" he said. I didn't understand what he was saying until I realized what I myself had said. "Osh, you are the most real thing in the world." "But I'm not your father," he said. "Of course, you are," I said. "But you're not my only father. And you're not my mother." "I know that," he said. But he turned back to his work until I said, "The letter you gave me. It said little feather. And there was a carving in one of the cottages on Penikese," I said, grabbing at his arm. "Osh, stop." Which he finally did, with a sigh. He turned the net inside out, dumping three sand eels into the pail and then **sluicing** in some water so they could breathe. He set it aside and sat down on a nearby rock. I sat next to him. The rock was cold and a little wet and not at all comfortable. "I found a carving of a lamb in the cottage, like the one on the grave marker but smaller," I said. "And next to it a carving of a little feather." Osh looked at me intently. "Like the one on your cheek?" he said. I nodded. "You think that means something." I nodded again, more slowly. "I do," I said.

(2,298 words)

Extracted from *Beyond the Bright Sea*, Corgi Children's. 64-68, 2017.

New Words

1. huddle ['hʌd(ə)l]

    v. crouch or curl up 蜷缩

    She would huddle in a chair, gazing in front of her, wandering off in the dizzy pursuit of a crack along the floor.

    她会蜷缩着坐在椅子上，目光追寻着前方地板上的一道裂缝，盯着看了又看，眼睛发花，出了神。

2. stern [stɜːn]

    n. the rear part of a ship 船尾

    The skipper pushed the boat hard, creating a broad white backwash at our stern.

    船长加快船速，使得船尾产生一片宽大的白色逆流。

3. wry [raɪ]

    a. humorously sarcastic or mocking 用反语表达幽默的；揶揄的

    Pinocchio looked at the glass, made a wry face, and asked in a whining voice, "Is it sweet or bitter?"

    匹诺曹看了一眼玻璃杯，做了个鬼脸，用哀怨的声音问："这是甜的还是苦的？"

4. stash [stæʃ]

    n. a secret store of valuables or money 藏匿处；藏匿物

    Police discovered a large stash of drugs while searching the house.

警方搜查这栋房子时发现里面藏有一大批毒品。

5. jerky [ˈdʒɜːki]

   n. meat (especially beef) cut in strips and dried in the sun 干肉条；熏肉条

   They packed boxes of butterscotch cupcakes, Sichuan-spiced beef jerky, and grapefruit marmalade.

   他们打包了几盒奶油糖果蛋糕、川味牛肉干和柚子果酱。

6. captive [ˈkæptɪv]

   n. a person who is confined; especially a prisoner of war 俘虏；猎获物

   Richard was finally released on February 4, one year and six weeks after he'd been taken captive.

   理查德经历了1年零6周的囚禁之后，终于在2月4日被释放。

7. leprosarium [ˌleprəˈseərɪəm]

   n. a place to quarantine people with leprosy 麻风病院

   The studies also found that the Lirenfang, founded in 556 AD by the Indian monk Narendraysas, was the earliest form of leprosarium in China.

   研究还表明，公元556年由印度来华僧人那连提黎耶舍建立的"疠人坊"是我国历史上首个麻风病院的雏形。

8. calamity [kəˈlæməti]

   n. an event resulting in great loss and misfortune 灾难；灾祸

   He described drugs as the greatest calamity of the age.

   他描述毒品是这个时代最大的灾难。

9. hurricane [ˈhʌrɪkən]

   n. a severe tropical cyclone usually with heavy rains and winds moving a 73-136 knots 飓风；爆发

   The insurance industry will pay out billions of dollars for damage caused by Hurricane Katrina.

   保险业将为卡特里娜飓风造成的损失赔付数十亿美元。

10. kale [keɪl]

    n. coarse curly-leafed cabbage 羽衣甘蓝

    In Denmark, a popular New Year's dish is sweetened kale cooked with cinnamon.

    在丹麦，一种很流行的新年食物就是甜甘蓝菜煮肉桂。

11. purr [pɜː(r)]

    v. make a low vibrating sound typical of a contented cat 发出喉音；猫发出呜呜声

    When interacting with her cubs, cheetah mothers purr, just like domestic cats.

    当猎豹妈妈和幼崽互动时，会像家猫一样呜呜叫。

12. frown [fraʊn]

    n. a facial expression of dislike or displeasure 皱眉；蹙额

The frown vanished away and gave place to an expression of evil satisfaction.

愁眉苦脸消失了,取而代之的是一种邪恶的满足。

13. bust [bʌst]

    v. ruin completely 弄坏;打碎

    Police busted up a counterfeiting racket in Miami last week.

    上周警方捣毁了迈阿密的一个伪造团伙。

14. bluff [blʌf]

    n. a high steep bank (usually formed by river erosion) 峭壁;悬崖

    One of the choppers set down on the beach below them, the other on the bluff above.

    一架直升机降落在他们下面的海滩上,另一架降落在他们上面的悬崖上。

15. pier [pɪə(r)]

    n. a platform built out from the shore into the water and supported by piles(用于供船上下客或装卸货物的)凸式码头

    In the years that followed, he lived a recurring cycle of drink and violence, often sleeping on the beach or under the pier in Brighton.

    接下来的几年里,他生活在酗酒和暴力的循环中,经常睡在布莱顿的海滩或码头下面。

16. leper ['lepə(r)]

    n. a person afflicted with leprosy 麻风病患者

    A Kiss to the Leper published in 1922 is Mauriac's first well-known book, and has been widely praised.

    1922年出版的《给麻风病人的吻》是莫里亚克的成名作,受到了广泛好评。

17. daft [dɑːft]

    a. informal or slang terms for mentally irregular(非正式)癫狂的;愚笨的

    The central bank's decision to raise the cost of borrowing earlier this year was daft.

    今年早些时候,中央银行提高借贷成本的决定是愚蠢的。

18. bandage ['bændɪdʒ]

    n. a piece of soft material that covers and protects an injured part of the body 绷带

    If the bleeding has stopped, cover the area with a bandage and take the person to the hospital.

    如果出血已经停止,用绷带包扎伤口并把病人送往医院。

19. scrape [skreɪp]

    n. an abraded area where the skin is torn or worn off 擦伤;擦痕

    She emerged from the overturned car with only a few scrapes and bruises.

    她从翻了的车里钻出来,只擦破一点皮,碰了几块淤青。

20. snore [snɔː(r)]

    v. breathe noisily during one's sleep 打鼾

    For those who snore only when sleeping on their back, placing a pillow under their upper back may encourage them to change sleeping position.

对于那些仅在仰卧时才会打鼾的人,在他们的上背部放一个枕头可以促使他们改变睡姿。

21. easel [ˈiːz(ə)l]

    n. an upright tripod for displaying something (usually an artist's canvas) 画架

    Here he made an easel of rough wood and put a great piece of gray paper on it and drew a picture.

    在这里他用粗糙原木做了一个画架,并在上面铺了一大张灰色的纸,画了一幅画。

22. solitude [ˈsɒlətjuːd]

    n. a state of social isolation 独处;孤独

    He enjoyed his moments of solitude before the pressures of the day began in earnest.

    他很喜欢一天的压力真正开始前的独处时刻。

23. scuttle [ˈskʌt(ə)l]

    v. move about or proceed hurriedly 急促奔跑;疾走

    Among the flowers the old women scuttled from side to side, like crabs.

    在花丛中,卖花的老婆子们急匆匆地横着跑来跑去,像螃蟹一样。

24. lurch [lɜːtʃ]

    v. move abruptly 突然改变(行为或态度)

    The state government has lurched from one budget crisis to another.

    州政府突然从一个预算危机陷入了另一个危机。

25. retrieve [rɪˈtriːv]

    v. get or find back; recover the use of 找回;收回

    The men were trying to retrieve the weapons left when the army abandoned the island.

    那些人正试图找回该军队撤离这个岛时留下的武器。

26. cinnamon [ˈsɪnəmən]

    n. tropical Asian tree with aromatic yellowish-brown bark(产桂皮的)樟属树木

    I hung on the cinnamon wood fence and watched the moon streak the waves silver.

    我爬上肉桂树篱笆,观赏着月色。月光洒在水面上,泛起道道银色波纹。

27. brusquely [ˈbruːskli]

    ad. in a blunt direct manner 唐突地;直率地

    After lobbying a visitor to his Oxford college too brusquely, he was eased out and took a job at another university which he despised.

    由于过于唐突地游说一位牛津大学的访客,他被牛津大学解雇,只得去了一所他原来不屑一顾的学校任教。

28. minnow [ˈmɪnəʊ]

    n. very small European freshwater fish common in gravelly streams 鲦鱼(一种小淡水鱼)

    At the edge of the woods there was a pond, and there a minnow and a tadpole swam among the weeds. They were inseparable friends.

森林边有个水塘,水塘里有一条小鲦鱼和一只小蝌蚪。它们是形影不离的好朋友。

29. strand [strænd]

   v. leave isolated with little hope of rescue 使困在(某处);使(船、船员、海洋生物)搁浅
   The National Guard says an armoury in Fairmont has opened to shelter stranded motorists.
   国民警卫队声称费尔蒙特的一个军务楼已经开放,以收容受困的汽车司机们。

30. sluice [slu:s]

   v. pour as if from a water channel 开闸放水;冲刷
   She swung her arm, the water sluiced down across the walk and washed the vomit into the gutter.
   她挥起胳膊,将水洒在人行道上,把我呕吐出的东西冲到排水沟里。

Phrases and Expressions

at the suggestion of 在……的建议下
fix up 修补;修理;解决;安顿
in no time 立刻;很快
put……out 出局
be obliged for 为……而感谢
hold……captive 俘虏某人
root cellar 储藏根块植物的地窖
bring……back to life 使复生;使复活;使脱离危险境地
bully someone into something 威胁某人干某事
inside out 彻底地;里面翻到外面

一、翻译策略/方法/技巧:创译法

什么是创译(Transcreation)?创译是 translation+creation 的合成词,是指将文本从一种语言转换为另一种语言,但是创译需要在转换语言的基础上实行再创作。在译文中可融入目标语言的市场文化、市场环境思想,更有效、更精准地向目标受众传达原文的中心思想。目前创译主要针对广告、市场营销、影视字幕、游戏等。通过译文融入目标市场的文化、风土人情、社会环境等以便提高消费者的接受度,从而刺激和吸引消费者。译者在创译时通常会不拘泥于源语言在语意与语音上的束缚,进行一定创造性的翻译,以求译语与源语在功能或效果上的对等。但是,创译法并不是纯粹的或天马行空的创作,它是基于源语在翻译时进行适当的拓展,只是赋予译者一定的"创意"空间。需要译者在翻译过程中融入更多的创造性,具备足够的文字掌控力。

创译,不仅仅是直接翻译或文本的本地化,创译者会专注于把握住原文想要表达的劝说或情感功能,以改写源语言让其适合目标语言受众的阅读需求。创译通常用于营销和宣传

资料,是一种以购买者为中心的目标语写作,从而实现有效传达信息,促成购买,传播形象。创译,作为一个新兴词,很多人还不了解它的重要性,但创译在我们的生活中其实随处可见。比如,星巴克的馥芮白(flatwhite)就是一个很好的例子,如果平铺直叙地翻译成"平白",是无法取悦星巴克的目标人群——那些小资生活情调追求者的。

还有一些品牌名也采用了创译的翻译方法,比如香奈儿(Chanel)、联想(Lenovo)、百事可乐(Pepsi-Cola)、帮宝适(Pampers)。再比如 Connecting people.这是一则诺基亚的广告,直译出来是"联系大众"。但若使用创译,可以翻译为:"科技以人为本",这样的广告词更能吸引消费者的注意。

另一种是运用汉字的"四字格"或七字结构法,言简意赅、意蕴深长。例如,Prepare to want one.(众望所归,翘首以待)——现代汽车,What's new Panasonic.(松下总有新点子)——松下电器。再如 A great way to fly.(新加坡航空,飞越万里,超越一切),Good to the last drop.(滴滴香浓,意犹未尽),Time is what you make of it.(天长地久),A diamond lasts forever.(钻石恒久远,一颗永流传)。

翻译,在这个信息全球化的时代,无疑是非常重要的,但是随着时代的发展,创译的重要性也在慢慢地显现并且服务这个时代。创译就是在翻译的基础上对内容进行微调,以使其与特定市场完美融合。换句话说,在保持原始内容的核心含义的基础上,可以以独特和创新的方式呈现,以引起读者的兴趣。创译需要在翻译的基础上进行再创作,翻译通过创译之后更加出彩动人,锦上添花,所以在一定基础上翻译与创译是相互影响、相互成就的。

二、译例与练习

Translation

1. When I said the word soup, Mr. Sloan finally smiled. "I would be obliged," he said, "for even an old ear of corn. Soup would be heaven." (Para.1)

我说"汤"这个字儿的时候,斯隆先生终于笑了一下。"给我一穗儿老玉米我就够感激的了",他说,"有汤喝那简直美上天了。"

【本句原文运用了比喻的修辞手法,被虐待的斯隆先生喝到一口汤,那感觉像是"天堂般的享受",这是语句本身的含义,但是这与文本目标读者的理解能力存在一定差距,即维索尔伦指出的"意义真空",为填补这种"意义真空",翻译时此处选择了夸张的修辞方法,将"天堂般的享受"夸张为"美上天了",能够突出作者情感,使中国读者产生对斯隆先生的同情,引起共鸣。】

2. He sounded a little like Mouse while he ate, purring and blinking as he bent low over the steam rising from his bowl. (Para.3)

斯隆先生吃东西时发出的声音和鼠儿差不多,他身子弓得很低,对着碗里冒出来的腾腾热气,眼睛一眨一眨的,喉咙咕噜咕噜的。

(阅读全文可知,Mouse 鼠儿是小主人公家里养的一只猫。)

3. "All I can tell you is that he was cruel. And more than cruel, for leaving me to die when

he might have set me free before he sailed off in my boat." (Para.4)

"但我只知道他是个残忍的家伙。何止残忍啊,他驾着我的船一走了之,把我一人留在岛上自生自灭,他本可以先放了我的。"

【本句译文中使用了两个四字格成语"一走了之"和"自生自灭",而非"走了、离开了"或者"让我等死"等这样的表达,它们有相同含义,但是由于语言具有变异性,译者可以将它们处理成其他形式的词语,例如此处使用的四字格,同时又因语言具有协商性,译者可以选择顺应语境的词语,此处"一走了之"和"自生自灭"既忠于原文语义,也顺应读者的语言能力。】

4. "I couldn't have chased him. And he busted up my radio so I couldn't call for help. No one would have come to Penikese for days, maybe weeks. He could have gotten clean away." (Para.4)

"我根本无法去追他。他还把我的无线电摔坏了,让我不能向外界求救。几天之内没人会来彭尼基斯岛,好几周都不会有人来。他可能早就跑得没影儿了。"

5. He paused, looking back. "I can tell you what he asked me," he said. "When we first landed on Penikese, he pulled me ashore and to the top of the bluff there by the pier and said, 'Where did she live?'" (Para.5)

他停下来,向身后看了看。"我告诉你,他问过我一个问题",他说道。"我们第一次在彭尼基斯岛登陆的时候,他把我拉上岸,拉到码头旁的悬崖顶上,说:'她住在哪里?'"。

6. Mr. Sloan shook his head again. "I think he was a bit daft. Or he just didn't understand that I hadn't been there when the others were." (Para.6)

斯隆先生又摇了摇头。"我觉得他有点蠢。也可能是他不知道,他们在那儿的时候我还没来呢。"

7. It was odd to spend the night alone on our island. Mouse tried to take Osh's place by saying very little and sitting quietly among his painting things. (Para.11)

独自一人在岛上过夜,感觉怪怪的。鼠儿想悄悄占着奥什的地盘,它几乎不作声,安静地趴在奥什的画旁边。

8. I lit a lantern when it got dark and carried it with me wherever I went, which wasn't very far in such a small place. (Para.11)

天黑的时候,我点了一盏灯笼,去哪儿我都提着它,这么大点儿地方,去哪儿都不远。

9. Our washtub, where we occasionally did a proper laundry, followed by a proper bath, was outside. So was our fishing gear and everything else. (Para.11)

我们的洗衣盆在外面,我们偶尔会在盆里像模像样洗回衣服,然后再像模像样洗个澡。我们的渔具和其他东西也在外面。

10. I sat by the cold fireplace and considered reading a book, but I'd read everything I had. I thought about practicing my sums, but that seemed a poor way to spend this unusual solitude. (Para.12)

我坐在冷冰冰的壁炉旁边,想看本书,但是家里的书我都看过了。我想要不练几道算数题,但这样打发少有的独处时间似乎也太惨了。

Exercises

1. Fill in the blanks with the proper given words, and then translate the sentences into Chinese.

  carving splash dumped wrecked intently enormous

1) They will examine small objects around the _____ craft, and see if they can identify human remains.

2) Weight is crucial in diving because the aim is to cause the smallest _____ possible.

3) He listened _____, but the stillness was profound and solemn—awful, even, and depressing to the spirits.

4) The country has made _____ strides politically but not economically.

5) We _____ our bags at the nearby Grand Hotel and hurried toward the market.

6) Blum flipped past an evocative sandstone _____ of a child and a pelican, and then admired a griffin plaque.

2. Translate the following sentences into Chinese.

1) It became difficult for Elizabeth even to think of more months of plodding study, when, sitting at her desk, she could picture the flowers and palms of Bermuda, its coral caves with floors of rippling water, or the lazy breakers tumbling in on some California beach.

2) Osh handed me a bushel basket and, armed with a pitchfork, began to harvest seaweed from the wrack line. When the basket was full, we each took a handle and lugged it up to the high rocks where we spread it to wait for rain. When it was washed and dried, we'd hoe it into the garden, food for our food.

3) He put an iron pan on the stove, added butter and spring onions, and, after a bit, tossed in the bass cheeks. The smell, the sizzle, brought Mouse running.

4) Osh gave her a bit of raw belly meat instead. She rolled onto her back and gripped the meat in both paws as she ate it.

5) I was a good sailor, but I knew that even a skiff could be a lot to handle in a strong wind, crosscurrents, shipping lanes.

6) But if I ever needed anything from the mainland, he sent for it with no fuss at all: pencils for my lessons, a new pair of winter boots when I outgrew the last, a new book now and then, medicine that Miss Maggie could not concoct on her own.

7) "There's weather coming," Osh said when I woke the next morning. "You may have a wet crossing home from the city." I walked outside and blinked at the blue sky, sheer yellow sunlight, small breeze.

8) On the crossing from Penikese, Mr. Sloan huddled in the stern, shivering and ducking at the mere suggestion of spray coming over the bow, and I felt guilty at the pleasure of a warm June afternoon while he suffered.

9) When a hurricane blew through, as one had just a year before—tearing off roofs and casting boats up on the shore—everyone suffered at least a little.

10) And when a ship wrecked in the Graveyard, people sometimes died—mariners or lifesavers or both—and everyone on the Elizabeths felt the stab of such disaster.

3. Translate the following sentences into English.

1) 他打冰、捕龙虾和卖画赚的钱都用来买我们自己不会种的东西啦。
2) 我身上又热又黏，双脚生疼，只想要一棵树，得一小片阴凉，或者有一片海就更好了，而且我饿得直想吃口袋里的午餐。
3) 他可没见过锚爪钩上来的那顶王冠，也没见过大银扣，我桂木盒子里的小金环就更不用说了。
4) 但是一想到他们把你绑在一艘破船里，推到海上，我就感到心痛。
5) 他把一个破海螺壳从一个恶魔手指上扯下来扔到一边。
6) 他把干草叉插进沙子里，用手捋了捋头发，然后坚定地看着我，嘴巴紧紧地闭着。
7) 我会驾驶小船、放养龙虾，还能救出一个快要饿死的捕鸟人，那我肯定也能自己坐轮渡去城里，再安全回来。
8) 我带着这些谜团进入了梦乡，第二天早晨起来，发现它们还在老地方等着我，丝毫没有进展。
9) 在房子外面，他停了下来，只拿了一张渔网和一个桶，然后就去了岩石底部的潮汐池，鲦鱼和沙鳗经常被困在那里。
10) 他把渔网翻过来，把三条沙鳗倒进水桶里，又灌了点水进去，这样鱼儿就可以在桶里呼吸了。

4. Choose the best paragraph translation. And then answer why you choose the first translation or the second one.

奥什好奇地看着我。"你以前一直想证明自己不是从彭尼基斯来的"，奥什说，"但你现在好像宁愿是从那儿来的，小乌鸦"。"我知道"，我答道，"是的啊。但这也没什么差别。我为什么会这样想呢？是想证明我的父母死于可怕的疾病吗？是想证明我的父母真的把我抛弃在一条漏水的小船上吗？我才刚出生啊！"奥什对我笑了笑，是那种苦笑。"他们那么做也是有苦衷的"，他说，"他们相信海浪会把你安全带走，把你安全送到别的地方。"

译文一：

Osh looked at me curiously. "You wanted to prove that you weren't from Penikese," he said, "but now you seem almost glad to think that you are, Crow." "I know," I said. "I am. But that doesn't make any sense. Why would I want that? To have parents who died from a terrible disease? To have parents who had sent me out to sea in a leaky skiff? When I was just a new baby?" Osh smiled at me, but it was a sad smile. "They had the best of all possible reasons for doing that," he said. "For trusting the sea to take you safely away and deliver you safely to a different shore."

译文二：

Osh looked at me curiously. "You always wanted to prove that you weren't from Penikese," said Osh, "but now you seem to prefer to be from there, Crow." "I know," I replied. "Yes. But it doesn't make any difference. Why do I think that? To prove that my parents died of some terrible disease? To prove that my parents really abandoned me in a leaky boat? I was just born!" Osh smiled at me, a wry smile. "They do it for a reason," he said. "They believe that the waves will carry you safely and send you somewhere else."

5．Translate the following passage into Chinese.

I was up and washed and dressed and out the door before daylight was more than a pale idea. I met Osh beyond the bass stands, coming home. He stopped at the sight of me. "Miss me, did you?" he said, almost smiling. I nodded. It was true. I had. But that was not all. "I think I know who my real parents are," I said. And the almost smile was gone, as if it had never been. He walked past me, past the bass stands, down over the bluff and across the sand bridge that was nearly dry for once, the tide dead low. "You'd best be along to Miss Maggie's", he said brusquely as I hurried to keep up with him. "The police will be there soon. You and Mr. Sloan can tell your tales together."

6. Translate the following passage into English.

"我在这儿工作时间不长，帮不了你"，她说，"但是病房里有几个人从那时候起就一直在这儿工作"。她给我指了育婴室的路。"你在那儿可以找到佩勒姆太太"，她说。"她现在帮忙照顾婴儿，但她以前在孤儿院工作。你先去问问她吧。如果真有人能回答你的问题，那就是她"。我穿过迷宫般的走廊，然后出了一扇门进入种着树的院子，接着走过院子和后街，走上人行道，踏上另一座大楼的台阶，这里不像圣卢克医院主楼那么大，我继续穿过门，经过走廊，拾阶而上来到一个病房，就听到了婴儿的哭声。

三、翻译家论翻译

他是"没头脑"和"不高兴"之父，他翻译了《夏洛的网》《安徒生童话》《柳林风声》《长袜子皮皮》《木偶奇遇记》《小飞侠彼得·潘》等世界儿童文学经典，总字数逾千万字。他集翻译家、作家、诗人、编辑家、出版家于一身。他是"中国儿童文学创作的先驱者"，他的翻译和创作对中国当代儿童文学发展产生了巨大影响，他的作品影响了几代中国儿童的成长。他就是任溶溶，一位文化智者，一位童心永驻的快乐"老顽童"。

任溶溶（1923—2022），本名任以奇，祖籍广东鹤山。1945年从上海大夏大学中国文学系毕业。1946年1月1日，他以易蓝的笔名在《新文学》杂志创刊号上发表了第一篇儿童文学翻译作品——土耳其作家萨德里·埃特姆的小说《黏土做的炸肉片》，从此开始了他的儿童文学翻译之路。

在七十多年的笔耕岁月中，任溶溶翻译了大量英语、俄语、日语及意大利语等多语种的儿童文学作品。任溶溶的译文通俗易读，亲切幽默，富有感染力。他善于发现儿童生活中充满童趣的语言、场面和情感体验，加以定格、放大、渲染，从而表现童年独特的生活情趣，恰如其分地做到了"紧贴儿童的心"。他曾如此表达初衷："我翻译许多国家的儿童文学作品，只希望我国小朋友能读到世界优秀的儿童文学作品，只希望我国小朋友能和世界小朋友一道得到快乐，享受好的艺术作品。"他把一个个妙趣横生的儿童文学人物带到中国孩子面前，为孩子们打造了一个五彩斑斓的文学世界。

在翻译工作中，任溶溶感到为儿童翻译书必须牢牢记住这是写中文，更注意中文，注意祖国语文的规范化。在他看来，儿童文学除了对儿童进行思想教育，并使他们获得艺术享受之外，还要向其进行语文教育。孩子正在学习语文阶段，一篇短文，一部长篇小说，都是向孩子进行语文教育，因此儿童文学工作者都要有语文修养。"我为什么搞儿童文学？因为儿童文学就好像在跟小孩子聊天、讲故事，我喜欢随便聊天，我用的文字也是大白话。"儿童文学作品最麻烦的是常有文字游戏，碰到这种情况就不能照字面译，要改成相应的、在中文里也有趣的东西，靠注释说明在原文里某字和某字谐音、某字语义双关等，就会使作品乏味。译者既要对得起读者，也要对得起作者，外国作家给儿童讲故事，不但要让他们听懂，而且听得有味道，改用中国话来讲，也同样要做到这一点。

"翻译无非是借译者的口，说出原作者用外语对外国读者说的话，连口气也要尽可能像。我总觉得译者像个演员，经常要揣摩不同作者的风格，善于用中文表达出来。"任溶溶说，自己是代替外国人用中国话讲他要讲的故事，YES 就是 YES，NO 就是 NO。他尽自己的力量，原作是怎样就翻译成怎样。

任溶溶翻译外国儿童文学作品，虽然小说、童话、剧本等无所不译，但最感兴趣的是译儿童诗。对于写诗，任溶溶说："诗要引人入胜，儿童诗最好从题目起就吸引孩子，诗的结尾又有回味。"他说："我翻译诗的过程是我学习的过程，我很有兴趣看一些成功的儿童诗人如何从生活中取材，又怎样巧妙地表现出来。这是为了提高自己的眼力和功夫，使自己也善于从我们的生活中取材，巧妙地表现。"

任溶溶曾多次获得重要儿童文学奖项，其中有"陈伯吹儿童文学奖"杰出贡献奖、"宋庆龄儿童文学奖"特殊贡献奖、中国翻译家协会"翻译文化终身成就奖"以及"上海文艺家终身荣誉奖"和"儿童工作白玉兰奖"。任溶溶曾说："与儿童文学结缘是我一生的幸运；为了繁荣儿童文学创作，我要不断地探索，不断地创新。"

（摘自：https://mp.weixin.qq.com/s?__biz=MjM5MDM1Mjk2Ng==&mid=2652935497&idx=1&sn=7fcdc02bbe5498aaa37d281c74a0ca86&chksm=bd9226608ae5af7633a8969823e7cd20212ff2d5e2cba21e8aee0d2ab27160109659f0446468&scene=27）

Text B

Taking a Ferry
By Lauren Wolk

Ideological and Political Education：保护海洋生态环境 建设美丽海湾

美丽海湾是美丽中国在海洋生态环境领域的集中体现和重要载体，建设美丽海湾也是加快建设海洋强国的必然要求和重点任务。

"十三五"以来，我国海洋生态环境保护取得显著成就，渤海综合治理攻坚战阶段性目标任务圆满完成，陆海统筹的近岸海域污染防治持续推进，"蓝色海湾"整治行动、海岸带保护修复工程等深入实施，海洋生态环境总体改善，局部海域生态系统服务功能明显提升。

当前，我国海洋生态环境保护面临的结构性、根源性、趋势性压力尚未得到根本缓解，海洋环境污染和生态退化等问题仍然突出，治理体系和治理能力建设亟待加强，海洋生态文明建设和生态环境保护仍处于压力叠加、负重前行的关键期，海洋环境污染形势依然严峻，近岸海域水质改善成效尚不稳固。

《"十四五"海洋生态环境保护规划》提出，到 2025 年，推进 50 个左右美丽海湾建设，形成秦皇岛湾、崂山湾、台州湾、马銮湾、大鹏湾、铺前湾等一批美丽海湾典范。

近年来，生态环境部门会同各有关部门和沿海地方加快推进美丽海湾建设，坚持环保为民，突出"美"的核心导向，编制印发了《美丽海湾建设基本要求》和《美丽海湾建设参考指标》等政策文件，让沿海地方能够准确把握总体思路和策略、核心目标和要求以及重点的努力方向。

坚持久久为功，擘画"实"的建设路径，在全国 18 000 千米海岸线及其毗邻的近岸海域，划定 283 个海湾，把"十四五"的各项目标任务逐步细化分解，精准落实到每一个海湾；坚持示范引领，探索"新"的建设模式，组织沿海地方征集了首批 8 个优秀提名的美丽海湾的案例，凝练了一批可复制、可推广、可借鉴的创新建设模式。

美丽海湾的基本内涵就是要让海湾符合"水清滩净、鱼鸥翔集、人海和谐"的建设目标要求，能够为公众提供优质的生态产品，为高质量发展夯实可持续的海洋生态环境基础。《美丽海湾建设参考指标（试行）》从系统性、科学性、导向性、可操作性的角度出发，研究提出了海湾水质优良比例、海洋生物保护情况、滨海湿地和岸线保护情况等反映海湾生态环境质量状况的建设指标，并鼓励各海湾因地制宜增设特色指标，为沿海地方建设美丽海湾提供了方向指引。

1. An hour later, I waited on the sand while the ferry passengers crossed the long boardwalk from the pier to dry ground. Lots of people had come for a holiday, a good shore dinner, and then home by dark. I watched them **parade** off the boardwalk and up the lane. Some of them looked at me curiously. One woman even said hello. And then they were gone, and the boardwalk free for those of us outward bound. I followed a few other islanders along the boardwalk toward the steamship. I knew them and they knew me, but no one said very much. At the **gangway**, the mate took my money, same as theirs. To him, I was just a skinny kid. I smiled on my way forward to the bow.

2. The breeze was strengthening as we pulled away from the pier. In the distance, mare's tails swept out of nowhere across the sky. They were often the first sign of a storm, the **frayed hem** of cloud heavy with weather, but they were a long way off, and the sky overhead was still a spotless blue as we steamed west.

3. At some point, I realized I was in unfamiliar waters, closer to the continent than the islands. Traveling by steam instead of by sail was odd, too. We went straight, regardless of the wind, at a steady speed, cutting through the waves instead of riding them, with but one purpose: to get there, which suited me fine that morning. I was anxious to reach the city.

4. But, as it turned out, I discovered something before we ever made land. As we neared the mouth of New Bedford Harbor, a **schooner** flying a long green **streamer** off its mainmast, its sails plump, its bow rising and dipping, came out through the channel toward us. We gave way a little, as all engine ships must to those under sail, but we passed close by, and I could see the sailors clearly. They were lined up at the **rail**, filling themselves with the sight of the home they might not see again for weeks, maybe months.

5. One of them looked like me. He was near the bow of his ship, I in the bow of the ferry, and for a long moment we were directly across from each other, staring. "Jason?" I called, but there was wind, and I could tell that he hadn't heard me. He cupped a hand behind one ear and leaned over the rail as far as he could, and I yelled again, more loudly, "Jason!" But I still couldn't be sure that he'd heard me. Without unlocking my eyes from his, I moved toward the stern, and he did the same as the two boats passed alongside each other, so for a minute—no more—we stayed as close as we could. And then, from the stern of the ferry, I stood and watched him in the stern of the schooner, the distance between us lengthening, and I waved. The sailor who looked like me waved back, but so did all the others lined up along the stern, nothing more than friendly. And then the ferry steamed into the harbor, the schooner **tacked** away, and the sailor was just a man.

6. "The Shearwater," I said aloud, so I would remember the name of that schooner and know it when it returned. I didn't move from that spot until I could no longer see the ship at all. Perhaps he was nothing to me. A stranger by now busy with the **rigging**, all of his thoughts tuned to the sea. How he looked, nothing more than a **coincidence**. How he looked, nothing more than something I wanted to see. Or perhaps he was still standing by the rail of the Shearwater, wondering about me. Perhaps even, like me, filled with an odd warmth that didn't weaken as he sailed away.

7. When the ferry tied up at the dock, I was tempted to stay on board and wait for the return trip. The sight of that sailor who looked so much like me might be a better answer than anything I would find in New Bedford. But the answer I wanted and the one I needed were two different things. I would do what I had come to do.

8. Now that I had reached the city, though, I was no longer so eager for it. Osh had been right. Even before I stepped foot onshore, I was startled by the place. It **reeked** of whale and waste. All along the dock, there were vast **pens** of barrels covered with seaweed and hundreds of **bales** of cotton waiting for the mills. Men everywhere, of every color, **stripped** to the waist, labored in the sun. Horses worked alongside engines alongside people alongside a harbor afloat with trash and oil. In the distance, factories smoked their pipes. Past the docks: tall buildings and long straight streets. Miss Maggie had told me about automobiles, and I had seen some pictures of them, but the ones I saw here—their **fumes** and rumble, the **blare** of their horns—amazed me. I didn't see a tree anywhere.

9. "You comin'?" a **deckhand** called to me, and I realized I was the only passenger still aboard. "I am," I said, hurrying past him and down the gangway onto the dock, where I stopped short. Nothing so far had alarmed me as much as the sight of a black dog, **boils** riding its back like a second spine, a rat as big as Mouse **clenched** in its teeth. No one else gave it a second look.

10. I didn't know where I was going, so I simply went. I hurried to catch up with the other islanders who had come for business of their own. "Hey," I called to one of them, a farmer, Mr. Cook. He turned. "Off on another grand adventure, are you, Crow? Saving the bird keeper wasn't enough for you?" "Do you know where the **orphanage** is?" I asked him. "Ain't no orphanage in New Bedford," he said. "Not anymore." And I cursed myself for thinking I could do this without any planning whatsoever. The city felt like a **fortress**, huge and unfriendly, and I had never felt so small. "But where it was—that's still there," he said, and I took hope. "What do you mean?" "Brought my mother here when she took sick. To the hospital. And just by there is the old building used to be the orphanage. Part of St. Luke's

now." He pointed away from the docks. "Go on up Union, left on County, right on Allen. And be quick, Crow. It's a long walk. And there's but one ferry back to Cuttyhunk today."

11. It felt odd to be wearing shoes in June, but I was glad to have them since the streets were sharp with bits of glass and metal, filthy with dust and **dreck**. Everyone walked so quickly here, nobody saying much to anyone else, all around them a cloud of **commotion**. The windows of the shops were full of things I couldn't imagine ever having, and the reflection of the passersby and the street full of cars made those glimpses of another life seem distant and insubstantial, like dreams.

(1,217 words)

Extracted from *Beyond the Bright Sea*, Corgi Children's. 64-68, 2017.

New Words

1. parade [pə'reɪd]

   v. walk ostentatiously（使）列队行进；游行

   More than four thousand soldiers, sailors, and airmen paraded down the Champs Elysées last Sunday morning.

   上周日上午，超过四千名士兵、水手和飞行员沿着香榭丽舍大街游行。

2. gangway ['gæŋweɪ]

   n. a temporary bridge for getting on and off a vessel at dockside 舷梯；进出通路

   The Navy normally docks at least five ships in Manhattan and Staten Island to celebrate New York's Fleet Week each May, lowering the gangway and letting the public come aboard.

   每年5月，为了庆祝纽约的舰队周，美国海军通常会在曼哈顿和斯塔顿岛停靠至少五艘军舰，放下舷梯，让公众上船。

3. frayed [freɪd]

   a. worn away or tattered along the edges 磨损的；毛边的；散口的

   The little mare was young, but thin, with legs planted wide apart and frayed ears.

   这匹小母马很年轻，但很瘦，两腿叉开，耳朵磨破了。

4. hem [hem]

   n. lap that forms a cloth border doubled back and stitched down 边，边缘；褶边

   Madeleine showered and put on her first spring dress: an apple-green baby-doll dress with a bib collar and a high hem.

   玛德琳洗了个澡并穿上了她的第一件春装：一条苹果绿的娃娃裙，带着围兜衣领和高高的褶边。

5. schooner ['sku:nə(r)]

n. sailing vessel used in former times 纵帆船

The owner of a schooner called the Silver Maid was signing up sailors for a voyage to China.

一条名叫"银姑"的纵帆船船主正在招雇水手,打算航行到中国去。

6. streamer ['striːmə(r)]

n. a long flag;often tapering 长旗;横幅;饰带

Colourful streamers and paper decorations had been hung from the ceiling.

五颜六色的飘带和彩纸饰品被挂在天花板上。

7. rail [reɪl]

n. a barrier consisting of a horizontal bar and supports 栏杆;扶手

She climbed the staircase cautiously,holding fast to the rail.

她紧握扶手,小心翼翼地爬上楼梯。

8. tack [tæk]

v. turn into the wind 抢风掉向;改变方针

At last we were beaten so far to the south that we tossed and tacked to and fro the whole of the ninth day.

终于,我们被风浪冲击得向南方飘去,在第九天,整整一天,船只忽前忽后地颠簸不停。

9. rigging ['rɪɡɪŋ]

n. gear consisting of ropes etc.supporting a ship's masts and sails 绳索;传动装置

A severe hurricane has done much damage to the rigging of the ship.

猛烈的飓风使这条船的帆缆严重破损。

10. coincidence [kəʊ'ɪnsɪdəns]

n. an event that might have been arranged although it was really accidental 巧合

By sheer coincidence,I met the person we'd been discussing the next day.

真是巧了,我在第二天遇见了我们一直在谈论的那个人。

11. reek [riːk]

v. smell badly and offensively 散发臭味

When the boats entered the narrow laneways between the houses,the water reeked of feces.

快艇开进房屋间的狭窄过道时,水中明显带有很多排泄物,散发出阵阵臭气。

12. pen [pen]

n. an enclosure for confining livestock 围栏;圈

We now only build pens with solid roofs as we find this type of pen more appealing to both the inhabitants and ourselves.

我们现在只兴建具有坚实屋顶的栏圈,因为我们发现这种类型的栏圈无论对鸟还是我们自己都更具吸引力。

13. bale [beɪl]

n. a large bundle bound for storage or transport 包;捆

The walls are made of tightly packed straw bales held together with bamboo pins and covered with fishing nets.

墙壁是一块块密实的草捆,用竹针结合在一起,上面盖着渔网。

14. strip [strɪp]

　　v. get undressed 脱衣服

　　One prisoner claimed he'd been dragged to a cell, stripped, and beaten.

　　一个犯人声称他曾被拖进一间牢房,被脱掉衣服殴打。

15. fume [fju:m]

　　n. a cloud of fine particles suspended in a gas 刺鼻(或有害)的烟气

　　Local residents needed hospital treatment after inhaling fumes from the fire.

　　当地居民吸入了大火的浓烟,需要入院治疗。

16. blare [bleə(r)]

　　n. a loud harsh or strident noise 刺耳的鸣响

　　Then, with a blare of trumpets and a banging of drums, the show began.

　　接着,随着刺耳的喇叭声和鼓的重击声,演出开始了。

17. deckhand ['dekhænd]

　　n. a member of a ship's crew who performs manual labor 甲板水手;普通水手;下级水手

　　He started as a deckhand on boats owned by Puglisi and Sarin, and eventually bought his own boats.

　　他一开始在普格利西和萨林的船上当甲板水手,后来自己买了船。

18. boil [bɔɪl]

　　n. a painful sore with a hard core filled with pus 疖;皮下脓肿

　　It's some sort of skin disease, flaking skin disease or other sorts of boils and skin states that seem to be associated, at least in the Israelite mind, with decomposition and death.

　　它是一种皮肤病,皮肤会一片片地剥落,或者是生疖和一些与皮肤有关的症状,至少在以色列人心目中,这种症状和腐烂、死亡联系在一起。

19. clench [klentʃ]

　　v. hold in a tight grasp 紧握;紧咬

　　The jaw is able to clench and chew because of the masseter muscle.

　　正是因为有咬肌的存在我们的下巴才能够进行咬合和咀嚼。

20. orphanage ['ɔːfənɪdʒ]

　　n. a public institution for the care of orphans 孤儿院

　　Angelina adopted Maddox when he was just seven months old after she visited a Cambodian orphanage while shooting her film *Lara Croft: Tomb Raider* in 2002.

　　2002 年,安吉丽娜在拍摄电影《古墓丽影》时访问了柬埔寨的一家孤儿院,并收养了当时只有七个月大的马多克斯。

21. fortress ['fɔːtrəs]

　　n. a fortified defensive structure 堡垒;要塞

Another interesting sight is the Akershus Fortress built in 1290, which is still used by the military.

另一个有趣的景点是建于1290年的阿克修斯堡垒,至今仍被军队使用。

22. dreck [drek]

n. something that is useless, of no value, or of very low quality 垃圾;污物;假货

For the foreseeable future, we will all continue to hate the industrial-scale dreck that dominates our diets, but will continue to eat it nevertheless.

在可预见的未来,我们都将继续讨厌主导我们饮食的工业规模的垃圾食品,但我们还会继续食用它们。

23. commotion [kə'məuʃn]

n. a disorderly outburst or tumult 骚动;暴乱

Noticing the commotion, she reached into her purse for a handful of hard candies and offered them to the kids in return for their good behavior.

她注意到骚动,从手包里掏出一把硬糖,给了表现好的孩子们。

Phrases and Expressions

pull away from 把某物从……拉开;使脱离

out of nowhere 不知打哪儿来;突然冒出来;莫名其妙地出现

heavy with 有大量的

mare's tails 马尾云

make land 到岸;看见陆地

give way 让路;撤退;倒塌;失去控制

tie up 绑好;缚牢;停泊

catch up with 赶上;追上

make a glimpse of 瞥一眼

一、翻译简析

1. In the distance, mare's tails swept out of nowhere across the sky. They were often the first sign of a storm, the frayed hem of cloud heavy with weather, but they were a long way off, and the sky overhead was still a spotless blue as we steamed west. (Para.2)

远处,不知从哪里冒出来的马尾云划过天际。马尾云通常是暴风雨来临的第一个征兆,是密布的乌云的钩钩边儿,但它们离我们很远,我们一路向西航行,头上的蓝天一尘不染。

【在翻译的过程中,完全按照原文的词汇和词典的释义来翻译是行不通的。英汉两种语言在文化背景和语法结构上存在着很大差异,因此我们必须根据具体的上下文进行灵活处理。本句翻译的难点是短语"the frayed hem of cloud heavy with weather",若按照词典释义,很容易误译为"云的磨损边缘沉重的天气",但仔细分析上下文可以看出这个短语是在描述马尾云的

形态和特点。汉语中,马尾云又叫钩卷云或者钩钩云,群众中流传着"天上钩钩云,地上雨淋淋"的谚语,所以将"frayed hem"译成"钩钩边儿",既能准确再现原文含义,又能呼应上文提到的作者的好心情。】

2. As we neared the mouth of New Bedford Harbor, a schooner flying a long green streamer off its mainmast, its sails plump, its bow rising and dipping, came out through the channel toward us. (Para.4)

我们接近新贝德福德港的入海口时,一艘双桅纵帆船通过航道向我们驶来,它的主桅上挂着一根长长的绿色彩带,船帆鼓鼓的,船头起起伏伏。

【本句的译文使用了 AABB 式叠词"起起伏伏"来形容船头在海上颠簸的状态,首先顺应了原文的内容,其次具有音韵美感,还能做到形象生动贴切,符合儿童文学的用语特征。】

3. The sailor who looked like me waved back, but so did all the others lined up along the stern, nothing more than friendly. And then the ferry steamed into the harbor, the schooner tacked away, and the sailor was just a man. (Para.5)

那个和我长得很像的水手也朝我挥了挥手,其他的水手在船尾排成一排,也都朝我挥了挥手,除了友好,别无其他。我们的渡船驶入港口,那艘纵帆船驶离,那个水手只是个过客。

【英译汉时,我们常常会遇到这样的词,它在词典上所有的词义都不符合句子的语境,如果照搬硬译的话,会使译文生涩难懂,甚至与原义相悖,造成误解。本句中的"man"就是一个这样的词,译成"人"或者引申成"普通人"都没法表达作者失望落寞的心情,她希望确认那个水手就是她要寻找的哥哥,所以将"man"引申译为"过客",才是最贴切的汉语词语。】

4. Nothing so far had alarmed me as much as the sight of a black dog, boils riding its back like a second spine, a rat as big as Mouse clenched in its teeth. (Para.9)

眼前的景象让我惊慌极了,我看到一条黑狗,背上长着一串儿疖子,看起来就像长了两条脊椎一样,它的牙齿之间正紧紧咬着一只和"鼠儿"一样大的老鼠。

【在翻译较复杂的长句时,有时使用词的引申、词性转换或者句子成分转换等翻译技巧还是不能确切表达原文的含义,这时就要突破原文的形式,改变句子的整体结构。本句是由主句和两个独立主格结构构成的长句,句中的三个状语成分较长,不容易安排,需要将它们分出来,处理成汉语短语或独立句。】

5. The windows of the shops were full of things I couldn't imagine ever having, and the reflection of the passersby and the street full of cars made those glimpses of another life seem distant and insubstantial, like dreams. (Para.11)

商店橱窗里的商品琳琅满目,超乎想象。橱窗玻璃映着街上人来人往,车辆穿梭,使这别样的生活看起来遥远陌生,如梦似幻。

【本句的译文运用了多个四字格,来形容新贝德福德城里的繁华,以及主人公小乌鸦对城市的距离感与陌生感,四字格的使用更使译文自然流畅,对仗工整,增强感染力,富有文采。】

二、参考译文

乘　渡　船

1. 一个小时后,我在沙滩上等着,渡船上的乘客穿过长长的木栈道,从码头走到干燥的地面上来。在海边吃顿丰盛的晚餐,天黑前就回家了。女士们身穿白色长裙,男士们则穿着汗衫。我看着他们一个接一个从木栈道上下来,走上小路。他们中的一些人好奇地看着我。有一位女士甚至还跟我打招呼呢。下船的乘客都走了,我们就可以经过木栈道上船了。我跟着其他几个岛民沿着木栈道向渡船走去。我认识他们,他们也都认识我,但大家都不怎么说话。在舷梯那儿,大副收走了我的船费,也收走了其他乘客的。对他来说,我只是个瘦小孩儿。想到即将踏上的旅程,我微笑着走向船头。

2. 船驶离码头,海上的风越来越大。远处,不知从哪里冒出来的马尾云划过天际。马尾云通常是暴风雨来临的第一个征兆,是密布的乌云的钩钩边儿,但它们离我们很远,我们一路向西航行,头上的蓝天一尘不染。

3. 不知什么时候,我意识到自己是在一片不太熟悉的海上,离大陆近些,离海岛远些。而且乘汽船和帆船的感觉也大不相同。我们迎着风,以稳定的速度勇往直前,穿过波浪,而不是乘风破浪,只有一个目的:到达新贝德福德。这正应了我那天早上的心情,我很想快点到达那座城市。

4. 但是,后来,我们上岸之前,我有了不一样的发现。我们接近新贝德福德港的入海口时,一艘双桅纵帆船通过航道向我们驶来,它的主桅上挂着一根长长的绿色彩带,船帆鼓鼓的,船头起起伏伏。所有轮船都得给帆船让道,我们的船稍稍为它让了路,两艘船近距离驶过,我能清楚地看到那条船上水手的长相。他们在栏杆边站成一排,他们的眼前是将要离开几个星期甚至几个月的家乡。

5. 他们其中一个水手跟我长得很像。他站在那艘船的船头附近,我也站在我们这条渡船的船头,有好长一段时间,我们正好面对面地凝视着彼此。"杰森?"我喊了一声,但有风,我看得出他没听见我叫他。他把一只手捂在一只耳后,身体尽可能地靠在栏杆上,想听清楚我说的话,所以我又更大声地喊:"杰森!"但是,我还是不确定他有没有听见。我目不转睛地盯着他的眼睛,两艘船相向而过的时候,我从船头跑到船尾,他也是,于是有那么一分钟——只那一分钟——我们尽量靠得很近。我们站在两艘船的船尾彼此望着,我们之间的距离越来越远,我朝他挥了挥手。那个和我长得很像的水手也朝我挥了挥手,船尾那一排其他水手也都朝我挥了挥手,除了友好,别无其他。我们的渡船驶入港口,那艘纵帆船驶离,那个水手只是个过客。

6. "海鸥号",我大声地喊出来,这样我就能记住那艘纵帆船的名字,等它回来的时候就能认出它来。直到完全看不见那艘船,我才挪了挪位置。也许他和我毫无关系。他只是个陌生人,现在正忙着扯帆索,心思都在大海上。他的长相只是巧合而已。他的样子,不过是我想看到的。也许他还站在"海鸥号"的栏杆旁,想着我呢。也许他甚至像我一样,心中充满了一种奇妙的温暖,这种温暖并没有随着船的远去而减弱。

7. 渡船停靠在码头的时候,我真想待在船上不下去,直接等着坐返程的船回去。与那个相貌酷似我的水手相遇,可能好过一切我在新贝德福德会找到的答案。但是,我想要的答案和我需要的答案是两码事儿。我还得老老实实地按原计划行事。

8. 既然我已经到了城里,我就不再那么心急了。奥什说的没错。还没上岸,这个小城就让我大吃一惊。它到处散发着鱼腥味和腐臭味。沿着码头,到处都是巨大的桶,上面覆盖着海草,还有几百包棉花,等着送往工厂。到处都是人,什么肤色的都有,他们光着上半身,在太阳底下干活儿。马也得干活儿,在发动机旁边干,在工人身边干,在漂着垃圾和油污的港口边干。远处,很多工厂的管道里冒着烟。走过码头,到处是高楼大厦和又长又直的街道。麦琪小姐给我讲过汽车的事,我也看过一些汽车的照片,但我在这里看到的那些汽车——它们刺鼻的烟气、隆隆的轰鸣声和刺耳的喇叭声——都让我震惊不已。而且,这里连一点树影儿都没有。

9. "你下船吗?"一个甲板水手问我,我这才意识到我是船上最后一个乘客了。"下船",我说着,匆匆从他身边经过,走下舷梯,来到码头上,但我突然停住了。眼前的景象让我惊慌极了,我看到一条黑狗,背上长着一串儿疖子,看起来就像长了两条脊椎一样,它的牙齿之间正紧紧咬着一只和"鼠儿"一样大的老鼠。对这些,其他人竟然视而不见。

10. 我不知道要往哪儿走,索性就走走看。我急忙赶上其他岛民,他们都是来办事儿的。"嘿",我对他们当中的一位农民库克先生喊道。他转过身说道,"又开始新的大冒险啦,是吗,乌鸦? 拯救一个护鸟员还远远不够吧?""你知道孤儿院在哪儿吗?"我问他。"新贝德福德没有孤儿院",他说,"没有啦"。我责备自己毫无计划,鲁莽行事。这个城市像一个巨大而冷漠的围城,我第一次觉得自己如此渺小。"但是以前有过——现在还在那儿",他说,我又看到了希望。"你这是什么意思呢?""我母亲生病那会儿,我带她来过这儿。去了那家医院。医院旁边的那个破楼以前就是孤儿院。现在是圣卢克医院的一部分啦。"他从码头指过去。"走到工会那儿,往左到郡政府,再往右到艾伦的店。要抓紧时间啊,小乌鸦。路远着呢! 今天回卡蒂杭克的船可只有一班呐!"

11. 六月天穿着鞋感觉怪怪的,不过还好有鞋,因为街上有很多尖锐的玻璃碴儿和金属屑,而且到处都是脏兮兮的尘土和垃圾。每个人都步伐匆匆,不多说话,置身于忙乱中。商店橱窗里的商品琳琅满目,超乎想象。橱窗玻璃映着街上人来人往,车辆穿梭,使这别样的生活看起来遥远陌生,如梦似幻。

Cultural Background Knowledge: Sea Fiction
海洋小说文化背景知识

英国因其独特的地理环境,有着悠久的海洋文化与文学传统,英国古今一大批作家都有着割舍不去的海洋情结,因此,英国海洋小说是英国文学之林中的一大景观。它如同促使它生长与进化的社会土壤一样,在历史的洪流中不断地改弦易辙,急剧演变。英国海洋小说,如其他类型的小说一样,不可避免地经历了一个从原始到成熟的发展过程。从整体来看,英国海洋小说从无到有,经历了雏形(18世纪以前)→成型(18世纪)→成熟(19世纪)→繁荣(20世纪),共400多年的发展历史。事实上,它的每一个发展阶段都同英国当时的社会、历史、政治、文化和经济息息相关。英国海洋小说的发展不是一个孤立或自发的文学现象,而是与英国的社会变化以及异域(尤其是欧洲各国)文化的繁荣昌盛彼此交融的。

  雏形期是"韵文叙事文学向散文叙事文学转型的时期"。而中世纪盎格鲁·撒克逊人的英雄史诗与雏形期的英国海洋小说血缘关系最近。成型期英国具有代表性的海洋小说有丹尼尔·笛福的《鲁宾逊漂流记》和《辛格顿船长》,文学大师乔纳森·斯威夫特的《格列佛游记》。成熟期英国具有代表性的海洋小说有查理·金斯莱的童话名著《水孩子》、罗伯特·路易斯·史蒂文森的《金银岛》和巴兰特的《珊瑚岛》等。繁荣期涌现了大量海洋小说家,比如约瑟夫·康拉德、弗吉尼亚·伍尔夫、威廉·毛姆、威廉·戈尔丁、约翰逊等。

  社会生活的日益发展丝毫没有使英国人对海外奇闻、荒岛探险减少兴趣。到了当代,英国社会已经发生了根本变化,第二次世界大战彻底改变了人们对世界、社会、人类的看法,岛屿成为表现人性的舞台。同时,自然本身也存在着善恶的矛盾,而孕育于自然之中的大善不仅能克服自然本身的恶,作为合乎人类意愿的超自然力量,它还能帮助人类战胜人性的恶,达到人性的更自然、更和谐的状态。英国海洋文学的长期持续繁荣是由其国家的社会政治背景所决定的,带有明显的社会功利性,在很大程度上反映了社会的变迁、盛衰及人们思想观念的演变,因而具有差异性和阶段性。英国航海类文学所取得的巨大成功,无疑是其他任何一个国家都望尘莫及的。虽然同一时期的西班牙、葡萄牙航海也比较盛行,但是出于地域文化的局限性,他们的文学并没有得到多大的发展。综观英国文学发展史,这一时期的文学作品受其地域文化的影响,无疑是最深刻的。航海热潮的疯狂盛行导致了文学的空前繁荣,而涌现出来的大量关于航海探险和部分作家亲身经历的游记叙述,又大大刺激了航海的盛行,这种航海与文学相互刺激并共同发展的局面,在世界文学史上是独一无二的。用勃兰兑斯的话说,英国作家一直是海洋景色"最佳的描绘者和解释者"。

  (摘自:郭海霞.英国海洋小说的起源与发展[M].外国语文,2012,28(S1):44.)

Unit Thirteen
Marine Science

海洋科学

Text A

Marine Science and Marine Robotics
By Ajay Menon and Brownell

Ideological and Political Education:促进海洋产业发展 强海兴国

海洋科学的研究领域十分广泛,它是19世纪40年代以来出现的一门新兴学科。2 000多年前,古罗马哲学家西塞罗曾经说过:"谁控制了海洋,谁就控制了世界。"

直到19世纪70年代,英国皇家学会组织的"挑战者"号完成首次环球海洋科学考察之后,海洋学才开始逐渐形成为一门独立的学科。20世纪五六十年代以后,海洋学获得大发展,形成一门综合性很强的海洋科学。海洋科学是研究海洋的自然现象、性质及变化规律,以及与开发利用海洋有关的知识体系。它的研究对象是占地球表面71%的海洋,包括海水、溶解和悬浮于海水中的物质、生活于海洋中的生物、海底沉积和海底岩石圈以及海面上的大气边界层和河口海岸带等内容。

随着海洋开发力度的日益加大,需要先进的海洋机器人进行水下操作来开发广袤的大洋。海洋机器人又称水下机器人,水下机器人已成为海洋开发的重要工具,我国已经能够生产和制造包括5G智慧海洋应用、高端海洋工程装备、海洋无人智能装备、无人无缆潜水器、无人机、无人船艇、应急救援机器人、海底搜索机器人、打捞作业机器人、水下工程机器人、水下考古机器人、海洋测绘机器人、水中摄影机器人、科学考察机器人、应急救援机器人、疏浚工程机器人等海洋科学勘探设备装置。

水下机器人是一种工作于水下的极限作业机器人,它的出现极大地拓展了人类探索海洋的广度和深度。中国的水下机器人研发始于20世纪70年代末期。经过全面调研、科学论证,中国科研人员完全依靠自主技术研制出中国第一台水下机器人,为21世纪万米深潜器的研发打下了坚实的基础。

"海人一号"的研制工作是中国科研人员完全依靠自主技术和立足于国内配套条件开展的,是中国水下机器人发展史上的一个重要里程碑。以蒋新松、封锡盛等为代表的中国第一代水下机器人专家为此做出了开创性贡献,中国水下机器人的制造由此掀开了波澜壮阔的宏伟篇章。

1. What is Marine Science? Marine science commonly is called **oceanography**. As these names may reveal, this branch of science deals with study of oceans. Professionals in this field are often called marine scientists or oceanographers, but they also may take titles that refer to their specialties. The topics that are covered by marine science can widely vary, including such things as ocean currents, sea floor geology, and the chemical composition of ocean water. Many people have only a vague understanding of marine science. One common misconception involves the use of the titles such as marine scientist and **oceanographer**. To a layperson, these may sound very specific. In reality, these titles hardly provide any information about what a person in this field does.

2. Marine science is so broad that it would require a lot of space to outline every possible career path. Many of the same components that are studied on land also are studied in the water. Marine biology, marine chemistry, and marine physics are three of the disciplines that fall into the category of oceanography. Within each of these disciplines there are numerous sub-categories in which a professional is likely to specialize. For example, within marine biology, one person could focus on plants while another focuses on **microscopic** organisms.

3. In some cases, oceanographers have majored in some type of marine science program. More often than not, however, these professionals majored in more basic programs such as Biology or Earth Sciences. Then, somewhere along the way, they **veered** off and began to concentrate on oceanography. People also tend to think that marine scientists carry out most of their duties in or on the water. This is a second misconception. A lot of the work done by such professionals typically is conducted in laboratories. Instead of wet suits and oxygen tanks, their gear commonly is composed of microscopes and computers. It is widely believed that the oceans affect many components of the natural system of the Earth. For example, oceans have been linked to the global climate. Marine life also is responsible for supporting part of the human food chain. As this is the case, a common objective of marine science is to draw relevance between the oceans and other parts of nature.

4. Marine science is often treated as a novelty science which commonly results in funding problems: a third misconception. This vast area of science can play a vital role in environmental conservation. It also may be a large contributor in the search for solutions to environmental problems such as global warming.

5. Marine Robotics. Oceans cover about 71% of the earth's surface and are responsible for several critical processes that sustain life on earth, such as producing oxygen, supplying food, and regulating weather patterns, and yet over 80% of the world's oceans are unexplored and

unmapped (National Oceanic and Atmospheric Administration). In part, this lack of exploration is because human divers can only reach a given depth before the pressure becomes too dangerous. Thus, to effectively understand and sustainably maintain the world's oceans, we turn to technology such as autonomous underwater vehicles (AUVs), which can operate without oxygen and reach depths previously unavailable to humans.

6. On the surface, while much of the ocean is still an undiscovered frontier to us, there is virtually no place on earth untouched by marine debris, which is **detrimental** to **aquatic** life, **chokes** ecosystems, and contaminates water. While sustainable efforts to prevent more debris from going into the water exist, their impact is limited; most pollution comes from debris already in the oceans, and that debris must somehow be removed. In an evaluation of marine litter robotic detection models, published by IEEE in 2023, researchers found that deep-learning-based object detection could **plausibly** detect marine debris in real-time, though the accuracy and interference rate of detection will differ depending on the model used. Moreover, an increase in the amount of available trash data would greatly benefit future research and detection model training. In solving this problem of marine debris detection (and eventual removal), the researchers also suggest the future investigation of multi-robot collaboration among multimodal robots.

7. In an effort to increase the amount of information about the ocean in general, Liquid Robotics, a subsidiary of the Boeing Company that focuses on developing marine technology, has developed the Wave Glider: a marine robot that sits on the surface of the water and can collect ocean data for up to a year. The Wave Glider can also **deploy** a **winch** to collect data at a depth, as well, such as subsea **acoustics**, water sampling, and fish tracking. Having a similar approach as Liquid Robotics, Openoceanrobotics USVs (Uncrewed or Unmanned Surface Vehicles) are also equipped with sensors, cameras, and communication devices so that they can capture information from anywhere on the ocean and have instant access to it. Harvesting energy from the sun, these boats travel nonstop for months, without producing any greenhouse gas emissions, noise pollution, or risk of oil spills. These boats can monitor oil spills, detect intentional dumping, and aid in the **cleanup** effort.

8. While not yet able to **delve** into the depths of the ocean to remove marine wastes, there are several projects already underway targeting the collection of floating pollution from the surfaces of our water systems. RanMarine, a company founded and operating in Rotterdam, the Netherlands, has created WasteShark: an autonomous surface vessel that can collect 500 kilograms of surface debris in a single deployment while producing zero greenhouse emissions itself. Similarly modeled, the company has also produced the DataShark, designed to autonomously collect and **collate** water quality health data from waterways in any environment.

9. Another operation, Clear Blue Sea, a nonprofit located in San Diego, California, is committed to innovating robotic solutions for removing plastic pollution from our water sources. The organization's current solution, a Floating Robot for Eliminating Debris (FRED), has four prototypes to date, each operating similarly to the WasteShark. FRED runs on solar power, producing zero emissions, which can be **customized** to operate in many different marine environments; and collects marine waste with brooms, a conveyer belt, and a collection bin. The collected waste, upon the robot's return to shore, gets sent to recycling centers for reuse.

10. Furthermore, the French start-up IADYS (Interactive Autonomous DYnamic Systems) has developed Jellyfishbot: a remote-controlled, electric-powered, trash-collecting robot that has been deployed in 15 French ports, removing plastic bags, bottles, and other debris from narrow and otherwise unreachable **nooks** in harbors where waste tends to accumulate. The start-up is also expanding the Jellyfishbot's capabilities to depolluting and decontaminating water systems by cleaning oil spills.

11. On a smaller, but no less impactful, scale, the University of Bristol in the United Kingdom developed the Row-bot in 2015: a project to create an autonomous robot that feeds off an organic matter in the dirty water it swims in, modeled after a water boatman insect. The Row-bot project aims to develop an autonomous swimming robot able to operate indefinitely in remote unstructured locations by **scavenging** its energy from the environment. When it is hungry, the Row-bot opens its soft robotic mouth and rows forward to fill its microbial fuel cell (MFC) stomach with nutrient-rich dirty water. It then closes its mouth and slowly digests the nutrients. The MFC stomach uses the bio-degradation of organic matter to generate electricity using bio-inspired mechanisms. When it has recharged its electrical energy stores, the Row-bot rows off to a new location, ready for another gulp of dirty water.

12. The concept of Stanford University's OceanOne, a **humanoid** diving robot, was born from the need to study coral reefs deep in the Red Sea, far below the comfortable range of human divers. It made its **debut** in 2023 in the Mediterranean Sea as it **retrieved** valuables from the La Lune shipwreck, which had not been disturbed since its sinking in 1664. The expedition to La Lune was OceanOne's maiden voyage, and based on its astonishing success, it's hoped that the robot will one day take on highly-skilled underwater tasks too dangerous for human divers, as well as open up a whole new realm of ocean exploration.

13. Crown-of-Thorns Starfish (COTS) is a **venomous** and **invasive** species, as these starfish feed on coral reefs and are responsible for an estimated 40% of the Great Barrier Reef's total decline in coral cover. To solve this, researchers at the Queensland University of Technology

in Australia developed the COTSbot and RangerBot, which can **inject** COTS with a fatal dose of bile salts, the same way teams of human divers also eliminate the starfish from the reef, with a 99% accuracy rate at detecting the starfish amongst the coral. Yet, the COTSbot was not designed to operate alone; rather, currently, it thins the field for human divers and has the potential to one day work in robotic swarms across the reef.

14. Researchers at Harvard University are also looking into the role that soft robotics can play in capturing delicate oceanic creatures such as jellyfish and releasing them without harm with a Rotary-Actuated Device, or RAD. The goal for the device is to one day be able to enclose the animal, collect cells and scan it to sequence its genome and print a 3D model back on the surface, and then let the animal go, all without greatly disrupting its natural state.

15. What is Remotely Operated underwater Vehicle (ROV)? A remotely operated underwater vehicle also called an ROV, is unmanned and usually **tethered** to the operator. It is an underwater robot that collects data about the underwater world about subsea structures or geological formations like hydrothermal vents. It uses a remote pilot and automated control technology, making it safe and convenient to operate. Observation-class ROVs are used for ocean exploration and provide images and high-definition video for research and study purposes. The uncrewed vehicle is fitted with additional equipment like water samplers, manipulator arms, etc. Modern ROVs have 8-hour battery life. An optical cable establishes a connection between the operator and the remotely operated vehicle, which enables its movement.

16. ROVs are incredibly complex and serve various purposes, from exploration and unmanned expeditions to research and sporting events. They are used by scientists, zoologists, botanists etc. Many industries, like aquaculture, agriculture, etc., use these devices for regular infrastructure inspections and repairs. Some of the salient features of ROVs, their usage, categorization, preparation, launching, operation, and shortcomings are presented.

17. What is a Remotely Operated Vehicle? An ROV is essentially a robot that can operate underwater. It works like a miniature submarine, but without the people using it from onboard. It works wirelessly or through a wired connection, although the latter is more common. Several subsystems together make up the incredibly complex ROV. The various components that form the broad operating mechanism of the ROV are: Electrical systems (wiring and **circuitry**), Mechanical structures, Sensors and appendages, and Task-specific structures. The base frame on which all these systems are supported is the skeleton of the ROV. It is made as light as possible to prevent additional weight and drag during motion.

18. It is covered with a Manifold structure to prevent accidental damage to the internal components. There are holder clips along its frame members to support wiring and other electrical circuit components. The skeleton is built to withstand impacts termed as "severe" and uses the **triangulation** principle of rigid mechanics. To understand how triangulation works, consider a square frame. Any force at a vertex would lead to **buckling** and collapse of the frame. Instead, a diagonal member is introduced, providing additional tension strength. By just adding a single member, the strength is increased across the frame. An X-joint is included in fragile areas, with two diagonal members for added strength in tension and compression.

19. Subsystems and Materials Used in ROV Design. The electrical systems refer to the wiring and circuitry that make up the heart of the ROV. The only reason they are so helpful and can be used in various fields is the highly robust and complex nature of the electronics outfitted on them. The primary components include the main motherboard and processing unit, where instructions are fed from a controller and then converted into a physical output. The controller is manually operated in most cases, and autonomous underwater vehicles (or AUVs) are scarce in this field owing to the various challenges presented. Signals from the controller input are either wired or wireless based on the level of innovation and design. It may also be task-specific. For instance, a wired robot may get entangled in an underwater wreck site.

20. On the other hand, a wireless signal may be blocked as the robot heads deeper into the wreckage. So, a careful evaluation of the working conditions and possible hazards should also be done. Once the signal reaches the onboard receiver, it is conveyed to the mechanical systems. The main system is **propulsion**, in the form of miniature and specialized marine propellers. They are powered by small servo motors housed in waterproof casings, the propellers number 2 between 3 and 5 blades per shaft. They can be operated in both clockwise and counter-clockwise directions to create steering. Some advanced ROVs in wreckage analysis and deep-sea exploration also have jointed propellers. These are rarely used and are more costly than their fixed counterparts. It is only required in situations where very high precision is required. Sensors and **appendages** form the very core of the ROVs functionality. The most common equipment includes cameras, depth gauges, temperature, and internal system sensors. The operator uses cameras since they must have eyes on their surroundings.

(2,265 words)

https://www.environmental-robotics.com/marine-robotics/

https://www.marineinsight.com/tech/what-is-remotely-operated-underwater-vehicle-rov/

New Words

1. oceanography [ˌəʊʃəˈnɒɡrəfi]

    n. the branch of science dealing with physical and biological aspects of the oceans 海洋学

    The importance of oceanography as a key to the understanding of our planet is seldom appreciated.

    海洋学是认识我们星球的关键,其重要性人们却很少理解。

2. geology [dʒiˈɒlədʒi]

    n. a science that deals with the history of the earth as recorded in rocks 地质学;地质状况

    He was a visiting professor of geology at the University of Georgia.

    他曾是佐治亚大学的地质学客座教授。

3. oceanographer [ˌəʊʃəˈnɒɡrəfə(r)]

    n. a scientist who studies physical and biological aspects of the seas 海洋学家;海洋研究者

    At the Pacific tsunami Warning Center, David Walsh, an oceanographer, said the earthquake created a 23-foot-high tsunami in Japan.

    在太平洋海啸预警中心,海洋学家大卫·沃尔什称,这次地震在日本造成了23英尺高的海啸。

4. microscopic [ˌmaɪkrəˈskɒpɪk]

    a. of or relating to or used in microscopy;too small to be seen except under a microscope; extremely precise with great attention to details 极小的,微小的;显微镜的,用显微镜的;非常仔细的,一丝不苟的;微观的

    Even today, microscopic meteorites continually bombard Earth, falling on both land and sea.

    即使是现在,微小的陨石仍在不断地撞击地球,然后落到陆地上和大海中。

5. veer [vɪə(r)]

    v. turn sharply;change direction abruptly;shift to a clockwise direction(使)转向;风向顺时针转;调转船尾向上风;转向;使顺风;使船尾向上风

    The ocean storm seems to veer away from you.

    大海的暴风雨似乎避着你而转向。

6. detrimental [ˌdetrɪˈment(ə)l]

    a. (sometimes followed by "to") causing harm or injury 有害的,不利的

    One day, I came across a book called *In Praise of Slowness* and realized that being busy is not only detrimental, but also has the danger of turning life into an endless race.

    有一天,我读到了一本名为《慢的崇拜》的书,意识到忙碌不仅有害,还有把生活变成一场无休止赛跑的危险。

7. aquatic [əˈkwætɪk]

    a. relating to or consisting of or being in water;operating or living or growing in water 水生的,水栖的;与水生动植物有关的;水的,水上的

Like most aquatic insects, mayflies moult as they grow.

像大多数水生昆虫一样，蜉蝣在成长过程中换羽。

8. choke [tʃəuk]

v. breathe with great difficulty, as when experiencing a strong emotion; be too tight; rub or press; wring the neck of; constrict (someone's) throat and keep from breathing 窒息，哽住，卡住；说不出话，哽咽；掐死，扼死；堵塞，塞满；(管子、渠道等)被阻塞，被塞满；失败，阻止，抑制，闷住，扑灭(火等)；窒息，哽住

The plant will soon choke ponds and waterways if left unchecked.

如不控制这种植物的生长，池塘和水道很快就要被阻塞。

9. plausibly [ˈplɔːzəbli]

ad. easily to believe on the basis of available evidence 似真地

Having bluffed his way in without paying, he could not plausibly demand his money back.

没付钱混了进去，他不大可能理直气壮地要回自己的钱。

10. deploy [dɪˈplɔɪ]

v. place troops or weapons in battle formation; to distribute systematically or strategically 部署，调度；利用

On June 7 Google pledged not to "design or deploy AI" that would cause "overall harm", or to develop AI-directed weapons or use AI for surveillance that would violate international norms.

谷歌在6月7日承诺不会"设计或部署"可能会造成"全面伤害"的人工智能，也不会开发人工智能制导武器，或将人工智能用于违反国际准则的监控。

11. winch [wɪntʃ]

n. lifting device consisting of a horizontal cylinder turned by a crank on which a cable or rope winds 绞车；曲柄

The main hoist winch and a system of elevators lift 2,340 tons of pipes and machinery through the 46 inches wide hole at the bottom of the boat.

主起吊绞车和升降机系统可通过船底46英尺宽的洞，提升2 340吨重的管道和机械。

12. acoustics [əˈkuːstɪks]

n. the study of the physical properties of sound 声学；音响效果，音质

Often the acoustics make it hard to identify where sounds are coming from.

这种声学效果通常会使得声音的来源难以辨别。

13. cleanup [ˈkliːnʌp]

n. a very large profit; the act of making something clean 清除；暴利；清扫工作

Figure 14 shows the dialog for adding a new cleanup service job.

图14显示了用于添加新的清理服务作业的对话框。

14. delve [delv]

v. turn up, loosen, or remove earth 钻研；探究；挖

He had a considerable Elizabethan library to delve in.

他拥有相当多的伊丽莎白时期的藏书供其钻研。

15. collate [kəˈleɪt]

　　v. compare critically; of texts; to assemble in proper sequence 核对,校对;校勘

　　We are here to collate the views of the general public and feed those back to the key decision makers within the organization.

　　我们会整理大众的普遍观点,并对其中主要的决定做出回复,也包括球迷组织提出的建议和意见。

16. customize [ˈkʌstəmaɪz]

　　v. make to specifications; make according to requirements 订制,改制(以满足顾主的需要)

　　Microsoft spokesman Paul Abrams said: "The firm spent about ＄5,000 on parts and components to customize the computer".

　　微软公司发言人保罗·艾布拉姆斯说,公司在订制计算机的零件和组件上花了约5 000美元。

17. nook [nʊk]

　　n. a sheltered and secluded place; an interior angle formed be two meeting walls 角落;隐匿处;核武器;凹处

　　Boxes are stacked in every nook and cranny at the factory.

　　箱子叠放在工厂的每个角落。

18. scavenge [ˈskævɪndʒ]

　　v. clean refuse from; collect discarded or refused material 以……为食;打扫;排除废气;清除污物

　　No squirrels are known to hunt vertebrates, or even scavenge meat.

　　没有发现有松鼠猎取脊椎动物,或者以肉为食。

19. gulp [gʌlp]

　　n. a large and hurried swallow; a spasmodic reflex of the throat made as if in swallowing 大口吞食,大口啜饮;一大口(饮入物);一大口(吸入的空气);吞咽,吸入

　　The brain then suddenly sends an emergency signal, telling the person to wake up and take in a big gulp of air.

　　然后大脑会忽然发出一个紧急信号,惊醒人们赶紧深呼吸一下。

20. humanoid [ˈhjuːmənɔɪd]

　　a. having human form or characteristics 像人的

　　n. an automaton that resembles a human being 类人动物

　　Japan has presented a humanoid robot, a cleaning robot and a porter robot in the Shanghai Expo.

　　在上海世博会上,日本推出了仿人机器人、清洁机器人和机器人搬运工。

21. debut [ˈdeɪbjuː]

n. the act of beginning something new; the presentation of a debutante in society 首次登台,(新事物的)问世

v. present for the first time to the public 初次登台,首次亮相;首次推出(某产品)

Their latest release is a worthy successor to their popular debut album.

继首张唱片大受欢迎之后,他们最新推出的专辑再获成功。

22. retrieve [rɪ'triːv]

v. get or find back; recover the use of; run after, pick up, and bring to the master; recall knowledge from memory; have a recollection 找回,收回;(狗等)衔回(物品、猎物);挽救,挽回;回忆,追忆

A pack of hunting dogs shot an Iowa man as he went to retrieve a fallen pheasant, authorities said.

当局说,当艾奥瓦州的一男子去取一只落下的野鸡时,一群猎犬追击他。

23. venomous ['venəməs]

a. extremely poisonous or injurious; producing venom; harsh or corrosive in tone 有毒的;恶毒的;分泌毒液的;怨恨的

In Kravis' venomous tone, he recognized the realization of his worst fears.

从克拉维斯恶毒的语调中他意识到他最害怕的事情发生了。

24. invasive [ɪn'veɪsɪv]

a. marked by a tendency to spread especially into healthy tissue; relating to a technique in which the body is entered by puncture or incision 扩散性的,侵入的;切入的,开刀的

Both native and exotic species can become invasive, and so they all have to be monitored and controlled when they begin to get out of hand.

本土物种和外来物种都有可能成为入侵物种,因此当它们开始失控时,我们必须对它们进行监视和控制。

25. inject [ɪn'dʒekt]

v. give an injection to; to introduce (a new aspect or element); force or drive (a fluid or gas) into by piercing 注射;(给……)添加,增加(某品质);投入(金钱或资源);(往物质、装置中)注入或射入(电流、粒子束等)

The US National Center for Atmospheric Research has already suggested that the proposal to inject ulphur into the atmosphere might affect rainfall patterns across the tropics and the Southern Ocean.

美国国家大气研究中心已经表示,向大气中注入硫的提议可能会影响整个热带和南大洋的降雨模式。

26. genome ['dʒiːnəʊm]

n. the ordering of genes in a haploid set of chromosomes of a particular organism; the full DNA sequence of an organism 基因组,染色体组

The platypus genome reveals the animal held onto genes for odor-detection.

鸭嘴兽的基因组还显示其带有气味感觉的基因。

27. tether ['teðə(r)]

v. tie with a tether(用绳或链)拴住

Judah Maccabee will tether his donkey to a vine, his colt to the choicest branch. And he will wash his garments in wine, his robes in the blood of grapes.

犹大·马加比把小驴拴在葡萄树上,把驴驹拴在美好的葡萄树枝上。他在葡萄酒中洗了衣服,在葡萄汁中洗了袍褂。

28. circuitry ['sɜːkɪtri]

n. electronic equipment consisting of a system of circuits 电路;电路系统;电路学;一环路

The company produced circuitry for communications systems.

这家公司为通信系统生产电路。

29. triangulation [traɪˌæŋgjuˈleɪʃn]

n. a trigonometric method of determining the position of a fixed point from the angles to it from two fixed points a known distance apart; useful in navigation 三角测量;三角形划分

There are three states of matter, three dimensions. Triangulation is how we get our bearings.

物质有三态,空间有三维,而三角测量是我们寻找自己坐标的方式。

30. buckle ['bʌk(ə)l]

v. bend out of shape, as under pressure or from heat; fold or collapse (使)弯曲,(使)变形;扣住,扣紧;屈服,让步;(人)精神崩溃

The winds were strong and huge numbers caused the bridge to buckle, but fortunately not to break.

强风和巨大的承受量导致桥梁弯曲,但幸运的是桥梁没有断裂。

31. propulsion [prəˈpʌlʃ(ə)n]

n. the act of propelling; a propelling force 推进;推进力

The nineteenth century witnessed a revolution in ship design and propulsion.

19世纪经历了船舶设计和推进上的一次重大变革。

32. appendage [əˈpendɪdʒ]

n. an external body part that projects from the body; a natural prolongation or projection from a part of an organism either animal or plant 附加物;下属;附器(如植物的枝叶和动物的腿尾)

Left atrial appendage has become a research hotspot for its special anatomic structure and function characteristics.

左心耳特殊的解剖结构、功能特点使其成为目前研究的热点。

Phrases and Expressions

major in 主修
more often than not 通常；多半
veer off 突然转向
in the search for 在寻找……
in part 部分地；在某种程度上
be detrimental to 有害于；对……不利
inject with 用……注射；增添……
get entangled in 陷入，卷入；纠缠
have an eye on 注意；监视

Terminology

bile salt 胆盐；胆酸盐；胆汁盐
manifold structure 流形结构

Proper Names

National Oceanic and Atmospheric Administration(NOAA) 美国国家海洋与大气管理局
autonomous underwater vehicles (AUVs) 无人水下自主航行器
IEEE 电气与电子工程师协会 (Institute of Electrical and Electronic Engineers)
USVs (Uncrewed or Unmanned Surface Vehicles) 无人或无人水面交通工具
Rotterdam 鹿特丹港市 (荷兰西南部港市)
Floating Robot for Eliminating Debris (FRED) 用于清除碎片的漂浮机器人
IADYS (Interactive Autonomous DYnamic Systems) 交互式自主动态系统
microbial fuel cell (MFC) 微生物燃料电池
Crown-of-Thorns Starfish (COTS) 棘冠海星
the Great Barrier Reef 大堡礁
Rotary-Actuated Device (RAD) 旋转式驱动装置
Remotely Operated underwater Vehicle (ROV) 遥控水下航行器

一、翻译策略/方法/技巧：直译法与意译法

　　直译(literal translation)和意译(free translation)是翻译中最常见的问题，也是最主要的两个翻译方法。直译是既保持原文内容，又保持原文形式的翻译方法或翻译文字。意译，也称为自由翻译，它是只保持原文内容、不保持原文形式的翻译方法或翻译文字。直译与意译相互关联、互为补充，同时，它们又互相协调、互相渗透，不可分割。

　　翻译是指在准确通顺的基础上，把一种语言信息转变成另一种语言信息的活动。英语和汉语是两种不同的语言。我们进行翻译时，必须掌握原作的思想、风格、事实、理论

与逻辑，同时也必须把原作的这些特征还原到译语中去。译语不必在数量和表现形式上与源语一致，但在内容方面要与源语保持一致。这些都是直译与意译应该遵从的共同原则。直译中，首先是忠实于原作的形式，其次是忠实于原作的内容，再次是翻译语言的流畅性和通俗性；而在意译中，忠实于原作的内容应放在第一位，翻译语言的流畅性和通俗性位居第二，但意译并不局限于原作的形式。可见，直译与意译都注重忠实于原作的内容。当原文结构与译语结构不一致时，仍字字对译，不能称为直译，是"硬译或死译"，即形式主义。凭主观臆想来理解原文，不分析原文结构，只看字面意义，编造句子也不能称为意译，是"胡译或乱译"，即自由主义。由此可见，直译和意译各有所长，可以直译就直译，不可以直译就采用意译，甚至双管齐下，两者兼施，才能兼顾到译文的表层结构和原文的深层意思。

直译与意译的对象可以是单词和词组，也可以是句子。直译既可以保持原作的思想和风格，又有助于传播原作的民族文化。

例1：

原文：How can I sue for what I so little deserve? I dare not presume—yet <u>Hope is the child of Penitence</u>.

译文：我怎么能奢求我根本不配得到的宽恕呢？我不敢妄想，<u>然而希望是悔过之子</u>。

分析：例1原文的画线部分是个含有修辞的句子，直译不仅能保持原作的特点，而且还可使读者接受原作的文学风格，因而译文保留了原文中的暗喻修辞，将其翻译成"希望是悔过之子"，而没有依据译者个人的思想和风格，将其翻译为"人只要悔过就还有希望"。

在翻译过程中，由于语言基础发生变化，相当多的句子不能采用直译来翻译，因此我们必须采取意译。

例2：

原文：ABSOLUTE.〔Reads.〕"As for <u>the old weather-beaten she-dragon</u> who guards you"—Who can he mean by that?

译文：爱博思路特：（读）"至于你的监护人，<u>那个饱经风霜的老母夜叉</u>"——他这么说是指谁？

分析：例2原文画线的部分含有一个具有文化色彩的词语"she-dragon"，如果我们把它直译成"老母龙"，中国读者肯定不能理解它的意思，因为在英语和汉语两种语言中龙的文化象征意义不同，所以译文采用灵活变通的方式，将其意译为"老母夜叉"。

直译与意译是翻译中的基础问题，各种作品的翻译都要涉及直译与意译。一般来说，法律文献、政治文献、学术文献、新闻报道要求字面对应准确性强，因而多用直译，而文学作品和电影剧本不求字面对应，只求高质量的阅读感受，因此多用意译。但"译无定法"，译者需要不断地进行理论和实践学习，才能恰如其分地把握好直译与意译的程度。

（摘自：冯庆华.实用翻译教程[M].上海：上海外语教育出版社，2010：43.）

二、译例与练习

Translation

1. The topics that are covered by marine science can widely vary, including such things as ocean currents, sea floor geology, and the chemical composition of ocean water. (Para.1)

　　海洋科学所涵盖的主题可能千差万别,包括洋流、海底地质和海水的化学成分等。

2. Within each of these disciplines there are numerous sub-categories in which a professional is likely to specialize. For example, within marine biology, one person could focus on plants while another focuses on microscopic organisms. (Para.2)

　　在每一个学科中,都有许多专业人员可能擅长研究的子类别。例如,在海洋生物学中,一个人可能专注于植物,而另一个人则专注于微观生物。

3. Oceans cover about 71% of the earth's surface and are responsible for several critical processes that sustain life on earth, such as producing oxygen, supplying food, and regulating weather patterns, and yet over 80% of the world's oceans are unexplored and unmapped (National Oceanic and Atmospheric Administration). (Para.5)

　　海洋覆盖了地球表面的71%,并负责维持地球生命的几个关键过程,如生产氧气、供应食物和调节天气模式,但世界上超过80%的海洋尚未被探索及还没有被绘制成地图(美国国家海洋和大气管理局)。

4. While sustainable efforts to prevent more debris from going into the water exist, their impact is limited; most pollution comes from debris already in the oceans, and that debris must somehow be removed. (Para.6)

　　虽然存在防止更多碎片进入水中的可持续努力,但其影响有限。大多数污染来自已存在于海洋中的碎片,这些碎片必须以某种方式清除。

　　【这是个由分号分开的转折关系与并列关系的复合句。翻译时先将While译为转折关系的"虽然……但……",分号后面的半句翻译为第二个句子。把长句断开来翻译,叫断句译法,在翻译中经常被使用。be removed是被动语态,翻译为汉语的主动语态,这是转态译法,翻译时要根据具体情况运用转态译法进行翻译。】

5. Having a similar approach as Liquid Robotics, Openoceanrobotics USVs (Uncrewed or Unmanned Surface Vehicles) are also equipped with sensors, cameras, and communication devices so that they can capture information from anywhere on the ocean and have instant access to it. (Para.7)

　　采用与液体机器人类似的方法,公海机器人无人船(无人驾驶的或无人水面交通工具)也配备了传感器、摄像头和通信设备,便于从海洋的各个地方获取信息并立刻接近它。

6. RanMarine, a company founded and operating in Rotterdam, the Netherlands, has created WasteShark: an autonomous surface vessel that can collect 500 kilograms of surface debris in a single deployment while producing zero greenhouse emissions itself. (Para.8)

　　莱恩海事公司是一家在荷兰鹿特丹成立并运营的公司,他们发明了能自动收集海洋垃圾的水上无人设备"垃圾鲨鱼":一种自主水面舰艇,可以在一次部署中收集500千克的水面碎片垃圾,同时本身不会产生温室气体排放物。

7. FRED runs on solar power, producing zero emissions, which can be customized to operate in many different marine environments; and collects marine waste with brooms, a conveyer belt, and a collection bin. (Para. 9)

清除碎片的漂浮机器人使用太阳能发电,零排放,可定制在多个不同海洋环境中的操作方式,用扫帚、传送带和收集箱收集海洋垃圾。

8. Furthermore, the French start-up IADYS (Interactive Autonomous DYnamic Systems) has developed Jellyfishbot: a remote-controlled, electric-powered, trash-collecting robot that has been deployed in 15 French ports, removing plastic bags, bottles, and other debris from narrow and otherwise unreachable nooks in harbors where waste tends to accumulate. (Para. 10)

此外,法国初创 IADYS 公司(法国交互式自主动力系统公司)开发了水母机器人:一种远程控制的电动垃圾收集机器人,已部署在法国 15 个港口,从港口狭窄和无法到达的角落里清除塑料袋、瓶子和其他垃圾,这些地方往往是垃圾堆积的地方。

9. The expedition to La Lune was OceanOne's maiden voyage, and based on its astonishing success, it's hoped that the robot will one day take on highly-skilled underwater tasks too dangerous for human divers, as well as open up a whole new realm of ocean exploration. (Para. 12)

对拉伦恩的考察是远程遥控深海作业机器人的处女航,基于这次意想不到的成功,人们希望机器人有一天能承担对人类潜水员来说过于危险的高技能水下任务,并开辟海洋探索的全新领域。

【这是个复杂长句,it 是形式主语,真正主语是 that 引导的主语从句。主语从句的主语是 the robot,谓语是 take on,宾语是 highly-skilled underwater tasks,句子的开头部分是状语,句子的后半部分"as well as open up……"是主语从句中的并列谓语。翻译时可采用直译法。对于如"OceanOne"的专有词汇,查阅后译为"远程遥控深海作业机器人",as well as 是"也;和……一样;不但……而且"的词意,这里译为"并"表达并列的意思。】

10. Researchers at Harvard University are also looking into the role that soft robotics can play in capturing delicate oceanic creatures such as jellyfish and releasing them without harm with a Rotary-Actuated Device, or RAD. (Para. 14)

哈佛大学的研究人员也在研究软机器人技术在捕捉水母等纤弱的海洋生物方面的作用,并利用旋转驱动装置将它们安全释放而不造成伤害。

Exercises

1. Fill in the blanks with the proper given words, and then translate the sentences into Chinese.

  detrimental gulp detect microscopic shaft aquatic

1) As we eat, specialized receptors in the back of the nose _____ the air molecules in our meals.

2) Manatees, _____ mammals inhabiting Florida's rivers and coastal waters, swim close to the surface and are frequently killed in collisions with boats.

3) Within samples collected from the solid substances lying beneath the ice they found fossils of _____ marine plants which suggest that the region was once open ocean not solid ice.
4) Brown grabs his breathing mask and takes a _____ of air.
5) The advantage of biological control in contrast to other methods is that it provides a relatively low-cost, perpetual control system with a minimum of _____ side-effects.
6) A crew was sent down the _____ to close it off and bail out all the water.

2. Translate the following sentences into Chinese.

1) The robot, also referred to as an ROV for remotely operated vehicle, has high-definition video capability and the ability to provide real-time feedback for explosive ordnance disposal divers, according to the release.
2) Fielding the ROV comes as both the Navy and Marine Corps are working to expand their use of unmanned platforms for various missions, including the Marines' Expeditionary Advanced Base Operations (EABO) and Littoral Operations in a Contested Environment (LOCE).
3) Using new sensors, submarine technology, and autonomous vehicles, humans have the opportunity to advance the public's understanding and appreciation of the ocean, much as the development of scuba equipment and underwater cameras allowed ocean explorer Jacques Cousteau to captivate the world 50 years ago.
4) Maritime transport is essential to the world's economy as over 90% of the world's trade is carried by sea and it is, the most cost-effective way to move goods and raw materials around the world.
5) Unmanned marine vehicles (UMVs) are generally classified into underwater and Surface Vehicles (SVs) with underwater vehicles currently taking the lion's share of the market because of their range of applications including subsea pipeline/infrastructure surveying and inspection, search and rescue to name a few.
6) Surface vehicles have also gained a widespread interest for applications ranging from oceanography, remote sensing, weapons delivery, force multipliers, environmental monitoring, surveying and mapping, and providing navigation and communication support to underwater vehicles.
7) Most of USVs rely on remote operator guidance for sending mission commands and to constantly overlook the vehicle's status either by direct observation or via a wireless video link.
8) Currently there are a number of companies producing different types of SVs not only for military establishments but also for industrial corporations, environmental institutions and government agencies.
9) There are other data collection methods such as satellite systems and networks of stationary and floating buoys used today by scientists and researchers; however there are

limitations to these systems, which include the range of buoy networks and the accuracy and cost of satellite systems.

10) Some common applications for environmental monitoring are water salinity, water temperature, CO_2 content in air and sea, barometric pressure, wave height, wind speeds, bathymetry (water depth), photographic observation, underwater acoustics, oil spill measurements and pollution measurements.

3. Translate the following sentences into English.
1) 一些任务功能的最佳应用包括可定制的样本收集(例如油样收集)、防御相关任务、危险清理、信标传输、运输、船只交通标记和鱼类跟踪。
2) 这些平台的耐用性、自主性和低运营成本(相对于研究船或海洋系泊船)为补充现有的公海和近岸海洋二氧化碳观测工作创造了相当大的可能性。
3) 团队一直在进行科学调查,以帮助美国国家航空和宇宙航行局更好地了解其他海洋世界可能支持生命的潜能,同时也在研究海洋探索过程,以了解如何执行远程科学任务,并简化未来的探索难度。
4) 该研究的长期目标是开发下一代低成本、高效的自主水面航行器,用于海洋、工业和环境调查。
5) 乔治强调说:"对洋流的不同理解、海洋哺乳动物的迁徙、不同的造船传统模式,以及持续航行的能力来说,太平洋上有许多不同的航海秩序。"
6) 美国海洋高速公路计划支持增加使用国家的通航水道,以缓解陆地拥堵,提供新的及有效的运输选择,并提高地面运输系统的生产能力。
7) 美国交通部长皮特·布蒂吉格说:"对美国海洋高速公路项目的投资有助于我们更快、更有效地将更多货物运送到美国人民手中来支持我们的供应链。"
8) 代理海事局局长露辛达·莱斯利说:"简而言之,拜登总统正在领导有史以来最大的联邦投资,促进我国港口和国内沿海服务的现代化,并改善我们的供应链和依赖这些供应链的美国人的生活水平。"
9) 卡斯卡斯基亚河是伊利诺伊州第二长河流,发源于伊利诺伊州中部的香槟市,河流长度超过300英里,在与密西西比河汇合处终止。
10) 卡斯卡斯基亚河自从成为通航水道以来,主要用于运输煤炭、洗涤石、矿渣、谷物和废金属等大宗商品。然而,这条水道上也运输4万~5万吨的成卷带钢,新租户一旦建成加工厂,预计将运送多达120万吨的成卷带钢用于加工与其他用途。

4. Choose the best paragraph translation. And then answer why you choose the first translation or the second one.
　　该项目指定将支持现有的轮渡服务,在密歇根州的卢丁顿和威斯康星州的马尼托沃克之间运送货车和乘客,横跨密歇根湖。渡轮服务由獾号(Badger)停泊,獾号是一艘有记录的美国船只,也是一艘历史悠久的汽车渡轮,由因特莱克海事服务公司拥有和运营。这项服务允许货运卡车(包括超大卡车)以及汽车、娱乐车、房车、摩托车和其他车辆,避免在伊利诺伊州芝加哥附近极其繁忙的南部路线上行驶。项目指定符合政府对美国就业的关注,包括工会海员就业,以及减少整个经济消费,特别是交通部门的碳排放。[另:Badger 獾州人(美国威斯康星州人的别称)]

译文一:

This project refers to an existing ferry service that transports both freight vehicles and passengers across Lake Michigan between Ludington, Michigan, and Manitowoc, Wisconsin. The ferry service is anchored by the Badger, a documented U.S. vessel and a historic car ferry that is owned and operated by Interlake Marine Services. The service allows freight trucks (including oversized trucks), along with cars, Recreational Vehicles, motorhomes, motorcycles, and other vehicles to avoid travel around the extremely busy southern route near Chicago, Illinois. The Project Designation aligns with the Administration's focus on American jobs, including trade union seamen's employments, and reducing consumption throughout the economy, particularly in the transportation sector.

译文二:

The Project Designation will support an existing ferry service that transports both freight vehicles and passengers across Lake Michigan between Ludington, Michigan, and Manitowoc, Wisconsin. The ferry service is anchored by the Badger, a documented U.S. vessel and a historic car ferry that is owned and operated by Interlake Marine Services. The service allows freight trucks (including oversized trucks), along with cars, ROVs, motorhomes, motorcycles, and other vehicles to avoid travel around the extremely busy southern route near Chicago, Illinois. The Project Designation aligns with the Administration's focus on American jobs, including union seafaring jobs, and the reduction of carbon emissions across the economy, particularly in the transportation sector.

5. Translate the following passage into Chinese.

On the other hand, a wireless signal may be blocked as the robot heads deeper into the wreckage. So, a careful evaluation of the working conditions and possible hazards should also be done. Once the signal reaches the onboard receiver, it is conveyed to the mechanical systems. The main system is propulsion, in the form of miniature and specialized marine propellers. They are powered by small servo motors housed in waterproof casings, the propellers number 2 between 3 and 5 blades per shaft. They can be operated in both clockwise and counter-clockwise directions to create steering. Some advanced ROVs in wreckage analysis and deep-sea exploration also have jointed propellers. These are rarely used and are more costly than their fixed counterparts. It is only required in situations where very high precision is required. Sensors and appendages form the very core of the ROVs functionality. The most common equipment includes cameras, depth gauges, temperature, and internal system sensors. The operator uses cameras since they must have eyes on their surroundings.

6. Translate the following passage into English.

《海洋科学前沿》发表严格的同行评议研究。到2050年,全球人口预计将达到90亿,很明显,传统的土地资源将不足以满足高质量生计所需的粮食或能源需求。因此,海洋正在成为未开发资产的来源,水产养殖、海洋生物技术、海洋能源和深海采矿等新型创新产业在以海洋为基础的蓝色经济快速增长的新时代下迅速发展。蓝色经济的可持续性密切依赖于我们如何减轻海洋生态系统面临的多重压力,这些压力与海洋工业经营规模和多样化及全球人类对环境的压力有关。因此,《海洋科学前沿》特别欢迎以海洋为基础解决出现新挑战的交流研究成果,包括提高预测和观测能力,了解生物多样性和生态系统问题,在地方和全球范围内,维持海洋健康的有效管理战略,以及提高可持续地从海洋获取资源的能力。

三、翻译家论翻译:钱钟书化境论

钱钟书(1910—1998),原名仰先,后改名钟书,字默存,号槐聚,笔名中书君,江苏无锡人。精通英、法、意、德、拉丁、西班牙等多种语言,中国现代学者、作家,被誉为"博学鸿儒""文化昆仑",与饶宗颐并称"南饶北钱"。1929年,钱钟书以英语满分的成绩考入清华大学外国语言文学系,然后在1937年获牛津大学艾克赛特学院副博士学位。1947年出版长篇小说《围城》。钱钟书学贯中西,对中国的史学、哲学、文学等领域有深入研究,同时不曾间断过对西方新旧文学、哲学、心理学等领域的研究,并以宏阔的世界性视野,创立了打通、参互和比较的独特治学方法,取得显著的学术成就,其中最有名的是近两万字的《林纾的翻译》,在国内外学术界都享有很高的声誉。

钱钟书提出了"化境"的概念,在翻译学界引起共鸣。"化境"是指语言转化成一种意境,让读者看到文字以外的镜像,给读者一个想象的空间。好的文章有优美的语言就会给人美的享受,同样的,也会给人一种艺术的享受。"化境"是指在某方面的成就达到一定水平和高度。

要真正做到"化境",翻译起来还是难而又难的,有的时候甚至难以达到。在翻译不同语言的转换过程中,文化的不同、思维习惯的差异,即使译者仔细揣摩,认真推敲,译文里不可避免地会出现译得不准确或生硬牵强的痕迹。"化境"贯穿翻译的整个过程,需要译者殚精竭虑地实现译文与原文在内容、风格与语境上的最大相似。

(摘自:https://wenku.baidu.com/view/71cbcecd730abb68a98271fe910ef12d2af9a9aa.html?_wkts_=1673749165359&bdQuery—%E9%92%B1%E9%92%9F%E4%B9%A6+%E5%8C%96%E5%A2%83)

Text B

Human-Robot Spaceflight Exploration
By Matthew J. Miller, Zara Mirmalek and Darlene S. S. Lim

Ideological and Political Education：奋斗者号下潜 展中国海洋科技成就

中国东濒太平洋,是海洋大国,发展海洋科学技术对于建设海洋强国、实现中华民族伟大复兴都具有重大意义。党和政府始终高度重视海洋科学技术发展,伴随着国家整体实力的提升,一代又一代海洋科技工作者不懈奋斗、艰苦求索,中国海洋科学技术事业取得了可喜的成就。

2012年以来我国的海洋科技进入全面加速发展期,为了促进南海及其周边海洋国家在海洋科技领域的务实合作,中国发起并实施了《南海及其周边海洋国际合作框架计划(2011—2015)》。该项计划聚焦南海及与其相连的印度洋和太平洋,重点推动南海及其周边国家共同关心的区域海洋可持续发展方面的合作,包括海洋与气候变化、海洋环境保护、海洋生态系统与生物多样性、海洋减灾防灾、区域海洋学研究、海洋政策与管理等六大领域。该项计划得到了印尼、泰国、柬埔寨、马来西亚、尼日利亚、巴基斯坦、斯里兰卡、瓦努阿图等20多个国家和有关国际组织的积极响应与参与,有力地配合和促进了21世纪海上丝绸之路建设。

2020年10月27日,"奋斗者号"在马里亚纳海沟成功下潜突破1万米达到10 058米,创造了中国载人深潜的新纪录。截至2021年12月5日,"奋斗者"号载人潜水器到达过全球海洋最深处,已完成21次万米下潜,有27位科学家到达过全球海洋最深处。2022年9月,中国全海深载人潜水器"奋斗者"号与4 500米级载人潜水器"深海勇士"号,在南海1 500米水深区域完成既定作业任务。这是中国首次投入两台载人潜水器进行联合作业。2022年11月27日,中国"探索一号"科考船搭载着"奋斗者"号全海深载人潜水器停靠在新西兰奥克兰皇后码头。

中国海洋科技事业不断向前发展,特别是党的十八大以来,海洋科学技术进入了跨越式发展期,海洋科技人才队伍呈"指数式"发展壮大,海洋科学研究能力和条件进一步优化提升。中国海洋科技人员总量超过10万人,其中,涉海中国科学院院士和中国工程院院士50多人,成为推动中国海洋科学技术发展的领军人才。广大海洋科技工作者为中国海洋科学技术发展做出了不懈努力和突出贡献。

1. In 2023, researchers from space and ocean communities began working together on studies of geological, biological and chemical phenomena at deep-sea **venting** sites while also participating as subjects of social science research on conducting ocean science and exploration via telepresence. The Systematic Underwater Biogeochemical Science and Exploration **Analog** (SUBSEA) program provided science operations and **ethnography** researchers the opportunity to examine the question, "Can the existing work domain of ocean science research using telepresence be used as an analog for informing future human-robot spaceflight low-latency telepresence (LLT) operations"?

2. The SUBSEA program included two ocean research cruises. The first of which was a twenty-day field program (Aug-Sept 2023) to the Lōihi **Seamount**, located off the coast of Hawaii's Big Island. A second cruise was performed on a sixteen-day field program (May-June 2023) in Gorda Ridge off the coast of Oregon. The 2023 cruise was studied to determine what conditions could be imposed for a 2024 cruise, which would run in accordance with analog research conditions for future human-robot spaceflight. This study presents the results from the 2023 research cruise that informed the specific flight-like conditions that were implemented for the 2024 research cruise.

3. A connection must first be made between the space flight and ocean exploration work domains. NASA is developing a long-term strategy for achieving extended human presence in deep space. This includes mission destinations such as **cis-lunar** space, the Moon, Near Earth Objects, the moons and the surface of Mars. A universal component of these **envisioned** future missions is the utilization of LLT operations. This involves unavoidable communication and data transmission latencies ranging from minutes to tens of minutes one-way light time (i.e. approx. 8-44 mins return) between crew (humans and robots) in deep-space locally controlling robotic assets and **Earth-bound** experts.

4. The impact of these communication delays combined with the potential system demands of LLT operations are a subject of NASA interest. On recent NASA Mars missions, scientists working with remote robots have had from one to several days to deliberate between sending and receiving data between Earth and Mars. Within future human-scale mission LLT operations, it is anticipated that decision-making timeframes will be significantly more compressed in order to affect actions by Earth-bound support to take advantage of having crew locally conduct LLT operations.

5. Examinations to date within the human spaceflight domain offer **hypothetical** applications of telepresence capability. Within the specific domain of scientific spaceflight telepresence, NASA

Goddard's former chief scientist James Garvin has stated, "There is a profound lack of real experience with low **latency** telepresence here on Earth, in geological field situations, with which to understand how to utilize the obvious benefits of this approach on the Moon, Mars, **asteroids**, or beyond... This experience gap limits our understanding of how to develop the engineering and technology capabilities required for using low latency telepresence in deep space field science." Therefore, the inclusion of telepresence capabilities in future human spaceflight missions remain **speculative** at best. Because human LLT operations are not common within the current spaceflight domain, these LLT hypotheses make them **susceptible** to the underspecification, ungrounded, and overconfidence properties of the envisioned world problem.

6. Human factors research emphasizes sociotechnical systems, context and communication in understanding the intersection of people, technology and work. Similar research **resides** within fields such as anthropology and cognitive systems engineering. A single disciplinary **approach** has yet to yield a robust account or guide to developing complex systems as work domains are constituted of multiple primary agents (e.g., humans, technology, geographic location) and interactive relationships. And while interdisciplinary research requires more time commitment, data collection and analysis are more robust.

7. Fashioning an analog spaceflight work domain within the ocean exploration work domain requires a definition of what it means for these two specific domains to be "analogous". The SUBSEA program neither created an actual future spaceflight work domain, nor attempted to presuppose and impose the components of a hypothetical future spaceflight domain in the ocean domain. This study approached the examination of what it means for ocean science and exploration to be an analogous work domain for human-robotic planetary exploration. For NASA spaceflight mission research, "analog" is widely used to describe environments on Earth that are similar to (analogous) planetary sites, e.g. gravity, scientific features.

8. For SUBSEA Cruise A, scientists studied seafloor fluid venting in Earth's deep ocean as it may relate to environments on other ocean worlds in the outer solar system that could host similar chemosynthetic ecosystems. A series of 10 dives over 16 days at sea was performed at the Lōʻihi Seamount, located off the coast of Hawaii's Big Island at more than 1 km beneath the ocean surface. This location was chosen because it hosts a distinct class of low temperature (<100 ℃) and shallow depth (hence, low pressure) fluid flow that might provide a particularly relevant scientific analog for seafloor **hydrothermal** conditions ($T = 50 \sim 200$ ℃; $p = 10 \sim 50$ MPa) inferred for Enceladus, one

of the 10 highest priority known ocean world candidates.

9. To address the research question of whether the **Nautilus** telepresence architecture could be **leveraged**, for the generation of a spaceflight LLT analog (Cruise B), the short answer is yes, but not necessarily in the form of what was originally proposed. At the onset of the SUBSEA project, the intent of Cruise B was to implement a communication latency, between ship and shore teams, close to the upper limit experienced between Mars-Earth (e.g., 15 minutes one way latency) within the existing Nautilus telepresence architecture. After Cruise A, based on assessments from various subteams, we found the technical implementation of this latency condition was not feasible, for Cruise B. Given this evaluation, the social science team developed a set of "flight-like" conditions, described below, both necessary (work modes modulating latency) and sufficient (communication tools and shifted work roles). These conditions were then **vetted** for feasibility within the existing sociotechnical system and for being robust to the inherent variability of dive opportunities that exist at sea. These conditions were **buttressed** by the addition of communication technologies and the reallocation of some team roles, that elevated and formalized what we found on Cruise A.

10. For Cruise B, we had the science team utilize a timeline for the planned dives that was separated into two distinct work modes, Mode 1 (M1) and Mode 2 (M2). Forty percent of the dives were allocated to M1; sixty percent of the dives were allocated to M2. During M1, scientists on shore did not have the ability to participate in real-time over audio channels with dive operations conducted on the ship. Instead, they relied on specific communication technologies (SUBSEA Dive Plan authored by shore, and Dive Recovery and Data Report authored by ship) for sending and receiving content, between the ISC and personnel on the Nautilus. These two products were reflectively developed based on Cruise A work practices and SUBSEA M2 goals.

11. During Cruise B M1, SUBSEA science team produced a daily SUBSEA Dive Plan for each dive and sent it to the ship. Once the dive was completed, the ship team produced a daily SUBSEA "Dive Recovery and Data Report" sent to shore. This communication **cadence** intentionally produced a temporal separation between requesting action (i.e., dive objective) and receiving confirmation of completion of action (i.e., dive objective). This was the intentional production of communication latency on the cycle of 24-hr exchanges between the two work groups. During M2, scientists on shore gained the ability to direct dive operations in real-time over audio and text channels in addition to the communication technologies described above. M2 communication

cadence included a combination of temporal separation (as defined in M1 with the communication technologies) and real-time discussion throughout dive operations.

(1,279 words)
Extracted from *Proceedings of the Human Factors and Ergonomics Society Annual Meeting*, vol.63,(1);292-296,November 1,2019.
https://journals.sagepub.com/doi/abs/10.1177/1071181319631217

New Words

1. vent [vent]
    v. give expression or utterance to;expose to cool or cold air so as to cool or freshen 表达,发泄(强烈感情,尤指愤怒);给……提供出口或排放口,排放
   Some probably vent their anger by throwing things, or perhaps pounding their lockers or kicking a nearby garbage can.
   一些人可能会通过扔东西或者踢门或者踢邻近的垃圾桶来发泄胸中的气恼。

2. telepresence ['teliprezns]
    n. the use of virtual reality technology to operate machinery by remote control or to create the effect of being at a different or imaginary location 思科网真;远程呈现(一种通过结合高清晰度视频、音频和交互式组件,在网络上创建一种独特的"面对面"体验的新型技术)
   Under the partnership, corporations looking to reserve telepresence spaces and facilities would be able to view room availability, book the meetings and view rates and restrictions.
   根据这一合作关系,需要预订"远程呈现"设备和场地的企业将可以浏览房间的可用性、预订会议、浏览价格及限制条件。

3. analog ['ænəlɒg]
    a. of a circuit or device having an output that is proportional to the input 模拟的;有长短针的
   The output can be more specific than just specifying the analog pin number, but the generic number allows flexibility because it does not tie down any one pin to any one job.
   输出可能比仅指定模拟管脚编号更为具体,但是本机编号灵活性更好,因为它不会将任何一个管脚束缚到某一项作业上。

4. ethnography [eθ'nɒɡrəfi]
    n. the branch of anthropology that provides scientific description of individual human societies 民族志;人种志;人种学
   Ways of objective material collection and on-the-sport research should be adopted in the research process so as to improve the efficiency of ethnography research method.

研究过程中应尽可能采取客观的资料收集方法及进行现场研究,以提高人种志研究方法的有效性。

5. seamount ['siːmʌunt]

   n. an underwater mountain rising above the ocean floor 海底山

   The team found that parts of the seamount's steep flanks were made of porous volcanic materials, leaving them weak and unstable.

   研究小组发现,海底山陡峭的侧翼局部由多孔的火山物质构成,这正是其脆弱和不稳定的诱因。

6. cis-lunar [sɪsˈluːnə]

   a. situated between the earth and the moon 地球和月亮之间的

   With its superior lift capability, the SLS will expand our reach in the solar system and allow us to explore cis-lunar space, near-Earth asteroids, Mars and its moons and beyond.

   凭借其卓越的提升能力,设定位置组号将扩大我们在太阳系的范围,并允许我们探索近月空间、近地小行星、火星及其卫星和更远的地方。

7. envision [ɪnˈvɪʒn]

   v. imagine; conceive of; see in one's mind 想象;预想

   Though the future we envision for all the world's children may not come easily, the founding of the United Nations itself is a testament to human progress.

   尽管我们为全世界儿童预想的将来可能不会轻易到来,但是联合国的成立本身就是对人类进步的证明。

8. earth-bound [ˈɜːθbaʊnd]

   a. confined to the earth 只在地球上的

   The young should be forward-looking in thoughts and earth-bound in actions.

   青年人在思想上应该高瞻远瞩,在行动上应该脚踏实地。

9. hypothetical [ˌhaɪpəˈθetɪkl]

   a. based primarily on surmise rather than adequate evidence 假设的;爱猜想的

   The key point to keep in mind here, however, is that these projects are real, not hypothetical exercises.

   但是,在此要记住的关键一点是:这些项目是真实的而不是假设的练习。

10. latency [ˈleɪtənsi]

    n. the state of being not yet evident or active 潜伏;潜在因素

    The average latency is something like 210 microseconds.

    平均延迟大约为210微秒。

11. asteroid [ˈæstərɔɪd]

    n. any of numerous small celestial bodies composed of rock and metal that move around the sun (mainly between the orbits of Mars and Jupiter) 小行星;海盘车;小游星

    In this case, the orbit of the rotating asteroid will slowly spiral in toward the sun.

在这种情况下,旋转小行星的运行轨道将缓慢地朝着太阳的方向盘旋。

12. speculative [ˈspekjələtɪv]

    a. not financially safe or secure 投机的;推测的;思索性的

    All trading can be divided into speculative trading and investment trading.

    一切买卖都可分为投机性买卖和投资性买卖。

13. susceptible [səˈseptəbl]

    a. (often followed by of or to) yielding readily to or capable of 易受影响的;易感动的;容许……的

    According to the researchers, the measles virus only needs a small number of susceptible children to cause outbreaks, so it is important that as many children as possible have immunity to the virus.

    科学家研究表明,麻疹病毒只需要一少部分易感儿童就能导致麻疹爆发,因此让尽可能多的儿童对麻疹有免疫力是很重要的。

14. reside [rɪˈzaɪd]

    v. make one's home in a particular place or community;live (in a certain place) 住,居住;属于

    Our perceptions about warming and cooling trends tend to depend upon where we happen to reside and the time frame we experience for reference.

    关于气候变暖和变冷的趋势,我们的看法往往取决于我们住在哪里以及我们所经历的时间范围。

15. approach [əˈprəʊtʃ]

    v. move towards;make advances to someone, usually with a proposal or suggestion 走进;与……接洽;处理;临近,逐渐接近(某时间或事件)

    All of this has helped me to organize many of my thoughts about how to approach the day with my kids.

    所有这些都帮助我整理了许多关于如何与孩子们度过一天的想法。

16. hydrothermal [haɪdrə(ʊ)ˈθɜːm(ə)l]

    a. of or relating to the action of water under conditions of high temperature, esp. in forming rocks and minerals 热液的;热水的

    Hydrothermal vents are cracks in the Earth's surface that occur, the ones we are talking about here are found deep at the bottom of the ocean.

    深海热泉是地球表面出现的裂缝,我们在这里提到的裂缝是在海底深处发现的。

17. nautilus [ˈnɔːtɪləs]

    n. cephalopod mollusk of warm seas whose females have delicate papery spiral shells 鹦鹉螺;鹦鹉螺号

    The Nautilus enhances ship safety during surfaced and submerged operations in crowded waters.

在拥挤的海面和水中航行时，鹦鹉螺显示器增强了船的安全性。

18. leverage ['li:vərɪdʒ]

    n. the mechanical advantage gained by being in a position to use a lever 手段，影响力；杠杆作用；杠杆效率

    v. supplement with leverage 利用；举债经营

    His position as mayor gives him leverage to get things done.

    他的市长身份使他有能力办成一些事情。

19. vet [vet]

    v. work as a veterinarian; examine carefully 审查，检查；诊疗

    Another pressing issue is the need to vet the integrity of the data the community provides.

    另一个紧迫的问题是，需要检查社区所提供的数据的完整性。

20. buttress ['bʌtrəs]

    v. reinforce with a buttress; make stronger or defensible 支持；用扶壁支撑（建筑物等）

    The sharp increase in crime seems to buttress the argument for more police officers on the street.

    犯罪率急剧上升似乎肯定了街上增加巡警的论点。

21. cadence ['keɪdns]

    n. (prosody) the accent in a metrical foot of verse; the close of a musical section 节奏；韵律；抑扬顿挫

    The part also meant memorizing endless monologues that needed to be delivered with Hoover's own breakneck cadence.

    这个角色还需要他背诵大段的独白，并且以胡佛独特的急促节奏表达出来。

Phrases and Expressions

off the coast of 距离……的海岸
on……missions 执行任务；完成使命
combine with 与……结合
at the onset of 在……开始的时候
remove from 除掉；移动

Terminology

space and ocean communities 空间和海洋群落

Proper Names

Systematic Underwater Biogeochemical Science and Exploration Analog (SUBSEA) 水下生

物地球化学科学与勘探模拟系统

low-latency telepresence (LLT) 低延迟网真

National Aeronautics and Space Administration(NASA)美国国家航空和宇宙航行局

Goddard 戈达德(姓氏)(1882—1945,美国物理学家,火箭工程学的先驱者)

Enceladus(希腊)恩克拉多斯(古希腊神话中的一位因为反对宙斯战败并被雅典娜埋葬在埃特那山下的癸干忒斯之一,与其他癸干忒斯一样是乌拉诺斯与盖亚之子。其英文名称 Enceladus 也是土卫二的名称)

International Space Congress(ISC) 国际宇宙空间大会

Exploration Ground Data System (EGDS) 勘探地面数据系统

一、翻译简析

1. In 2023, researchers from space and ocean communities began working together on studies of geological, biological and chemical phenomena at deep-sea venting sites while also participating as subjects of social science research on conducting ocean science and exploration via telepresence.(Para.1)

  2023 年,来自太空和海洋群体的研究人员开始共同研究深海排气点的地质、生物和化学现象,同时,作为社会科学研究对象,研究人员通过远程呈现进行海洋科学探索。

  【这是个较长的复合句,句式结构不是很复杂。在汉译时,按照汉语语序,时间状语放在前面,介词短语修饰主语 researchers,译为定语"……的",宾语后的介词短语 at deep-sea venting sites 译为定语"深海排气点的"的具体译法。在翻译过程中,用具体的词进行翻译,从而降低语言差别,使译文产生与原文同样的效果。】

2. This involves unavoidable communication and data transmission latencies ranging from minutes to tens of minutes one-way light time (i. e. approx. 8-44 mins return) between crew (humans and robots) in deep-space locally controlling robotic assets and Earth-bound experts.(Para.3)

  这不可避免地涉及通信和数据传输延迟,在当地控制机器人资产的外太空机组人员(人类和机器人)与地球专家之间会出现几分钟到几十分钟的单向光时(即大约 8～44 分钟返回)延迟。

  【这个句子难翻译在于宾语的修饰语很长。先翻译 between……and…… 然后再译 ranging from……由小及大。词汇中的 locally 译为"当地",deep-space 译为"外太空",light time 译为"光时"更可取一些。因此,英译汉时,常常需要调整原文的各个成分,使译文被读者接受。】

3. Within future human-scale mission LLT operations, it is anticipated that decision-making timeframes will be significantly more compressed in order to affect actions by Earth-bound support to take advantage of having crew locally conduct LLT operations.(Para.4)

  在未来人类规模的低延迟网真操作中,预计决策时间框架将大大压缩,影响到只在地球上支持的运转方式,便于利用当地人员进行低延迟网真作业的优势。

【这是一个含有形式主语的主语从句。在汉译时,把句子断开来翻译,先翻译从句的主语部分 decision-making timeframes,然后翻译目的状语 to affect actions……更为可取一些,就是我们常用的断句的翻译方法。】

4. After Cruise A, based on assessments from various subteams, we found the technical implementation of this latency condition was not feasible, for Cruise B. (Para.9)

基于来自各个子工作组的评估,在巡航 A 之后,我们发现对于巡航 B,这种延迟条件技术无法实现。

【这个句子不长,汉译时,subteam 译为"子工作组",按照汉语的表达习惯把介词短语 After Cruise A 放在"基于……的"后面翻译。for Cruise B 的介词短语翻译成汉语"发现"的宾语,其宾语是宾语从句 the technical implementation of this latency condition was not feasible。】

5. Instead, they relied on specific communication technologies (SUBSEA Dive Plan authored by shore, and Dive Recovery and Data Report authored by ship) for sending and receiving content, between the ISC and personnel on the Nautilus. (Para.10)

相反,他们依靠特定的通信技术(海底潜水计划由岸上授权编写,潜水恢复和数据报告由船上授权编写)在国际宇宙空间大会和鹦鹉螺号船上人员之间发送和接收电子信息。

【这个句子的翻译难点在于括号中的专有名词的翻译。在汉译时,把括号中的 authored by shore 译为"由岸上创始人编写"做汉语的后置定语,修饰"水下潜水计划"。因此,英译汉时,恰当使用翻译策略,使译文被读者接受。】

二、参考译文

人机航天探索

1. 2023 年,来自太空和海洋群落的研究人员开始共同研究深海排气点的地质、生物和化学现象,同时,作为社会科学研究对象,研究人员通过远程呈现进行海洋科学探索。水下生物地球化学科学系统和勘探模拟计划为科学操作和民族志研究人员提供了检验的机会,"使用远程呈现海洋科学研究的现有工作领域是否可以用作未来人机航天低延迟网真呈现的操作模拟"?

2. 海底项目包括两次海洋巡航研究。第一次是为期 20 天的实地考察项目(2023 年 8 月至 9 月),地点位于夏威夷大岛海岸外的深水海底火山。第二次巡航在俄勒冈州海岸的高架桥进行了为期 16 天的实地项目(2024 年 5 月至 6 月)。研究人员对 2023 年的巡航进行了研究,以确定 2024 年的巡航设置条件,2024 年的巡航将根据未来人机航天的模拟研究条件进行巡航。这项研究介绍了 2023 年巡航研究的结果,这些结果为 2024 年特定飞行的巡航实施提供了依据。

3. 首先,必须在空间飞行和海洋探测领域之间建立联系。美国国家航空和宇宙航行局正在制定一项长期战略,以扩大人类在深空的生存条件,这包括任务目的地,如顺月空间、月球、近地物体、月球和火星表面。这些设想的一项未来任务组成部分是利用有限运输进行作业。这不可避免地涉及通信和数据传输延迟,在当地控制机器人资产的外太空机组人员(人类和机器人)与地球专家之间会出现几分钟到几十分钟的单向光时(即大约8~44分钟返回)延迟。

4. 这些通信延迟的影响与低延迟网真操作的潜在系统需求相结合是美国国家航空和宇宙航行局感兴趣的话题。在近期的美国国家航空和宇宙航行局火星任务中,与远程机器人一起工作的科学家们有一到几天的时间来考虑在地球和火星之间发送和接收数据。在未来人类规模的低延迟网真操作中,预计决策时间框架将大大压缩,影响到只在地球上支持的运转方式,以便于利用当地人员进行低延迟网真作业的优势。

5. 在人类航天领域内确定日期的测试提供了远程监控功能的假设应用程序。在科学航天飞行远程呈现的特定领域内,美国国家航空和宇宙航行局戈达德的前首席科学家詹姆斯·加文表示:"在地球上,在地质领域的情况下,对于如何利用这种方法在月球、火星、小行星或其他地方的明显好处,我们非常缺乏真正的经验。这种经验差距限制了我们在外太空领域如何开发与使用低延迟网真呈现的工程和技术能力的理解。"因此,在未来的人类航天任务中,包含的网真呈现能力充其量只是推测。由于人类的低延迟网真操作在当前的航天领域中并不常见,这些低延迟网真假设容易受到预期世界的不规范、无扎实基础与过度自信的影响。

6. 人为因素的研究强调社会技术系统,以人、技术与工作交集的环境与沟通。类似的研究存在于人类学和认知系统工程等领域。由于工作领域是由多个主要代理机构(例如人、技术、地理位置)和交互关系组成,单一的学科方法尚未产生稳健账户或引导开发复杂系统。虽然跨学科研究需要投入更多的时间,但数据收集和分析更加可靠。

7. 在海洋探索工作领域中形成模拟航天工作领域需要定义这两个特定领域的"类似性"。海底项目既没有创建一个实际的未来航天工作领域,也没有试图在海洋领域预先假设和强加一个假设的未来航天领域的组成部分。这项研究探讨了海洋科学探索作为人类-机器人行星探索的类似工作领域意味着什么。在美国宇航局的航天任务研究中,"模拟"广泛应用于描述地球上与(类似的)行星地点相似的环境,例如重力、科学特征。

8. 对于海底巡航 A,科学家们研究了地球深海的海底流体排气,因为它可能与太阳系外其他海洋世界的环境有关,这些海洋世界可能拥有类似的化学合成生态系统。在夏威夷大岛海岸 1 千米以下的深水海底火山,在 16 天内进行了 10 次潜水。之所以选择这一地点,是因为它拥有独特的低温($<100\ ℃$)和浅深度之后低压流体流动,可能为海底热液条件提供特别相关的科学模拟($T=50\sim200\ ℃;p=10\sim50\ MPa$)推测的土卫二,是已知的 10 个优先级最高的海洋世界候选之一。

9. 为了解决是否可以利用鹦鹉螺号远程网真架构的研究问题,对于太空飞行低延迟网真模拟(巡航 B)的生成,简洁的答案是肯定的,但这不一定是以最初提议的形式提出。在海底项目开始之初,巡航 B 的目标是实现现有的鹦鹉螺号在远程网真架构内实现船舶和岸上团队之间的通信延迟,接近火星与地球之间的上限(例如,单程延迟 15 分钟)。基于来自各个子工作组的评估,在巡航 A 之后,我们发现了巡航 B,这种延迟条件技术无法实现。鉴于这个评估,社会科学团队开发了一组"类似飞行"的状态,如下所述,既是必要的(工作模式调节延迟),也是充分的(通信工具和转换的工作角色)。然后在现有社会技术系统中,对这些条件下的可行性进行了审查,并对海上存在的潜水机会的内在可变性进行了稳健分析。这些条件加强了通信技术和一些团队角色重新分配的支持力度,这提升并证实了我们在巡航 A 中发现的一切。

10. 对于巡航 B,我们让科学团队利用计划潜水的时间表,将其分为两种不同的工作模式,模式 1(M1)和模式 2(M2)。40%的潜水分配给 M1,60%的潜水分配给 M2。在 M1 期间,岸上的科学家无法通过音频频道实时参与在船上进行的潜水操作。相反,他们依靠特定的通信技术(海底潜水计划由岸上授权编写,潜水恢复和数据报告由船上授权编写)在国际宇宙空间大会和鹦鹉螺号船上人员之间发送和接收电子信息。这两款产品根据巡航 A 的工作实践和海底 M2 的目标进行开发。

11. 巡航 B 在 M1 期间,海底科学团队为每次潜水制订了每日海底潜水计划,并将其发送到船上。潜水完成后,船舶团队每天都会制作一份海底"潜水恢复和数据报告",并将其发送到岸上。这种通信节奏有意地在请求动作(即潜水目标)和接收动作完成确认(即潜水目标)之间产生了时间间隔。这是有意在两个工作组之间的 24 小时交换周期内产生的通信延迟。在 M2 期间,除了上述通信技术外,岸上的科学家还获得了通过音频和文本渠道实时指导潜水作业的能力。M2 通信节奏包括时间间隔(在 M1 中定义的通信技术)和整个潜水作业中进行实时讨论。

Cultural Background Knowledge: Marine Science
海洋科学文化背景知识

海洋科学主要研究全球海洋的自然现象、性质及其变化规律,以及海洋的开发利用问题。海洋科学的研究对象包括海水、溶解和悬浮于海水中的物质、生活于海洋中的生物、海底沉积和海底岩石圈,以及海面上的大气边界层和河口海岸带等各种现象。海洋是人类社会可持续发展的重要领域,"海洋强国"战略对于我国未来的发展具有极为重大而深远的意义。

古代人类在生产活动方面积累了很多有关海洋的知识,也获得了关于海洋的经验和见解。公元前 7 世纪至公元前 6 世纪,出生于爱奥尼亚米利都城的西方哲学之父古希腊泰勒斯认为,大地是浮在茫茫大海之中的。公元前 4 世纪,古代先哲古希腊的亚里士多德在《动物志》中描述和记载了爱琴海沿岸的 170 余种动物。从 15 世纪资本主义兴起之后,人类对海洋有了更多的了解。在西方人称为地理大发现时代的 15 世纪至 16 世纪,意大利航海家克里斯托弗·哥伦布于 1492—1504 年间先后 4 次横渡大西洋到达南美洲。葡萄牙人伽马于 1498 年从大西洋绕过好望角经印度洋到达印度。1519—1522 年葡萄牙人麦哲伦完成了人类第一次环球航行。1768—1779年英国人库克船长3次进行海洋探险,首先完成了环南极航行,并最早进行了科学考察,获取了第一批关于大洋深度、表层水温、海流及珊瑚礁等海洋资料。19世纪到20世纪中叶是海洋科学的奠基与形成时期,既表现在海洋探险逐渐转向为对海洋的综合考察,海洋研究不断深化,关于海洋的研究成果众多并形成了多个海洋理论体系。20世纪中叶至今是现代海洋科学时期,海洋科学得以迅速发展,逐步进入现代海洋科学的新时期。

21世纪是海洋发展战略的世纪,21世纪是"海洋科学的新世纪"。发展海洋科学、繁荣海洋经济、保护海洋环境,是新时代赋予我们的伟大历史使命。

Unit Fourteen
Ocean Exploration

海洋探索

Text A

International Ocean Exploration
By Jesse H. Ausubel and Paul G. Gaffney II

Ideological and Political Education：提高深海探测能力 助力海洋强国建设

海洋是人类可持续发展的战略资源宝库，蕴藏着丰富的渔业资源、能源、矿产等数不尽的宝藏。到目前为止，人类对海洋的探测十分有限，仅为5%左右。究其原因，主要是因为深海探测装备未能达到深海探索的要求。

从古至今，中国人探索海洋的脚步从未停止。600多年前，明朝航海家郑和7次出海，历经28年，远航30多个亚非国家，完成了史无前例的航海活动。郑和的宝船是当时世界上最可靠、最大的帆船，堪称有史以来海上最大的交通工具。据专家称，明朝海军在其全盛时期，实力可能超过了任何历史时期的任何其他亚洲国家，也远远超过了任何当代欧洲国家，甚至超过了它们的总和。近年来，中国积极研发深海探测装备，取得了一项又一项海洋科技突破。此前亮相的蛟龙系列潜水器，和后来亮相的海牛二号探测器，都是具有代表性的作品。2018年，中国自主研发的深海载人潜水器"深海勇士"号是全世界同一级别深海下潜作业时间最长的潜水器。"深海勇士"号4 500米成功海试，标志着中国深海技术装备由集成创新向自主创新的历史性转变，成为继美国、法国、俄罗斯和日本之后拥有4 500米载人深潜技术的国家。2020年，中国的"奋斗者"号成功探索被称为地球第四极的深逾万米的马里亚纳海沟，成为继美国之后第二个抵达此处的国家。2021年4月，"海牛Ⅱ号"海底大孔深保压取芯钻机系统刷新了世界深海海底钻机钻探深度，填补了我国深海海底钻机装备的空白，标志着我国在这一技术领域达到世界领先水平。2021年9月，海上风电新型桩-桶复合基础研发及其工程应用，促进了我国海上风电工程技术的研发升级，填补了国内外相关技术发明的空白。2021年11月，国产全平台远距离高速水声通信机突破全球最高指标，打破了国外封锁，实现了技术超越。2021年12月，我国完全自主研发的海洋环流数值模式"妈祖1.0"，实现了"中国芯"对"欧美芯"的替代，展现了我国在海洋预报领域的高水平和自立自强的能力。另外，我国自主完成北极高纬密集冰区国际首次大规模海底地球物理综合探测，创造了多项国际新纪录，赢得了国际科学界广泛赞誉。我国主持制定的首项海洋调查国际标准于2021年正式出版发布，为国际社会贡献中国方案并参与全球海洋治理。

全球还有95%的海洋有待人类前去探索，哪个国家拥有更先进的深海探测装备，就能拥有更多的先机。中国人民将继续用万丈豪情和骄人智慧在深蓝之下书写震动世界的中国传奇。更深、更广、更远，强大的深海探测能力将引领中国向海洋强国加快迈进。

1. In 2022, Public Law 111-11 formally established a U. S. national ocean exploration program. The law assigns the federal lead for ocean exploration to the National Oceanic and Atmospheric Administration (NOAA) but urges participation of other agencies with **oceanographic** capabilities. The following three sections offer highlights of Breakout Group discussions **consolidated** by campaign area.

2. Arctic. The Arctic groups discussed potential future ocean exploration campaigns in the (mostly) American Arctic from the North Slope **littoral** to approximately 80 degrees north **latitude**. An area of shallow and deep water mostly ice-covered in winter with a marginal ice zone (MIZ) and ice edge year-round, it is tectonically and geologically not well understood, and its changing physical oceanography demands characterization. The Arctic, with its **harsh** weather, persistent ice (even in recent years), unavailability of ship support except for short periods in summer, and long distance from **robust** shore support redefines ocean exploration. Ocean exploration here must be characterized by "duration" not "one-stop shopping." Autonomous sensors must be deployed when the weather and ice allow. **Retrieval** of the same sensors also depends on environmental friendliness. A ship is not readily available to **hover** nearby in case of trouble, nor is technical support available from a nearby laboratory. Explorers must plan for campaigns with duration of 8 to 12 months or longer. Hence, exploration in the Arctic **verges** on observation and can offer a transition to it.

3. Arctic exploration is currently and will continue to be enabled by autonomous underwater vehicles, mobile ice **implanted buoy** networks, bottom-mounted instruments, and ASVs in the open water and the MIZ. Conveniently, autonomous vehicles and sensors developed for Arctic exploration will be useful elsewhere: "if it will work in the Arctic it will work anywhere." Such an exploration/observation network of autonomous devices will have to be supported by a previously installed and complementary network of navigation (since GPS will not be available), refueling/repowering, and communications **nodes**. Surface ships simply cannot carry out these functions throughout most of the year. Such a support infrastructure may be (a) mobile and ice-based with instruments hanging into the water column;(b) mobile and surface ocean buoys-based;(c) bottom-mounted with or without **tendrils** rising through the water column;or (d), likely, combinations of all of the above.

4. Future exploration vehicles and instruments will need to employ adaptive sampling software wherein an instrumented **drone** can learn as it goes, find or change its way and sample where it finds the water column or bottom most interesting. Simply "**mowing the grass**," gaining full sensor coverage, then identifying new or changing features is not practical in the Arctic.

5. It will always be expensive and **hazardous** to explore in the Arctic because of weather, ice, and distance, and also the environmental sensitivities of polar ecosystems. Moreover, much Arctic exploration is carried out in darkness. These factors heighten the importance of partnerships in general and in particular with industry and international friends (e.g., Canada, First Nation peoples, Denmark, NATO). For the technical, safety, and environmental reasons already stated, and because the area has potential for international competition, is home to several endangered species, and is commonly used for traditional purposes by First Nation peoples, exploration campaigns will require extensive planning.

6. Technologies of note in the next decade: ① Long-duration autonomous vehicles smart enough to devise exploration plans on the fly; ② Smaller, autonomous devices that need less power even assuming high-capacity batteries and recharging capability; ③ Pre-**deployed**, cheap (perhaps nonrecoverable) network(s) to support navigation, communications/data dumps and refueling or recharging; ④ Ice-hardened vehicles and instruments that do not get crushed by moving ice in any season.

7. Southeast U.S. Atlantic **Bight**. The Southeast U.S. Atlantic (SEUS) Bight groups separately discussed an ocean exploration campaign in the SEUS. Unlike the other regional campaigns in the 2022 Forum, the federal government is already planning a SEUS campaign, for the period 2022—2023, with funding commitments under consideration and ship time in negotiation among Bureau of Ocean Energy Management (BOEM), USGS, and NOAA, and thus the groups considered the near term as well as 2025—2030. The area to be explored stretches south from the Baltimore Canyon to the Blake Plateau. This plan, while covering a very large area, focuses on **discrete** areas for high-resolution exploration. To generalize, the SEUS planners are most interested in identifying **seeps** and deep coral and sponge habitats; canyon areas are expected targets.

8. In accordance with the joint-agency plan, BOEM may issue a request for proposals (RFP) in early 2022. The discussants noted that, so far, the RFP is not asking for proposals that include archeological exploration activities. In view of the commitment of ship time, this seems a lost opportunity in an area that was a major seaway for early settlement of the Americas.

9. While the SEUS area has been "home base" for ocean research and survey activities for decades, it highlights several issues.

10. Although the area has been widely surveyed bathymetrically, present high-resolution **bathymetry** and **acoustic** imagery and information do not **suffice** to **discern** seeps,

sponge, and coral communities and archeological **artifacts**. Greater attention could be given to high-resolution bathymetry, at least in high interest areas, by launching swarms of AUV/ASVs equipped with multibeam sonars. Here the ship not only collects bathymetric data, but can also serve as a launch platform and mother ship for autonomous vehicles that can significantly increase bathymetric coverage at lower cost. Increasingly, too, these autonomous vehicles could be launched from shore on long-endurance missions. In areas like SEUS, where the shore is within reasonable range, shore launched autonomous vehicles could **frugally** serve to decrease reliance on ships.

11. As one thinks about using emerging autonomous vehicles or present-day remotely operated underwater vehicles (ROVs) in areas of suspected seeps and deep sponge/coral habitats, one needs either higher-resolution bathymetry to help prevent costly AUV/ROV collisions with **craggy canyon** features and direct the vehicles toward suspected or **hypothesized** targets, or the deployment of adaptive, deep-diving autonomous vehicles that can make decisions on fly to avoid danger, search for the most interesting features, and find recharging, communication, and navigation nodes.

12. Assuming an unusually rich general baseline of information in this particular campaign area, the campaigners have an opportunity to compare archived data with newly collected higher-resolution data. Determining how much we really did or didn't know and what has changed could help better design future campaigns. To complete such comparative analysis, campaign sponsors need to establish a "dedicated fund" for a pre-campaign effort to gather **archival** oceanographic/bathy metric data in the areas of highest interest, at all resolutions and wavelengths, and make that data available to explorers for operational planning and post-campaign comparative analysis. One group suggested establishing a campaign "joint information center" to marshal archival and new data, and communicate with those ashore. Both groups stressed the need to invest in campaign data management and visualization.

13. This campaign would be less about exploring for the first time, and more about trying new tools, bringing new technologists to the cruise, looking at exploration in different ways, and comparing past observations with new findings at higher resolution. In short, one might design it as a scientific control for exploration.

14. The notion of a campaign designed with multiple platforms and multiple sensors in mind, even in the 2020—2025 timeframe, should be an incentive for greater OE collaboration. The SEUS campaign already brings three federal agencies together. The opportunity now exists to attract partners and cosponsors from the private sector.

Moreover, the already planned campaign follows current exploration practices and is an area with some general, baseline data. What better opportunity to involve individuals with new ideas or specific talents to witness an expedition or try out a technology? Such guest riders may not be from the usual oceanographic community. They might be medical technology developers or citizen scientists, for example.

15. Campaigns such as SEUS can bring more platforms and players together. It offers good opportunity to **integrate** telepresence on more platforms and with more participants ashore and incentivize cross communication and data/imagery sharing among platforms when two are in the campaign area at the same time. To the present, telepresence for ocean exploration has centered on one vessel of discovery but in the 2020s working synchronously from multiple platforms will surely become normal.

16. Technologies of note over the next decade: ① Exploration ships and ships of opportunity as "deployers" of autonomous vehicles; ② Long duration autonomous vehicles smart enough to adapt exploration plans on the fly; ③ **Swarms** of autonomous vehicles equipped with multibeam sonars; ④ Take advantage of geography and use shore-launched autonomous vehicles when possible; ⑤ Pre-campaign dedicated funding to assemble data already collected; ⑥ Establish a campaign "joint information center"; ⑦ Develop (little development is required) and invest in mobile telepresence units available to vessels when engaged in the campaign and in integration of telepresence teams; ⑧ Find ways to connect an exploration campaign with other exploration or science projects in the same general area in the same time frame. For example, a SEUS exploration campaign might integrate with the National Ocean Partnership Program's ADEON project, which aims to develop passive acoustic sensing in the same region.

17. Gulf of Mexico. The Gulf of Mexico **plenary** presentation had offered two target areas for the possible campaign, deep and shallow. The two Gulf groups chose to divide their efforts.

18. One group focused on the deep central Gulf in two sub-areas outside of any nation's EEZ, the so-called Doughnut Holes. EEZs from the United States, Mexico, and Cuba nearly converge, but leave two largely unexplored gaps that are roughly 3,000 meters deep. Not only is the pristine ocean bottom in these sub-areas interesting, so too is the **prevalence** of the Gulf's important Loop Current and the eddies it sheds westward into U.S. and Mexican oil and gas drilling zones.

19. The other group discussed shallower Gulf marine sanctuaries. Out for public comment is a NOAA proposal to extend the Flower Garden Banks Marine Sanctuary, located in

the Northern Gulf, eastward from Texas toward Louisiana. At the same time, better understanding similarities and linkages between the shallow Florida Keys Marine Sanctuary and the largely unexplored national marine park on Cuba's western tip attracted great interest. This discussion unexpectedly extended west along a traverse to include characterizing the Yucatan Channel as a conduit between the Caribbean and the Atlantic through which water enters and exits the Gulf.

20. Both groups stressed the international opportunity that presents itself when exploring "America's Sea", the Gulf of Mexico. In every scenario the discussants see benefit in partnering with Mexico and Cuba in campaign planning and actual expedition **excursions**. Advanced technology brought to the Gulf for demonstration during an exploration expedition would need to comply with international technology-transfer restrictions. A second common point was the potential for rich submerged archeological and historical discoveries.

21. Like the groups discussing the other geographies, the Gulf discussants urged higher-resolution bathymetry and backscatter information to characterize biological colonies and define bottom geology. They see this need being gradually addressed over time by undersea and surface autonomous vehicles. The shallower waters closer to shore offer an attractive opportunity for citizen science in the next decade. Borrowing from the SEUS discussion, shore-launched autonomous vehicles can be a bonus.

22. One technology not discussed by the other groups can fit nicely in the Gulf's geography if politics allow; namely, the dual use of undersea communication cables to host oceanographic sensors in key Gulf straits. In deeper waters, for example, when addressing the Doughnut Holes, one Gulf group called for fitting more marine mammals and other large animals with instruments as a way of supplementing gliders and other autonomous and remotely operated mechanical vehicles.

23. These approaches could be particularly exciting in shallow water campaigns in the Gulf over the next decade. The group that chose to focus on the shallower area moved away from the classic exploration model where an explorer goes to a fixed spot and attempts full characterization and imaging of bottom biology and geology and the surrounding water column. The group seized on the unknowns of water flow in and out of the Gulf and chose a less standard exploration model where the exploration "follows the water" through the Yucatan Channel. Ideas that emerged included using the rich collection of potential vessels of opportunity (commercial shipping, cruise ships, fishing vessels, and oil/gas industry vessels) as platforms to launch autonomous vehicles, moving over time toward "swarms" of cheaper/smaller/enduring/disposable sensors. Both Gulf

groups want and foresee better battery life on smaller vehicles and free-floating sensors.

24. Technologies of note over the next decade: ① Cross-strait cable systems hosting oceanographic sensors; ② Autonomous vehicles with advanced multibeam sonars; ③ Cheaper, smaller, enduring, disposable sensors that can be deployed in swarms; ④ Vehicles and sensors easily deployable by untrained crews on ships of opportunity as an extension of the expendable bathy **thermograph** (XBT) concept; ⑤ Mobilization of citizen scientists with instruments that employ a wider variety of sensors: imagers, chemical sensors, eDNA samplers and sequencers, etc. ; ⑥ Sensors fitted to large marine mammals and other large animals; ⑦ Continued development of biosampling and collection devices that consider the fragility of the samples and ambient environment in which they live; ⑧ Passive acoustic monitoring (PAM) to determine biodiversity Levels.

(2,213 words)

https://oceanexplorer.noaa.gov/national-forum/media/noef-2016-report.pdf

New Words

1. oceanographic [əʊʃɪənəʊˈgræfɪk]
   a. 海洋学的;有关海洋学的(等于 oceanographical)
   The 12-day mission is the 23rd for Atlantis, which is named for an oceanographic research vessel.
   这次为期12天的使命是"亚特兰蒂斯号"航天飞机的第23次执行任务。"亚特兰蒂斯号"是一艘以海洋命名的科研飞船。

2. consolidate [kənˈsɒlɪdeɪt]
   v. make firm or secure; strengthen; unite into one; bring together into a single whole or system 使巩固,使加强;合并,统一
   Judge Charles Schwartz is giving the state 60 days to disband and consolidate Louisiana's four higher education boards.
   查尔斯·施瓦茨法官给该州60天解散且合并路易斯安那的4个高等教育委员会。

3. littoral [ˈlɪtərəl]
   a. of or relating to a coastal or shore region 沿海的;海滨的
   The new movement and the modern climate promoted ancient karst further to grow, forming the modern littoral karst.
   新构造运动和现代天气促进古岩溶进一步发育,形成现代滨海岩溶。

4. latitude [ˈlætɪtjuːd]
   n. the angular distance between an imaginary line around a heavenly body parallel to its equator and the equator itself 纬度;纬度地区

He noted the latitude and longitude, and then made a mark on the admiralty chart.

他记下经度和纬度,然后在海图上做了标记。

5. harsh [hɑːʃ]

   a. unpleasantly stern;severe;sharply disagreeable;rigorous(环境)恶劣的,艰苦的;严厉的,残酷的

   The harsh conditions in deserts are intolerable for most plants and animals.

   沙漠中的恶劣条件对大多数植物和动物来说是无法忍受的。

6. robust [rəʊˈbʌst]

   a. physically strong;strong enough to withstand or overcome intellectual challenges or adversity;rough and crude 强健的,强壮的;(系统或组织)稳固的,健全的;(物体)结实的,坚固的;(观点等)强烈的,坚定的

   It was a typically robust performance by the Foreign Secretary.

   这是外交大臣典型的有信心的表现。

7. retrieval [rɪˈtriːv(ə)l]

   n. the act of regaining or saving something lost (or in danger of becoming lost); the cognitive operation of accessing information in memory 找回,取回;(计算机系统信息的)检索;恢复,挽回

   In a sense, forgetting is our brain's way of sorting memories, so the most relevant memories are ready for retrieval.

   从某种意义上说,遗忘是大脑对记忆进行分类的一种方式,这样联系最紧密的记忆随时可以被检索出来。

8. hover [ˈhɒvə(r)]

   v. hang in the air; fly or be suspended above; move to and fro; hang over, as of something threatening, dark, or menacing; be undecided about something; waver between conflicting positions or courses of action 翱翔,盘旋;徘徊,守候;处于不稳定状态;上下波动,左右摇摆

   No one really knows how many different faces someone can recall, for example, but various estimates tend to hover in the thousands, based on the number of acquaintances a person might have.

   例如,没人确切地知道一个人究竟能记住多少张不同的面孔,但是依据一个人可能相识的人数来判断,各类评估倾向于几千张上下。

9. verge [vɜːdʒ]

   v. border on;come close to 趋向,接近;濒临;处在边缘

   As global markets continue to see-saw on the verge of another recession, central banks across the world face a difficult balancing act of monetary policy.

   随着全球市场继续在另一场萧条的边缘摇摆不定,全球各国央行都面临着艰难的货币政策平衡行动。

10. implant [ɪmˈplɑːnt]

    v. become attached to and embedded in the uterus; fix or set securely or deeply; put firmly in the mind(尤指医学)植入,移植;灌输(观点或态度);把……嵌入,埋置

In a Mass at St Peter's, he prayed for God to "implant peace in our hearts".

圣彼得大教堂的一场弥撒中,教宗向上帝祈祷"在我们心中广植和平之信念"。

11. buoy [bɔɪ]

n. a float attached by rope to the seabed to mark channels in a harbor or underwater hazards 浮标,航标;救生圈

Flotation collars are used to buoy space capsules that land in the sea.

漂浮套管用于使降落于海面的航天舱浮于海面。

12. node [nəʊd]

n. any thickened enlargement; protect from impact; the source of lymph and lymphocytes 结点;节点;网点;结,结节;茎节;(尤指人体关节附近的)硬结

All of these are transparent to you, as you still end up with a node that has the name you want.

所有这些对于您来说都是透明的,最终得到的是一个具有您所希望的名称的节点。

13. tendril ['tendrəl]

n. slender stemlike structure by which some twining plants attach themselves to an object for support 卷须;蔓;卷须状物

The morning glory climbs the trunk with its tendril.

牵牛花用卷须攀着树干。

14. drone [drəʊn]

n. an aircraft without a pilot that is operated by remote control; someone who takes more time than necessary; someone who lags behind(遥控的)无人驾驶飞机(或导弹)

The drone can know when to bring a cup of coffee by studying a person's personal information.

通过研究一个人的个人信息,无人机可以知道什么时候该带来一杯咖啡。

15. mow [məʊ]

v. cut with a blade or mower 刈,割(草)

He continued to mow the lawn and do other routine chores.

他继续给草坪割草,并做其他日常杂务。

16. hazardous ['hæzədəs]

a. involving risk or danger 危险的,有害的;碰运气的

The report calls for a ban on the import of hazardous waste.

这篇报道呼吁禁止危险废弃物的进口。

17. deploy [dɪ'plɔɪ]

v. to distribute systematically or strategically; place troops or weapons in battle formation 部署,调度;利用

It also pledged not to deploy AI whose use would violate international laws or human rights.

它还承诺不会部署违反国际法或人权的人工智能。

18. bight [baɪt]

n. a broad bay formed by an indentation in the shoreline; a loop in a rope; a bend or curve (especially in a coastline) 海湾,绳圈;曲线

The threatened Australian sea lion is found only in the Great Australian Bight, which arcs around the southern shore of the continent.

危险的澳大利亚海狮只在环绕欧洲大陆南海岸的大澳洲湾出没。

19. discrete [dɪˈskriːt]

a. constituting a separate entity or part 分离的

A new study suggests that the conversational pace of everyday life may be so brisk it hampers the ability of some children for distinguishing discrete sounds and words.

一项新的研究表明,日常生活中的对话节奏可能过于轻快,以至于阻碍了一些孩子区分不同的声音和单词的能力。

20. seep [siːp]

n. a leak through small openings(油、水)渗出地表的地方

A seep is a moist or wet place where water, usually groundwater, reaches the earth's surface from an underground aquifer.

渗漏是指水(通常是地下水)从地下蓄水层到达地表的潮湿或潮湿的地方。

21. bathymetry [bəˈθɪmətri]

n. measuring the depths of the oceans 海洋测深学;深度测量法

Laser sounder bathymetric system is of great importance to be capable of charting coastal bathymetry rapidly and cheaply.

激光海洋测深仪是一种快速便宜的海深勘测系统,对绘制沿海区域海图有重大意义。

22. acoustic [əˈkuːstɪk]

a. of or relating to the science of acoustics(爆破性水雷或其他武器)能为声波引爆的;(装置,系统)利用声能运作的;声音的,听觉的

The mechanism of this instability is based on the cycle of acoustic waves between their corotation radius and the shock.

这种不稳定性的机制是基于声波在其同心圆半径和冲击之间的循环。

23. suffice [səˈfaɪs]

v. be sufficient; be adequate, either in quality or quantity 足够,足以;满足……的需求;有能力

Joseph's top priority became getting his hands on enough grass-fed milk to keep customers satisfied, since his own 64-cow herd wasn't going to suffice.

约瑟夫的首要任务是弄到足够的草饲牛奶来满足顾客的需求,因为他自己64头牛的牧群还不足以满足。

24. discern [dɪˈsɜːn]

v. detect with the senses(艰难地或努力地)看出,觉察出;了解,认识

It is often difficult to discern how widespread public support is.

了解公众支持的广泛程度常常是困难的。

25. artifact [ˈɑːtɪfækt]

n. a man-made object taken as a whole(尤指有文化价值或历史价值的)人工制品,历史文物;非自然存在物体,假象

The Clovis point may be the most analyzed artifact in archaeology.

克洛维斯矛头可能是考古学中被分析得最多的手工艺品。

26. frugally [ˈfruːɡəli]

ad. in a frugal manner 节约地;节省地

The more we learn how to live frugally, the more satisfied with our lives.

我们越是学会了过简朴的生活,就越容易对自己的生活感到满足。

27. craggy [ˈkræɡi]

a. rocky and steep; having hills and crags 崎岖的;多峭壁的

The crowd swarmed ashore and soon the forest distances and craggy heights echoed far and near with shouts and laughter.

人群向岸上涌去,不久,树林中、崎岖的高崖处都回荡着他们的喊叫声和笑声。

28. canyon [ˈkænjən]

n. a ravine formed by a river in an area with little rainfall(两边为峭壁、谷底通常有溪流的)峡谷

The Grand Canyon is one of the natural wonders of the world.

科罗拉多大峡谷是世界自然奇观之一。

29. hypothesize [haɪˈpɒθəsaɪz]

v. to believe especially on uncertain or tentative grounds 假设,假定

To explain this, they hypothesize that galaxies must contain a great deal of missing matter which cannot be detected.

为了解释这一点,他们假定银河系一定包含了大量无法探测到的不明物质。

30. adaptive [əˈdæptɪv]

a. having a capacity for adaptation 适应的,有适应能力的

Societies need to develop highly adaptive behavioural rules for survival.

社会要生存需要建立有高度适应性的行为准则。

31. archival [ɑːˈkaɪvəl]

a. of or relating to or contained in or serving as an archive 档案的

This image is a composite of archival Hubble data taken with the WideField Planetary Camera 2 and the Advanced Camera for Surveys.

这幅图像是由哈勃存档数据合成而来,这些数据来自宽视野2号行星照相机和高级测量照相机。

32. integrate [ˈɪntɪɡreɪt]

v. make into a whole or make part of a whole; become one; become integrated; calculate the integral of; calculate by integration(使)合并,成为一体;(使)加入,融入群体;(使)取消种族隔离;求……的积分;表示(面积、温度等)的总和,表示……的平均值

He didn't integrate successfully into the Italian way of life.

他没有成功融入意大利的生活方式中去。

33. swarm [swɔːm]

　　n. a moving crowd;a group of many insects 一大群(移动中的昆虫);(移动着的)一大群人;(多指发生在火山附近的)地震群;(天文)一大群小型天体同时在空中出现

　　Stroll around any Mexican city for a while and you will notice a background hum like a swarm of angry bees.

　　在任何一个墨西哥城市转一圈,你都会注意到周围的嗡嗡声,就像一群愤怒的蜜蜂。

34. plenary ['pliːnəri]

　　a. full in all respects 充分的;全体出席的

　　The meeting today is of special significance because it marks CD's 1,000 th plenary meeting.

　　今天召开的是裁谈会第 1 000 次全会,具有特殊意义。

35. prevalence ['prevələns]

　　n. the quality of prevailing generally;being widespread 流行,盛行

　　The prevalence of several silkworm diseases has led to a decline in silk products. Especially in France, the silk industry has never recovered.

　　几种桑蚕疾病的流行导致了丝绸产品下滑。特别在法国,丝绸工业再也没有恢复。

36. scenario [sə'nɑːriəu]

　　n. a postulated sequence of possible events;a setting for a work of art or literature;an outline or synopsis of a play (or, by extension, of a literary work) 设想,可能发生的情况;(电影、戏剧等的)剧情梗概;(艺术或文学作品中的)场景

　　The worst-case scenario is that an aircraft will crash if a bird destroys an engine.

　　如果小鸟毁坏了一部发动机,最坏的情形是飞机坠毁。

37. excursion [ɪk'skɜːʃn]

　　n. a journey taken for pleasure;wandering from the main path of a journey 短途旅行,远足;涉猎;移动;游览团,远足队;离题;偏移,偏差

　　Another pleasant excursion is Matamoros, 18 miles away.

　　另一个怡人的短程旅游是 18 英里外的马塔莫罗斯。

38. thermograph ['θɜːməʊˌgrɑːf]

　　n. a thermometer that records temperature variations on a graph as a function of time 温度记录器;热录像仪

　　He invented thermograph and uncomplicated telescope and microscope.

　　他发明了温度计以及简单的望远镜和显微镜。

Phrases and Expressions

an area of 一个……的区域;一个……的地区
be available to 可被……利用或得到的
in darkness 在黑暗中
be home to 为……的所在地

in short 总之；简言之
comply with 照做，遵守
over time 随着时间的过去；超时
in and out of 在……内外
fit to 使适应；调整到

Terminology
80 degrees north latitude 北纬 80 度
open water 公海；开阔水面；无冰水面
eDNA 不同生物的基因组 DNA 的混合[类似于对环境的生物多样性调查，方法是检测环境样品中尽可能多的 DNA 序列，与数据库比对分析它们所属的物种分类信息，最终鉴定在这个环境中生活的所有物种，这个方法又被称为 DNA 宏条形码（DNA metabarcoding）]

Proper Names
the National Oceanic and Atmospheric Administration (NOAA) 国家海洋和大气管理局
marginal ice zone (MIZ) 边缘冰区
the Baltimore Canyon 巴尔的摩峡谷
the Blake Plateau 布莱克高原
request for proposals (RFP) 征求建议书
AUV/ASVs 自主式水下/航空交通工具（Autonomous Underwater/Space Vehicle）
remotely operated underwater vehicles (ROVs) 遥控水下航行器
EEZ 专属经济区（Exclusive Economic Zone）
the Doughnut Holes 甜甜圈漏洞（指加入美国联邦医疗保险的人不得不支付某种药物费用的情况，因为这种药物的费用介于他们的医疗保险计划所支付费用的两个水平之间）
cruise ships 游轮，游艇
XBT 投弃式温深仪（expendable bathythermograph）
Passive acoustic monitoring (PAM) 被动声学监测
biodiversity Levels 生物多样性水平

一、翻译策略/方法/技巧：归化

  归化的翻译方法，初见于 19 世纪的德国思想家与哲学家弗里德里希·丹尼尔·施莱尔马赫（Friedrich Daniel Ernst Schleiermacher，1768—1834）在《论不同的翻译策略》中的精确总结，颇近似于异化和归化："翻译的方法不外乎两种：要么尽可能不去打扰作者，让读者向作者靠拢（异化）；要么尽可能不打扰读者，让作者向读者靠拢（归化）"。
  晚清民初之际，中国翻译文学的主流策略是归化。20 世纪 30 年代，留日前期的鲁迅裹挟于归化的时代主潮之中，在他的前期译文中，如《月界旅行》《斯巴达之魂》等译作，域外文学文化儒家化、道家化乃至法家化的痕迹相当明显，经过鲁迅的改译已很难见到翻译的痕迹。鲁迅先生已经开始较系统地阐述"归化"与"异化"，他认为"凡是翻译，必须兼顾着两面，一则当然力求其易懂,一则保存着原作的风姿,但这保存,却又常常和易懂相矛盾……"。译文符合译入语文化规范则属归化。

那么,"归化"指什么呢?归化翻译法旨在尽量减少译文中的异国情调,为目的语读者提供一种自然流畅的译文。美籍翻译研究学者劳伦斯·韦努蒂(Lawrence Venuti)对"归化"与"异化"进行了系统的讨论,英语单词"domestication"指"归化"。在他看来,所谓"归化",就是指译者在翻译过程中采用一种透明的、流畅的风格,从而最大限度地降低目标语读者对外国文本的陌生感。韦努蒂指出,归化翻译法是英美文化社会中占主导地位的翻译策略。

归化翻译法通常包含以下几个步骤:① 谨慎地选择适合于归化翻译的文本;② 有意识地采取一种自然流畅的目的语文体;③ 把译文调整成目语篇体裁;④ 插入解释性资料;⑤ 删去原文中的字面表层意义,寻找贴近目标语的译文;⑥ 协调译文和原文中的观念与特征。

我们从下面的例子来看归化翻译法:

例1:爱德华·萨义德撰写的书名 On Late Style: Music and Literature Against the Grain 的翻译

台湾学者做了"归化"的处理,译为《论晚期风格:反常合道的音乐与文学》,"反常合道"为苏东坡的诗学理论,把它作为典故用在了书名翻译上是归化的准确处理。

例2:弱水三千,只取一瓢

原文:Driving back to my home in Sherbrooke Street was an endless journey of destructive emotions and thoughts. In a truck-stop restaurant, I sat staring at a glass of cheap red wine. Of all the gin joints in all the towns in all the world, she walks out of mine.

归化翻译:在我开车回到布鲁克大街的路上,我陷入了无尽的悲伤之中。随后,我来到一家汽车旅馆,端着一杯廉价红酒出神,弱水三千,终究我已不是她的那一瓢水了。

例3:Once the wife of a parson, always the wife of a parson.

归化翻译:嫁鸡随鸡,嫁狗随狗。

例4:One swallow does not make a summer.

归化翻译:一燕不成夏。

例5:The fox may grow grey, but never good.

归化翻译:江山易改,本性难移。

例6:New brooms sweep clean.

归化翻译:新官上任三把火。这句话的字面意思是"新扫把扫得干净",比喻"新任职的人干得好"。和汉语中的"新官上任三把火"有异曲同工之妙。

(部分摘自:刘敬国,何刚强.翻译通论[M].北京:外语教学与研究出版社,2011:112-113.)

二、译例与练习

Translation

1. An area of shallow and deep water mostly ice-covered in winter with a marginal ice zone (MIZ) and ice edge year-round, it is tectonically and geologically not well understood, and its changing physical oceanography demands characterization. (Para.2)

浅水区和深水区在冬季大多被冰覆盖,全年都有边缘冰区(MIZ)和冰缘,这种现象在构造和地质学上尚无法得到解释,而且其不断变化的海洋学上的外在特性也需要明确。

2. Such an exploration/observation network of autonomous devices will have to be supported by a previously installed and complementary network of navigation (since GPS will not be available), refueling/repowering, and communications nodes. (Para. 3)

这种自主设备的探索或观察网络必须得到先前安装的互补导航网络(因为 GPS 不可用)、加油或重新供电和通信节点的支持。

3. Future exploration vehicles and instruments will need to employ adaptive sampling software wherein an instrumented drone can learn as it goes, find or change its way and sample where it finds the water column or bottom most interesting. (Para. 4)

未来的勘探车辆和仪器将要采用自适应采样软件,其中装有仪表的无人机可以边走边学习,自主寻找或改变其路径,并在它发现最感兴趣的水柱或底部的位置进行采样。

4. For the technical, safety, and environmental reasons already stated, and because the area has potential for international competition, is home to several endangered species, and is commonly used for traditional purposes by First Nation peoples, exploration campaigns will require extensive planning. (Para. 5)

出于已加以说明的技术、安全和环境原因,且该地区具有国际竞争的潜力,是几种濒危物种的家园,并且通常被原住民用于传统用途,勘探活动将需要进行广泛的规划。

5. Unlike the other regional campaigns in the 2022 Forum, the federal government is already planning a SEUS campaign, for the period 2022—2023, with funding commitments under consideration and ship time in negotiation among Bureau of Ocean Energy Management (BOEM), USGS, and NOAA, and thus the groups considered the near term as well as 2025—2030. (Para. 7)

与 2022 年论坛中的其他区域活动不同,联邦政府正在规划 2022 年至 2023 年期间的 SEUS 活动,这项活动得到了海洋能源管理局(BOEM)、美国地质勘探局(USGS)和美国国家海洋和大气局(NOAA)的资助。他们还考虑了近期以及 2025—2030 年的规划。

6. As one thinks about using emerging autonomous vehicles or present-day remotely operated underwater vehicles (ROVs) in areas of suspected seeps and deep sponge/coral habitats, one needs either higher-resolution bathymetry to help prevent costly AUV/ROV collisions with craggy canyon features and direct the vehicles toward suspected or hypothesized targets, or the deployment of adaptive, deep-diving autonomous vehicles that can make decisions on fly to avoid danger, search for the most interesting features, and find recharging, communication, and navigation nodes. (Para. 11)

当人们考虑在疑似渗漏和深层海绵或珊瑚栖息地区域使用新兴的自动驾驶车辆或远程操纵潜水器时,需要更高分辨率的水深测量,以防止自主式水下交通工具或远程操纵潜水器与崎岖的峡谷地貌发生碰撞,还需将车辆引向可疑或假设的目标,或部署自适应、深潜的自动驾驶车辆。这些车辆可以在飞行中做出决策以避免危险,还可以检索那些最引人注目的特征,并找到充电、通信和导航节点。

【这个句子结构十分复杂,属于典型的英文长句。英语中常用的表示句子成分之间主次关系的语法手段几乎全都出现了,集中体现了英语句子侧重形合的语言特点。在翻译英文长句时,最首要的是要分清句子内部的主次关系,也就是说要将主干和次要部分界定清晰。英语长句的次要部分结构往往比较复杂,接下来就需要将诸个次要部分之间的逻辑关系搞清楚,这也往往是翻译中的难点。句子的主干部分其实很简短"one needs either higher-resolution bathymetry to"。动词不定式之后并列了两个动词"help"和"direct"。动词"help"后面的宾语是"collisions",而"direct"后面的宾语比较复杂一些,分别是"vehicles"和"deployment"。名词词组"adaptive, deep-diving autonomous vehicles"后面又接用了一个定语从句来具体说明这种自动驾驶车辆的主要功用。总体来看,次要部分依旧是采用了常见的"套句"的方式来组织的。搞清楚了主次部分以及次要部分之间的逻辑关系,接下来就需要将这些成分拆解成对应的中文小句,然后根据它们之间的逻辑关系和中文的表达习惯重新进行组织。上述分析、理解和调整的过程也是翻译英语长句时的一般方法。最后一点需要特别说明的是,该句中有很多专业术语,不要望文生义、表面化翻译,一定要通过查阅专业资料,译为最标准的译文。这也是在翻译此种应用文时需要注意的问题。】

7. To complete such comparative analysis, campaign sponsors need to establish a "dedicated fund" for a pre-campaign effort to gather archival oceanographic/bathy metric data in the areas of highest interest, at all resolutions and wavelengths, and make that data available to explorers for operational planning and post-campaign comparative analysis. (Para.12)

为完成此类比较分析,活动赞助商需要为活动前的工作建立"专项基金",以收集最感兴趣区域所有分辨率和波长的海洋学或水深测量档案方面的数据,并将这些数据提供给探险者用于运行规划和运转后的比较分析。

【该句在结构上稍显复杂之处在于句子的后半段,也就是"a pre-campaign effort to"后面的部分。这一部分实际上是两个动词的并列结构,分别是"gather"和"make"。作者又在两个动词后面分别使用了名词词组、介词词组及形容词词组,从而使得整个句子显得比较复杂。翻译过程中,还要注意"make"一词的处理。习惯上比较容易译成"使……"。其实这是一种"黔驴技穷"式的翻译方法。虽不能说是误译,但也绝算不上佳译。实际翻译过程中,可根据原文语境适当做些转换,像参考译文中,就将其译成"将……提供给",既不落俗套,又通顺达意,两全其美。】

8. This campaign would be less about exploring for the first time, and more about trying new tools, bringing new technologists to the cruise, looking at exploration in different ways, and comparing past observations with new findings at higher resolution. (Para.13)

此次活动将不再是第一次探索,而是更多地尝试新工具、将新的技术人员带到邮轮上,并以不同的方式看待探索,还要将过去的观察结果与更高分辨率的新发现进行比较。

9. It offers good opportunity to integrate telepresence on more platforms and with more participants ashore and incentivize cross communication and data/imagery sharing among platforms when two are in the campaign area at the same time. (Para.15)

它提供了一个很好的机会,可以将远程呈现集成到更多平台上,让更多参与者上岸,并在两个平台同时位于活动区域时激励平台之间的交叉通信和数据或图像共享。

10. In deeper waters, for example, when addressing the Doughnut Holes, one Gulf group called for fitting more marine mammals and other large animals with instruments as a way of supplementing gliders and other autonomous and remotely operated mechanical vehicles. (Para. 22)

在更深的水域，例如在解决"甜甜圈漏洞"问题时，一个海湾组织呼吁为更多的海洋哺乳动物和其他大型动物配备仪器，作为对滑翔机和其他自主和遥控机械车辆的补充。

Exercises

1. Fill in the blanks with the proper given words, and then translate the sentences into Chinese.

autonomous   redefine   infrastructure   in accordance with   collaboration   sanctuary

1) The task facing medicine in the twenty-first century will be to _____ its limits even as it extends its capacities.

2) Simon Blackmore, who researches agricultural technology at Harper Adams University College in England believes that fleets of lightweight _____ robots have the potential to solve this problem.

3) After the fall of Rome, the first European society to regulate behaviour in private life _____ a complicated code of etiquette was twelfth-century Provence, in France.

4) The government built the bridge directly from the airport to the Songdo International Business District, and the surface _____ was built at the same time as the new airport.

5) It will decide on the Weddell Sea _____ proposal at a conference in Australia in October, although a decision on the peninsula sanctuary is not expected until later.

6) Ordinary search techniques proved unsatisfactory, so McKee entered into a _____ with Harold E. Edgerton, professor of electrical engineering at the Massachusetts Institute of Technology.

2. Translate the following sentences into Chinese.

1) Simulation results show the stability of the MAUV adaptive formation control algorithm and its significance in the field of MAUVS formation research and many ocean exploration applications.

2) As a mobile ocean platform, and owing to its high positioning capacity and good operation stability, the self-elevation drilling platform takes a leading role in continental shelf exploration.

3) Through other activities, NOAA Ocean Exploration contributes to the protection of ocean health, understanding of climate change impacts on the ocean and our planet, sustainable management of our nation's marine resources, acceleration of our national

economy, and the building of a better appreciation of the value and importance of the ocean in our everyday lives.

4) So in order to overcome the defect of the exploration vehicle with man, it is necessary to develop an intelligent ocean environment resource exploration unmanned surface vehicle.

5) From the viewpoint of resource demand, ocean multi-wave exploration is a useful technology to resolve the difficult problem just as blur zoon, structure distortion and gas pollution etc.

6) In recent years, the field such as underwater navigation ocean exploration locating of underwater object and underwater communication develop very fast.

7) South China Sea is a marginal sea in western Pacific Ocean, and has very good accumulation conditions and exploration prospect for natural gas hydrate.

8) As a new kind of autonomous underwater vehicle (AUV), an underwater glider has valuable applications in oceanographic survey or in ocean exploration.

9) A teleoperator is a kind of robot which be widely used in space exploration, ocean exploration, working in dangerous environments and so on.

10) It researches the design and development of ocean environment exploration unmanned surface vehicle system based on ARM.

3. Translate the following sentences into English.

1) 我们可以在遥远的冰冷卫星上探测到地下海洋，甚至可以探测到相对论引力在太空中产生的微小涟漪。

2) 从毕达哥拉斯到赫伯特·斯宾塞，每个人都做到了，尽管科学通常探索的是一片海洋，它更愿意将其视为统一体或宇宙，并称之为秩序。

3) 由于您的肺部在太空或水下感受到的压力不同，液体通气在探索宇宙深处或海洋深处时可能很有用。

4) 据全球海洋探索基金会称，"地球上大约四分之三的火山活动实际上发生在水下"。

5) 海底仪器监测鲸鱼叫声、浮游生物密度、温度、酸度、氧合和各种化学浓度。这些传感器可以安装在剖面浮标上，这些浮标在大约1 000米的深度自由漂移。

6) 让我们一起探索星空，征服沙漠，根除疾病，探索海洋深处，鼓励艺术和商业。

7) 坚持创新驱动发展，在人工智能、量子信息、集成电路、生命健康、脑科学、农业、航天科技、地球深海探测等领域实施一批战略项目。

8) 我们一贯反对任何其他国家在中国管辖的海域进行油气勘探开发活动，希望有关外国公司不要卷入南海争议。

9) 英国石油公司勘探和开采石油的总执行官道格·萨特斯表示，他的公司正用可远程操作的机械来尝试弄清到底有多少石油已泄漏到海洋中。

10) 和太空探索一样，深海探索也需要新的仪器和技术。太空是寒冷的真空，而海洋深处虽然很冷，但压力很大。

4. Choose the best paragraph translation. And then answer why you choose the first translation or the second one.

随着海洋信息化的发展和各类探测设备的不断进步,海洋数据体量已经呈现出爆炸性增长的状态,海洋领域已经进入了大数据时代。世界各国都在积极推动海洋领域的发展,建立起覆盖全国甚至全球的海洋观测网络,通过形式多样的探测设备获取海洋实时的数据,形成数量庞大的海洋数据库。庞大的数据体量向从海量数据中挖掘信息的技术提出了更高的挑战,如何提取更有用的信息,并将其应用于海洋科学的各个领域,已经成了未来海洋科学发展的重点问题。

译文一:

With the development of Marine informatization and the continuous progress of all kinds of detection equipment, the volume of Marine data has shown explosive growth, and the marine field has entered the era of big data. All countries in the world are actively promoting the development of the marine field, establishing a nationwide and even global marine observation network, obtaining real-time marine data through various forms of detection equipment, and forming a large number of marine databases. The huge amount of data poses a higher challenge to the technology of mining information from massive data. How to extract more useful information and apply it to various fields of Marine science has become a key issue in the development of marine science in the future.

译文二:

As scientists invented various detection equipment with the development of marine information and the continuous progress, the volume of marine data has shown an explosive growth, and the marine field has entered the era of big data. All countries in the world are actively promoting the development of the marine field, and they are establishing an ocean observation network covering the whole country and even the whole world, acquiring real-time ocean data through various forms of detection equipment, and forming a huge number of marine databases. The huge volume of data poses a higher challenge to the technology of mining information from massive data. So how to extract more useful information and apply it to various fields of marine science has become a key issue in the future development of marine science.

5. Translate the following passage into Chinese.

Assuming an unusually rich general baseline of information in this particular campaign area, the campaigners have an opportunity to compare archived data with newly collected higher-resolution data. Determining how much we really did or didn't know and what has changed could help better design future campaigns. To complete such comparative analysis, campaign sponsors need to establish a "dedicated fund" for a pre-campaign effort to gather archival oceanographic/bathy metric data in the areas of highest interest, at all resolutions and wavelengths, and make that data available to explorers for operational planning and post-campaign comparative analysis. One group suggested establishing a campaign "joint information center" to marshal archival and new data, and communicate with those ashore. Both groups stressed the need to invest in campaign data management and visualization. This campaign would be less about exploring for the first time, and more about trying new tools, bringing new technologists to the cruise, looking at exploration in different ways, and comparing past observations with new findings at higher resolution. In short, one might design it as a scientific control for exploration.

6. Translate the following passage into English.

海洋占地球总面积的70%，占地球生存空间的90%以上。事实上，在广袤的海洋世界里，海洋影响陆地上的气候和天气现象，大量海洋动植物是人类重要的食物来源。海洋在世界各大洲和国家之间航行，世界上大约一半的人口生活在沿海地区。因此，了解世界海洋对我们极为重要。地球海洋的变化将直接影响我们在陆地上的生活，因此我们有必要在早些时候发现这种变化，除了增加我们对海洋的了解以求安全和经济利益之外，探索洋底也将满足人类好奇心和对探索未知的渴望。

三、翻译家论翻译

梁实秋，号均默，原名梁治华，字实秋，著名学者、散文家、翻译家、文学评论家。作为翻译家，他一生笔耕不辍，译作不断，一生坚持将西方一流文学名著介绍到中国，是我国第一位独立完成《莎士比亚全集》翻译的翻译家。

梁实秋1903年1月生于北京，1915年考入清华留美预备校，1923年8月赴美留学，专攻英语和英美文学。1926年夏回国于南京东南大学任教，先后任暨南大学、青岛大学、北京大学、北京师范大学等校外文系教授、系主任。1948年移居香港，1949年到台湾。1987年11月3日因心脏病病逝于台北。

梁实秋在工作之余，积极投身文学翻译工作。1928年年初，他与叶公超、徐志摩等共同创办杂志《新月》，并有大量译作在该杂志上发表，其中著名的有《阿伯拉与哀绿绮思的情书》和罗伯特·彭斯的叙事诗《汤姆·奥桑特》。1958年，他翻译了《沉思录》。1967年，梁实秋终于完成了《莎士比亚全集》的全部翻译工作，包括37部戏剧和3部诗作，共计400余万字。此外，梁实秋的著名译著还包括《英国文学史》及《世界名人传》。

梁实秋的翻译理论著述不多，其翻译思想散见于其散文及20世纪30年代的翻译论战。此外，在翻译莎翁巨著的艰苦岁月里，他不断从实践中总结翻译经验，为后来的译者提供了宝贵的经验借鉴。

1. "信"与"顺"辩证统一的翻译标准

一方面，梁实秋强调"信"，即"存真"，充分忠实原文，尊重原作者。他指出，"翻译的目的是要把一件作品用另一种文字忠实表现出来，给不懂原文的人看"。另一方面，梁实秋也强调"顺"的重要。"顺"即"通顺"，他认为，译者可以适时跳出原文句法，使译文更加自然流畅。

梁实秋指出,译者应兼顾忠实与通顺,在"信"与"顺"之间做好权衡,尽力再现作品原貌。

2. 谨慎的翻译选择与严谨的翻译态度

首先,梁实秋坚持译介世界一流名著。翻译涉及的第一个问题是翻译文本的选择。他曾说:"愚以为有学术性者,有永久价值者,为第一优先。"其次,翻译版本选择上,梁实秋反对转译。梁实秋曾形象地把原著比作美酒,他认为,译作无论多好,与原著相比,总仿佛是"掺了水或透了气的酒一般,味道多少变了,如果转译,气味就更大了"。此外,梁实秋坚持翻译与研究结合的严谨翻译态度。文学翻译不仅是语言的转换,更与文化紧密相连,翻译态度是保证译作质量的前提。

3. 异化和归化结合的莎翁翻译策略

在语言与文化层面,他更多采用异化策略。例如,翻译人名、地名等专有名词时,他不赞成误导读者的中国本土式翻译,主张"音译"。若人名体现人物某种特征时,他会适当归化,在翻译的人名中传达该信息,便于读者了解人物的特征。又如,莎剧中有很多双关语、俚语和俗语,几乎无法翻译,他采用异化策略,然后辅以注释,采用异质文化丰富汉语。梁实秋充分考虑读者,采用异化和归化结合的翻译策略,使译文更加流畅,符合本国读者表达习惯,提高译作的可读性和欣赏性。

(摘自:https://zhuanlan.zhihu.com/p/447571711)

Text B

The Ocean Unexplored and Unprotected
By Emily Petsko

Ideological and Political Education：探海格局形成 展中国海洋探索成就

经过多年努力，中国业已形成了由"潜龙""海龙""蛟龙"构建的"三龙"探海格局，并且不断提升到新水平，折射出中国海洋科技装备领域突飞猛进的发展，为人类认识海洋、开发、利用海洋作出了重大贡献。

1. "潜龙"纵横海底平原和山川

潜龙三号长3.5米、高1.5米、重1.5吨，外形像一条橘黄色的胖鱼。这条"胖黄鱼"主要应用于深海资源环境勘查，从类别上来说属于"自主无人潜水器"，具备强大的探测功能。它除了能够进行海底微地貌成图、温盐深探测、甲烷探测等之外，还具备浊度探测、氧化还原电位探测、海底照相以及磁力探测等诸多探测功能。作为"自主无人潜水器"，潜龙三号是名副其实的水下智能机器人，不仅可以灵活自如地畅行水下，而且凭借其先进的探测装备，完成多样复杂的科考任务。

2. "海龙"海底取样手到擒来

海龙系列潜水器是一种遥控水下机器人，擅长局部作业、定点精细探测，可以进行大功率、长时间的水下工作，在海底硫化物矿区勘查中被广泛使用，典型应用包括矿区调查、矿区成图、勘查取样。海龙Ⅲ是6 000米勘查取样型无人缆控潜水器，最大作业水深6 000米，作业功率170马力，具备海底自主巡线能力以及更强的推力、高速和重型设备搭载能力，以支持搭载多种调查设备和重型取样工具，达到了世界先进水平，在深海探测领域具有广阔的应用前景，可成为海洋观测、深海取样、获取并传递深海信息的良好作业平台。海龙11 000万米级深海无人遥控潜水器，设计最大工作深度1.1万米。随之开发的万米深水电池、可加工浮力材料、陶瓷浮力球、光纤微细缆、多芯贯穿件等，均为中国首创成果的部件。

3. "蛟龙"志在载人全海深无禁区

在中国深潜器"三龙"中，最引人瞩目的莫过于"蛟龙"了，不仅在于其体积显得更大，外形与众不同，更在于其有"载人"功能，需要潜航员深海驾驶，可以给科学家群体等提供进入深海开展科考活动的难得机遇。"蛟龙家族"中最耀眼的"明星"无疑是蛟龙号，它是一艘由中国自行设计、自主集成研制的载人潜水器，也是"863"计划中的一个重大研究专项，其设计最大下潜深度为7 000米，是目前世界上下潜最深的作业型载人潜水器。

1. Scientists have successfully photographed a black hole, landed **rovers** on Mars, and sent spacecraft to the dark side of the moon. Yet, one of the last unknown frontiers, and one of the most **deceptively** familiar, is on our very own planet. More than 80% of the ocean remains unexplored. And because it's difficult to protect what we don't know, only about 7% of the world's oceans are **designated** as marine protected areas (MPAs). With this in mind, we explain why a body of water that covers most of Earth's surface is also one of the most vulnerable, and least understood, places in the universe.

2. Under Pressure. One of the biggest challenges of ocean exploration comes down to physics. Dr. Gene Carl Feldman, an **oceanographer** at NASA's Goddard Space Flight Center, explains that the ocean, at great depths, is characterized by zero visibility, extremely cold temperatures, and crushing amounts of pressure. "In some ways, it's a lot easier to send people into space than it is to send people to the bottom of the ocean," Feldman told Oceana. "The intense pressures in the deep ocean make it an extremely difficult environment to explore."

3. Although you don't notice it, the pressure of the air pushing down on your body at sea level is about 15 pounds per square inch. If you went up into space, above the Earth's atmosphere, the pressure would decrease to zero. However, if you went diving or **hitched** a ride in an underwater vehicle, those forces would start to stack up the further down you went. "On a dive to the bottom of the Mariana Trench, which is nearly 7 miles deep, you're talking about over 1,000 times more pressure than at the surface," Feldman said: "That's the **equivalent** of the weight of 50 jumbo jets pressing on your body." Of course, human-occupied submersibles aren't the only way to explore and study the ocean. We can even learn some lessons from space. Feldman specializes in satellite technologies that record the color of the ocean as a means of measuring the distribution and abundance of phytoplankton, which can change rapidly and even double in a day.

4. When these technologies were first used in the late 70s, satellites were able to capture detailed images of the ocean within minutes, while it would take a ship 10 years of continuous sampling to collect the same number of measurements, according to Feldman. With that said, some things are better measured in the water, however difficult it may be to get there.

5. Bearing Witness. Ocean exploration technologies have come a long way. Floats and drifters, devices that rely on ocean currents to carry them while they collect data, have been **complemented** in recent years by an ever-sophisticated fleet of underwater vehicles. This can include human-occupied vehicles (HOVs), remotely-operated ones (ROVs),

and autonomous and hybrid ones. In a recent webinar hosted by the Woods Hole Oceanographic Institution, film director and ocean enthusiast James Cameron called for an "all-of-the-above" approach, but also **highlighted** the value of continuing to **plunge** people into the great unknown. In 2012, Cameron set a record when he visited the Mariana Trench, the deepest part of the ocean, in a "vertical torpedo" sub. "I call it bearing witness," Cameron said. "There's something very exciting about being physically present and using all of your senses. Plus, you can come back and tell the story, and that engages an audience. The most important aspect of exploration, in my mind, is coming back and telling the tale."

6. Oceana uses a combination of technologies on its **expeditions**, which have charted previously unexplored waters, including areas off of Southern California, several **seamounts** in the Canary Islands, and a deep trench south of Malta. Another seamount near Morocco, which was also previously unexplored, led to the discovery of a deep-sea coral reef, the only one of its kind that is still growing in the Mediterranean Sea. In Europe's biodiversity hotspots, expeditions resulted in the first-ever record of a live brown-snout spookfish, as well as the documentation of two starfish species, one black coral, and one stony coral that were previously thought to only live in the Atlantic Ocean.

7. And in Chile, following Oceana expeditions which recorded the rich and unique marine life in the Desventuradas and Juan Fernández islands, the government was persuaded to make those places marine parks. A multiple-use MPA was also established in the Caleta Tortel **commune** following multiple Oceana expeditions. While the benefits are undeniable, expeditions are expensive, and the lack of detailed maps and data make them all the more challenging. Ricardo Aguilar, the leader of Oceana's expeditions in Europe, said they can't bank on **bathymetric** information-which can serve as a guide to an area's underwater **terrain**, because in most cases it doesn't exist. "We only have good information on less than 5% of the world's oceans, and maybe sparse information on another 10%," Aguilar said, "Therefore, how can we protect areas where we have no clue what is there?"

8. Herein lies the catch-22: We need exploration to collect more information, but many agencies around the world are reluctant to fund projects where there are too many unknowns. "There seems to be an increasing trend towards avoiding risk at all costs, which means that you often have to prove that you know all the answers before you can even start your investigation," Feldman said. "But this is the wrong approach. Science is not only about having the answers. Science is really about asking the questions."

9. Evidence, not Excuses. Oceana has protected nearly 4 million square miles of ocean to date, and expeditions have been crucial to this success. Expeditions yield photographs, video **footage**, scientific data, and narratives that can all be used to **bolster** the case for new or expanded protections. "By exploring previously unexplored areas, we have been able to discover new species and new habitats, but also to identify vulnerable habitats or threatened species that were protected 'on paper,' but because nobody knew they could be found in these places, there were no measures to effectively protect them," Aguilar said. "One of the most common excuses governments have used for not taking action is the lack of information to choose which areas to protect and how to manage them. Also, the opposition of different stakeholders to creating new MPAs was due to this lack of data."

10. Oceana **endorses** plans to protect 30% of the ocean by 2030, a goal known as 30×30. While the world still has a long way to go, continued ocean exploration can supply the evidence needed to protect the ocean and the many resources it provides.

11. Even though data is useful, Feldman says a **compelling** case for MPAs can still be made when the **reverse** is true. Because scientists do not have a complete understanding of how one change to the ocean affects the entire ecosystem, and which of those changes may be the tipping point that causes collapse, it is common sense that areas should be designated for protection and further research. "Since we don't know how all the pieces fit together, setting aside areas where we just say, 'We're leaving these alone' or 'We're going to have minimal **intervention**' is perhaps the safest thing to do until we know better," Feldman said. "The idea of setting aside areas that are environmentally important and unique is probably a really smart move until we can get smarter about how we manage the ocean."

(1,249 words)

https://oceana.org/blog/why-does-so-much-ocean-remain-unexplored-and-unprotected/

New Words

1. rover [ˈrəʊvə(r)]
    n. someone who leads a wandering unsettled life; a planetary detector 漫游者，流浪者；探月车，(行星等)探测器
    The top priority is no longer freeing the rover from the sandpit, but getting its solar panels pointed more to the sun.
    目前的第一要务已经不是让探测器脱离沙坑了，而是让太阳能板更偏向太阳的方向。
2. deceptively [dɪˈseptɪvli]
    ad. in a misleading way 看似；不像看上去那么……

This deceptively simple empathic interaction which we have been discussing has many and profound consequences.

我们一直在讨论的看似简单的共情互动,其实有许多深刻的影响。

3. designate ['dezɪgneɪt]

　　v. assign a name or title to; give an assignment to (a person) to a post, or assign a task to (a person) 把……定名为,把……描述为;任命,指定;标明,标示

There are efforts under way to designate the bridge a historic landmark.

人们在努力把这座桥定为历史地标。

4. oceanographer [ˌəʊʃə'nɒɡrəfə(r)]

　　n. a scientist who studies physical and biological aspects of the seas 海洋学家;海洋研究者

At the Pacific tsunami Warning Center, David Walsh, an oceanographer, said the earthquake created a 23-foot-high tsunami in Japan.

在太平洋海啸预警中心,海洋学家大卫·沃尔什称,这次地震在日本造成了23英尺高的海啸。

5. hitch [hɪtʃ]

　　v. travel by getting free rides from motorists; connect to a vehicle; to hook or entangle 搭便车(旅行),搭顺风车;拴住,套住,钩住

China's first planetary probe, the tiny Yinghuo-1 orbiter, will also hitch a ride to Mars with Phobos-Grunt.

中国的第一个行星探测器,小小的"萤火一号"轨道飞行器,也将搭乘"福布斯-格朗特"号飞船前往火星。

6. stack [stæk]

　　v. load or cover with orderly piles; arrange in piles 使成整齐的一堆;使成叠(或成摞、成堆)地放在……

He ordered them to stack up pillows behind his back.

他命令他们在他背后堆放一些枕头。

7. equivalent [ɪ'kwɪvələnt]

　　a. equal in amount or value 等同的,等效的

　　n. a person or thing equal to another in value or measure or force or effect or significance etc. 对等的人(或事物),对应的人(或事物)

If they want to change an item in the budget, they will have to propose equivalent cuts elsewhere.

如果他们想要改变预算中的一个款项,必须得提出其他等值的削减。

8. complement ['kɒmplɪmənt]

　　v. make complete or perfect; supply what is wanting or form the complement to 补充,补足

He insists there is a vast and important difference between the two, although they complement each other.

他坚持认为,尽管两者相辅相成,但两者之间存在着巨大且重要的差异。

9. webinar ['webɪnɑː(r)]

n. web seminar 网络研讨会;在线会议

An on-the-spot workshop and a workshop webinar on the same topic, which one would you prefer?

如果有主题相同的现场讲座和网络讲座,您更愿意参加哪一个?

10. highlight ['haɪlaɪt]

v. move into the foreground to make more visible or prominent 突出,强调

You should highlight what you have to offer the company in the application letter, such as a specific skill or experience.

你在求职信里应该突出你能为公司提供什么,比如特定的技能或经验。

11. plunge [plʌndʒ]

v. thrust or throw into; cause to be immersed; engross (oneself) fully (使)突然向前倒下(跌落);猛推,猛插;投入(液体中以使淹没);(使)投身,(使)突然开始从事

She was about to plunge into her story when the phone rang.

她刚要开始大谈她的经历,电话响了。

12. expedition [ˌekspə'dɪʃ(ə)n]

n. a journey organized for a particular purpose; an organized group of people undertaking a journey for a particular purpose 远征,考察;探险队,考察队

Samples recovered from the expedition revealed important differences in chemical composition and fossil distribution among the sediment layers.

探险队采集的样品揭示了沉积物层之间化学成分及化石分布上的重大差异。

13. seamount ['siːˌmaʊnt]

n. an underwater mountain rising above the ocean floor 海底山

After clearing his way through the seaweed on the surface, he swam down a few meters to the seabed, where he spied a seamount covered with a great number of sea cucumbers.

穿过海水表面的海藻以后,他下潜好几米游到了接近海底的地方。在那儿,他看到了一个海底山,山上全是海参。

14. commune ['kɒmjuːn]

n. a body of people or families living together and sharing everything; the smallest administrative district of several European countries 社群,群体;(尤指欧洲一些国家的)最小行政区

She lives in one of the poorest villages in the commune and makes a living growing rice and sweet potato.

她居住在这个地区最为贫困的一个村庄里,靠种植水稻和地瓜生活。

15. bathymetric [ˌbæθɪ'metrɪk]

a. of or relating to measurements of the depths of oceans or lakes 测深的;等深的

Laser sounder bathymetric system is of great importance to be capable of charting coastal bathymetry rapidly and cheaply.

激光海洋测深仪是一种高效且低廉的海深勘测系统,对绘制沿海区域海图有重大意义。

16. terrain [təˈreɪn]

    n. a piece of ground having specific characteristics or military potential 地形,地势;领域

    He knows the terrain of this locality like the back of his hand.

    他对这一带的地形了如指掌。

17. sparse [spɑːs]

    a. not dense 稀少的,稀疏的;简朴的

    The deep sea typically has a sparse fauna dominated by tiny worms and crustaceans, with an even sparser distribution of larger animals.

    深海通常有稀疏的动物群,以微小的蠕虫和甲壳类动物为主,体型较大的动物分布更少。

18. footage [ˈfʊtɪdʒ]

    n. film that has been shot 一组(电影,电视)镜头

    The police replayed footage of the accident over and over again.

    警察一遍又一遍地重放事故的片段。

19. bolster [ˈbəʊlstə(r)]

    v. support and strengthen 增强,激励;巩固(地位),加强;改善,改进

    Bolster your knowledge of business areas you're less familiar with.

    多学习一些你不熟悉的商业知识。

20. endorse [ɪnˈdɔːs]

    v. give support or one's approval to; guarantee as meeting a certain standard (公开)赞同,认可;宣传,吹捧

    This is an admirable goal, and one I wholeheartedly endorse.

    这是一个令人钦佩的目标,我非常赞同。

21. compelling [kəmˈpelɪŋ]

    a. tending to persuade by forcefulness of argument; driving or forcing 令人信服的,有说服力的;引人入胜的,扣人心弦的;非常强烈的,不可抗拒的

    Factual and forensic evidence makes a suicide verdict the most compelling answer to the mystery of his death.

    自杀判决,在事实和法庭举证的支持下,成为解开他自杀之谜的最佳答案。

22. reverse [rɪˈvɜːs]

    v. change to the contrary 逆转,彻底改变(决定、政策、趋势等);颠倒,反转

    n. a relation of direct opposition 相对,相反(the reverse);逆向,逆转;反面,背面

    Although I expected to enjoy living in the country, in fact the reverse is true.

    尽管我原以为会喜欢乡村生活,但实际情况正好相反。

23. intervention [ˌɪntəˈvenʃ(ə)n]

    n. a policy of getting involved in the affairs of others or other countries, so as to alter or hinder an action, or through force or threat of force 干预,干涉,介入;调停,斡旋

    Military intervention will only aggravate the conflict even further.

    军事介入只会使冲突加剧。

Phrases and Expressions

with……in mind 考虑到……;把……记在心里
a body of 一片,大量,一团
come down to 归根结底,可归结为;实质上是
amounts of 大量的
rely on 依靠,依赖
call for 要求;需要;提倡;邀请
set a record 创造纪录
bank on 指望;依赖
at all costs 无论如何,不惜任何代价
be crucial to 对……至关重要
set aside 留出;驳回,撤销;不顾
leave……alone 听其自然;置之不顾;不管;不打扰

Terminology

jumbo jet 大型喷气式客机
phytoplankton 浮游植物
ocean currents 海流,洋流
torpedo sub 鱼雷潜艇
coral reef 珊瑚礁
spookfish 长吻银鲛
starfish 海星
the tipping point 引爆点;临界点

Proper Names

marine protected areas (MPAs) 海洋保护区
NASA's Goddard Space Flight Center 美国宇航局戈达德太空飞行中心
Mariana Trench 马里亚纳海沟
human-occupied vehicles (HOVs) 载人车辆
remotely-operated vehicles (ROVs) 远程操控车辆
the Woods Hole Oceanographic Institution 伍兹霍尔海洋研究所(美国马萨诸塞州)
Canary Islands 加那利群岛
Malta 马耳他(欧洲岛国)
Mediterranean Sea 地中海
Atlantic Ocean 大西洋
the Desventuradas and Juan Fernández islands 德斯文图拉达斯岛和胡安·费尔南德斯岛
Caleta Tortel 卡列塔托尔特尔(智利南端的一个港口城市)

一、翻译简析

1. Feldman specializes in satellite technologies that record the color of the ocean as a means of measuring the distribution and abundance of phytoplankton, which can change rapidly and even double in a day. (Para. 3)

　　费尔德曼专门研究卫星技术,用来记录海洋的颜色,以此作为测量浮游植物分布和丰富程度的一种手段,这些海洋里的浮游植物可以在一天内迅速变化甚至翻倍。

　　【这个句子是由一个 that 引导的限制性定语从句和一个 which 引导的非限制性定语从句构成,中间还有一个 as a means of 引导的方式状语,翻译时如果把整个长句结构直接转换成汉语,译文会显得臃肿不堪。汉语多短句,英语多长句,这时,我们需要考虑采用拆分的方法把句子进行拆解。这里我们把第一个 that 引导的定语从句拆解出来,翻译时不把它翻译成前置定语,然后把 as a means of 引导的方式状语也单独拆解出来,按照原句的语序,只需变成短小的汉语短句,然后逐层释放信息即可,这样就能符合汉语的思维习惯。另外,as a means of 是介词短语,翻译时处理成动词短语"作为手段"。】

2. Floats and drifters, devices that rely on ocean currents to carry them while they collect data, have been complemented in recent years by an ever-sophisticated fleet of underwater vehicles. (Para. 5)

　　海洋勘探设备,如浮子和漂流器,依靠洋流在海上漂流来收集数据,近年来又有新的、越来越复杂的水下航行器也加入了海洋勘探队伍。

　　【这个句子是由一个 that 引导的定语从句和一个 while 引导的时间状语从句构成,而且原句是一个被动句。汉语中很少使用被动句,因此翻译时要进行句型转换处理,想办法把被动句变成主动句。这时,我们需要考虑把 complemented 进行转换,原句强调的是海洋勘探设备的迭代更新,因此,选词时,译者就选用了一个"加入队伍",这样就避免了使用被动句。另外,原句使用了限制性定语从句,翻译处理时译者没有把定语进行前置翻译,在对句子进行处理时,译者把原来的定语从句和状语从句进行重新整合,捋清原句逻辑层次,按照符合汉语的思维习惯把句子化成短句,逐层给出新信息。】

3. Oceana uses a combination of technologies on its expeditions, which have charted previously unexplored waters, including areas off of Southern California, several seamounts in the Canary Islands, and a deep trench south of Malta. Another seamount near Morocco, which was also previously unexplored, led to the discovery of a deep-sea coral reef, the only one of its kind that is still growing in the Mediterranean Sea. (Para. 6)

　　海洋环保组织在海洋勘探中使用了多种技术,对以前从未探索过的水域进行地图绘制,这些水域包括南加州海域、加那利群岛的几座海山以及马耳他南部的深海沟。摩洛哥附近的另一座海山以前也从未被勘探过,这一发现还让他们找到了一处深海珊瑚礁,这也是地中海中唯一一处仍在生长的深海珊瑚礁。

　　【这个句子由两个 which 引导的非限制性定语从句和一个同位语构成,句子结构不太复杂。英语多使用名词,而汉语则多用动词。因此,英译汉时,更多要将名词转换成汉语的动词,从而使静态叙述转换成动态叙述。在处理本句时,译者把 led to the discovery 处理成了"让他们找到了",这样更符合汉语习惯。另外,在最后一句同位语的处理上,译者在翻译过程中,将同位语转换成了同位语从句,人为地增加了"这也是",突出强调了这一发现的重

大意义,跟原文行文风格匹配。】

4. And in Chile, following Oceana expeditions which recorded the rich and unique marine life in the Desventuradas and Juan Fernández islands, the government was persuaded to make those places marine parks. A multiple-use MPA was also established in the Caleta Tortel commune following multiple Oceana expeditions. (Para. 7)

  海洋环保组织勘探队还记录了在德斯文图拉达斯岛和胡安·费尔南德斯岛上生活的丰富而独特的海洋生物,这件事最终说服了智利政府将这些地方建成海洋公园。海洋环保组织经过多次勘探后,在智利南部的沿海村庄卡莱塔托尔建立了一个允许多用途的海洋特别保护区。

  【这个句子是由一个现在分词和一个 which 引导的定语从句及两个被动语态构成,翻译时如果把这种结构直接转换成汉语,译文会英语化。这时,我们需要考虑如何把被动转换成主动。虽然原文中主语是政府,但是结合上下文尤其前面的背景,译者看到这是海洋环保组织说服智利政府设定保护区域的一个成功案例,遵循这个思路,译者就把时间状语进行了移植,译成了政府的前置定语,为了规避使用被动语态,译者把智利政府变成宾语,把前面海洋环保组织的发现变成主语,为了避免跟前文重复,这里译者把前文中谈到的海洋环保组织的发现用"这件事"代替,使用了一个外位语结构,既突出了要点,又符合汉语思维习惯。】

5. Even though data is useful, Feldman says a compelling case for MPAs can still be made when the reverse is true. Because scientists do not have a complete understanding of how one change to the ocean affects the entire ecosystem, and which of those changes may be the tipping point that causes collapse, it is common sense that areas should be designated for protection and further research. (Para. 11)

  尽管数据在说服过程中相当有用,但同时费尔德曼也说,当数据不足时,仍然可以拿出令人无法拒绝的理由来说服相关单位建立海洋特别保护区。海洋的某个变化会影响到整个生态系统,该变化如何影响生态系统以及哪些变化可能成为导致崩溃的临界点,关于这一点,即使是科学家也并不能完全了解,因此应该指定区域进行保护和进一步研究是常识。

  【这个句子由一个 even though 引导的让步状语从句,when 引导的时间状语从句,because 引导的时间状语从句,两个 how 和 which 引导的宾语从句,以及 it 作形式主语引导的主语从句构成,翻译时捋清句子逻辑关系至关重要,这里 when the reverse is true 是一个理解上的难点,通过阅读上下文,发现 when the reverse is true 对应的是 Even though data is useful,因此,处理时,译者就对这层隐含的对比关系进行了显化处理,翻译成"当数据不足时",a compelling case for MPAs can still be made 对应的是上文中海洋环保组织拿出铁的数据说服智利政府设定保护区一事,因此翻译时就译成了"拿出令人无法拒绝的理由来说服相关单位建立海洋特别保护区",适当地增译了一些背景知识,以便读者更好地理解原文。】

二、参考译文

亟待保护、尚未被探索的海洋

1. 科学家们成功地拍摄到了黑洞,探测器也顺利在火星上着陆,飞船也完成了月球飞越,到达了月球的背面。然而,人类最后未知(也是我们自认为非常熟悉,其实是非常陌生)的边界之一,还是我们人类赖以生存的星球。在这个星球上,超过80%的海洋目前仍处于尚未被探索的状态。也因为对其不了解,想要对其进行保护做起来很难。目前,世界上大约只有7%的海洋被划为海洋保护区。知道了这一点,我们就能解释为什么覆盖地球大部分表面的海洋是宇宙中最脆弱也是我们最不了解的地方之一。

2. 压力是潜入深海的最大挑战。海洋探索之所以困难,究其原因还是跟物理相关。在美国宇航局戈达德太空飞行中心工作的海洋学家吉恩·卡尔·费尔德曼博士说,深海中温度极低,能见度为零,有着超强的水压。"在某些方面,将人送到太空比将人送到海底要容易得多",费尔德曼这样告诉海洋环保组织。"深海中的巨大压力使其成为一个极其难以探索的环境。"

3. 你或许没有注意到,在地表接近海平面的地方,我们全身的皮肤每一平方英寸都承受着大约15磅的压力。如果向上进入太空,也就是进入地球大气层上方,我们身体肌肤承受的压力就会降低到零。然而,如果我们去潜水或乘坐水下航行器进入深海,这时压力就会开始增大,进入海底越深,身体承受的压力就越大。"在潜入将近7英里深的马里亚纳海沟底部时,我们身体所承受的压力就会是地表海平面的1 000倍以上",费尔德曼说,"这相当于50架巨型喷气式飞机碾压在身上的重量"。当然,人类乘坐的潜水器并不是探索和研究海洋的唯一途径。我们甚至可以从探索太空中获取一些经验。费尔德曼专门研究卫星技术,用来记录海洋的颜色,以此作为测量浮游植物分布和丰度的一种手段,这些海洋里的浮游植物可以在一天内迅速变化甚至翻倍。

4. 费尔德曼还说,这些技术首次在20世纪70年代后期使用,卫星在几分钟内就能捕捉到海洋的详细图像。然而,如果我们用普通船只来收集同样的信息,则需要10年的连续采样才能收集相同数量的测量结果。话虽如此,有些深海数据还是得在深海测量才更好,无论到达那里有多困难。

5. 这是海洋探索历史的见证。海洋勘探技术已经取得了长足的进步。海洋勘探设备,如浮子和漂流器,依靠洋流在海上漂流来收集数据,近年来又有新的、越来越复杂的水下航行器也加入了海洋勘探队伍。这些水下航行器包括载人水下航行器、远程操作航行器以及自主水下航行器和混合动力航行器。在伍兹霍尔海洋研究所最近主办的网络研讨会上,电影导演,也是海洋爱好者詹姆斯·卡梅伦呼吁在海洋探索时采取"上述所有"方法,该研讨会还强调了继续探索未知领域的重要价值。2012年,卡梅伦创造了历史,他乘坐一艘名叫"垂直鱼雷"的潜艇,访问了海洋最深处的马里亚纳海沟。"我称之为见证",卡梅伦说,"亲身在场并用你所有的感官来感受海洋深处是非常令人兴奋的。另外,你还可以回来讲述这次经历,听众会很感兴趣的。对我来说,探索最重要的一点是我可以回来跟人讲述这次离奇的经历"。

6. 海洋环保组织在海洋勘探中使用了多种技术,对以前从未探索过的水域进行地图绘制,这些水域包括南加州海域、加那利群岛的几座海山以及马耳他南部的深海沟。摩洛哥附近的另一座海山以前也从未被勘探过,这一发现还让他们找到了一处深海珊瑚礁,这也是地中海中唯一一处仍在生长的深海珊瑚礁。在欧洲的生物多样性热点地区,勘探队还首次记录了一条活的棕色鼻子长吻银鲛,两种海星物种,一处黑珊瑚和一处石珊瑚,这些物种以前被认为只生长在大西洋。

7. 海洋环保组织勘探队还记录了在德斯文图拉达斯岛和胡安·费尔南德斯岛上生活的丰富而独特的海洋生物,这件事最终说服了智利政府将这些地方建成海洋公园。海洋环保组织经过多次勘探后,在智利南部的沿海村庄卡莱塔托尔建立了一个允许多用途的海洋特别保护区。尽管勘探带来的好处是不可否认的,对海洋进行勘探也是相当昂贵的,缺乏详细的地图和数据使勘探过程更是困难重重。海洋环保组织在欧洲勘探队的负责人里卡多·阿吉拉尔表示,他们不能完全依靠这些水深信息——只能将其视为一个地区水下地形的指南,因为在大多数情况下,这些海底水深信息并不存在。"人类只对连5％都不到的海洋掌握了完整而准确的信息,也许还有另外10％,我们也或多或少知道一些",阿吉拉尔说,"可以说,我们对这些海洋区域几乎一无所知,保护又从何说起?"

8. 这就是第22条军规式的黑色幽默:我们需要探索来收集更多信息,但世界各地的许多机构都不愿意资助有太多未知数的项目。"似乎有一种不惜一切代价避免风险的趋势,这就意味着在开始启动勘探调查之前,你就必须证明你知道所有的答案",费尔德曼说,"但这种方法是错误的。科学不仅仅要有解决问题的答案。相对于解决问题,提出问题更加重要。"

9. 海洋勘探给我们提供证据,而不是找理由的借口。迄今为止,海洋环保组织已经保护了近400万平方英里的海洋,海洋勘探对取得如上成就至关重要。通过勘探,我们就能获得照片、视频、科学数据和论证叙事方式,这些都有助于说服相关单位设定新的保护区域或扩大保护范围。"通过勘探以前未探索的地区,我们就会发现新的物种以及新的栖息地,这样做还有助于识别哪些栖息地生态环境脆弱或哪些物种正在遭受威胁,这些栖息地或受威胁的物种往往只是在纸面上受到保护的,因为没有人能找到这些栖息地或物种,所以也就没有措施对它们进行有效的保护",阿吉拉尔说,"政府不采取行动的最常见借口之一是缺乏足够信息来选择保护哪些地区以及如何进行管理。此外,不同利益相关者反对创建新的海洋保护区也是出于数据缺乏。"

10. 海洋环保组织支持到2030年要完成保护30％海洋的倡议,这一目标也被称为"30×30"目标。虽然距离这一目标的实现还有很长的路要走,但持续的海洋探索可以为人类提供所需的各种数据,从而更好地保护海洋以及保护海洋资源。

11. 尽管数据在说服过程中相当有用,但同时费尔德曼也说,当数据不足时,仍然可以拿出令人无法拒绝的理由来说服相关单位建立海洋特别保护区。海洋的某个变化会影响到整个生态系统,该变化如何影响生态系统,以及哪些变化可能成为导致崩溃的临界点,关于这一点,即使是科学家也并不能完全了解,因此应该指定区域进行保护和进一步研究是常识。"我们也不知道所有因素如何组合叠加产生影响,因此需要留出特定区域供我们来研究,直到我们对此有更多了解,这时我们只需说'我们不管这些'或者'我们将进行最少干预',可能是最安全的做法",费尔德曼说,"把这些对环境来说极其重要或者环境相对独特的区域单独划出进行研究和保护,这种做法在我们学会如何更明智管理海洋之前,可能是非常明智的举动。"

Cultural Background Knowledge: Ocean Exploration
海洋探索文化背景知识

伴随深海探测技术的发展，人类深入认识深海的时代正在来临。500年前达·芬奇设计潜水服，150年前凡尔纳写《海底两万里》，当时的科学幻想如今正在成为现实。

从科学角度看，探索深海能够帮助人类深入了解海洋的奥秘、地球的奥秘。水深超过2 000米的深海，占据地球表面的3/5，无论是温室气体排放的归宿，还是气候长期变化的源头，都要追溯到海水深层。不仅如此，海底是距离地球内部最近的地方：大陆地壳平均35千米厚，大洋地壳则为7千米。揭示板块运动的规律、窥探地球内部的真相，也要到深海底部进行探索。

从经济角度看，深海蕴藏着丰富的矿产、油气和生物资源。目前，海洋石油产量占世界石油产量的30%，高居世界海洋经济首位，其中发展最快的是深水油田。近年来，全球重大油气发现，70%来自水深超过1 000米的水域。海底有待开发的资源非常丰富，现在还只是起步阶段。比如海底的微生物新陈代谢极其缓慢，生殖周期在千年以上，但人类尚不知如何利用其"长寿基因"；太平洋一片深海黏土所含的稀土元素可供人类使用几十年，但开采利用技术尚待研发。深潜、深钻、深网是当今探索深海奥秘的三大手段，即深潜科学考察、国际大洋钻探和国家海底科学观测网建设。深潜是直观的深海探索，但在空间和时间上都存在局限性。深潜最深只能到海底，从海底往下得靠钻探，这就是深钻；深潜的运行时间只能以小时计，想要长期连续观测就得将传感器放到海底，联网观测，这就是深网。深潜、深钻和深网，共同担起深海探索的技术重任。

目前，我国已建立起"三深"格局，深海科考进入快速发展期。我国从2005年起开始推进海底观测网的建设，2009年建设近岸的实验观测站。此后，又在南海北部进行了大量深水海流和沉积过程的长期观测。2017年，我国国家海底科学观测网正式被批复建立，将在我国东海和南海分别建立海底科学观测系统，从海底向海面进行全天候、实时和高分辨率的多界面立体综合观测。近10年来，建设中的海底科学观测网除了光缆联网的设备外，还有着大量无线联网的活动观测平台，包括自主水下航行器、水下滑翔机、海底爬行车等。

Unit Fifteen
Marine Cultural Industry

海洋文化产业

Text A

Maritime Cultural Landscape

By Dan Atkinson and Alex Hale

Ideological and Political Education：整治发展无序状态 推动产业高质量发展

随着海洋强国战略的推进,我国沿海各地对于海洋资源开发利用的重视程度越来越高,但是,海洋文化产业属于新生事物,从目前的发展状况来看主要存在五个方面的短板。第一,海洋文化资源开发效率不高。目前我国各地对海洋文化资源的开发基本上处于无序状态,甚至同一地区之间都缺乏协调配合,造成文化产品和产业项目的重复建设和互相克隆,导致资源浪费。第二,空间布局有待优化。由于缺乏统筹规划和协调机制,目前我国沿海城市文化产业发展各自为政,内部联系松散,存在产品同质化竞争、产业项目重复建设等问题,资源集约化配置程度不高,制约跨地区的产业合作和要素自由流动。第三,发展新动能有待释放。由于缺乏现代科技和创意元素的积极融入,我国海洋文化资源优势尚未高效开发利用,各地基本以滨海旅游和低附加值的工艺品为主要开发对象,亟待形成海洋文化新产品、新业态、新模式。第四,品牌效应有待发挥。由于各地同质化问题和发展新动能不足,我国海洋文化产品主题化不明显,特色不突出,独创性不高;海洋文化品牌杂乱,形象零散,对海洋强国战略建设的支撑作用尚未完全发挥。第五,产业体系有待完善。文化和旅游及相关产业融合深度不够,没有形成完整的海洋文化产业链和有效延展的价值链,产业集聚度、开放度、外向度、创新力、附加值、综合质量和效益均需提升。

我国海洋文化产业发展,需要着力解决当前阶段存在的短板,合理开发海洋文化资源,统筹规划海洋文化产业发展布局,加快海洋文化资源与科技、创意融合,打造海洋特色文化品牌,实施系列重大工程项目,推动海洋文化产业高质量发展。

1. The Maritime Cultural **Landscape**. The term "maritime culture" grew out of a broader understanding of not only the use of the sea by humans, but the attendant structures, cultural identifiers and associations made between people and **seafaring**. For example, Britain would claim to be a seafaring nation as a result of ship-building traditions on the Clyde and elsewhere, world-wide maritime exploration and the prestige of the Royal Navy. Norse culture is similarly considered maritime due to extensive sea voyages, raiding and the colonization of land such as Greenland. Today, people driven in cars along roads and over bridges, are in contact with the sea in quite a different way. They often treat the sea more as part of a leisure canva on which to play beside on holiday, dive into and explore, and sail on, rather than remember that it is still relied on to move cargo, for example. However, coastlines are still referred to as landmarks and now nations increasingly consider underwater resources and oil fields, and not fishing alone, as an important part of national wealth.

2. Westerdahl has subsequently broadened his definition of maritime cultural landscape as: the archaeological concept combining sea and land would be the maritime cultural landscape. It means that the starting point for the subject of maritime archaeology is maritime culture. If the **holistic** approach proposed by this document and the concept of "Source to Sea" is to be developed, this demands that an even more wide-ranging view be taken, which **encompasses** modern as well as historic popular culture, to re-define and broaden the definition of marine and maritime cultural landscapes.

3. There can be no doubt that the maritime and marine historic environment in Scotland enjoys an enviable status with regard to the broad and varied resource located off its coasts, along its coastline, within its estuaries, and inter-connected to the network of inland waters. This research framework was developed in a period of legislative change, and one where there is a growing awareness of the maritime and marine historic resource. As a nation, and as members of the international community, there exists a need to fulfill the obligations **conferred** for the better understanding, management and conservation of our maritime cultural heritage. An important way in which this goal can be approached is through this cross-sector, **thematic** research framework document.

4. European marine cultural heritage obligations such as those set out in the United Nations Convention on the Law of the Sea (UNCLOS) 1982, the European Convention on the Protection of the Archaeological Heritage (Revised) 1992, (the Valletta Convention) and the UNESCO Convention on the Protection of the Underwater Cultural Heritage 2001 have influenced the exciting developments for the protection of the resource within the UK; notably the UK Marine and Coastal Access Act 2009, and more significantly in the case of this

framework, the Marine (Scotland) Act 2010. **Statutory** bodies and NGOs have worked hard to provide a voice for the maritime and marine historic environment and to ensure there is a place for the historic environment in the new legislation. This is an encouraging start, and hopefully one around which a research framework can grow and develop.

5. In March 2009, Historic Scotland published a discussion article, *Towards a Strategy for the Marine Historic Environment* as a result of a wide-ranging consultation. The article set out the challenges and opportunities that, at that time, lay ahead with regard to the marine historic environment, not only at a national strategic level, but also in regional and local contexts. In addition, the "Desirable Outcomes" section also considers how future strategies and initiatives can be measured, indicating areas where this framework can be influential in helping shape how we approach marine and maritime research at a strategic national and regional level. Examples include areas such as "Challenges and Future Directions"-an identified theme within this framework and one where useful cross-referencing will benefit the development of a long-term and sustainable framework. By effective integration of the objectives of the Marine (Scotland) Act 2010, and those highlighted in the discussion article, the framework can provide a useful basis for academic and voluntary sector research interests. It is also a useful curatorial tool in helping guide national and local government when making decisions with regard to the priorities for the effective management and understanding of the maritime and marine historic environment.

6. In addition, this research framework document can also help influence relevant policies and legislation that exist for other areas. This includes **terrestrial** policies and plans and those of Museums and Galleries concerning the effective management of their maritime material culture and monuments such as historic ships and vessels. Indeed, this inclusive cross sector approach is embodied in the **overarching** theme throughout this framework, namely "From Source to Sea". As it will become clear, this approach aims to promote an overarching, holistic and integrated mechanism upon which all areas of the archaeological discipline, at all levels, can actively contribute to understanding of the maritime and marine historic environment. In this respect, it is also important to employ a pro-active position in ensuring full co-operation with trans-boundary research frameworks, such as those in the rest of the UK and beyond.

7. From Source to Sea: the scope and **remit** of the Marine and Maritime Panel. This **contiguity** of archaeological relationships from the palaeo-seabed to upland rivers led the panel to term its remit "From Source to Sea", **encapsulating** the interrelationship of all aspects of human activity that ultimately link archaeological sites to the maritime zone. This wide-ranging definition of the marine and maritime resource requires a holistic approach: the panel addressed the subject

in thematic terms, but these themes ranged in type and definition, some geographical (e. g. Coastal Hinterlands, Inland Waters), some site-specific (Ships and Vessels) and some methodological (Submerged Landscapes). This diversity is both a reflection of the difficulties of setting remit boundaries discussed below, but also a reflection of the diversity of approaches taken to the archaeology of the marine and maritime environment.

8. The first task facing each of the ScARF (the Scottish Archaeological Research Framework) thematic panels is also the largest and most problematic: that of defining the scope of the remit of the panel. In the case of the Marine and Maritime panel, this task was particularly difficult: arguably, with such a high ratio of coastline to land area and with nowhere further than around 80 km from the sea, all of Scotland can be considered maritime. Unlike other panels, the scope of the Marine and Maritime panel was not restricted to any particular chronological range, nor to any geographical region. In defining the remit, the panel necessarily took an inclusive approach to the maritime resource, so that shipwrecks form only one element of the material record of people's interaction with the sea and the coastal zone. All aspects of past human activity whether directly related to the exploitation of the marine environment, or simply located in a maritime setting, were considered to be under the panel's jurisdiction. Such a view leads archaeology of all types to be relevant to this research framework and therefore this document is cross-cutting and inter-leaves with the other ScARF panel research documents. Settlement sites of the earliest hunter-gatherers, prehistoric **inundated** landscapes, medieval **fortifications**, historic shipwrecks, fish-traps, modern harbors and vessels still afloat, are all considered part of our surviving maritime heritage. Furthermore, the direct linkage of inland waterways to the coast as "**arteries**" of communication, trade and transport, means that the archaeology and history of Scotland's extensive **lochs**, rivers and canals cannot be excluded-hence the introduction of the "source to sea" approach.

9. Past Achievements, Future Directions. It is often a characteristic of new research agendas that the achievements of past research initiatives are portrayed as incomplete or inadequate, while new directions are listed as the routes to understanding. This, however, would not be an accurate portrayal of the history or future of marine and maritime archaeology in Scotland. Archaeological studies of human interaction with the sea have a long history in Scotland, while maritime research in the traditional sense such as underwater excavations and surveys of shipwrecks has been a long standing strength. The primary aim of this panel was to identify, **collate** and summarize the achievements of this previous research as a premise for the identification of the most productive avenues of future research. The changes of 2008 to 2012 (including the Marine [Scotland] Act 2010 and the formation of Marine Scotland) in the marine and maritime sector in Scotland have been positive advances and change is continuing

apace: this will be reflected in the changing understandings set out in this, and future, frameworks.

10. In order to build a holistic approach to the marine and maritime environment, a number of approaches need to be explored, preferably in combination and through collaboration. This includes adopting a sea-oriented perspective. Approaching marine and maritime landscapes from the perspective of the sea, both metaphorically and physically by boat, enables researchers to experience seascapes as people would have done, prior to the automobile age. The researcher has to understand landing and launching places, navigation aids, currents, tides and winds, and where sources of certain resources, such as fresh water, can be found. This approach demands an understanding or a source of local knowledge and experience, which is often lost or rarely encountered. However, historic maps, charts and local knowledge combined can lead to fruitful insights into the maritime geography and history of an area.

11. Landscape-scale approaches demand limits and islands are particularly suitable as study areas in this respect. Similarly waterways are readily defined by watershed and other factors, whereas sea routes are infinite in variability and the boundaries of study may be more usefully defined by the commercial, political or economic objectives of the navigation. At the other end of the geographical spectrum, site-scale researches of Scotland's marine and maritime cultural landscapes focus on **exemplars** such as log-boats, shipwrecks, fish-traps, navigation aids, cleared landing places, **vernacular quays** and buildings associated with fishing and boats, all of which may lead out into the wider issues alluded to above. Consideration of the land from the sea also gives the advantage of perceiving the land as a transient surface, one that is constantly changing. The present day shore line is but one boundary in an infinite and ever-changing environment that humans have interacted with and adapted to throughout history.

12. Finally, artifact and literary study approaches look at cultural objects and events as indicative of connections, both physical and psychological, with the sea, from fishing weights and **sextants** to place names and sea **shanties**, but also to 'associations' such as Capstan navy cut cigarettes, launched in 1894 and still sold today under the same brand name. This product uses a **capstan** as a logo, an image **redolent** in contemporary culture of the "manly" virtues of strength and teamwork, bonded to the peerless prestige of the pre 1914 Royal Navy.

13. It is an important aim of this panel to highlight the wealth and diversity of Scotland's maritime archaeological resource, but central to this aim must be the move away from the stereotypical view of "Maritime Archaeology" as a specialist sub-division of the mainstream discipline. For many reasons, considering maritime archaeology as a specialism is misleading

and particularly so, perhaps, in Scotland where marine and maritime culture has pervaded all aspects of human activity in all periods. It is intended that this framework document be used alongside those of the other panels, with the research directions outlined here aligning and complementing those of the other thematic and chronological subject areas.

(1,929 words)
http://scarf.scot/thematic

New Words

1. landscape [ˈlændskeɪp]

 n. a large area of countryside, especially in relation to its appearance 乡间,野外;(尤指乡村的)风景,景色

 The cathedral dominates the landscape for miles around.
 大教堂在方圆数英里的乡间高高耸立着。

2. seafaring [ˈsiːfeərɪŋ]

 n. seafaring industry 航海业;海上航行

 The trading and seafaring skills of the Phoenicians resulted in a network of colonies, spreading westwards through the Mediterranean.
 腓尼基人的贸易和航海技能造就了一个殖民地网络,并通过地中海向西扩展。

3. holistic [həˈlɪstɪk]

 a. dealing with or treating the whole of something or someone and not just a part 整体的,全面的

 The medical world is now paying more attention to holistic medicine which is an approach based on the belief that people's state of mind can make them sick or speed their recovery from sickness.
 医学界现在越来越重视整体医学,这个方法是基于这样的信念,人们的精神状态可以使他们生病或加速他们从疾病中恢复。

4. encompass [ɪnˈkʌmpəs]

 v. to include different types of things 包含,包括(尤指很多不同事物)

 The project will encompass rural and underdeveloped areas in China.
 这项工程将覆盖中国的农村和不发达地区。

5. confer [kənˈfɜː(r)]

 v. to give an official title, honor, or advantage to someone 授予,赋予

 An honorary doctorate was conferred on him by Columbia University.
 哥伦比亚大学授予他荣誉博士学位。

6. thematic [θɪˈmætɪk]

 a. relating to or based on subjects or a theme 主题的,主旋律的;题目的

 In her study, the author has adopted a thematic rather than a chronological approach.

作者在她的研究中采用了按主题而非按时间顺序的方法。

7. statutory ['stætʃətri]

   a. decided or controlled by law 依照法令的,法定的;法令的

   The FCC has no statutory authority to regulate the Internet.

   美国联邦通信委员会没有规范因特网的法定权力。

8. terrestrial [tə'restriəl]

   a. relating to the earth 地球的,地球上的;陆栖的,陆生的

   Thousands of dams built since the mid-19th century have "completely altered the planet's terrestrial plumbing", he said.

   他说自19世纪中叶以来修建的几千座水坝已经"彻底改变了地球的地表径流"。

9. overarching [ˌəʊvər'ɑːtʃɪŋ]

   a. most important, because of including or affecting all other areas 首要的;支配一切的;包罗万象的

   The overarching question seems to be what happens when the U.S. pulls out?

   中心的问题看来是当美国撤出时会发生什么?

10. remit [rɪ'mɪt]

    n. the area that a person or group of people in authority has responsibility for or control over 职权范围;控制范围;许可权

    The remit of this official inquiry is to investigate the reasons for the accident.

    这次官方调查旨在查清事故发生原因。

11. contiguity [ˌkɒntɪ'gjuːəti]

    n. the fact of being next to or touching another, usually similar thing 邻近,接触

    The north wing of the house is in immediate contiguity to the kitchen.

    房子的北翼与厨房直接相连。

12. encapsulate [ɪn'kæpsjuleɪt]

    v. to express or show the most important facts about something 简要描述,概括

    It was very difficult to encapsulate the story of the revolution in a single one-hour documentary.

    将革命的前后经过浓缩成一部一小时的纪录片是很难的。

13. inundate ['ɪnʌndeɪt]

    v. to flood an area with water (洪水)淹没

    If the dam breaks it will inundate large parts of the town.

    如果水坝决堤,这个城镇的大部分地区都将被淹没。

14. fortification [ˌfɔːtɪfɪ'keɪʃn]

    n. strong walls, towers, etc. that are built to protect a place 碉堡,防御工事

    The fortifications of the castle were massive and impenetrable.

    城堡的防御工事巨大厚实,坚不可摧。

15. artery ['ɑːtəri]

    n. one of the thick tubes that carry blood from the heart to other parts of the body 动脉;干线

He had an operation last year to widen a heart artery.

他去年动了一次扩张心脏动脉的手术。

16. loch [lɒk]

n. in Scotland, a lake or inlet of the sea or ocean （苏格兰）湖；（狭长的）海湾

They went till Inverness via Loch Ness.

他们从尼斯湖走到因弗内斯。

17. collate [kəˈleɪt]

v. to bring together different pieces of written information so that the similarities and differences can be seen 核对，校对；校勘

The photocopier will collate the documents for you.

影印机会为你整理文件顺序。

18. apace [əˈpeɪs]

ad. quickly 飞快地，迅速地；急速地

The project is coming on apace (= advancing quickly).

这个工程进展神速。

19. exemplar [ɪɡˈzemplɑː(r)]

n. a typical or good example of something 模范，榜样；标本

It is an exemplar of a house of the period.

这是那个时期的房屋典范。

20. vernacular [vəˈnækjələ(r)]

a. in architecture, a local style in which ordinary houses are built（建筑）民间风格的

This does not imply copying vernacular architecture, but rather that new buildings should be in keeping with it.

这并不意味着要抄袭传统建筑，而是使新建筑与之和谐。

21. quay [kiː]

n. a long structure, usually built of stone, where boats can be tied up to take on and off their goods 码头

A number of fishing boats were moored to the quay.

很多渔船系泊在码头。

22. sextant [ˈsekstənt]

n. a device used on a ship or aircraft for measuring angles, such as those between stars or that between the sun and the earth, in order to discover the exact position of the ship or aircraft 六分仪（航海定向仪器）

Imagine trying to deduce this with the naked eye, a sextant and little else.

试想一下在没有六分仪和其他仪器的帮助下，仅用肉眼观测就来推断的困难吧。

23. shanty [ˈʃænti]

n. a small house, usually made from pieces of wood, metal, or cardboard, in which poor people live, especially on the edge of a city 棚屋，简陋小屋；a song that sailors sang in the past while they were working on a ship 水手号子，船夫曲

These people are given tin sheets to build shanty towns—but no compensation.

这些贫民能够获得一些锡板建造简陋的棚屋小镇,但并没有获得其他补偿。

24. capstan ['kæpstən]

n. a machine with a spinning vertical cylinder that is used, especially on ships, for pulling heavy objects with a rope 绞盘;起锚机;主动轮

Sailors hove up the cable with windlass or capstan.

水手们用绞盘或起锚机把缆绳拉起来。

25. redolent ['redələnt]

a. having or emitting an odor or fragrance; aromatic 有……的强烈气味;芬芳的,香的; suggestive;reminiscent 使人想到的,使人联想起的

The album is a heartfelt cry, redolent of a time before radio and television.

这张专辑唱出了人们的心声,将人们带回到没有收音机和电视的时代。

Phrases and Expressions

as a result of 由于

be in contact with 与……有接触

refer to……as 把……称作;把……当作;称为

with regard to 至于;就……而言;至于,就

set out 宣布,陈述

lie ahead 在前面;即将发生;即将来临

in this respect 在这一点上;在这一方面;从这个方面来说

be restricted to 局限于;仅限于;被限制在

be relevant to 与……有关;和……相关;与……相关

portray……as 把……描绘成

prior to 此前;在……之前;先于

allude to 影射,暗指;间接提到

Proper Names

the United Nations Convention on the Law of the Sea (UNCLOS)《联合国海洋法公约》

the European Convention on the Protection of the Archaeological Heritage (Revised)《欧洲保护考古遗产公约》(修订版)

一、翻译策略/方法/技巧:异化

异化是指译者在翻译时故意保留源语文本中的某些异质性(foreignness),以此打破译入语的种种规范。与归化相比较,异化倾向于源语文化和原文作者。或者说异化的翻译更注重体现源语文本在语言和文化上的差异性,尽可能多地保留源语文化的特色和作者的独特表达方式,使得读者能够领会到原作的风貌,有一种身临其境的感觉。其实,在实际翻译过程中,译者在具体异化策略的选择上,事实上受到很多因素的影响,如作者意图、文本类型、译者水平、翻

译目的、读者对象、大的翻译环境等。而且,尽管异化与归化是相互对应的两种翻译策略,但是两者在翻译中不仅不是矛盾的,而且是互为补充的,文化移植本来就是需要不同的方法和模式。根据不同情况的需要,译者既可以采用归化的原则和方法,也可以采用异化的原则和方法。至于在译文中必须保留哪些源语文化,怎样保留,哪些源语文化的因素必须做出适当调整以适应译入语文化,都是可以根据实际情况来加以选择的。对于译者来说,重要的是在翻译过程中要有深刻的文化意识,即意识到两种文化的异同。

试看下列译例,可以更加深入理解异化原则和方法:

原文:抽刀断水水更流,举杯浇愁愁更愁。

译文:Drawing sword, cut into water, water again flows.
　　　Raise up, quench sorrow, sorrow again sorrow.

原文是唐代大诗人李白的诗,出自《宣州谢朓楼饯别校书叔云》。英语译文则出于美国意象派大诗人埃兹拉·庞德之手。尽管他的译文第一眼看上去似乎有些不伦不类之感,但是有一点必须承认的是他的译文还是最大程度上保留了中文侧重意合的特点,还有中国古典诗词习惯将诸多意象叠加在一起的审美效果,这些处理手法都十分典型地体现了异化的原则和方法。这也让外国读者能一窥中国古诗的审美情趣和语言特色。当然,也必须看到异化背后英语原文所付出的牺牲。这也是翻译中十分常见的状况。

原文:In the country of the blind, the one-eyed is king.

译文:山中无老虎,猴子称大王。

如果将原文这句英文谚语直译的话,可以译成"盲人国里独眼称王"之类的译文。从归化和异化的角度来看,这应该属于一种典型的归化策略,也就是几乎照搬了原文的语言形式而未做适应译语表达习惯的改变。相比之下,参考译文则是一种完全的异化策略的运用,因为完全搬用了译语中与原文意思相同的谚语,等于是一种套用式的翻译。如果说上文的译例是让读者能够充分体会到中文的审美情趣和语言特色的话,那么第二个译例通过恰当地使用异化策略完全符合了译语读者的阅读和认知习惯,既通俗又易懂,不存在理解上的障碍。

(摘自:姜倩,何刚强.翻译概论[M].上海:上海外语教育出版社,2016:214.)

二、译例与练习

Translation

1. The term "maritime culture" grew out of a broader understanding of not only the use of the sea by humans, but the attendant structures, cultural identifiers and associations made between people and seafaring. (Para.1)

"海洋文化"一词不仅源于对人类使用海洋的更广泛了解,还源于随之产生的各种结构、文化标识以及人们与航海之间联系的更广泛了解。

2. If the holistic approach proposed by this document and the concept of "Source to Sea" is to be developed, this demands that an even more wide-ranging view be taken, which encompasses modern as well as historic popular culture, to re-define and broaden the definition of marine and maritime cultural landscapes. (Para.2)

要实践该文件提出的整体方法和"从源头到海洋"的概念,就需要采取包含现代和历史流行文化在内的更广泛的观点,重新定义海洋和海洋文化景观。

【英语多用被动语态,而汉语常用主动语态。因此英译汉时,我们需要采用转态译法,把英语的被动句转换成汉语的主动句。另外,句子中"broaden the definition of"与前文的"an even more wide-ranging view"和"re-define"同义,若直译出来,会显得繁杂啰嗦。再者,出于汉语语法和表达习惯的需要,句子中的连词"if"和代词"this"也没有必要直译出来,而是改译成"要……,就需要……"的结构。因此,我们需要采用省译法把"broaden the definition of""if""this"省略,从而使译文准确、简洁、通畅、地道。】

3. There can be no doubt that the maritime and marine historic environment in Scotland enjoys an enviable status with regard to the broad and varied resource located off its coasts, along its coastline, within its estuaries, and inter-connected to the network of inland waters. (Para. 3)

毫无疑问,苏格兰由于在沿海地区、海岸线沿线、河口内以及与内陆水域连接处拥有广泛多样的资源,因而在海洋和海洋历史环境方面享有令人羡慕的地位。

4. The article set out the challenges and opportunities that, at that time, lay ahead with regard to the marine historic environment, not only at a national strategic level, but also in regional and local contexts. (Para. 5)

该文件不仅在国家战略层面,而且在区域和地方层面,阐述了当时在海洋历史环境方面面临的挑战和机遇。

5. As will become clear, this approach aims to promote an overarching, holistic and integrated mechanism upon which all areas of the archaeological discipline, at all levels, can actively contribute to understanding of the maritime and marine historic environment. (Para. 6)

显而易见,这个措施旨在建立一个全面、整体和综合的机制,并在此基础上,考古学科的各个领域在各个层面都能积极促进对海洋和海洋历史环境的了解。

6. This contiguity of archaeological relationships from the palaeo-seabed to upland rivers led the panel to term its remit "From Source to Sea", encapsulating the interrelationship of all aspects of human activity that ultimately link archaeological sites to the maritime zone. (Para. 7)

由于从古海床到高地河流的考古关系存在着连续性,涵盖了人类活动各个方面的相互关系,最终将考古遗址与海洋区域联系了起来,因此,该小组将其职权范围命名为"从源头到海洋"。

7. Furthermore, the direct linkage of inland waterways to the coast as "arteries" of communication, trade and transport, means that the archaeology and history of Scotland's extensive lochs, rivers and canals cannot be excluded—hence the introduction of the 'source to sea' approach. (Para. 8)

此外,作为通信、贸易和运输的"动脉",内陆水道与海岸有着直接的联系。这意味着不能忽略对苏格兰广泛的湖泊、河流和运河的考古以及它们的历史——因此引入了"从源头到海洋"的方法。

8. The changes of 2008 to 2012 (including the Marine [Scotland] Act 2010 and the formation of Marine Scotland) in the marine and maritime sector in Scotland have been positive advances and change is continuing apace: this will be reflected in the changing understandings set out in this, and future, frameworks. (Para. 9)

2008—2012年,苏格兰海洋和海事部门的改革(包括发布《2010年苏格兰海洋法案》和建立苏格兰海洋局)取得了积极进展,改革还在如火如荼地进行着:这将反映在本框架和未来框架所阐述的不断变化的理解中。

【苏格兰海洋局是苏格兰政府核心部门的一部分,负责管理苏格兰的水域。它成立于2009年4月1日,是苏格兰领先的海洋管理组织,其职能包括海事管理、渔业研究与服务、渔业保护等。】

9. This product uses a capstan as a logo, an image redolent in contemporary culture of the "manly" virtues of strength and teamwork, bonded to the peerless prestige of the pre 1914 Royal Navy. (Para.12)

这款产品使用绞盘作为标志,这一形象在当代文化中重现了"男子汉"力量和团队合作的美德,与1914年前皇家海军无与伦比的威望紧密相连。

【英语中一个用作状语或定语的词组或者名词词组经常包含丰富的语义。有时,要把原文词组的内容连同其形式一起翻译成地道的汉语几乎不可能。这时,我们需要采用转句译法,将其转译成汉语中的一个句子。这个句子中需要把同位语"an image……teamwork"转译成句,从而使译文更符合汉语表达习惯。】

10. For many reasons, considering maritime archaeology as a specialism is misleading and particularly so, perhaps, in Scotland where marine and maritime culture has pervaded all aspects of human activity in all periods. (Para.13)

出于许多原因,将海洋考古学视为一门专业是有误导性的,尤其是在苏格兰。因为在苏格兰,海洋和海洋文化已经渗透到各个时期人类活动的各个方面。

Exercises

1. Fill in the blanks with the proper given words, and then translate the sentences into Chinese.

　　vernacular　　collate　　exemplar　　thematic　　contiguity　　holistic

1) This layer can contain any set of externally acquired _____ models.
2) The vast majority of the population used their own regional _____ in all aspects of their lives.
3) The _____ problems that Milton is attempting to tackle are written into the very grammar and the syntax of the poem.
4) These sites _____ publicly available information and present it in a user-friendly way.
5) Academic inquiry, at least in some fields, may need to become less exclusionary and more _____.
6) Scientists want to investigate the relation between xerophthalmia occurrence and smut _____.

2. Translate the following sentences into Chinese.

1) Repeatedly, archaeological research on landing sites draws upon the equivalence between a naturally suitable coast and a landing site/harbour.
2) The main objective of ocean museums is making certain objects in the collection visible

or, on the contrary, leaving them invisible.

3) Coastal communities have been widely explored by various cultural researchers, all of whom have applied and developed their own approaches to interpreting landscapes.

4) Despite the diversity of maritime cultural studies worldwide, terrestrial and marine areas are still frequently studied in isolation.

5) The Norwegian experience seems to be to some extent transferable to other areas in Northern Europe and the North Atlantic.

6) In this way, it was more or less a natural choice to take a remarkable ship-find that was discovered in Gredstedbro.

7) This paper explores an innovative approach to the analysis of maritime cultural landscapes in an Australian Colonial setting by engaging wider definitions of cultural landscape.

8) If such characteristics can be defined, what are the possibilities of seeing any of these traits in marine cultural industry?

9) The program is international and introduces postgraduate students with different national and educational backgrounds to the various aspects of the marine cultural industry.

10) Maritime archaeology has been viewed as peripheral in the archaeological discussion.

3. Translate the following sentences into English.

1) 布局和加快海洋文化产业发展，对于我国海洋强国战略和文化强国战略的整体推进都具有深远的意义。

2) 海洋文化资源依附于沿海地区人民在与海洋相关的生产和生活实践中形成的观念、风俗、习惯、规范等思维及行为方式。

3) 创意在产业发展全过程中的融入至关重要，这也是海洋文化产业可持续发展的关键因素。

4) 文化产业发展对自然资源的依赖度低，利用和开发的主要是文化资源。

5) 从社会意义上看，海洋文化产业发展有助于满足人民群众的美好生活需求，提升人们的幸福感、满足感。

6) 中国历来重视国民海洋意识的培养和海洋资源的利用，并依托海洋资源拓展经济增长空间。

7) 在政府的推动和支持下，舟山群岛逐渐形成了海洋文化与民俗业、海洋工艺品业、海洋艺术业、海洋文化会展业、滨海休闲旅游业为核心的海洋文化产业集群。

8) 海洋文化产业发展，需要政府通过战略制定、政策出台、法律保障、资金投入提供引导和保障，为海洋文化产业发展提供健康的环境。

9) 沿海地区美轮美奂的景观可以成为文化产品中故事的线索和发生地，增添文化产品的审美价值和艺术魅力。

10) 海洋文化产业属于新生事物，从目前的发展状况来看主要存在五个方面的短板。

4. Choose the best paragraph translation. And then answer why you choose the first translation or the second one.

绵长的海岸线上重要的节点城市对周边地区和相关产业具有明显的辐射带动作用,通过整体规划、合理布局、整合资源,构建海洋文化产业发展的节点和枢纽城市,有助于形成层次多样、内容丰富的区域性海洋文化产业集群。品牌的文化力决定着品牌的持续竞争力及发展深度,几乎都不约而同地选择打造海洋文化产业品牌,让其成为当地的支柱产业,借助品牌的辐射力和联动效应,带动相关产业发展,实现海洋文化产业集群化,形成完整的海洋文化产业体系。

译文一:

The important node cities on the long coastline have obvious radiating and driving effects on surrounding areas and related industries. Through overall planning, rational layout, and integration of resources, the construction of node and hub cities for the development of the marine cultural industry will help form a rich regional marine cultural industry cluster featured by the variety of levels and contents. The cultural power of the brand determines the sustainable competitiveness and development depth of the brand. Judging from the successful cases of the development of the marine cultural industry, almost all of them choose to build the brand of the marine cultural industry and make it a local pillar industry. With the help of the radiation and linkage effect of the brand, it will drive the development of related industries and realize the clustering of the marine cultural industry, forming a complete marine cultural industry system.

译文二:

The important node cities on the long coastline have obvious radiating and driving effects on surrounding areas and related industries. Through overall planning, rational layout, and integration of resources, the construction of node and hub cities for the development of the marine cultural industry will help form a rich regional marine cultural industry clusters' variety of levels and contents. The cultural power of the brand determines the sustainable competitiveness and development depth of the brand. With the judgment of the successful cases of the development of the marine cultural industry, almost all of them choose to build the brand of the marine cultural industry and make it a local pillar industry. With the help of the radiation and linkage effect of the brand, it will drive the development of related industries and realize the clustering of the marine cultural industry, forming a complete marine cultural industry system.

5. Translate the following passage into Chinese.

The Maritime Cultural Landscape. The term "maritime culture" grew out of a broader understanding of not only the use of the sea by humans, but the attendant structures, cultural identifiers and associations made between people and seafaring. For example, Britain would claim to be a seafaring nation as a result of ship-building traditions on the Clyde and elsewhere, world-wide maritime exploration and the prestige of the Royal Navy.

Norse culture is similarly considered maritime due to extensive sea voyages, raiding and the colonization of land such as Greenland. Today, people driven in cars along roads and over bridges are in contact with the sea in quite a different way. They often treat the sea more as part of a leisure canvas on which to play beside on holiday, dive into and explore, and sail on, rather than remember that it is still relied on to move cargo, for example. However, coastlines are still referred to as landmarks and now nations increasingly consider underwater resources and oil fields, and not fishing alone, as an important part of national wealth.

6. Translate the following passage into English.

我国有1.8万千米的海岸线,面积在500平方米以上的岛屿为6 536个,沿海地区经济发达,蕴藏着内涵丰富、价值独特的海洋文化资源。海洋文化从性质上来看,既是历史现象,也是社会现象,是基于海洋物质文化和精神文化而创造出的文明形态,是诸多涉及海洋文化要素历经传承的凝练。海洋文化资源是沿海地区语言、习俗、服装、文学、艺术、饮食、生产方式、生产技术、生活方式等文化要素的集合,人类生活和社会进步都依赖于海洋文化资源,海洋文化产业发展必然要依托海洋文化资源。

三、翻译家论翻译

傅雷"翻译精神和美感思想"。1951年,傅雷在《高老头》的重译本序言中,开宗明义地指出:"以效果而论,翻译应当像临画一样,所求的不在形似而在神似……各种文学各有特色,各有无可模仿的优点,各有无法补救的缺陷,同时又各有不能侵犯的戒律。像英、法,英、德那样接近的语言,尚且有许多难以互译的地方;中西文字的扞格远过于此,要求传神达意,铢两悉称,自非死抓字典,按照原文句法拼凑堆砌所能济事"。同时,傅雷首次提出"重神似 be alike in spirit,不重形似 be similar in form or appearance"。从此,"重神似不重形似"的翻译主张便引起了翻译学者们的高度重视,深深影响了翻译理论的发展,并在我国文学翻译领域中成为核心理念。不过,也需要说明,首先提出"神似"说的并不是傅雷,其实在20世纪二三十年代,陈西滢、曾虚白等人便拿临画和翻译作比,就有过"注重神似"的说法,只是由于种种原因,没有被流传开来。然而傅雷的这一重提,自然与其在翻译上的成就分不开,最终使其逐渐形成了一派学说。

"神似说"的主要独特之处在于,用文艺美学的视角去把握文学翻译,把翻译活动纳入美学的范畴。而且,"神"与"似"这一对概念早在我国古典美学就提出。"我国古代美学家把审美对象分为'神'与'形'两部分,'神'即精神、内容,或事物发展变化的内在因素;'形'即形体、形质"。之后的诗文理论受我国古典美学"尚情""尚意"的审美倾向的影响,"神似"逐渐重于"形似",注重传神便成为诗文美学的主流。傅雷早年对艺术史有过研究,熟知我国古典美学和绘画诗文领域中的"形神论",因此将其借用来讨论文学翻译问题。

下面,我们来看几个例子:

例1:There is something of magic appeal in the rush and movement of "boom" town, just a clatter of hammers and saws, rounds of drinks and rolls of money.

译文:一个"繁荣"的城市在那熙熙攘攘、川流不息的尘嚣中,自有一种迷人的魅力,到处都在大兴土木,到处都在觥筹交错,到处都是成沓的钞票。

在这个译本中,译文似乎是多用了三个"到处",这种增词并不是译者随意而为,而是依据对原句深刻的理解,为了表达的需要而做的处理。三个"到处"增加了译句的表现力,而且几个成语用得非常贴切。

例2:He was a fool for danger.

译文:他是个天不怕地不怕的人。

需要一提的是,在文学翻译中要做到绝对的"信""传神"是比较难的。因为源语和译语在文化内涵、思维方式等方面有较大差异,很难做到翻译完全对等。我们只有通过反复阅读原文,根据作者字面形象去揣测作者的心思,理解原文字、词、句的深层内涵,并把信息传递给译者才能使译语读者准确地理解原文作者的思想感情。死译,字当句对地翻译,容易让译文读者一头雾水,不知所云。傅雷是我国在翻译理论与实践两方面都可独树一帜的少数翻译大师之一。"傅译"自成一派,着重传神,在我国翻译史上占有极其重要的地位,而其"神似说"对于从事翻译理论及翻译实践人士均有重大借鉴意义。

(摘自:金圣华.傅雷与他的世界[M].北京:生活·读书·新知三联书店,1997:97.)

Text B

Maritime Cultural Heritage Linked to Women

By Maria Grazia Cantarella

Ideological and Political Education：加快发展海洋文化产业 维护国家主权

关心海洋、认识海洋、经略海洋、建设海洋强国。进入21世纪,海洋在国家经济发展格局和对外开放中的作用更加重要,在维护国家主权、安全、发展利益中的地位更加突出,在国家生态文明建设中的角色更加显著,在国际政治、经济、军事、科技竞争中的战略地位也明显上升。经过多年发展,我国海洋事业总体上进入了历史上最好的发展时期,海洋作为高质量发展战略要地的地位日益凸显。在实现第二个百年奋斗目标的新征程上,我们必须进一步关心海洋、认识海洋、经略海洋,协同推进海洋生态保护、海洋经济发展和海洋权益维护,推动我国海洋强国建设不断取得新成就。

《中华人民共和国国民经济和社会发展第十四个五年规划和2035年远景目标纲要》提出:"积极拓展海洋经济发展空间"。加快建设海洋强国,要坚持陆海统筹,促进海洋经济高质量发展。眼下,海洋经济正在成为国民经济新增长点,在扩大内需、破除资源瓶颈、加快新旧动能转换等方面发挥着不可替代的作用。保护海洋生态环境是永续利用海洋资源的基础,我们必须坚持保护与开发并重,保护好海洋生态环境。像对待生命一样关爱海洋,努力实现经济效益与生态效益双赢,才能为子孙后代留下一片碧海蓝天。

当前,我国主权利益、安全利益、发展利益在海洋方向上日趋重合。2022年6月17日上午,我国完全自主设计建造的第三艘航空母舰"中国人民解放军海军福建舰"下水命名,映照着人民海军由弱到强的发展历程,也宣示了维护国家主权、安全、发展利益的坚定决心。加快建设海洋强国,要统筹发展和安全,不断提高维护海洋权益和海洋安全的综合能力。身处经济全球化时代,一片海洋将彼此连通,没有与世隔绝的孤岛。应对海上安全问题,开展海洋全球治理,需要各国携手同行。从提出海洋命运共同体理念到推进21世纪海上丝绸之路建设,从开展海上渔业合作和资源共同开发到配合国际社会打击各种非法渔业活动,我国不断为维护国际海洋秩序、增进海洋繁荣福祉贡献智慧与力量。树立合作共赢理念,广袤海洋必将为人类带来更丰厚的馈赠。

世界最长的跨海大桥港珠澳大桥、国内首个海上数字智慧应用工程"海联网"、世界领先的智慧海洋海底无线通信组网等,都是我国海洋建设的伟大成就。不久前播出的纪录片《蓝海中国》,翔实介绍了海洋领域超级工程和大国重器,折射出我国发展海洋事业的硬核实力。放眼未来,海洋事业潜力无限、大有可为,必将为经济社会发展注入不竭动力,支撑中国号巨轮劈波斩浪、扬帆远航。

1. Maritime cultural heritage is all heritage, tangible or intangible, linked to the ocean, fishing practices, objects found on shipwrecks, old vessels, etc. It is a witness of the relationship between coastal communities, the maritime environment and the landscape itself. This heritage is essential for the creation and strengthening of their cultural **identity**. Throughout Europe and the world, what examples of maritime cultural heritage can be especially linked to women? Let's explore the coast and history to find out……

2. Long Cultural Century-year Old Traditions. In the framework of 2022's European Heritage Days, European cultural practices and heritage connected to the maritime and coastal landscape and to fishing traditions can be **highlighted**. It is also relevant to notice that 2022's theme focused on **sustainability** and sustainable heritage. In fact, when thinking about, for instance, fishing practices and traditions appear that they are more environmentally sustainable and are being rediscovered in order to safeguard the environment and protect natural resources because they are usually more respectful of the nature's **cycles** and have survived for generations.

3. Maritime cultural heritage expands from national and local maritime museums which tell the story of past powerful maritime power, such as in the United Kingdom, through the exhibition of maps, **manuscripts, navigational instruments**, ship models; to local cultural practices that are directly linked to the ocean or the coastal landscape. Many such examples and projects working to preserve this heritage can be found, for instance, in the EU-funded project "PERICLES", which specifically works and has worked towards the preservation, safeguarding and enhancement of coastal spaces in order to enable sustainable usage of maritime and coastal cultural heritage. In this framework, projects have been implemented in Estonia, Malta, Scotland and Ireland, Denmark, Brittany and other places.

4. Portuguese Women Communities Knitting for Fishermen. These European Heritage Days brought to light several stories, projects and practices connected to the maritime sector. Among these, an interesting approach to traditional **hand-knitted** fishermen clothes, made by women in Póvoa de Varzim, Portugal, has produced an artistic project which puts together the traditional practice of **embroidery** of fishermen's clothes with **literary** texts about the figure of Penelope, Ulysses' wife. This project, started in 2021, tries to highlight and celebrate the traditions and the heritage produced by locals and connected to the practices of the fishing town by shedding light on the central figures of women in this town, and how this community has always been fundamentally **matriarchal**. Even if it's the figure of women waiting on land, it still shows the unvalued unpaid and unrecognized yet often central work of women related to fisheries activities.

The project also aimed at **reviving** this tradition and at encouraging local people to know more about their own heritage and celebrating it.

5. Other practices elsewhere bring together cultural heritage, fishing traditions and the role of women: how heritage can make evolve gender **norms** and **vice versa**.

6. For instance, the story of a centuries-old male-dominated tradition opening up to fisherwomen is very much worth mentioning and interesting. The place is Oostduinkerke, on the coast of Belgium on the North Sea, and the practice is horseback **shrimp** fishing. A tradition that has been recognized as representative of the intangible cultural heritage of humanity by UNESCO in 2013, this practice has commonly been associated with men and passed down from father to son. In 2015, though, several local associations representing these fishermen also allowed a fisherwoman to join the team, effectively putting an end to the "all men's club". Previous to that, there were only 17 fishermen remaining practicing this type of fishing. Nele Bekaert, the first woman to be recognized as horseback **shrimper**, says that it is thanks to the inscription of the practice on the UNESCO intangible list that she could be admitted and recognized on the same level of her male colleagues. In fact, the Convention for the Safeguarding of the Intangible Cultural Heritage clearly states that "consideration will be given solely to such intangible cultural heritage as is **compatible** with existing international human rights instruments, as well as with the requirements of mutual respect among communities, groups and individuals, and of sustainable development". This implies, among other things, the respect of gender equality.

7. Furthermore, gender norms may change gradually as does intangible cultural heritage, and for this reason it may happen that cultural practices that were once **confined** to the domain of a specific gender, change over time to include also other gender groups. This story shows how even long-standing traditions and norms that have been always considered the "usual" can change over years or generations. This is by definition what is understood to be intangible cultural heritage which is not by chance also called "living heritage", because it can change and **transform** over time, and this includes **acquiring** a more open and inclusive dimension, for example, by opening up to other gender groups.

8. Korean Women Shellfish Divers. Other traditions linking together traditional fishing practices with women are to be found outside Europe, for instance, in Korea, in particular on the island of Jeju. In this place, women are the **protagonists** of a long-standing tradition which consists in free diving in the ocean to gather shellfish such as **abalone** and sea **urchins**. The women involved in this practice are called Haenyeo, and this tradition too has been **inscribed** in the UNESCO intangible cultural heritage of

humanity list. Women have a central role in this community and this practice has allowed them over the centuries to gain an important social status. The practice has always been passed down from mother to daughter, however, nowadays younger generations try to find different occupations and move out of the island. For this reason, now the remaining Haenyeo who still dive to fish are sometimes over 80 years old. Furthermore, this tradition is also considered environmentally sustainable, as shown by the fact that this type of fishing has been done for centuries and has allowed women to be at the **forefront** of decision-making in natural resources management.

9. All over the world there are many practices, traditions, cultural heritages that involve fishing or are connected to the coastal and maritime landscape. It is important to highlight this type of heritage not only because it represents the identity and history of a community, but also because it can teach something and can bring to light the role of specific groups or people. In the realm of fishing practices, those that come from the past and have been practiced for centuries, have usually environmentally and sometimes socially sustainable characteristics that could be useful to **replicate** elsewhere. In other cases, as shown, these traditions may have **undergone** changes that have further highlighted their importance and significance by being accessible to more social and gender groups. In this context, the UNESCO Convention for the Safeguarding of the Intangible Cultural Heritage and its list, as well as other projects and instrument can be a vehicle through which these practices are highlighted and if necessary updated or improved in certain aspects.

(1,170 words)

https://womeninseafood.org

New Words

1. identity [aɪˈdentəti]

n. the distinct personality of an individual regarded as a persisting entity; the individual characteristics by which a thing or person is recognized or known; an operator that leaves unchanged the element on which it operates 身份,本体;个性,特性;同一性,一致;恒等运算,恒等式

The suspect had refused to give any details of his identity and had carried no documents on his person.

犯罪嫌疑人拒绝交代任何有关他身份的细节,身上也没带任何证件。

2. highlight [ˈhaɪlaɪt]

v. apply a highlighter to one's cheeks or eyebrows in order to make them more prominent; move into the foreground to make more visible or prominent 突出,强调;用亮色突出;挑染

Please highlight any terms that are unfamiliar to you.

请把你们不熟悉的用语都标示出来。

3. sustainability [səˌsteɪnə'bɪləti]

　　n. the property of being sustainable 持续性，能维持性

　　When choosing what products to buy and which brands to buy from, more and more consumers are looking into sustainability.

　　在选择购买产品和品牌时，越来越多的消费者开始关注可持续性。

4. cycle ['saɪk(ə)l]

　　n. a periodically repeated sequence of events; the unit of frequency; one hertz has a periodic interval of one second; a wheeled vehicle that has two wheels and is moved by foot pedals; an interval during which a recurring sequence of events occurs 循环，周期；整套，系列；自行车，摩托车；自行车骑行；一段时间

　　We need to break the vicious cycle of violence and counterviolence.

　　我们需要结束暴力和反暴力的恶性循环。

5. manuscript ['mænjuskrɪpt]

　　n. handwritten book or document; the form of a literary work submitted for publication 手稿，原稿；手抄本，手写本

　　These odd assertions were interpolated into the manuscript sometime after 1,400.

　　这些零散的主张是 1 400 年后被补充进手稿里的。

6. navigational [ˌnævɪ'geɪʃənl]

　　a. of or relating to navigation 航行的，航运的

　　Many water scooter operators are inexperienced and ignorant of navigational rules, which increase the potential for accidents.

　　许多水上摩托车驾驶员缺乏经验，且对航行规则一无所知，这增加了发生事故的可能性。

7. instrument ['ɪnstrəmənt]

　　n. the means whereby some act is accomplished; a device that requires skill for proper use; any of various devices or contrivances that can be used to produce musical tones or sounds 仪器；器械；乐器；促成某事的人（或事物），手段；受利用（或控制）的人，工具；文据，正式法律文件

　　Mr. Fahrenheit, Daniel Gabriel Fahrenheit was a German instrument maker.

　　华勒斯先生，也就是丹尼尔·加布里尔·华伦海特是一位德国仪器制造师。

8. hand-knit [hænd nɪt]

　　v. protect from impact 手工编织

　　I don't have patience to hand-knit a gorgeous sweater like my knitting friends.

　　我没有耐心能够像我织毛衣的朋友们那样织出一件华丽的毛衣。

9. embroidery [ɪm'brɔɪdəri]

　　n. decorative needlework; elaboration of an interpretation by the use of decorative (sometimes fictitious) detail 刺绣技法，刺绣活儿；绣花，刺绣品；夸张之词，渲染成分

Craft Resources also sells yarn and embroidery floss.

"手工之家"也售卖纱线和绣花丝线。

10. literary ['lɪtərəri]

　　a. of or relating to or characteristic of literature; knowledgeable about literature; appropriate to literature rather than everyday speech or writing 文学的,文学上的;书面的;爱好文学的,从事文学研究(或写作)的

　　She has published more than 20 books including novels, poetry and literary criticism.

　　她已出版了 20 多本书,其中包括小说、诗歌和文学评论。

11. shed [ʃed]

　　v. get rid of; pour out in drops or small quantities or as if in drops or small quantities; cause or allow (a solid substance) to flow or run out or over; pour out in drops or small quantities or as if in drops or small quantities 去除,摆脱;(植物)落(叶),(动物)蜕(皮),脱(毛);脱掉(所穿衣物);掉落(货物);挡水,防水;射出,发出(光);流,洒,落(泪)

　　Within two years, Leah shed the safety of her accounting job and made the switch complete.

　　两年间,莉娅摆脱了会计工作的安全感,完成了转型。

12. matriarchal [ˌmeɪtri'ɑːkl]

　　v. characteristic of a matriarchy 母系氏族的;女家长的

　　The animals live in matriarchal groups.

　　这些动物以母系氏族的方式生活。

13. revive [rɪ'vaɪv]

　　v. cause to regain consciousness; give new life or energy to; be brought back to life, consciousness, or strength; return to consciousness (使)复原,(使)复苏;重新唤起,重新记起;重新使用,使复兴;重新上演

　　The Chancellor of the Exchequer has indicated that he plans to revive a scheme put forward last year by the International Monetary Fund, which has not yet provided any relief.

　　这位英国财政大臣表示,他计划重启国际货币基金组织去年提出的一项计划,该计划迄今尚未提供任何帮助。

14. norm [nɔːm]

　　n. a standard or model or pattern regarded as typical; a statistic describing the location of a distribution 社会准则,行为规范;标准,平均水平;常态,平常事物

　　From now on, when someone asks you how your life is, try responding with words like "exciting and fun" instead of the cultural norm that says "busy".

　　从现在开始,当有人问起你的生活如何时,试着用"令人兴奋和有趣"这样的词来回答,而不是用"忙碌"这种文化套话来作答。

15. vice versa [ˌvaɪs'vɜːsə; ˌvaɪsi'vɜːsə]

　　ad. with the order reversed 反之亦然

　　A recurring theme is that as circumstances change, different sorts of leaders are required; a leader who thrives in one environment may struggle in another, and vice versa.

一个反复出现的主题是,随着环境的变化,人们需要不同类型的领导者;在一种环境中应对自如的领导者可能在另一种环境中艰难挣扎,反之亦然。

16. shrimp [ʃrɪmp]

n. small slender-bodied chiefly marine decapod crustaceans with a long tail and single pair of pincers; many species are edible; disparaging terms for small people 虾,小虾;(非正式)矮子,无足轻重的人

Paella's ingredients include rice, olive oil, roasted rabbit, chicken, shrimp, red and green peppers, onions, and plenty of garlic.

海鲜饭的原料包括米饭、橄榄油、烤兔肉、鸡肉、虾、红青椒、洋葱和大量的大蒜。

17. shrimper [ˈʃrɪmpə(r)]

n. a person engaged in shrimping 捕虾之人

Eric Drury is a shrimper but knows that his business is not the only one threatened.

埃里克·特鲁里是一位捕虾者,但他也知道他的生意不是唯一受到威胁的。

18. compatible [kəmˈpætəb(ə)l]

a. able to exist and perform in harmonious or agreeable combination; (of a couple) existing together harmoniously; capable of being used with or connected to other devices or components without modification 可共存的;可和睦相处的;兼容的;与……一致的

They uncovered clues to what was going wrong by researching a fascinating subject: how birth order affects not only your personality but also how compatible you are with your mate.

他们通过研究一个有趣的课题发现了问题的线索:出生顺序不仅影响你的个性,还影响你和伴侣的和谐度。

19. confine [kənˈfaɪn]

v. restrict; place limits on (extent or access); prevent from leaving or from being removed; deprive of freedom; take into confinement 限制,局限;防止……扩散;关押,监禁;使离不开(或受困于床、轮椅等)

Risk does not, of course, confine itself only to the troubled bits of the globe.

当然了,风险不会仅仅局限于地球上的一些动荡不安的角落。

20. transform [trænsˈfɔːm]

v. convert (one form of energy) to another; change in outward structure or looks; change or alter in form, appearance, or nature 使改观,使变形,使转化;变换(电流)的电压;(数)变换(数学实体)

Number words and their written forms transform our quantitative reasoning as they are introduced into our cognitive experience by our parents, peers and school teachers.

当我们的父母、同龄人和学校老师把数词及其书面形式引入我们的认知经验时,它们改变了我们的定量推理方式。

21. acquire [əˈkwaɪə(r)]

v. come into the possession of something concrete or abstract; gain knowledge or skills;

come to have or undergo a change of (physical features and attributes)获得,得到;学到,习得;患上(疾病);逐渐具有,开始学会

Less inherently interpersonal subjects, such as math, could acquire a social aspect through team problem solving and peer tutoring.

像数学这样本身就较少涉及人际关系的学科,可以通过团队解决问题和同伴辅导获得社交方面的知识。

22. protagonist [prə'tægənɪst]

n. a person who backs a politician or a team etc.; the principal character in a work of fiction(戏剧、电影、小说等的)主人公;(比赛、斗争中的)主要人物,主要参与者;(政策、运动的)倡导者,拥护者

Rebecca West wrote a good many novels and one in particular called The Return of the Soldier, the protagonist of which is also a traumatized war victim.

丽贝卡·韦斯特写了很多小说,其中有一本叫《军士返乡》,书中的主人公也是一个饱受创伤的战争受害者。

23. abalone [ˌæbə'ləʊni]

n. any of various large edible marine gastropods of the genus Haliotis having an ear-shaped shell with pearly interior 鲍鱼

Abalone shell, as a kind of organic gemstone material, shows the beautiful and unique iridescence. Abalone can be used to produce blister pearl.

作为一种有机宝石材料的鲍鱼贝壳,具有美丽而独特的晕彩,鲍贝还能用来生产具有同样晕彩的贝附珍珠。

24. urchin ['ɜːtʃɪn]

n. ocean urchin; poor and often mischievous city child 海胆;刺猬;顽童,淘气鬼

The bright red color of a fire urchin warns predators that its spines are poisonous. Sea otters are the major predators of sea urchins.

火海胆鲜红的颜色警告捕食者它的刺有毒。海獭是海胆的主要捕食者。

25. inscribe [ɪn'skraɪb]

v. write, engrave, or print as a lasting record; mark with one's signature; carve, cut, or etch into a material or surface 题写;题献;铭记;雕

In the very books in which philosophers bid us scorn fame, they inscribe their names.

一些哲学家们在书中大谈视虚名为粪土,然而就是在这些书上却赫然写着他们的大名。

26. forefront ['fɔːfrʌnt]

n. the position of greatest importance or advancement; the leading position in any movement or field; the part in the front or nearest the viewer 重要位置,最前沿;(思考、关注的)重心

The results of these early efforts are as promising as they are peculiar, and the new nature-based AI movement is slowly but surely moving to the forefront of the field.

这些早期努力的结果越独特就意味着它们越有希望,而尽管以自然为基础的新人工智能发展缓慢,但肯定正在走向该领域的前沿。

27. replicate [ˈreplɪkeɪt]
    v. reproduce or make an exact copy of; make or do or perform again; bend or turn backward 重复,复制;(遗传物质或生物)自我繁殖,自我复制;复证(实验或试验以得出一致的结果)
    Today's top violin makers can pretty much replicate all the physical attributes of a Cremonese violin.
    今天的顶级小提琴制造商几乎可以复制克雷莫纳小提琴的所有物理属性。

28. undergo [ˌʌndəˈgəʊ]
    v. go or live through; go through (mental or physical states or experiences) 经历,经受
    The pufferfish must be prepared a special way, by licensed chefs who undergo rigorous training, to be enjoyed.
    河豚须由经过严格训练的有执照的厨师用一种特殊的方式烹饪才能食用。

Phrases and Expressions

be linked to 与……连接;与……有关联,与……有联系
be essential for 对……必不可少的
in the framework of 在……框架中
think about 认真考虑
survive for 为……生存
bring to light 揭露;发现;公开
put together 放在一起;组合;装配
shed light on 阐明;使……清楚地显出
be worth doing 值得做……
be recognized as 被认为是……,被公认为……
pass down 传下来;遗传
previous to 在……以前
thanks to 由于,幸亏
be compatible with 一致;适合;与……相配
in the realm of 在……领域里

Proper Names

2022's European Heritage Days 2022 欧洲文化遗产日
Póvoa de Varzim 葡萄牙北部的波华市
UNESCO 联合国教科文组织(United Nations Educational, Scientific, and Cultural Organization)

一、翻译简析

1. In fact, when thinking about, for instance, fishing practices and traditions it appears that they are more environmentally sustainable and are being rediscovered in order to safeguard

the environment and protect natural resources because they are usually more respectful of the nature's cycles and have survived for generations. (Para. 2)

事实上,仔细考虑,捕鱼习俗和传统似乎更具有环境可持续性,因为它们通常更遵循自然的周期规律,并且世代延续,为了保护环境和自然资源,人们正在重启这些习俗和传统。

【英语的逻辑顺序一般是先主后从,先果后因。本句原文把信息重心放在前面,然后在句末补充说明原因,译文部分遵循汉语先因后果的逻辑顺序,对语序做了调整,把原因放在主句的两个并列结构中间,按照事物的自然逻辑发展顺序排列信息,这样更符合汉语读者的思维习惯。】

2. Maritime cultural heritage expands from national and local maritime museums which tell the story of past powerful maritime power, such as in the United Kingdom, through the exhibition of maps, manuscripts, navigational instruments, ship models; to local cultural practices that are directly linked to the ocean or the coastal landscape. (Para. 3)

国家和地方海事博物馆,通过展览地图、手稿、航海仪器、船模,讲述过去强大海上强国的故事,英国的海事博物馆就是如此。海洋文化遗产从国家和地方海事博物馆扩展到与海洋或沿海景观直接相关的地方文化习俗。

【本句原文的主句为"Maritime cultural heritage expands from national and local maritime museums to local cultural practices"。主句的两个宾语各有一个定语从句对它们进行修饰,其中"which"引导的定语从句较长,如果按照原文的结构和语序进行翻译,会让译文读者感到费解,因此需要调整语序并进行断句,使"which"引导的定语从句单独成句放在前面,并把从句中的状语调整到谓语之前。】

3. Even if it's the figure of women waiting on land, it still shows the unvalued unpaid and unrecognized yet often central work of women related to fisheries activities. (Para. 4)

即使女性给人的印象是在陆地上等待渔船归来,这仍然表明与渔业活动有关的女性从事的工作不受重视,没有报酬,不被认可,却往往是核心工作。

【本句翻译的难点是状语从句"Even if it's the figure of women waiting on land"的处理。首先分析结构,"it"做形式主语指代动名词短语"waiting on land",可以理解为"waiting on land is the figure of women"。然后选择词义,"figure"作为与人相关的名词在词典里有"身材""人影""人物""人像"等中文释义,有"alternative names for the body of a human being""a model of a bodily form""a well-known or notable person""the impression produced by a person"等英文释义,根据文章的上下文和逻辑关系,这里只能选择"the impression produced by a person",翻译为"给人的印象"。】

4. This is by definition what is understood to be intangible cultural heritage which is not by chance also called "living heritage", because it can change and transform over time, and this includes acquiring a more open and inclusive dimension, for example by opening up to other gender groups. (Para. 7)

根据定义,这就是所谓的非物质文化遗产,它也被称为"活遗产",这并非偶然,因为它可以随着时间的推移而变化和转变,这包括获得更开放和更包容的维度,例如向其他性别群体开放。

【本句翻译的难点是定语从句"which is not by chance also called 'living heritage'"的处理。首先,这个定语从句有些复杂,需要翻译成后置的独立分句。其次,如果不改变原句的结构将其翻译成"它也不是偶然地被称为'活遗产'",读起来别扭,不妨将"not by chance"从原来的位置上提出来,放到后面单独处理,翻译成"它也被称为'活遗产',这并非偶然",这样才能顺应汉语的表达习惯。】

5. In this place, women are the protagonists of a long-standing tradition which consists in free diving in the ocean to gather shellfish such as abalone and sea urchins. (Para. 8)

在那里,有个悠久的传统,女性是这个传统的主角,她们在海里自由潜水,捕捞鲍鱼和海胆等贝类。

【英语和汉语在句子结构上存在差异,英语多长句,汉语多短句,英语多从句,汉语多分句,英语多代词(包括 that、which 等关系代词),汉语多名词。根据这些差异,翻译时按照汉语陈述事实的逻辑习惯,将主句拆分成两个分句,定语从句也拆分成分句。】

二、参考译文

与女性有关的海洋文化遗产

1. 海洋文化遗产是所有与海洋、捕鱼活动、沉船上发现的物品、旧船只等有关的有形或无形遗产。它见证了沿海社区、海洋环境和景观本身之间的关系。这种遗产对于创建和加强它们的文化特征至关重要。在欧洲和世界各地,有哪些海洋文化遗产特别与女性有关?让我们一起探索海岸和历史,把它们找出来。

2. 悠久的百年文化传统。在 2022 年欧洲遗产日的框架内,与海洋和沿海景观以及渔业传统相关的欧洲文化习俗和遗产可以得到凸显。同样值得注意的是,2022 年欧洲遗产日的主题聚焦于可持续性和可持续遗产。事实上,仔细考虑,捕鱼习俗和传统似乎更具有环境可持续性,因为它们通常更遵循自然的周期规律,并且世代延续,为了保护环境和自然资源,人们正在重启这些习俗和传统。

3. 国家和地方海事博物馆,通过展览地图、手稿、航海仪器、船模,讲述过去强大海上强国的故事,英国的海事博物馆就是如此。海洋文化遗产从国家和地方海事博物馆扩展到与海洋或沿海景观直接相关的地方文化习俗。许多这样的典型和项目都致力于保护这种遗产,例如欧盟资助的"伯里克利"项目,该项目专门致力于保护、保卫和优化沿海空间,以实现海洋和沿海文化遗产的可持续利用。在这一框架下,多个项目已在爱沙尼亚、马耳他、苏格兰和爱尔兰、丹麦、布列塔尼等地实施。

4. 葡萄牙妇女社群为渔民编织衣物。这些欧洲遗产日展示了与海事部门相关的几个故事、项目和习俗。其中,葡萄牙波瓦-迪瓦尔津的妇女手工编织了传统渔民服装,她们采用的方法很有趣,形成一个艺术项目,该项目将渔民服装刺绣的传统工艺与文学作品里描述尤利西斯妻子佩内洛普形象的文字结合在一起。该项目始于 2021 年,通过揭示该城女性的核心形象以及为什么这个社群从根本上来说一直是母系社会,来彰显和弘扬当地人创造的与渔村习俗有关的传统和遗产。即使女性给人的印象是在陆地上等待渔船归来,这仍然表明与渔业活动有关的女性从事的工作不受重视、没有报酬、不被认可,却往往是核心工作。该项目还有一个目的就是恢复这一传统,并鼓励当地人更多地了解和弘扬自己的文化遗产。

5. 其他地方也有将文化遗产、渔业传统和女性作用结合在一起的习俗,涉及文化遗产如何推动性别规范的演变,反之亦然。

6. 例如,几个世纪以来男性主导的传统向女性渔民开放的故事非常值得一提,也很有趣。故事发生在位于比利时北海沿岸的奥斯特敦刻尔克城,在那里人们骑马捕虾。2013年,联合国教科文组织将这一传统认定为人类非物质文化遗产的代表,这种捕虾方式通常与男性有关,由父亲传给儿子。然而,2015 年,代表这些男性渔民的几个地方协会也允许一名女性渔民加入团队,有效地结束了那个"全男性俱乐部"。在此之前,只有 17 名渔民还在继续这种捕虾方式。纳里·贝卡尔特是第一个被承认为"捕虾骑手"的女性,她说,多亏这种捕虾方式被列入联合国教科文组织非物质文化遗产名录,她才能得到与男同事相同的承认和认可。事实上,《保护非物质文化遗产公约》明确规定:"只考虑符合现有国际人权文书,符合社群、团体和个人相互尊重和可持续发展要求的非物质文化遗产。"除了其他方面,这意味着尊重两性平等。

7. 此外,就像非物质文化遗产一样,性别规范可能会逐渐发生变化,因此,曾经局限于某个特定性别领域的文化习俗可能会随着时间的推移而发生变化,也会接纳其他性别群体。这个故事表明,即使是长期以来人们"习以为常"的传统和规范,也会随着岁月或世代的流逝而发生变化。根据定义,这就是所谓的非物质文化遗产,也被称为"活遗产",这并非偶然,因为它可以随着时间的推移而变化和转变,这包括获得更开放和更包容的维度,例如向其他性别群体开放。

8. 韩国女性潜水捕捞贝类。在欧洲以外也可以找到其他将捕鱼习俗与女性联系在一起的传统,例如在韩国,特别是在济州岛。在那里,有个悠久的传统,女性是这个传统的主角,她们在海里自由潜水,捕捞鲍鱼和海胆等贝类。参与这种劳动的女性被称为"海女",联合国教科文组织也将这一传统列入了人类非物质文化遗产名录。女性在社群中起着核心作用,几个世纪以来,这种习俗使她们获得了重要的社会地位。这种习俗一直是母亲传给女儿的,然而,如今的年轻一代试图寻找不同的职业,并搬出这座岛屿。因此,现在还在潜水捕鱼的海女,有的已经 80 多岁了。此外,这一传统也被认为在环境上是可持续的,事实表明,潜水捕鱼已经持续了几个世纪,使女性处于自然资源管理决策的最前沿。

9. 世界各地有许多涉及渔业、沿海和海洋景观的习俗、传统和文化遗产。强调这种类型的遗产很重要,不仅因为它代表了一个社会群体的特征和历史,而且因为它可以教会人们一些东西,可以揭示特定群体或人的作用。在捕鱼习俗方面,那些从过去到现在已实行了几个世纪的捕鱼方式通常具有环境可持续性,有时具有社会可持续性,在其他地方进行复制可能是有用的。在其他情况下,如上文所示,这些传统可能经历了变化,使更多的社会群体和性别群体能够接触到这些传统,从而进一步突出了它们的重要性和意义。在此背景下,联合国教科文组织《保护非物质文化遗产公约》及其名录,以及其他项目和文书,可以作为强调这些习俗的载体,并在必要时在某些方面更新或改进这些习俗。

Cultural Background Knowledge: Marine Cultural Industry
海洋文化产业文化背景知识

  海洋文化产业是当代海洋文化建设的重要载体，是海洋文化自信的重要体现，是国家文化建设和海洋战略的重要组成部分，对于推动海洋文化传播、促进海洋经济与科技文化融合发展、服务海洋外交、提高国民海洋意识、捍卫海洋权益、推进海洋建设、发展海洋文明具有独特而重要的作用。

  人类文化精神的创新、扩散方式，以及社会文化产品的生产和消费方式，因一个时代生产力和生产关系的发展变化而发展变化，同时又为社会的维系和发展提供重要动力。在人类经历了农耕文明、工业文明，正在步入智能时代的今天，现代海洋文化产业作为一种文化精神新的创造方式和文化消费的新方式，其地位和作用将会越来越突出。

  目前美、日、英等海洋强国海洋文化产业发展已经相当成熟，海洋文化产品的生产和输出，极大地推进了西方国家海洋战略和海洋文化传播，形成了世界文化流通中的西方海洋文化话语体系，强化了全球海洋发展和国际政治阐释权，海洋文化产业发展已经成为西方国家海洋战略和文化战略的重要组成部分。

  从历史发展看，西方海洋战略思想历经了早期偏重于海权争夺的军事阶段，到重视国家整体实力建设的综合战略阶段，再到把文化软实力建设作为国家海洋战略重要组成部分的文化战略三个阶段。

  1890年，美国学者马汉在《海权论》中首次把海权与国家战略、民族兴衰联系在一起，提出了海洋战略系统理论。1911年，英国学者科贝特出版了《海洋战略的若干原则》一书，构建起国家、海洋、海军三级战略体系，各级战略相互支撑，政治、经济、军事、文化各要素形成一个相互关联的有机整体。1982年《联合国海洋法公约》诞生，确立了"群岛国"、200海里"专属经济区"、350海里"大陆架延伸"等新概念，"海权"的外延进一步扩大，"战略"一词进一步从军事概念演化为综合性概念，海洋战略成为涉及政治、经济、文化、法律、社会、军事等领域的综合体系。20世纪末，美国学者约瑟夫·拉皮德提出"文化回归"概念，以约瑟夫·奈、塞缪尔·亨廷顿、温特以及平野健一郎为代表的学者们进一步丰富了海洋战略的文化内涵。

  中国不仅拥有漫长的海岸线、辽阔的海洋面积和极其丰富的海洋自然资源，还有历史悠久、内涵丰富的海洋文化特色资源和独立于世的海洋文化精神财富。发展中国海洋文化产业，我们有足够的文化自信。在中国海洋科技水平和海洋经济发展水平持续提高的基础上，充分利用海洋文化资源，发掘中国海洋文化精神内涵，打造中国海洋文化产业创新经济带，走经济、文化融合发展之路，通过"科技+""文化+""互联网+"的方式创新海洋经济形态，对于增强中国的文化自信、推动海洋经济升级发展和"21世纪海上丝绸之路"建设将发挥出巨大作用。

参考答案

Unit One

Text A
二、译例与练习
Exercises

1. Fill in the blanks with the proper given words, and then translate the sentences into Chinese.

1) regulatory
肾上腺素不直接作用于大脑,怎么会对大脑功能产生调节作用呢?

2) initiate
被交谈的人与主动发起谈话的人有着同样积极的体验。

3) accommodation
当地政府将立即为多达三千名无家可归的人提供临时住所。

4) dynamic
近年来巴塞罗那已经成为地中海地区最具活力、最为繁荣的城市之一。

5) incorporate
技术变革被认为是利益相关方之间谈判的结果,这些利益相关方试图将自己的利益纳入机制的设计和配置中。

6) extract
厨房里有一台机器,只要管家用拇指按下小按钮两百次,半小时就能榨出两百个橙子的汁。

2. Translate the following sentences into Chinese.

1) 现有的海上通信、运输和旅游业迅速扩张,围绕海洋的领土主张和信息需求也在迅速增长。

2) 在维持社会和国家之间的知识流动以及社会文化交流方面,海洋发挥了关键作用。

3) 共享这些资源有助于维持区域政治和经济稳定,从而有助于区域海洋和环洋地区附近的民生福祉。

4) 长期以来沿海人民有许多共同的社会准则,其中许多准则已被编入现代法律。这些海事行为准则旨在改善海上安全和福祉,便利旅行和商业往来。

5) 经过几个世纪的贸易往来,美洲、欧洲和非洲航海民族之间形成了一系列联系,因而逃离国家暴力和劳动力剥削的逃亡者结成了松散的联盟。

6）在许多当代文化中，海滩和海边与家庭度假、童年记忆、浪漫情调和晚年团聚有着紧密的联系。

7）知道如何划船、修理发动机、穿越海浪或用矛刺鱼，这些都是海事职业中让人充满自豪感的实用技能。

8）渔民们为自己工作或与朋友和家人一起工作，他们很看重这种独立和自由，不愿意在室内工作，受时间表的约束，向老板汇报，所以即使有更赚钱的工作，他们也常常强调他们需要独立，并选择留在渔场。

9）海洋和海岸激发了视觉和创意艺术的灵感，至少从第一次洞穴绘画创作开始，人类就感到有审美表达的需要。

10）那些不属于海洋的人，也可能获得与海洋融为一体的幸福感。

3. Translate the following sentences into English.

1) While the ocean is the place where you go to do your job, it may also be where you feel most free, most in control of your own destiny, most competent and most valued by others.

2) Sharks in the Pacific islands were imbued with spiritual powers, considered as ancestor guardians or gods who offered protection from the unpredictable forces of the ocean.

3) In the last 50 years, there has been rapid growth in new ocean industries such as mariculture, deep ocean drilling for hydrocarbons and minerals, desalinization and offshore wind farms.

4) Wind energy has traditionally been used for voyaging at sea but is now increasingly used to generate electricity for land-based human activities through offshore wind farms.

5) The ocean is seen as an underutilized source of raw materials for contemporary societies.

6) Shipping and fishing have long been contributors to generating wealth and jobs and supporting livelihoods in coastal and island economies.

7) These societies began to connect, some 5,000 to 3,000 years ago, when traders learned to use the monsoon to trade across the ocean rather than along coasts.

8) Populations in lockdown have flocked to beaches when allowed to do so, causing concerns that a second wave of COVID-19 infections would manifest in the Northern Hemisphere summer.

9) As thoughts turn to how to rebuild economies and restart social life, the coming months will provide opportunities to reinforce how important 'blue spaces' are to people and to ensure that people have access to them for their well-being.

10) Coastal populations have been growing about twice as fast as national growth rates, and population densities there are twice the world's average.

4. Choose the best paragraph translation. And then answer why you choose the first translation or the second one.

译文二比较好。

译文一:在措辞和结构上都过于拘泥于原文的词汇和结构,译文有些生硬。

译文二:措辞和结构都很灵活,能够准确再现原文含义,符合译入语语言习惯。

5. Translate the following passage into Chinese.

主观幸福感也受到焦虑的影响,心理学家确定了六种存在焦虑:身份、幸福、孤独、生命的意义、自由和死亡。在我们与海洋的关系中,或者更广泛地说,在我们与自然的关系中,无论这种关系是有关职业的、居住的、消费的还是娱乐的,所有这些焦虑都可以面对或缓解(或两者都可以)。我们注意到,生活与海洋资源利用密切相关的群体(渔民、海员、土著人民、海洋旅游和娱乐专业人员)与海洋有着复杂的多维关系,这些关系往往是深度精神层面的,而且强烈地反映了他们的社会和文化身份。

6. Translate the following passage into English.

The Chinese central authority referred to the 'blue economy' in the 13th five-year plan in 2016, but contemporary policy on China's ocean economy dates to the start of the 'opening and reform' period in the late 1970s. Blue economic development in China accelerated around the turn of the 21st century. Since China ratified the UN Convention of the Law of the Sea (UNCLOS) in 1996, the state has established numerous exclusive economic zones, called for 'implementing ocean development' and issued various five-year plans for ocean economic development. The focus has culminated with the explicit goal of becoming a 'maritime power', possessing military defence capabilities, a strong ocean economy and advanced marine science and technology. In 2019, Premier Li Keqiang summarized the state vision of China's blue economy as to 'vigorously develop the blue economy, protect the ocean environment, and construct a great maritime nation'.

Unit Two

Text A

二、译例与练习

Exercises

1. Fill in the blanks with the proper given words, and then translate the sentences into Chinese.

1) consisting of

石油主要是指原油和天然气,可能来自沉积在海底的有机物。

2) with respect to

传记作者往往围绕着主题在两种立场间跳跃摇摆。

3) entity

他们就是通过削弱你对自己智力的信心,让你觉得外部实体才是值得信赖的。
4) perpetual
它与我们常见健康恐慌相比如何,比如人们在合成运动场上对铅的恐慌?
5) originating
源自冰原的冰山在数百万年间的降雪中最终在陆地上得以形成。
6) marginal
这个特别法庭是为社会中的主流人群设立的,而不是为了无足轻重的小人物。

2. Translate the following sentences into Chinese.
1) 进入海洋的大多数污染物都来自沿海及腹地的人类活动。
2) 多数情形下,海洋中的深水循环都受到水流的控制,这些水流都来自大西洋、红海和南极洲的洋流。
3) 海床上的主要分布物是海底山,这是一种陡峭而平顶的海底山峰。
4) 日本暖流除了会影响天气之外,还会影响到气候,尽管这种影响在成百上千万年的长时段中依旧不够清晰。
5) 深海生物已经进化出了一些绝佳的觅食机制,因为这些地区的食物极为稀缺。
6) 颇具讽刺意味的是,当前那些引发环境问题的,同那些原本用来防止自然界过度损耗和过度开发的,竟是同一种东西。
7) 海洋颜色的变化最初是因为悬浮在水中的生物体在类型和数量上的改变。
8) 蒸发和降水决定了某一特定地区海洋的含盐量,而太阳能则是两者背后的决定性因素。
9) 若将海洋最深处的积水量考虑在内,静水压力是影响深海生物最为重要的环境因素之一,这一观点应该不足为奇。
10) 这是因为比起温水,冷水的溶氧力更强,并且极深之水一般都源自极地浅海。

3. Translate the following sentences into English.
1) The general characteristics of the deep sea can be listed as follows: high pressure, nonluminousness, low temperature, high salinity, high oxygen content and abundant sediments.
2) The first massage transmitted by the fresh oceanic intellectuals concerns eternal alteration within ocean.
3) The eruption made by the submarine volcano was caused by the continuous accumulation of underground energy, rather than some individual occurrence.
4) The area of submarine world explored by human till now only accounted for 5%, while 95% still remained unknown.
5) There occurred numbers of massive crustal disturbances formulating abundant coastal ranges.
6) Although the present deep diving technology seems quite clumsy, through it we can take a glance at mysterious landscape in the deepest ocean.
7) After confirming the depth of the ocean floor human tended to look over its true colors.

8) Most submarine volcanoes locate around the area of the plate movements, which is called mid-oceanic ridge.

9) Pacific Ocean, stretching 15,900 km in maximum length from the south to the north, and 19,000 km in maximum width from the east to the west, covers an area of 18,134.4 square km.

10) Oceanic ridge lying in the mid-Atlantic ocean, its moving direction parallels to Atlantic east-and-west coastline, distributing in the shape of "S".

4. Choose the best paragraph translation. And then answer why you choose the first translation or the second one.

译文一较好。

译文一：这篇译文通过使用短语、分词和各类从句将原文译成语义相对独立的三个语义单元，并分别以三个结构较为复杂的句子加以表现。第一个语义单元通过并列 by 介词短语，以及穿插使用定语从句和名词短语，将印度洋不同时代和文化背景下的名称交代的一清二楚。因为大量使用了 by 介词短语有效地避免了同一主语的无谓重复，符合英文讲求简练的语言特点。同时，因为使用了名词性短语和定语从句又主次分明地将整个句子的语义层次清晰地表达了出来。第二个语义单元总体上看是一个以 not until 开头的倒装句，符合英语惯用倒装句的特点。为了凸显句内不同的语义层次，先后使用了名词短语和现在分词短语，整个句子显得主次分明、错落有致。

译文二：这篇译文在用词上同第一篇译文相比并无太大差异，原文语义标的也十分准确。然而，在句子结构方式上与第一篇译文有很大的不同。不难看出，这篇译文通篇均以英语简单句为主要语义表达单位，尤其是在表述印度洋的名称变化时，将原文译成了三个独立的简单句。相较于第一篇译文的处理方式，这样的译文不仅读起来颇有累赘之感，更为重要的是未能充分体现出英语重"形合"的造句思维，依旧带有汉语重"意合"的色彩。因此，尽管第二篇译文无论是在语法上还是在遣词造句上都无错处，但是称不上是一篇符合英语语言思维方式的好的译文。

5. Translate the following passage into Chinese.

海水的特性取决于纯水和溶解物的性质。海水中的固体溶解物有两个来源。一部分产生于陆上岩石的风化作用，随雨水流入大海，另一部分则来自地球内部。大部分溶解物都是通过深海热泉口进入海洋的，少部分则先是由火山进入大气，再随雨雪流入海洋。海水的化学成分几乎无所不有，但是水中大部分溶解物的成分仅由几种离子组成。这一点让人颇感惊奇。实际上，海水中98%的溶解物都是由6种离子组成。其中钠和氯约占到85%，因此海水都是咸味的。海水中含盐对居于其中的生物影响颇大。比如，大多数的海洋生物若居于淡水之中便会死亡。不仅如此，即便含盐量有些微妙的变化都会对诸种生物产生不利影响。

6. Translate the following passage into English.

Compared with modern ocean, primitive ocean is far from huge in terms of scale evidenced by the estimation that its water volume only accounts for 10% of the one of modern ocean. Later, the infusion of constitution water deposited within earth's interior expanded primitive ocean into present magnificent modern ocean. Seawater in primitive ocean is not bitter and salty like the one in modern ocean. Mineral salt in modern ocean has been increasing year by year as pouring into submarine world from land through hydrological circulation in nature, while organic molecules in primitive ocean are proved to be more abundant. The organic molecules taking shape in chemical evolution of primitive atmosphere are carried into ocean by rainwater, and submerge to the mid-coat promptly thus avoiding the damage of ultraviolet rays caused by deficiency of ozonosphere in primitive atmosphere.

Unit Three

Text A

二、译例与练习

Exercises

1. Fill in the blanks with the proper given words, and then translate the sentences into Chinese.

1) merchant

近年来商船队的数量大为缩减。

2) navigation

今天,在我们所生活的世界中,全球定位系统、数字地图和其他导航应用程序都可以在智能手机上使用。

3) maritime

这两项活动都以航海为基础,这是腓尼基人从他们的海上前辈,克里特岛的米诺斯人那里发展起来的一种能力。

4) oceangoing

国际贸易上,新奥尔良平均每年有5 000艘远洋货轮在此停靠,超过40个国家在这座城市设有领事馆。

5) route

研究者们正试图通过一条间接途径获取同样的信息。

6) encounter

为什么有些学生遇到困难就放弃了,而另外那些技能上并无更多过人之处的学生却继续努力学习?

2. Translate the following sentences into Chinese.

1) 这是当时最重要的贸易路线之一,满载丝绸的商队经常通过这条通道,使它成为历史上和经济上重要的通道。

2) 随着海上丝绸通道越来越出名,丝绸之路贸易开始变得多余。在隋朝,新的丝绸港口如南海出现了。

3) 自从中国和世界其他国家之间建立了丝绸贸易的海上联系以来,一些国家开始对丝绸贸易感兴趣。

4) 最热门的丝绸贸易路线始于中国南部地区的港口,包括吴、魏、齐、鲁等地区。由于海上通道的紧密性,通过这些港口进行进出口丝绸贸易非常容易。

5) 在整个中世纪,丝绸的流行一直持续着,拜占庭对丝绸服装的制造有详细的规定,说明了丝绸作为典型的皇家面料的重要性,也是王室收入的重要来源。

6) 丝绸生产知识是非常宝贵的,尽管中国皇帝努力保守秘密,但它最终还是传播到了国外,首先是印度和日本,然后是波斯帝国,最后在公元6世纪传到了西方。

7) 此外,皇帝问了许多问题,问他们是否知道这个秘密,僧侣们回答说,某些蠕虫是丝绸的制造者,本能迫使它们一直工作。

8) 中世纪早期,随着来自阿拉伯半岛的水手们在阿拉伯海和印度洋之间开辟了新的贸易路线,这一网状贸易规模得到拓展。

9) 通过展示出的各种路线,商人们从世界各地的陆地和海上运输各种各样的货物。

10) 基督教、伊斯兰教、印度教、琐罗亚斯德教和摩尼教以同样的方式传播,因为旅行者吸收了他们遇到的文化,然后把它们带回了家乡。

3. Translate the following sentences into English.

1) Through the Maritime Silk Road, silks, china, tea, and brass and iron were the four main categories exported to foreign countries; while spices, flowers and plants, and rare treasures for the court were brought to China.

2) China's internal trade and commercialization were highly developed from the early days of the People's Republic of China, with the total value always greatly exceeding the value of international trade.

3) It was the second Silk Route. Its waters and islands straits were as the sands and mountain passes of Central Asia; its ports were like the caravanserais.

4) Quanzhou (in Fujian province) is considered by many experts to be the terminus of the old Maritime Silk Road. It was arguably the most important port in the Nan Hai trade until at least the Yuan Dynasty.

5) Under traditional shipping technologies, without modern maps and navigation aids, the waters of the Nan Hai were extremely dangerous. Chinese ships have crossed the dangerous seas of the Nan Hai for more than 2,000 years.

6) From the collapse of the Roman Empire up to at least the sixteenth century, China's level of commercialization, urbanization, technology, and culture was much ahead of that in Europe.

7) The Chinese government's policy of the "New Silk Road by Land and Sea" has the development of infrastructure and commercial relationships at its core.

8) In many important ways, China's newly enunciated policy thus builds on the history of ancient trade networks and cultural interaction between China and Central and Southeast Asia.

9) Infrastructure building, in order to support commerce and foster social stability, was a foundation-stone of China's own long-term prosperity over the course of more than 2,000 years.

10) The lands around the Nan Hai and those from farther afield along the Maritime Silk Road sent tribute missions to China's rulers from as early as the Han Dynasty.

4. Choose the best paragraph translation. And then answer why you choose the first translation or the second one.

译文一好。

译文一：在遣词造句上，译文内容没有偏离原文。句式通顺，词汇和表达方式都与原文一一对应，地名词汇的翻译也准确地道。虽然是直译，但是有着无可替代的价值。

译文二：直观上看是直译，但不能完全表达原文的具体含义，句子语法也有错误。

5. Translate the following passage into Chinese.

蚕丝是一种起源于中国古代的纺织品，由蚕在结茧时产生的蛋白质纤维制成。根据中国传统，养蚕是为制丝，蚕业是在公元前2700年左右发展起来的。丝绸被认为是一种价值极高的产品，是中国朝廷专门用来制作布料、窗帘、旗帜和其他尊贵物品的。在大约3 000年的时间里，它的制作技术在中国是一个严格保密的秘密。依据封建王朝的法令，任何向外国人透露其制作过程的人都将被处以死刑。湖北省的古墓可以追溯到公元前4世纪和3世纪，其中包括第一批完整的丝绸服装，以及出色的丝绸作品，包括锦缎、纱布和刺绣丝绸。

6. Translate the following passage into English.

The oldest of the Maritime Silk Road routes exist between southern China and islands of Indian Ocean and South Pacific Coast. By now, these oldest trade routes for silk had reached Korea, Silla, Japan, India and Persia, covering a major portion of North, South and South East Asia. Others of the oldest maritime trade routes of silk exist between China and Coast of Persian Gulf and Red Sea including Kuilong in Indian peninsula, Sumatra, Orr Island and Gulf of Siam and Vietnam coast. By the end of the Yuan dynasty, more than 220 countries had become a part of trade routes starting with this marine Silk Road. In the early Qing Dynasty, on the basis of many routes in the Ming Dynasty, the North American route, the Russian route and the Oceania route were explored.

Unit Four

Text A

二、译例与练习

Exercises

1. Fill in the blanks with the proper given words, and then translate the sentences into Chinese.

1) renovate

经过仔细的调查,委员会决定翻修这位伟大作家曾经住过的老房子。

2) increase

对于一个日产量从未超过1 100万桶的国家来说,如此大规模的产能增长是非同寻常的。

3) surmount

跨文化研究的目标之一就是克服障碍,寻求文化差异和文化认同。

4) fragment

她什么都读,对新闻的每一个片段都细细品味。

5) colonial

这艘商船的迅速发展使北方殖民经济多样化,并使其更加自给自足。

6) coexist

新闻网站可以与报纸共存,还能带来可观的利润。

2. Translate the following sentences into Chinese.

1) 《海洋民俗记录》的工作始于1986年7月,当时我在佛罗里达州东北海岸的梅波特渔村测试记录技术。

2) 虽然关于海洋传统的例子大多来自佛罗里达州,但记录文化资源的技术可以应用于许多其他海洋领域。

3) 记录海洋民俗有两个主要目的:一是促进对海洋文化传统的了解,无论是在海洋、河流、湖泊,还是溪流附近,都能找到关于那里人们传统生活的相关信息;二是为非专业人员提供识别和记录一般海事惯例的指南。

4) 1989年,南佛罗里达州民俗历史博物馆完成了一项关于佛罗里达群岛民俗生活的调查。

5) 考古学告诉我们,原始的美学物品往往是有用的东西,要么是生产工具,要么是通过猎取动物获得的。

6) 海洋行业的历史角色和独具特色的当地文化造就了海洋职业风俗,并且使得这些风俗一直保持下去。

7) 佛罗里达州拥有绵长的海岸线和丰富的湖泊、河流,因而孕育了休闲垂钓和商业捕鱼的传统。

8) 海洋行业使佛罗里达州成为一个在航海、航运和海产品生产等领域具有重要战略意义的地区。

9) 因为一个关于拉丁美洲文化的《佛罗里达民俗项目》,我要对迈阿密戴德县民俗生活进行调查。

3. Translate the following sentences into English.

1) The value evaluation is the precondition of the reasonable development and using of the resources of the maritime culture.
2) China has a long history of Marine culture, and the tidal culture occupies a large proportion.
3) Maritime community traditions along America's shorelines include occupational and recreational folklife, water-to-table foodways, and folk art.
4) Zhoushan fishermen paintings have a strong local folk art style and contain special marine culture amorous feelings.
5) It's a whole heritage site and is crammed with narrow, cobbled streets and steep-roofed medieval buildings.
6) Anyone who has traveled through Europe no doubt recalls its impressive landmarks, museums and historic sites.
7) About 100 ships, seven of them from Chinese, have been attacked by Somali pirates since the beginning of the year.
8) I said that smoking should be banned, but she objected saying that the tobacco industry is a very important source of government revenue.
9) Pottery seems to have been invented in different places at different times right across the world.
10) When purchasing perishable food items, I look for those that have the longest "use by" date, even if I intend to consume them immediately.

4. Choose the best paragraph translation. And then answer why you choose the first translation or the second one.

相比较而言，译文一更好。

"妈祖"是中国的海洋女神，对于中文以外的读者是一个陌生的概念，因此，一般在翻译中要添加限制性说明。译文一中，第一句增加了"in China"，起到了限定作用。而译文二中没有，这就有可能让译文读者在初读文字时有种莫名其妙的感觉。另外，原文"被联合国教科文组织列入《人类非物质文化遗产代表作名录》"，其中"列入"在译文二中翻译为"include"，不是很准确。中国文化中的"祭祀"活动，除了杀生供奉神灵，更主要的是敬拜。译文一将其翻译为"sacrificial and worship"更符合中国祭祀文化的内涵，在这一点上，译文一也好于译文二。

5. Translate the following passage into Chinese.

巴卡达瑞斯与尼维斯的利奇群落和圣乔治港遗址的比较结果，进一步证实了海盗营地特征的突出性，其中包括烟斗比例过高，碗占据多数，酒具和实用器皿缺乏，以及陶瓷器皿和形式的多样性普遍不足。这些差异代表了海湾人与尼维斯人不同的物质文化选择。虽然还需要做更多的工作来确定真实的海盗模式，但本文所提出的模型显然能够突出考古记录中海盗活动的特征。这些特征可以在未来进行测试。巴卡达瑞斯是迄今为止发掘出的唯一一个明显与海盗有关的遗址，而关于此课题的研究工作又很少，没有多少关于海盗的网站可以调查。少数几艘挖掘出来的海盗沉船还陷于伦理和鉴定的争论。然而，一个更大的问题是，之前在研究"海盗"材料时已经提到过的,如何将他们与其他海上团体区分开来?

6. Translate the following passage into English.

  Zheng He (1371—1433) was a navigator and diplomat during the Ming Dynasty (1368—1644). Born in Yunnan Province, in southwest China, the influential historical figure was the son of a hajji, a Muslim who had made the pilgrimage to Mecca. His family claimed to descend from an early Mongol governor of Yunnan and a descendant of King Muhammad of Bukhara. His original family name, Ma, was derived from the Chinese rendition of Muhammad. When he was 10 years old, Yunnan was re-conquered by the newly established Ming Dynasty. The young Ma He, as he was then known, was among the boys who were captured and sent into the army as orderlies. Placed under the command of the Prince of Yan in 1390, Ma He had distinguished himself as a junior officer, skilled in war and diplomacy. The Prince of Yan revolted against his nephew, Emperor Jianwen, and took the throne in 1402. After the Prince of Yan became Emperor Yongle, the war-devastated economy of Ming Dynasty was soon restored and the Emperor then sought to display his naval power overseas. Selected by the emperor to be commander-in-chief of the missions to the Western Seas, he set sail in 1405, commanding 62 ships and 27,800 men.

Unit Five

Text A

二、译例与练习

Exercises

1. Fill in the blanks with the proper given words, and then translate the sentences into Chinese.

1) involvement

  外交部已断然否认与此有任何牵连。

2) sustainability

  能源和可持续发展专家表示,我们未来能源需求的答案很可能来自大量的压缩,包括传统和替代性的压缩。

3) ground

  他把汽车排挡踩得嘎嘎作响。

4) flourish

  随着植物越来越茂盛,你需要将它们移植到更大的花盆中。

5) harpoon

  鱼叉绳必须很长,还要小心地装在船上,以备鲸鱼潜水时能方便地放出。

6) worldview

  鲁迅的短篇小说问世之初因其语言风格和其中的观点被视作异类。

2. Translate the following sentences into Chinese.

1) 深海通常有稀疏的动物群,以微小的蠕虫和甲壳类动物为主,更大的动物分布更少。
2) 当这个不速之客搭乘了一艘从墨西哥湾钻井平台出发的潜艇后,这种深海爬行动物吓了石油工人一跳。
3) 一些未来学家设想纳米技术也可以用于在小型潜水艇里探索深海,甚至可以用来发射装有微型仪器的手指大小的火箭。
4) 在苏格兰,野生鲑鱼的数量已经减少,这是由于不受控制的深海和沿海捕捞、污染,以及对鲑鱼栖息地的各种其他威胁引起的。
5) 深海动物依靠颗粒物质作为食物,这些颗粒物质最终来自光合作用,下降的排放口会使平流的作用相形见绌。
6) 这一样本包括曾经柔软的硬化沉积物的卵石、深海淤泥以及石膏颗粒和火山岩碎片。
7) 那么为什么深海金属硫化物矿的商业化开发要等这长时间?
8) 论文作者指出,全球气候的变化可以从很多方面影响深海食物供应。
9) 除了海水渗入土壤之外,海洋也在酸化,导致珊瑚白化。
10) 由于深海拖网捕捞是最近才出现的,所以它造成的损害目前还是有限的。

3. Translate the following sentences into English.

1) Hawaii whispers romance in the lap of the ocean on the beach and in the sweet sound of a love song.
2) These organisms absorb carbon dioxide from the atmosphere and take it to the bottom of the ocean when they die, where it stays for thousands of years.
3) They crunch numbers that simulate the processes that drive Earth's climate, like incoming sunlight and the circulation and composition of the ocean and atmosphere.
4) Natural disasters such as the earthquake that triggered the Indian Ocean tsunami in 2004 have also caused reef loss.
5) California is constrained by its mountains and the ocean, to say nothing of the demands of environmentalists keen to preserve its remarkable natural beauty.
6) Only in 2005, following the Indian Ocean tsunami, did it adopt an agreement obliging members to help one another in natural calamities.
7) One big factor in the warm temperatures was a powerful El Nino, a natural warming of Pacific Ocean surface temperatures that affected the climate from 2015 to 2016.
8) The ocean and seas surrounding the islands are deep blue and many of New Zealand's cities lie on a bay and have a natural deep harbour.
9) When the designer first visited this site, he thought the key issues for this project were how to intertwine, dissolve, cut and take in elements of this natural environment, such as ocean, greenery, sky and wind.
10) The Ocean and seas surrounding the islands are deep blue and many of New Zealand's city lie on a bay and have a natural deep harbour.

4. Choose the best paragraph translation. And then answer why you choose the first translation or the second one.

译文二更好。

译文一：翻译过于字面逐字逐句解释，不够灵活。

译文二：原因是在理解原文的基础上，翻译表达更符合英文的表达习惯，因此译文二更合理。

5. Translate the following passage into Chinese.

夏威夷本土文化。土著居民与自己的土地、领土和资源有着深刻的联系。数百年来，夏威夷原住民与这些岛屿的自然环境共同进化，并积累了对祖先土地和海洋的深刻知识和理解。今天，夏威夷原住民仍然把环境作为夏威夷文化和世界观的主要来源和基础。根据夏威夷宇宙学，夏威夷原住民与自然界有着独特的亲缘关系。库穆里波是夏威夷的一首（上帝）创造天地的圣歌，描绘了宇宙力量爆发为运动和热量，导致自然元素激发创造，如同由黑夜逐渐积累进化而来的生命形态。从第一个时代的珊瑚虫开始，库穆里波在第二个时代宣布了鲸鱼的存在。"Hānau ka palaoa noho i kai"（夏威夷语），出生的是生活在海洋中的鲸鱼。最终，人类从这个共同的起源诞生。

6. Translate the following passage into English.

There are diverse habitats in the deep sea, such as abyssal plains, seamounts, abyssal abysses, hydrothermal vents, cold springs, and whale colonies. Different habitats breed different ecosystems, among which the seamount ecosystem has the highest biodiversity and is home to almost all kinds of animals, from the most primitive microorganisms to the highest mammals. The seamount is thought to be a habitat for deep-sea life. About 2,000 species have been found from the seamount, but the actual number of species "living" in the seamount may be much higher. Because of the layered distribution of Marine life, with each species distributed in a specific water layer, the three-dimensional structure of a seamount allows it to accommodate organisms from different water layers.

Unit Six

Text A

二、译例与练习

Exercises

1. Fill in the blanks with the proper given words, and then translate the sentences into Chinese.

1) sustainable

在巴黎召开了一个旨在推进所有国家可持续发展的大型国际会议。

2) mitigate

会谈已经持续了几十年，然而，到目前为止，还不清楚如何缓解旅游业对这个岛屿的影响。

3) generate

去年修建的这座风力发电站发的电或许足以供 2 000 个家庭之用。

4) implemented

政府在新闻发布会上公布说国民医疗保健制度的改革将于明年实施。

5) perspective

由于地理位置不同,德国对俄罗斯局势的看法与华盛顿大相径庭。

6) vulnerable

游客更容易受到攻击,因为他们不知道城里哪些地方不该去。

2. Translate the following sentences into Chinese.

1) 海洋覆盖了地球表面的71%,是全球生命支持系统的一个基本组成部分,也是资源的宝库、环境的重要调节器。

2) 中国是一个发展中的沿海大国。中国高度重视海洋的开发和保护,把发展海洋事业作为国家发展战略,加强海洋综合治理,不断完善海洋法律制度,积极发展海洋科学技术和教育。

3) 这份长达一百八十页的文件,有三百余条,并有八个附件。它涉及所能想到的每一个与海洋有关的问题,从岛屿的定义,到对在淡水生长而在海洋产卵的鱼类的管辖权,都做了明确的规定。

4) 孟加拉国坚持认为:一直以来国际间都是使用自然延伸原则作为主要标准,来认定200海里外大陆架权利的归属问题,他们认为缅甸的大陆板块和孟加拉湾的海床正好处于两个板块边界地带,两个板块之间是不连续或者中断的,因此,根据自然延伸原则,缅甸不应享有该区域的大陆架权利。法庭对此说法没有给予支持。

5) 最近,毛里求斯和塞舌尔联合提交《联合国大陆架界限公约》,将马斯卡伦高原的专属经济区又扩大了 39.6 万平方千米。

6) 人们认为,当前正处于经济增长和政治成熟的阶段,与此同时,全球和地区发展也带来了更多的机遇,如果他们把握得当,上述种种有利条件将推动他们进入下一个发展阶段。

7) 印度洋地理位置相对开阔,附近大国数量很多,但都不是占主导地位的超级大国,进而也就不涉及战略问题和力量平衡问题,这与西太平洋和中国附近海域的状况有很大不同。

8) 环境资源是旅游的主要吸引点,因此需要合理规划和充分保护这些环境资源,使其成为旅游依托。

9) 海南岛和台湾岛的面积差不多,那里有许多资源,有富铁矿,有石油和天然气,还有橡胶和别的热带亚热带作物。海南岛好好发展起来,是很了不起的。

10) 海洋资源维持着全世界约 30 亿人的生计,其中绝大多数人生活在发展中国家。但这些人的生计目前正受到威胁,由于污染和过度捕捞等人类活动,海洋及其维持生命的能力正处于严重危险之中。

3. Translate the following sentences into English.

1) This program facilitates dialogue and coordination between countries and organizations. It also catalyzes partnerships, and builds on lessons learned.

2) From the SOI Global Dialogue with Regional Seas Organizations and Regional Fishery Bodies to the SOI Training of Trainers program, the various SOI activities aim to achieve a balance between the conservation and sustainable use of marine and coastal biodiversity, at global, regional, national, and local levels.

3) This event will outline the path to implementation, focusing on showcasing efforts and mechanisms to support the restoration of coral reefs to sustainable financing for the effective conservation, protection and restoration of coral reefs.

4) In this landmark year for the ocean, ocean action has been advancing on multiple fronts. The 2022 UN Ocean Conference held in Lisbon generated hundreds of new voluntary commitments and a political declaration aimed at scaling up ocean action based on science and innovation for the implementation of Sustainable Development Goal 14.

5) Since the late 1970's, environmental studies undertaken along with mineral exploration activities in the international seabed area have resulted in remarkable gains in knowledge of deep-sea biodiversity.

6) Sustainable ocean action is action for biodiversity. The goals in the Ocean Panel's shared agenda directly align with 85% of the targets of the Convention on Biological Diversity post-2020 Global Biodiversity Framework.

7) This event will explore the challenges and opportunities to move from knowledge to action in marine and coastal ecosystems in the context of the post-2020 Framework.

8) This interagency and cross-sectoral panel will highlight U.S. efforts at multiple scales to implement ocean solutions to mitigate the effects of climate change on marine biodiversity.

9) Over the last three decades, biodiversity has plummeted due to human caused stressors such as pollution, development, overexploitation and climate change and the ocean is not exempt from these losses, with rapid species decline, degrading iconic landscapes like coral reefs.

10) Urgent and comprehensive action is needed to halt biodiversity loss and restore and protect important marine ecosystems such as coral reefs, mangroves and salt marshes to ensure their continued existence.

4. Choose the best paragraph translation. And then answer why you choose the first translation or the second one.

1) 景点介绍属于外宣材料，画面感是吸引游客的重要手段，外宣翻译应该突出读者中心地位，要从读者的感受方面组织安排信息，拉近与读者之间的距离，让读者产生立马出发预定行程的冲动。外宣材料的呼唤功能要在翻译过程中有所体现。这点在译文一中有所体现，译文二这点做得不好。

2) 景点宣传中的核心信息应该是景点名称，应该以景点作为中心词进行信息展开，应该以景点名作为句子主语，突出行为主体，也起到宣传该景点的作用。译文一这点做得很好，然而，译文二的介绍文字在句式上则没有体现出该景点的主体地位，反而让其他次要信息居于句首,导致喧宾夺主、重心偏离。除此之外,译文二版本逻辑把控也不好。缺乏整体逻辑把控,因此造成译文整体结构或逻辑不恰当。只有把握核心信息、找准句子重心,才能使译文主次分明,更好地传达信息。

综上两点所述,译文一相比之下更好。译文一无论在受众意识还是信息呈现方面都进行调控和把握,因此译文一质量更好。

5. **Translate the following passage into Chinese.**

鉴于海岛目前生态系统所面临的挑战,如何做好环境管理对于海岛的可持续发展来说至关重要。由于岛上土地资源稀缺,产生了各种土地使用问题,如何平衡资源利用,在旅游业、农业和其他产业的用地之间如何进行权衡,哪些土地可划作生态保护区,这些都是管理者需要解决的问题。环境资源是旅游的主要吸引点,因此需要合理规划和充分保护这些环境资源,使其成为旅游依托。坎林,在他对塔斯马尼亚旅游的研究中,提出了对公园或保护区的九种不同划分方式(国家公园、国家保护区、自然保护区、野生动物保护区、保护区、自然休闲区、区域保护区、历史遗址和私人保护区)。他还指出,划分方式还可能包括指定禁止游客进入的区域,以及明确哪些区域必须由导游和翻译陪同。书中还指出,拥有独特野生动物的岛屿也必须采取相应的措施来保护这些特有种群。他还提出,就像在塞舌尔一样,如果岛上居民能够意识到野生动物所具有的经济价值,那么旅游业实际上还可以演绎成反偷猎机制的一部分,或者成为对动物进行保护管理的引擎。另外,制定相关的政策确保陆地和海洋不受污染也是必要的。他还说,特别对于小岛屿来说,制订计划来规范岛屿对垃圾废物的管理,规定对垃圾废物如何进行回收,也至关重要。此外,如有必要,海岛还应该通过激励计划鼓励使用替代燃料来源(风能、太阳能、地热等),来助力旅游目的地,使其发展更具可持续性。

6. **Translate the following passage into English.**

Coastal and marine tourism covers a wide range of activities taking place on the seashore or in the deep oceans, all of which are designed for the purpose of exploring, sightseeing, entertainment, sports and recuperation. The ocean is vast, thus its great potential for tourism needs to be further explored. Sea air, which contains a large amount of iodine, high concentration of oxygen, ozone, sodium carbonate and bromine, and very little dust, is very conducive to human health and thus is suitable for various tourism activities. Compared with land tours, sea tours will evoke distinctive emotional cues. Sea tours usually center around such traditional events as admiring the sun slowly rising above the horizon and setting under the horizon. Included in sea tour itineraries are also a wide spectrum of water sports from swimming in the sea to deep sea diving or snorkeling, from fishing to windsurfing or sailboarding, from racing boats to kayaking etc. Cruise ships are the main means of transportation for marine tourism. There are hundreds of luxury cruise ships in the world today, which not only provide accommodation and food for tourists, but also a variety of services and entertainment facilities.

Unit Seven

Text A

二、译例与练习

Exercises

1. Fill in the blanks with the proper given words, and then translate the sentences into Chinese.

1) take a toll
   然而,香烟烟雾是否确实对儿童的大脑产生负面影响,或者是否是其他东西所起的作用,尚不明确。

2) suspend
   如果你超过了信贷限额,我们有权暂停或取消你的账户。

3) accelerate
   政府想要加速机构改革,寻找新的整顿国家的方法。

4) on a par with
   这个水上公园将与周围那些最好的公共游泳场所一样好。

5) generate
   我们打算让所有人对我们生活的环境产生个人责任感。

6) innovation
   为了拓展业务,我们在团队内必须提倡独创性,激发创造力,鼓励创新,提高专业管理技能。

2. Translate the following sentences into Chinese.

1) 海洋资源越来越被视为解决未来几十年地球面临的多重挑战所不可或缺的一种资源。

2) 在这次涵盖了海商法、海洋资源评估和海运经济等主题的会议上,与会者也表达了对中国市场的重视。

3) 鱼类产品提高了鱼类的经济价值,这使得依靠渔业加工和出口的国家能够充分利用自己的水产资源,获得最大的经济利润。

4) 在很多地区,由于风力太小而不足以支撑风力涡轮机和风力发电厂,使用太阳能或者地热成为上乘之选。

5) 沿海地区是陆地和海洋之间的过渡区,其特点是生物多样性程度非常高,此地区包括地球上一些最丰富却又最脆弱的生态系统,比如红树林和珊瑚礁。

6) 海洋工业与海洋生态系统的相互依存性,加之海洋健康遭受到的日益严重的威胁,使人们越来越意识到需要对海洋采取综合管理方法。

7) 在全球范围内,尽管难以准确评估非法、不监管和不报告捕捞(IUU)的确切规模,但是其对鱼类资源造成了严重影响。

8) 旅游业的长期中断,对于国内经济依赖旅游业的国家有重大后果。

9) 科学对于实现全球可持续性和海洋的充分管理至关重要,因为它能够加深我们对海洋资源和海洋健康的理解和监测，并帮助我们预测海洋状况的变化。

10）海洋科技的重要性不言而喻，仍然需要把它保持在最前沿，以应对海洋健康恶化加速、气候变化和海洋经济活动加速所带来的挑战。

3. Translate the following sentences into English.

1) Effective fisheries management requires improved sea surveillance and monitoring as well as the availability of science and data on fisheries.
2) These examples contribute to a common understanding of what constitutes sustainable activities across ocean-based sectors.
3) Alongside the mainstreaming of a more sustainable use of ocean resources across ocean-based industries, specific actions are required to conserve and restore marine ecosystems.
4) In addition to cross-sector approaches, cross-country approaches may be needed due to the transboundary nature of natural marine assets.
5) Marine Protected Areas have been cited to increase the general health of the marine ecosystem through mitigating harmful activities in the oceans such as mining and overfishing, thus enabling healthy oceans.
6) In a desperate economic climate, Peter, like countless others, saw the sea as his only viable alternative.
7) Studies suggest that the fin trade still accounts for a significant proportion of the estimated 70 to 100 million sharks fished globally each year.
8) Sustainability should remain a crucial factor in decision making surrounding the ocean economy.
9) Ocean entwines the two-thirds of surface of the earth. From that perspective, blue waters of the oceans are deemed as the bloodstream of this planet earth.
10) Product, price, promotion, and place, classical points in traditional marketing, also apply to marketing aquaculture products.

4. Choose the best paragraph translation. And then answer why you choose the first translation or the second one.

译文一：翻译几乎是词对词、句对句直译过来的，未能恰当使用翻译策略和技巧，因而未能脱离源语与译语形式对应的束缚。译文篇章结构不够清晰，句间逻辑关系稍弱，行文不够紧凑有力，遣词造句仍有较大提升空间。

译文二：译文逻辑清晰，层次分明，用词准确，充分发挥了英语形合的语言特点。例如第一句通过使用"the reasons……may vary……"和紧随其后"from……to……"结构，使得译文充分尊重了英语尾重原则和重形合原则，显化了句际间逻辑关系。

综上所述，译文二比译文一更胜一筹。

5. Translate the following passage into Chinese.

手工渔业涉及数百万从业者,尤其是女性。这使其成为发展中国家另一个重要的海洋产业,但也面临一些共同的挑战。这些国家往往缺乏收获后的设施,如干燥设备、制冰厂和冷藏设备。此类设施是增加海鲜产品价值和获得更好价格所必需的,同时也是减少手工渔业收获后损失所必需的。当没有冰的港口缺乏储存设施时,渔民有时倾向于以更便宜的价格卖掉剩下的鱼,或者只能面临渔获物的变质。粮食及农业组织估计,每年全球约有35%的渔货物遭到损失或浪费。因此,整个鱼类生产系统的经济发展高度依赖于加强捕捞后的加工管理,以及探索进一步的可持续捕捞措施(如认证和生态标签)。展望未来,过度捕捞、气候变化、沿海污染、生物多样性丧失以及非法、未报告和无管制捕捞的影响将对海鲜生产造成损害,因为它们加剧了手工渔业的固有挑战。一些国家将越来越需要更有效的海洋保护和可持续渔业管理战略,以重建种群实现营养安全。

6. Translate the following passage into English.

Marine and coastal tourism are highly dependent on the quality of natural ecosystems to attract visitors, as they rely on the recreational value of beaches and clean waters. Yet unmanaged tourism is contributing to ecosystem degradation and fragility, jeopardizing it's the sector's own economic sustainability. Climate change vulnerability is also a risk for countries that rely on tourism, seen for instance in coral reefs bleaching events, with the greatest risk in small island states where tourism is the largest sector of the national economy. Other challenges affecting the sector include beach degradation resulting from sand harvesting, mangroves deforestation and an ever-growing coastal population that pressures coastal ecosystems.

Unit Eight

Text A
二、译例与练习
Exercises

1. Fill in the blanks with the proper given words, and then translate the sentences into Chinese.

1) accentuate

在某些情况下,节约能源往往会使情况更加严重。

2) get involved in

部队明确表示不愿卷入镇压示威活动。

3) omnipotent

但是谷歌的回击恰是在这个曾经无所不能的软件巨头显得很弱势的时候。

4) prototype

即使它是独创性的原型,它也是非常令人不安和有害的东西。

5) precarious

我觉得要在华盛顿这个地方出版一份这么理性中立的报纸根本不现实。

6) treacherous

虽然理查德是一个血腥暴力的人,但他过于冲动,既不奸诈,也不残忍成性。

2. Translate the following sentences into Chinese.

1) 该雕塑的灵感来自美丽的硅藻家园,它经常在一种圆形容器中利用硅酸盐生成一种用于装饰的几何形状的有机物。

2) 尽管可以做出如胸腔一样的结构简单并且具有不错的独立功能的身体部分,然而随着时间的推移,它也可以随着骨骼框架的增长而增长。

3) 通过这个项目,他不仅想创造一尊让人赏心悦目的雕塑,而且还希望为海洋生物和珊瑚创造一个可持续的栖息地。

4) 榕树的根部有很深的凹槽,为鱼类、藻类、珊瑚和其他海洋生物提供了完美的繁殖地和栖息地。

5) 集美丽、绚丽和坚韧于一身是一种令人羡慕的特质,而章鱼就拥有这一特质。

6) 这位艺术家认为,在海底放置一尊仙人掌雕塑将提供一种独特的并置效果,还可为人们的诠释提供多重空间。

7) 《渴望》这件作品展现的是一个年轻女孩仰望海底世界奇观的侧影。

8) 从喉舌处冒出的是一串气泡,其中许多气泡将作为学生个人设计的框架,并使雕塑成为鱼类栖息地。

9) 作为一名自学成才的海洋艺术家,葛弗利的作品总是挑战艺术规范,突破固有界限,并不断变化。

10) 随着时间的流逝,这些间隙将成为活牡蛎和其他海洋物种的栖息地。

3. Translate the following sentences into English.

1) Oceanic art or maritime art is any form of art depicting or inviting major inspiration from ocean (i.e. painting, printmaking and sculpture).

2) With the advent of landscape art in the Renaissance, the so-called ocean landscape became an important part of the work. But until later it was the pure ocean views.

3) With romantic art approaching, the sea and the coast were taken as the subject of creation by many landscape painters, and works without ships for the first time became commonplace.

4) Ships and their representations have appeared in art since prehistoric times, but navy did not appear until the Middle Ages, only when it began to become a special type of art.

5) For the painter Franco Salas Borquez, the silent ocean, as an idea, was a representation of the eternal landscape.

6) Marine painting differs from other painting subjects in that it only focuses on the main motive of the sea, and is difficult to draw boundaries from other fields.

7) There is a real sea view, Journey to St. Julian and St. Martha, but two pages were burned in 1904, only surviving in black and white photographs.

8) The Marine Art and Science Museum integrates marine cultural and artistic resources, such as marine paintings, marine maps, and marine models.
9) Along with mythology and the praise of rulers, map painting was an important part of marine painting in this century.
10) With Romantic art, the sea and the coast were filled by many experts of landscape painters, and for the first time works without boats became common.

4. Choose the best paragraph translation. And then answer why you choose the first translation or the second one.

译文一较好。

译文一：之所以相对较好，主要是因为它在处理原文句子时体现出来的英语侧重"形合"的语言特点，并且在实际操作中也十分恰当。原文可以分作三个相对独立的语义单位，分别对应的是段落中的三个中文句子。而这三个语义单位都可以通过从句、分词和短语等语言手段转译成三个英语复合句，外在语言模式上达到了齐整的效果，这是符合英语的语言特点和思维方式的。在翻译完第一个中文句子的主干部分后，可以将"荷兰黄金时代"这个重复出现的短语作为写从句的关节点，再将最后一部分翻译成现在分词用以补充说明当时海景画的内容和重要性。相较于译文二，译文一在第二个和第三个语义单位的衔接上加入了短语"in a contrast"，此处增译法的使用可谓十分恰当。因为前后文描述的分别是东西方在海景画创作上的显著差异，两者恰好是对照关系。加入的"in a contrast"正准确地体现了这种潜在的逻辑关系，同时也加强了前后译文之间的衔接性，使译文变得更加连贯。最后译文一在处理第三个语义单位时，亦通过连词"while"将前后句连缀成一个复合句，进一步体现了英语"形合"的语言特点。

译文二：这篇译文在用词上同第一篇译文相比并无太大差异，原文语义表达得也十分准确。然而，在句子结构方式上与第一篇译文有很大的不同。不难看出，这篇译文通篇基本上以英语简单句为主要语义表达单位，尤其是在处理第一个语义单位时，未能抓住"荷兰黄金时代"这个关节点，使得整个英文句子显得略微松散。相较于第一篇译文的处理方式，这样的译文不仅读起来颇有累赘之感，更为重要的是未能充分体现出英语重形合的造句思维，依旧带有汉语重意合的色彩。因此，尽管第二篇译文无论是在语法上还是在遣词造句上都无错处，但是与译文一相比较还是略逊一筹。

5. Translate the following passage into Chinese.

《添砖加瓦》这件艺术品可以说实现了艺术家扎卡里·朗大约一年前的构想。当时他想建造一个金属雕塑，为生命的诞生提供一个栖息之地。在他的脑海中，这应该是一件美丽的不锈钢结构，它大胆、坚固、生命力旺盛，但又微妙地平衡着并努力维持着生命。扎卡里可以看到许多变化的角度和空间，让丰富多彩的海洋生物得以展示和安置在宽大（在海洋尺度上看起来很小）的位置上。这些微妙地保持着平衡的"砖瓦"让人想起生命的脆弱，但通过一段时间的关注也可以发现即便是星球上一些最为脆弱和重要的生物体照样可以茁壮成长。这也是为了能让人类真正去关心、关注那些丰富多彩和令人惊叹的海洋生物，并愿意为它们奉献力量。这件艺术品的诞生之地在美国中部的俄克拉荷马城。之所以安置在这里，是希望即便是那些内陆国家和地区的人们也能在观念上做出改变。扎卡里也希望通过这件艺术品可以在中美洲引发热议,因为那里的人们似乎对今日海洋所面临的诸多问题更加麻木无知。他想告诉世人的是,无论身在何处,只要你对现有的东西发挥创意,就可以成为解决方案的一部分。

6. Translate the following passage into English.

Ivan Konstantinovich Aivazovsky (1817—1900), a Russian romantic painter, is recognized as the greatest marine art master in the world. Aivazovsky was born in 1817 in an Armenian family located in the Crimean Black Sea port of Feodosia. Aivazovsky is a painter mainly based on romanticism, and his works also have a style of realism, usually showing large-scale dramatic scenes, and mainly covering romantic struggles between humans and the sea, and the so-called "blue ocean" and cityscape. In his early creation, the expression of color was extremely abundant. Until the last two decades of his artistic career, the expression of color began to become delicate and steady, when he created a series of seascape works in silver tones.

Unit Nine

Text A

二、译例与练习

Exercises

1. Fill in the blanks with the proper given words, and then translate the sentences into Chinese.

1) interpretation

然而,在该修正案存在的头 80 年里,最高法院对该修正案的解释背叛了这种平等的理想。

2) jurisdiction

因此,当各省省长聚集在尼亚加拉大瀑布城,像往常那样不停地抱怨时,他也应该开始在他们的管辖范围内做一些有利于他们的预算和病人的事情。

3) implicit

大型企业往往在稳定和安全的环境下更舒适地经营,它们的管理机构往往促进现状的发展,抵制变革中隐含的威胁。

4) withdraw

尽管发达国家的人口、工业产出和经济生产率还在飙升,但人们从含水层、河流和湖泊取水的速度已经放缓。

5) legislation

他们认为,宗教和政治应该明确分开,且他们普遍反对人道主义立法。

6) infrastructure

迅速发展中的国家的政府将废物最小化的思想纳入目前正在规划、设计和建造的运输基础设施和储存设施。

2. Translate the following sentences into Chinese.

1) 几千年来，舟山群岛渔民与大海相处，捕鱼为业，创造了丰富多彩的海洋文化。

2) 热带水库中，罗非鱼和白鲢等是主要养殖种类，它们的混养是热带水库典型渔业模式。

3) 汉晋时期西南地区渔业活动是西南地区人民的主要文化活动之一。

4) 内容包括渔业生产直接消耗系数、价值构成、国民经济效益分析、各种养殖类型的经济指标比较等。

5) 两国金融、经贸、地热、渔业、人文等领域合作与交流进展顺利，双方在国际及地区事务中也保持良好沟通与协调。

6) 因此，内陆水产品人工养殖比海水产品人工养殖前景更广阔。

7) 但是由于在人类现代史上北冰洋大部分地区始终是终年冰雪覆盖，因而它也是各大洋中最疏于勘查的。

8) 在美国移民史中大西洋扮演着一个特别的具有构成性影响的角色——对大多数美国人的祖先而言，大西洋是一条通往美好生活的危险走廊。

9) 这是一本生动活泼的书，讲述了太平洋斜坡的历史，以及太平洋如何成为美国的一个湖泊。

10) 他的成名一方面得益于他的捕鲸史激发了赫尔曼·梅尔维尔创作《白鲸记》的灵感，另一方面也来自他在《光荣之海》中似乎有理的主张：海洋，而不是西部，才是美国的第一前线。

3. Translate the following sentences into English.

1) The formation of fur clothing culture is because of lower productivity and special fishing and hunting mode of production at that time.

2) It is considered that there are relationships between agro-farming culture and fishery's culture. They are not isolated and statically.

3) The true cultural heart and soul of Alaska's fisheries, however, is salmon.

4) As an island nation, fishing is important to the local economy, a part of the local culture.

5) One of the island's most popular cultural attractions is the Kalokoeli Fishpond, where ancient Hawaiians once practiced a remarkably sophisticated form of aquaculture.

6) Honghu wetland owns much economic value in aquatic production, shipment, tourism and so on, with its social value in aesthetics, education and scientific research.

7) Pond fish culture is the most important aquaculture model in China. In 2004, pond fish production accounts for 70.36% of total freshwater aquaculture production.

8) The present invention belongs to the field of aquaculture feed technology, in particular, it relates to an additive premixed material for low-salinity culture of Fanna prawn.

9) The invention relates to an on-line water quality monitoring method and an on-line water quality monitoring system of aquaculture ponds, which belong to the field of pond culture.

10) Aquaculture studies concentrate on the selection, culture, propagation, harvest and marketing of domesticated fish, shellfish and Marine plants, both freshwater and saltwater.

4. Choose the best paragraph translation. And then answer why you choose the first translation or the second one.

译文二更好。

译文一:译文语言必须通顺,符合规范,用词造句应符合本民族语言的习惯,要用民族的、科学的、大众的语言,以求通顺易懂。不应有文理不通、逐词死译和生硬晦涩等现象。而这版翻译过于生硬机械,单一地追求词汇的对应而忽视了语法的准确。

译文二:此版翻译更佳,原因是译文在忠实原文的基础上,表述清晰简洁,语法准确,语言清晰、精练、严密。因此这个翻译更好一些。

5. Translate the following passage into Chinese.

非法、不报告和不管制捕鱼以及支持非法、不报告和不管制捕鱼的渔业相关活动继续严重破坏和威胁渔业的可持续性、沿海社区的生计和海洋经济。非法、不报告和不管制捕鱼在很大程度上是看不见的,它使以证据为基础的渔业管理的存量评估复杂化,同时使守法的渔民在资源和市场上面临不公平竞争。此外,非法、不报告和不管制导致重要的税收损失。非法、不报告和不管制捕鱼还可能威胁粮食安全,例如,在依赖当地海产品的地区和社区,非法、不报告和不管制捕鱼将鱼类从当地市场转移,并可能由于非法产品的错误标签而造成食品安全风险。它有时也与争夺稀缺资源和有争议水域的冲突有关;跨国犯罪活动;以及对强迫劳动的剥削。虽然非法、不报告和不管制捕鱼的负面影响众所周知,但非法、不报告和不管制捕鱼可以从政府对渔业的支持中受益。本报告评估了如何避免这种情况。

6. Translate the following passage into English.

The concept of culture can be divided into broad sense and narrow sense. In the narrow sense, culture refers to the spiritual wealth of social ideology, while in the broad sense, culture refers to the sum of material wealth and spiritual wealth created by human beings in social and historical practice. Fishery is the way of production in which human beings directly ask for food from nature in the early stage, and it is the earliest production behavior of human beings. At that time, relying on mountains and eating means hunting, while relying on water and drinking means fishing. In other words, the ancestors relied on water, utilized the natural reproduction and vitality of aquatic organisms, and obtained aquatic products through labor, which we call fishery. The industrious and simple ancestors not only created a rich material civilization, but also created a brilliant fishing culture.

Unit Ten

Text A

二、译例与练习

Exercises

1. Fill in the blanks with the proper given words, and then translate the sentences into Chinese.

1) sacrificed

基本观点是,众生皆平等,虽然不得已之时为了大众的福祉,个体还是可以被牺牲的。

2) feminine

我们生来就有极其不同的基因倾向,这种倾向被社会强化为"男子气"或"女性化"。

3) encountered

抑郁症是一种以显著的而持久的心境低落为主要临床特征的常见的情感性精神疾病,已成为现代社会的常见病、多发病。

4) innate

身体的排泄物、死亡和腐烂的气味都可能是危险或疾病的迹象,它们都会触发我们与生俱来的厌恶感。

5) assumed

希腊人认为,语言结构与思维过程之间存在着某种联系。这一观点在人们尚未认识到语言的千差万别以前就早已在欧洲扎下了根。

6) assessment

她根据监管人的评价和孩子的自测,对每个孩子的同情心程度以及犯错之后感受负面情绪的倾向进行了评分。

2. Translate the following sentences into Chinese.

1) 如果您要创业,您需要合理的财务建议和强有力的计划——只靠空中楼阁白日做梦是不行的。

2) 大多数时候,为高中生演奏古典音乐就像把有价值的东西给了不识货的人。但每隔一段时间就会有几个孩子欣赏它。

3) 他加入了我们的党,结果却是一只披着羊皮的豺狼,因为他是被极右分子派来混在我们的团体之中的。

4) 安德森女士总是将班上的学生分开,让调皮的男生坐在前面好看管他们。

5) 科学是讲求实际的。科学是老老实实的学问,来不得半点虚假,需要付出艰巨的劳动。

6) 伊朗的核计划如同一把达摩克利斯之剑,始终悬在美国头上,令其寝食不安。

7) 不要边视频边吃红烧肉来逗引我啦!这简直是对不能回家过年的人的折磨。

8) 随着互联网热潮的到来,他在毕业后成了一名精明的投资者,投资的每家公司都获利颇丰。

9) 她赢了这场官司，但得不偿失，因为她得支付高额诉讼费。

10) 库存品已经变质，而且全都搞混了，因此经理决定快刀斩乱麻，把它们当作废品出售，然后重新储藏新的物资。

3. Translate the following sentences into English.

1) There is a current television craze for dance contests featuring celebrities not known for their terpsichorean talents.

2) Sadly, his reforms opened up a Pandora's box of domestic problems.

3) It is unfair that historians always attribute the fall of kingdoms to Helen of Troy.

4) Mr. Jones made a long speech at the meeting. Everyone thought it was a Penelope's web.

5) The project, which seemed so promising, turned out to be a Pandora's box.

6) I wonder why they argued in the dark. The broken car seems an apple of discord.

7) It is only through hardships and disasters that a friendship of Damon and Pythias can manifest itself.

8) The Russians have long regretted selling Alaska to the Americans for a mess of pottage.

9) All the tickets have been sold for the singer's performance in Paris this week—the public clearly believes that this will be her swan song.

10) Although they know it is not easy to reform the Augean stables of the society, they are still trying to do it.

4. Choose the best paragraph translation. And then answer why you choose the first translation or the second one.

两个版本的译文各有优缺点。

首先，在专有名词的翻译方面，即含有较深文化意义的人名、神名、地名的翻译方面，译文一采取了原文优先的策略，即以异化为主，几乎所有的有具体名称的神祇都用拉丁字母化的现代汉语拼音翻译，甚至包括"吴姬天门"这样的地形名称，这样的好处是能保持和原文专有名词指代的一致，给熟悉中国文化的人以方便，整体上没有脱裂感，利于源语国文化的传播；不足是过于死板机械，专有名词通篇都采用这种翻译方式，虽翻译难度大大降低，但对英文世界的非专业读者而言，会觉得枯燥难懂，要全部弄明白这么多专有名词及其关系，会是一件头疼的事。而译文二在专有名词的翻译方面，正好和译文一相反，译者将所有的专有名词都翻译了过来，而且用的是归化的方式翻译过来，即对专有名词里的每个字都加以理解并用英文表达出来。这种翻译方式，会让句子活起来，增加读者的兴趣，激发他们的想象，这一尝试会给人以耳目一新的感觉，利于目的语国读者的接受，但这样做最大的问题是隔断了学习中国文化或试图了解中国文化的人与源语文化在专有名词上一致的普遍的认定。

其次,在对段落中上下句子之间的逻辑关系的处理方面,译文一中把"系昆之山"和"共工之台"放在一个句子里,句式雷同,没有能够突出共工之台与下句的联系。然而译文二却把共工之台之句单独翻译成一句,突出了共工之台的重要性。在"射者不敢北乡"一句,译文二较为简省,没有表达出射者不敢面对的方向在北面这层意思,同时射者应该是不敢把箭向北射,而不是单独的射者不敢面对共工之台,译文一则把这两层意思都表达了出来,不过缺失了不敢北向,其实就是不敢面对坐落在北面的共工之台这层暗含的意思。翻译这句话的关键就是对共工身份的背景知识的掌握,这样才能很好地处理上下句子之间的逻辑关系。

5. Translate the following passage into Chinese.

纵观所有人类历史和古代社会,妇女在促进社会进步、保护资源、教育青年等方面都扮演了非常重要且积极的作用,妇女的存在确保了社会的持续发展,也保障了人类的文化遗产和历史遗产能够得到传承。尽管在古代文献中鲜有文字明确描述妇女所发挥的积极作用,但在神话中,妇女所隐含的象征意义却让女性发挥的作用和在历史上的影响不言自明。例如,作为文明摇篮的环地中海地区就有丰富多样的女性象征表达,这些象征表达就反映了过去社会的社会经济结构以及海洋如何对这些古代社会命运产生影响。值得注意的是,水最初是女性象征主义的一部分。地中海沿海地区的女性元素与水之间的联系可以追溯到史前时代,因为如下水生特征,比如海洋灾难和自然现象(海啸、洪水、暴风雨和降雨、岛屿和沿海地区的淹没、海岸侵蚀和海岸被侵犯/回归、海流、地峡和海峡、潮汐和漩涡)都与人类生活和文明进步(即航海、考古天文学,以及通过海上通信网络打造的社会经济联系,或者战争和地缘政治冲突)高度相关。妇女通过交流和教育帮助提升年轻人的认知,并且在补救以及修复由环境因素或人为因素给人类造成的损害方面发挥着重要作用。在妇女的教导以及引导下,青年在成年后,就能更好地承担起领导角色,从而有助于减轻环境危害给人类社会及其资源造成的威胁和影响。

6. Translate the following passage into English.

Although ancient Chinese mythology has neither relatively complete plot and mythological figures nor systematic genealogy, they do have distinct features of oriental culture, among which the spirit of esteeming virtue is particularly significant. When compared with Western mythology, especially Greek mythology, this spirit of esteeming virtue is even more prominent. In Western mythology, especially Greek mythology, the criteria for judging whether a god is good or not are mostly the god's wisdom and strength, while in ancient Chinese mythology, the criterion lies in morality. This way of thinking is deeply rooted in Chinese culture. For thousands of years, this spirit of esteeming virtue has affected people's comments on historical figures and expectations of real people.

Unit Eleven

Text A

二、译例与练习

Exercises

1. Fill in the blanks with the proper given words, and then translate the sentences into Chinese.

1) namely

电动汽车过去曾遇到例如行驶里程有限和充电站紧缺的问题,因而限制了人们对电动汽车的使用。

2) issued

该部门已经发布了初步的环境影响评估,总体看来是支持该项目的。

3) coherent

创新科学会产生新的命题,使不同的现象以更连贯的方式相互关联。

4) turnaround

政府现在为本国汽车行业提供支持,向其提供贷款并监督其重整计划的推行。

5) stimulate

虽然尚未得到证实,但专门用于刺激小脑的体育锻炼可能会成为主流干预措施。

6) undertake

为了保护历史,希腊政府组建了一个委员会来承担雅典卫城的恢复工作。

2. Translate the following sentences into Chinese.

1) 尽管城市的物质和非物质价值以一种非常复杂和互动的方式联系在一起,但遗产的概念和当前的保护措施不可避免地吸引了公众的注意力。

2) 随着工业化的兴衰,滨水区的功能发生了历史性的改变。现在很多工业都在迁离滨水区,留下了大量废弃的工业用地,其中大部分毗邻脆弱的滨水区生态系统。

3) 保护历史身份和绿地管理是实现城市历史公共空间社会可持续性的城市重建战略的主要途径。

4) 作为提升城市活力和文化特色的积极因素之一,城市滨水区的港口遗产在城市保护和城市重建方面至关重要,为城市提供了与水边重新连接的机会。

5) 建造和改造这些场地意味着技术和自然力量之间的持续较量,需要在施工过程中开发和应用最合适的方法。

6) 历史港口空间是社会生活的主要支柱之一,对经济、社会、建筑和象征性方面都有影响。

7) 尽管港口遗产具有无可争议的文化、历史和技术价值,但在城市中构成我们城市滨水区的港口建筑却被遗忘了,这似乎是自相矛盾的。

8) 近年来,在城市可持续发展理念的引领下,人们重新认识了滨水区,并意识到这些地区有创造无与伦比的再生机会的潜力——开发新的用途、重新激发地方活力,并对曾经被遗弃或边缘化的港口遗产产生归属感。

9) 因此，港口作为促进城市海上贸易发展的门户是不可或缺的，许多沿海城市的起源和繁荣都归功于水线运输和贸易，而这类城市和港口通常在功能和空间上密切相关。

10) 水边与城市交汇的独特位置，自古以来就是城市发展初期许多故事的发源地。

3. Translate the following sentences into English.

1) Given the relocation of port-related activities, the outskirts of port cities often become derelict areas where urban restructuring occurs.

2) Historically, the first waterfront redevelopment projects dated from the 1960s when cities like Boston, Baltimore, and San Francisco set the trend.

3) Convincing residents of the benefits of waterfront redevelopment and its possible link to port activity is often not an easy task.

4) Most cities have understood that the industrial heritage of the port is something that should be reinforced rather than eradicated.

5) The strong and close relationship between the port and the city, which is often a consequence of historical circumstances, has become disrupted due to the negative impact of the port on the urban environment.

6) If well exploited, the historic port cities not only serve to consolidate the identity of the city but also bring economic benefits which help to sustain the city as a whole.

7) The study concludes with a reminder that regeneration is about more than the restoration of fabric, involving as well the capture of the spirit of historic port cities.

8) As defined by UNESCO, cultural heritage includes tangible monuments, sites, and objects and intangible traditions and living expressions inherited and passed on through the generations.

9) There is widespread recognition that the legacy of the past should influence future development but how this is to be achieved and what benefits it brings vary from place to place.

10) The historic culture of the port city, events, landmarks, existing architecture and nature should be utilized to give the waterfront redevelopment character and meaning.

4. Choose the best paragraph translation. And then answer why you choose the first translation or the second one.

译文二比译文一更胜一筹。

译文一：译文几乎是一个逐字逐句翻译的过程，缺乏翻译的技巧，并且存在用词不够准确的情况。因此，译文略显臃肿啰唆，逻辑关系不够清晰。

译文二：翻译过程中忠实于原文的前提下，运用了恰当的翻译方法和技巧，用词简洁凝练，句法结构严谨明了，行文符合英语重形合的特点。

5. Translate the following passage into Chinese.

该区域的道路分为三个层次。第一层是该区域及其边界的主要行车道。例如，位于南侧的"正门"里德道(Reid Road),在建筑立面、街道设施和环境绿化方面都进行了全面升级。它将该地区与南部的高档住宅区连接起来,还创建了反映该地区特色的入口街道。第二层是该地区的公共优先通行权道路,包括北码头街和南码头街。它们位于开发建筑的背面,用于缓解内部车辆交通,并为垃圾拾取等相关的市政服务提供空间,确保运河两侧主要行人通道的安全、舒适和连续性。第三层是一条围绕运河港口池塘的完整而连贯的步行街。该设计将运河作为整个区域公共空间的中心,从东侧的港口池尽头到西侧的小威尼斯,创造了一条环绕运河的连续路径。

6. Translate the following passage into English.

Many cities are undergoing major redevelopment of older derelict waterfront areas, intending to turn them into commercial, cultural, tourist, or upscale residential areas. Cities are using waterfront redevelopment to revitalize urban areas and bolster local economies with new jobs, better housing, improved community amenities, and added tourism opportunities. The redevelopment of the waterfront has also become an opportunity to rekindle relationships between the city and the port. In some cases, waterfront redevelopment is largely an aesthetic undertaking where the extent to which the projects can revitalize urban areas remains unproven. However, even then, they represent a shift from industrial to post-industrial land use.

Unit Twelve

Text A

二、译例与练习

Exercises

1. Fill in the blanks with the proper given words, and then translate the sentences into Chinese.

1) wrecked

他们将检查失事船只周围的小物件,看看能否辨认出人类遗骸。

2) splash

体重对于跳水运动至关重要,因为跳水的目标是尽可能不激起水花。

3) intently

他聚精会神地听着,但周围一片寂静,深沉而肃穆——甚至可怕,令人沮丧。

4) enormous

这个国家在政治上取得了巨大进步,但在经济上却没有。

5) dumped

我们把包扔在附近的格兰德旅馆,急匆匆地向市场赶去。

6) carving

布卢姆匆匆走过一个令人回味的砂岩雕塑，上面雕刻着一个孩子和一只鹈鹕，然后欣赏了一块狮鹫牌匾。

2. Translate the following sentences into Chinese.

1）还要再苦读几个月？伊丽莎白甚至连想都不敢想。坐在书桌前，她可以想象出百慕大的鲜花和棕榈树、珊瑚洞底的层层涟漪，或者是加利福尼亚海滩上翻滚着的慵懒细浪。

2）奥什递给我一个篮子，他手拿一把干草叉，开始沿着退潮后的海边收海藻。篮子装满了，我和奥什一人握着一个把手把它抬到高处的岩石上，我们把海藻摊在那儿，等雨水把它们冲刷干净。海藻洗净晾干后，我们会把它们锄到菜园地里做肥料。

3）奥什在炉子上架起铁锅，在里面放了点黄油和小葱，过一会儿把鲈鱼片放进去。煎鱼片香喷儿喷儿的，发出了滋滋声，鼠儿立刻就跑过来了。

4）奥什给了它一块生的鱼腹肉。它打了个滚儿，仰面躺在地上，用两个小爪子抱着鱼肉吃起来。

5）我很会驾船，可我知道在航道上遇见风暴和逆流，连条小船都很难驾驭。

6）但如果我需要内岛外的东西，奥什也会毫不犹豫地叫人捎给我：学习用的铅笔啦，一双合脚的新冬靴啦，时不时买本书啦，麦琪小姐调不成的药啦。

7）"要变天了，"第二天我刚醒来，奥什就说，"你从城里回家的路上可能会赶上雨。"我走到门口，眯眼看着蓝汪汪的天空，沐浴着金灿灿的阳光，吹着缕缕微风。

8）从彭尼基斯岛回来的船上，斯隆先生蜷缩在船尾，颤抖着，一看到船头有浪花要溅过来，他就低头躲避。六月的午后送来暖意，而我却因这愉悦的心情感到惭愧，因为斯隆先生可正在遭罪呢。

9）飓风呼啸而过的时候，每个人多多少少都会遭殃，就像去年的那场飓风——掀翻了屋顶，搁浅了船只。

10）船只在格瑞亚德海域失事的时候，有时会有人丧命，要么是水手，要么是救生员，要么都有，伊丽莎白群岛上所有的居民都能感受到灾难的恐怖。

3. Translate the following sentences into English.

1) The money he earned from cutting ice or trapping lobsters or painting his pictures went for things we couldn't grow ourselves.

2) I was hot and sticky, my feet hurt, I wanted just one tree and a small patch of shade—or the ocean, better yet—and I was hungry for my pocket lunch.

3) But he had not seen the crown that had come up on an anchor fluke, or the giant silver buckle, or my little gold ring in the cinnamon box.

4) But it breaks my heart to think of them tying you into that old skiff and pushing you out onto the tide.

5) He ripped an old conch shell off a hank of dead man's fingers and tossed it aside.

6) He stuck his pitchfork in the sand, ran his hands through his hair and then looked at me

steadily, his mouth tight.

7) I could sail a skiff, wrangle a lobster, save a birdman from starving to death and surely I could ride a ferry to the city and back again.

8) Then I took its mysteries to bed with me and found them waiting in the morning, right where I'd left them.

9) Outside the house, he stopped only to grab a bait net and a pail before heading for the tidal pools at the base of the rocks where minnows and sand eels often found themselves stranded.

10) He turned the net inside out, dumping three sand eels into the pail and then sluicing in some water so they could breathe.

4. Choose the best paragraph translation. And then answer why you choose the first translation or the second one.

译文一比较好。

译文一：语言丰富具体，结构准确灵活，在语言风格和语境关系上忠实于原文。

译文二：用词稍显平淡无味，结构基本对照原文，没能突显儿童文学作品的特点。

5. Translate the following passage into Chinese.

天刚蒙蒙亮的时候，我就起床洗漱，穿戴好出门了。我在椴树林那儿刚好遇见了回家的奥什。他一看见我就停下来了。"想我了，是吗？"他几乎笑着说。我点点头。是的，我一整晚都很想他，但我还想了别的事情。"我想我知道我真正的父母是谁了"，我说。奥什脸上近乎微笑的表情不见了，就好像他从来没笑过。他经过我身边，走过椴树林，走下峭壁，跨过沙桥。沙桥这次几乎干涸了，潮水跌得很低。"那你最好去麦琪小姐家看看，"我疾步快走跟上他时，他冷不丁蹦出这一句。"警察马上就到。你可以和斯隆先生一起把情况说清楚。"

6. Translate the following passage into English.

"I haven't worked here long enough to help you," she said, "but there are a few people in the wards who've been here since then." She gave me directions to the nursery. "You'll find Mrs. Pelham there," she said. "She helps with the babies now, but she used to work in the orphanage. Start with her. If anyone can answer your questions, she can." I made my way through a maze of corridors, out through a door into a yard with, finally, trees, across the yard and the street beyond and up the walkway to the steps of another building, not as big as St. Luke's proper, through the door, down a corridor, and up a flight of stairs to a ward where I could hear babies crying.

Unit Thirteen

Text A

二、译例与练习

Exercises

1. Fill in the blanks with the proper given words, and then translate the sentences into Chinese.

1) detect

当我们吃东西的时候,鼻子后部的特殊感受器会检测到我们食物中的空气分子。

2) aquatic

海牛是生活在佛罗里达河流和沿海水域的水生哺乳动物,它们游到海面附近,经常在与船只相撞时被撞死。

3) microscopic

在从冰下的固体物质收集的样本中,他们发现微型海洋植物的化石,这表明这个区域曾经是开阔的海洋而不是坚硬的冰。

4) gulp

布朗抓起他的呼吸面罩,猛吸了一口气。

5) detrimental

与其他方法相比,生物防治的优点是它提供了一种成本相对较低、有害副作用最小的永久性防治系统。

6) shaft

一队工作人员被派下去关闭竖井,并把所有积水舀出来。

2. Translate the following sentences into Chinese.

1) 根据发布的消息,该机器人也被称为遥控机器人,具有高清视频功能,并能够为潜水员处理爆炸物时提供实时反馈。

2) 海军和海军陆战队在各种作战任务中使用无人平台系统,包括海军陆战队的远征高级基地作战和在有争议环境中的濒海作战。

3) 就像50年前,海洋探险家雅克·库斯托使用水下设备和水下摄像机征服世界一样,利用新的传感器、潜艇技术和自动驾驶船舶,人类有机会帮助公众了解海洋与欣赏海洋。

4) 海上运输对世界经济至关重要,因为世界上90%以上的贸易是通过海运进行的,它是在世界各地运输货物和原材料的最具成本效益的方式。

5) 无人海上航行器通常分为水下航行器和水面航行器,由于水下航行器的应用范围包括海底管道/基础设施测量和检查、搜索和救援等,目前水下航行器占据了最大的市场份额。

6) 在海洋、遥感、武器投放、力量倍增器、环境监测、测绘以及为水下航行器提供导航和通信支持等应用领域,人们对水面航行器也产生了广泛的兴趣。

7) 大多数无人水面舰艇依靠远程操作员的指导来发送任务指令,并通过直接观察或通过无线视频链接不断监测舰艇的状态。

8) 目前,有许多公司生产不同类型的安全阀,不仅用于军事机构,而且可用于工业公司、环境机构及政府机构。

9) 科学家和研究人员目前还使用其他数据收集方法,如卫星系统以及固定浮标和浮动浮标网络等。然而,这些系统也有局限性,包括浮标网络的范围以及卫星系统的精度和成本。

10) 环境监测的一些常见形式有:海水盐度、水温、空气和海洋中的二氧化碳含量、气压、浪高、风速、测深(水深)、摄影观察、水声、溢油测量和污染测量。

3. Translate the following sentences into English.

1) Some of the best applications for task performance include customizable sample collection (e.g. Oil sample collection), defense related tasks, hazard clean up, beacon transmission, transportation, boat traffic markers and fish tracking.

2) The durability, autonomy, and low operational cost (relative to research ships or oceanographic moorings) of these platforms create tremendous potential for supplementing existing open-ocean and near-shore marine carbon dioxide observing efforts world-wide.

3) The team has been conducting scientific investigations to help NASA better understand the potential that other ocean worlds could have to support life, while also studying the oceanographic exploration process as a way to understand how to conduct remote science missions and to streamline future exploration.

4) The long-term goal of this research is to develop future generations of low cost, high-efficiency ASV for oceanographic, industrial and environmental surveys.

5) George emphasized that there are numerous Pacific systems of navigation, different understandings about ocean currents, sea mammal migrations, different boat building traditions, and varied capacity to continue voyaging.

6) The America's Marine Highway Program supports the increased use of the nation's navigable waterways to relieve landside congestion, provide new and efficient transportation options, and increase the productivity of the surface transportation system.

7) "Investments in the America's Marine Highway Program help us move more goods more quickly and more efficiently to the American people, supporting our supply chains," said U.S. Secretary of Transportation Pete Buttigieg.

8) "Put simply, President Biden is leading the largest-ever federal investment in modernizing our country's ports, and our domestic coastwise services, and improving both our supply chains and the lives of Americans who depend on them," said Acting Maritime Administrator Lucinda Lessley.

9) The Kaskaskia River is the second-longest river in Illinois, originating in central Illinois around Champaign, Illinois and terminating at its confluence with the Mississippi River, a distance of more than 300 miles.

10) The Kaskaskia River has been predominately used to ship bulk commodities of coal,

scrubber stone, slag, grain, and scrap metal since it was established as a navigable waterway; however, 40,000-50,000 tons of unitized coiled steel are also moved on this waterway with a new tenant expected to ship up to 1.2 million tons of coiled steel for processing and other uses once it constructs its processing plant.

4. Choose the best paragraph translation. And then answer why you choose the first translation or the second one.

译文二较好。

译文一：这个段落的开头与结尾的句子在翻译上出现了翻译腔，语法上也有错误。其他句子还是比较通顺的。地名词汇的翻译也比较地道准确。

译文二：整个段落的译文语法、专有词汇都准确，专有词汇的翻译客观、准确。最重要的是译文准确，即"信"又"达"又"雅"。

5. Translate the following passage into Chinese.

当机器人深入残骸时，无线信号可能会被阻断。因此，还应仔细检查评估工作条件和可能出现的危险。一旦信号到达机载接收器，就会被传送到机械系统。主要系统是推进系统，采用微型和专门的船用螺旋桨。它们由装在防水外壳中的小型随动系统电机提供动力，2号螺旋桨每轴3到5个叶片。它们可以在顺时针和逆时针方向操作，以创建转向。一些用于残骸分析和深海勘探的先进水下机器人也采用了联合螺旋桨。它们很少被使用，而且比固定的同类产品更昂贵，只在非常高精度的情况下才需要。传感器和附件构成了水下机器人功能的核心，最常见的设备包括摄像头、深度计、温度和内部系统传感器。操作员使用摄像机，因为他们必须用眼睛观察周围的环境。

6. Translate the following passage into English.

Frontiers in Marine Science publishes rigorously peer-reviewed research. With the human population predicted to reach 9 billion people by 2050, it is clear that traditional land resources will not suffice to meet the demand for food or energy, required to support high-quality livelihoods. As a result, the oceans are emerging as a source of untapped assets, with new innovative industries, such as aquaculture, marine biotechnology, marine energy and deep-sea mining growing rapidly under a new era characterized by rapid growth of a blue, ocean-based economy. The sustainability of the blue economy is closely dependent on our knowledge about how to mitigate the impacts of the multiple pressures on the ocean ecosystem associated with the increased scale and diversification of industry operations in the ocean and global human pressures on the environment. Therefore, Frontiers in Marine Science particularly welcomes the communication of research outcomes addressing ocean-based solutions for the emerging challenges, including improved forecasting and observational capacities, understanding biodiversity and ecosystem problems, locally and globally, effective management strategies to maintain ocean health, and an improved capacity to sustainably derive resources from the oceans.

Unit Fourteen

Text A

二、译例与练习

Exercises

1. Fill in the blanks with the proper given words, and then translate the sentences into Chinese.

1) redefine

21世纪医学面临的任务将是重新定义其限制,即使它扩大了其功能。

2) autonomous

西蒙·布莱克摩尔在英国哈珀亚当斯大学研究农业技术,他认为轻型自主机器人有可能解决这个问题。

3) in accordance with

罗马帝国灭亡后,在欧洲,第一个按照复杂的礼节规范私人生活行为的社会,是12世纪法国的普罗旺斯。

4) infrastructure

政府建造了从机场直达松岛国际商务区的桥梁,并在建造新机场的同时完善了地面基础设施。

5) sanctuary

10月份在澳大利亚召开的一次会议上将先对威德尔海洋保护区的提议做出决定,然后才轮到针对半岛保护区的提议。

6) collaboration

普通的搜索技术不能令人满意,所以麦基与麻省理工学院电子工程教授哈罗德·E.埃杰顿合作。

2. Translate the following sentences into Chinese.

1) 仿真结果表明,多水下机器人编队自适应控制算法有较好的稳定性,对于多水下机器人编队研究及多种海洋探测应用具有重要的意义。

2) 自升式钻井平台属于海上移动式平台,由于其定位能力强且作业稳定性好,在大陆架的勘探开发中居主力军地位。

3) 通过其他活动,美国国家海洋和大气局的海洋探索有助于保护海洋健康,了解气候变化对海洋和我们星球的影响,可持续管理我国海洋资源,加速我国国民经济,以及在我们日常生活中更好地认识海洋的价值和重要性。

4) 为了克服载人监测船的缺陷,有必要开发一种智能化的海洋环境资源探测机器人。

5) 从我国油气资源来看,海上多波地震是解决海底模糊带、构造变形和气污染等问题的有效技术。

6) 近年来，水下导航、海洋探测、水下目标定位以及水下通信等领域迅猛发展，这些领域常常使用水声信号作为信息传输的载体。
7) 中国南海是西太平洋最大的边缘海之一，具备良好的天然气水合物成藏条件和勘探前景。
8) 水下滑翔器作为一种新型水下机器人系统，对于海洋环境监测与资源探测具有重要应用价值。
9) 遥控机器人是一类非常实用的机器人，广泛应用于空间探测、海洋探测、危险环境作业等领域。
10) 它研究的是基于非同步反应模式的海洋环境探测机器人系统的设计与开发。

3. Translate the following sentences into English.

1) We can detect subsurface oceans on far-off icy moons, maybe even detect tiny ripples in space due to relativistic gravity.
2) From Pythagoras to Herbert Spencer, everyone had done it, although commonly science had explored an ocean which it preferred to regard as Unity or a Universe, and called Order.
3) Because of the difference in pressure your lungs feel in space or underwater, liquid ventilation could be useful while exploring the far reaches of the universe or the depths of the ocean.
4) According to the Global Foundation for Ocean Exploration group, about "three-quarters of all volcanic activity on Earth actually occurs underwater".
5) Undersea instruments monitor whale songs, plankton density, temperature, acidity, oxygenation, and various chemical concentrations. These sensors may be attached to profiling buoys, which drift freely at a depth of about 1,000 meters.
6) Together let us explore the stars, conquer the deserts, eradicate disease, tap the ocean depths and encourage the arts and commerce.
7) It will pursue innovation-driven development and implement a number of strategic projects in the fields of artificial intelligence, quantum information, integrated circuits, life and health, brain science, agriculture, aerospace science and technology, and deep Earth and ocean exploration.
8) We oppose any other country's oil and gas exploration activity in the waters under China's jurisdiction and hope relevant foreign company do not involve itself in the South China Sea dispute.
9) BP's chief operating officer of exploration and production, Doug Suttles, said the company was using remote operative vehicles (ROVs) to try to find out how much oil was leaking into the sea.
10) Like space exploration, deep sea exploration requires new instruments and technology. While space is a cold vacuum, the ocean depths are cold, but highly pressurized.

4. Choose the best paragraph translation. And then answer why you choose the first translation or the second one.

译文一优于译文二。

译文一：在进行科技文章翻译时，要弄清科技文章的文体特征，以及英汉两种语言在科技文章表达上的不同，不拘泥于原文单词或句子的语法特征，对词性、句子成分、句子类型等进行转换，使得译文通顺流畅，符合各自语言习惯，避免出现文理不通、逐句死译和生硬。此版译文在忠实原文体裁的基础上，表述得通顺自然，语言简洁，比较合理。

译文二：科技英语的又一大特点就是简洁。因此，译者在翻译科技文章的过程中，应尽量避免冗长、累赘的句子，通过使用灵活多变的句法结构，如独立主格结构、名词化结构等方式体现该文体简洁清晰的风格。而译文二运用了大量汉语表达习惯的句式，不太符合原文科技文体的语言特色。

5. Translate the following passage into Chinese.

假设在这个特定的竞选领域有一个异常丰富的信息基线，竞选者就有机会将存档数据与新收集的高分辨率数据进行比较。确定我们到底知道多少，不知道多少，以及改变了什么，可以帮助我们更好地设计未来的活动。为了完成这种比较分析，运动赞助商需要设立一项"专用基金"用于运动前的工作，以收集各种分辨率和波长的最感兴趣的领域的海洋学/水深测量档案数据，并将这些数据提供给探索者，以便进行业务规划和运动后的比较分析。一个小组建议建立一个运动"联合信息中心"来收集档案和新数据，并与岸上的人交流。两家机构都强调需要在竞选数据管理和可视化方面进行投资。此次活动将不再是第一次探索，而是更多地尝试新工具、将新的技术人员带到邮轮上，并以不同的方式看待探索，还要将过去的观察结果与更高分辨率的新发现进行比较。简而言之，人们可以将其设计为探索的科学控制。

6. Translate the following passage into English.

Oceans account for 70% of the Earth's total area and more than 90% of its living space. In fact, in the vast ocean world, the sea affects the climate and weather phenomena on land, and a large number of marine plants and animals are important food sources for humans. The ocean sails between the continents and countries of the world, and about half of the world's population lives in coastal areas. Therefore, it is extremely important for us to understand the world's oceans. Changes in the Earth's oceans will directly affect our lives on land, so it is necessary for us to detect such changes early on. In addition to increasing our understanding of the ocean for security and economic benefits, exploring the ocean floor will also satisfy human curiosity and desire to explore the unknown.

Unit Fifteen

Text A

二、译例与练习

Exercises

1. Fill in the blanks with the proper given words, and then translate the sentences into Chinese.

1) exemplar

该层可以包含任意系列的外部获得的范例模型。

2) vernacular

绝大多数人在生活的各个方面都使用他们自己的地方方言。

3) thematic

弥尔顿试图解决的主题问题被写进了诗歌的语法和句法中。

4) collate

这些网站整理了所有公开资料,并以方便用户的形式将其呈现出来。

5) holistic

至少在部分领域,学术调查可能需要减少排他性,变得更全面一些。

6) contiguity

科学家们打算探讨干眼症与煤尘接触之间的关系。

2. Translate the following sentences into Chinese.

1) 自然条件优越的海岸与着陆点或港口之间的等效性在对着陆点的考古研究中被反复利用。

2) 海洋博物馆的主要目标是让某些藏品可见或完全隐于视线之外。

3) 各类文化研究人员在沿海地区进行了广泛的探索,他们纷纷运用自己的方法来解读景观的意义。

4) 尽管世界范围内的海洋文化研究多种多样,但陆地和海洋区域的研究仍然互无关联。

5) 挪威的经验似乎在某种程度上可以运用到北欧和北大西洋的其他地区。

6) 这样一来,选择在格雷斯德布罗发现的一艘特殊船只,或多或少是一个自然的选择。

7) 本文通过拓展文化景观更为广泛的内涵,探索了一种分析澳大利亚殖民地环境中海洋文化景观的全新方法。

8) 如果这些特征可被定义,那么在海洋文化产业中看到这些特征的可能性有多大?

9) 该项目是面向全球的,旨在向不同国家和教育背景的研究生介绍海洋文化产业的各个方面。

10) 海洋考古学一直被视为考古学讨论的外围领域。

3. Translate the following sentences into English.

1) The layout and acceleration of the development of the marine cultural industry has far-

reaching significance for the overall promotion of our country's strategy of strengthening the country by ocean and the strategy of strengthening the country by culture.

2) Marine cultural resources are attached to the ideas, customs, habits, norms and other ways of thinking and behavior formed by people in coastal areas in the production and life practices related to the ocean.

3) The integration of creativity in the whole process of industrial development is very important, which is also a key factor for the sustainable development of the marine cultural industry.

4) The development of cultural industries is less dependent on natural resources, and cultural resources are mainly used and developed.

5) From a social perspective, the development of the marine cultural industry helps to meet people's needs for a better life and enhance people's sense of happiness and satisfaction.

6) China has always attached great importance to the cultivation of national marine awareness and the utilization of marine resources, and has expanded the space for economic growth relying on marine resources.

7) With the promotion and support of the government, Zhoushan Islands has gradually formed a marine cultural industry cluster with marine culture and folk customs, marine handicrafts, marine art, marine cultural conventions and exhibitions, and coastal leisure tourism as the core.

8) The development of the marine cultural industry requires the government to provide guidance and protection through strategy formulation, policy introduction, legal guarantees, and capital investment to provide a healthy environment.

9) The beautiful landscapes in coastal areas can become the clues and places where stories in cultural products take place, adding to the aesthetic value and artistic charm of cultural products.

10) The marine cultural industry is a new thing. Judging from the current development status, there are mainly five shortcomings.

4. Choose the best paragraph translation. And then answer why you choose the first translation or the second one.

译文一较好。

译文一：相较于译文二，译文一的主要长处在于它在具体翻译过程中，恰当而准确地反映了英语侧重形合的语言特点，这一点尤其在翻译信息比较丰富的源语文本时需要注意。总体上来看，这段英语译文综合使用了英语中常有的实现形合的语言手段，包括从句、现在和过去分词、各类短语（主要是介词、名词和形容词短语）和独立主格结构等。这样组织语言信息之后，整个英语文段就显得主次分明，错落有致，做到了既准确，又地道。

译文二：尽管在用词上和译文一没有显著差别，但是在处理句子的主次关系上，有几处处理得不甚恰当。具体来说，翻译"形成层次多样、内容丰富的区域性海洋文化产业集群"

时，译文二处理成"a rich regional marine cultural industry clusters' variety of levels and contents"。这样的话，整个句子在表述"区域性海洋文化产业集群"与"层次多样、内容丰富"之间的所属关系显得十分累赘。这样如译文一一样用一个动词过去分词词组后置修饰就显得很协调。再就是翻译"从海洋文化产业发展的成功案例来看"时，译文二处理成"With the judgment of the successful cases of the development of the marine cultural industry"。这样翻译尽管语法上没什么问题，但是翻译成介词短语的话，原文中"判断"的动作性不够。再者考虑到下文也出现了类似的表达"with the help of"，为了避免紧挨着两个句子的重复性表述，可以译成"Judging from the successful cases of the development of the marine cultural industry"。

5. Translate the following passage into Chinese.

海洋文化景观。"海洋文化"一词不仅源于对人类使用海洋的更广泛理解，还源于随之产生的各种结构、文化标识以及人们与航海之间的联系的更广泛理解。例如，由于克莱德河和其他地方的造船传统、世界范围的海上探索和皇家海军的声望，英国会声称自己是一个航海国家。由于广泛的海上航行和掠夺以及对格陵兰岛等地方的殖民化，北欧文化同样被认为是海洋文化。然而，如今驾车行驶在道路和桥梁上的人们却以一种截然不同的方式与大海接触。他们通常将大海更多地视为休闲规划中的一部分，可以在度假时在上面玩耍、深潜探索和航行，而不会认为仅用于运输货物时才有意义。然而，海岸线仍然被称为地标，现在各国越来越多地将水下资源和油田，而不仅仅是渔业，视为国家财富的重要组成部分。

6. Translate the following passage into English.

Our country has a coastline of 18,000 kilometers and 6,536 islands with an area of more than 500 square meters. The coastal areas are economically developed and contain marine cultural resources with rich connotations and unique values. In terms of nature, marine culture is not only a historical phenomenon, but also a social phenomenon. It is a form of civilization created based on marine material culture and spiritual culture, and it is the condensed inheritance of many elements related to marine culture. Marine cultural resources are a collection of cultural elements such as language, customs, clothing, literature, art, diet, production methods, production technologies, and lifestyles in coastal areas. Human life and social progress depend on marine cultural resources. The development of marine cultural industries must be relying on marine cultural resources.

参考文献

[1] 陈福康.中国译学理论史稿[M].上海:上海外语教育出版社,2000.
[2] 陈宏薇.高级汉英翻译[M].北京:外语教学与研究出版社,2009.
[3] 陈晓舟.功能对等视角下的英语长句翻译[D].长沙:湖南师范大学,2012.
[4] 冯庆华.实用翻译教程:英汉互译[M].3版.上海:上海外语教育出版社,2010.
[5] 郭富强.英汉翻译理论与实践[M].北京:机械工业出版社,2004.
[6] 何刚强.笔译理论与技巧[M].北京:外语教育与研究出版社,2009.
[7] 黄友义.坚持"外宣三贴近"原则,处理好外宣翻译中的难点问题[J].中国翻译,2004,25(6):27-28.
[8] 黄忠廉.翻译方法论[M].上海:华东师范大学出版社,2019.
[9] 姜倩,何刚强.翻译概论[M].2版.上海:上海外语教育出版社,2016.
[10] 金圣华.傅雷与他的世界[M].北京:生活·读书·新知三联书店,1997.
[11] 黎昌抱.王佐良翻译风格研究[M].北京:光明日报出版社,2009.
[12] 李明秋.点、线、面三位一体的水产英语特点与翻译策略[J].大连海事大学学报(社会科学版),2013,12(1):111-114.
[13] 李欣.我国海洋型民俗文化的独特魅力[J].人民论坛,2017(13):240-241.
[14] 林语堂.林语堂名著全集:第十九卷-语言学论丛[M].长春:东北师范大学出版社,1994.
[15] 刘敬国,何刚强.翻译通论[M].北京:外语教学与研究出版社,2011.
[16] 刘宓庆.新编汉英对比与翻译[M].北京:中国对外翻译出版公司,2006.
[17] 罗新璋,陈应年.翻译论集[M].2版.北京:商务印书馆,2009.
[18] 钱锺书.谈艺录[M].北京:中华书局,1984.
[19] 沈复.浮生六记:汉英对照绘图本[M].林语堂,译.北京:外语教学与研究出版社,1999.
[20] 唐瑾.中国近代翻译文学概论[M].武汉:湖北教育出版社,2005.
[21] 田传茂,杨先明.汉英翻译策略[M].上海:华东师范大学出版社,2007.
[22] 汪榕培.汪榕培学术研究文集[M].潘智丹,整理.上海:上海外语教育出版社,2017.
[23] 王秉钦.20世纪中国翻译思想史[M].天津:南开大学出版社,2004.
[24] 曾诚.实用汉英翻译教程[M].北京:外语教学与研究出版社,2002.
[25] 张春柏,王大伟.英语笔译实务:三级[M].北京:外文出版社,2017.
[26] 张培基.英汉翻译教程[M].上海:上海外语教育出版社,2018.
[27] 庄绎传.翻译漫谈[M].北京:商务印书馆,2015.
[28] HATCH H E. Material culture and maritime identity:identifying maritime subcultures through artifacts[M]//The Archaeology of Maritime Landscapes. New York:Springer,2011:217-232.

[29] HATIM B, MASON I. Discourse and the Translator [M]. London: Longman Group United Kingdom, 1993.

[30] POMBO P. Tales of ocean, migration and memory in ancestral homespaces in Goa [M]//The Palgrave Handbook of Blue Heritage. Cham: Palgrave Macmillan, 2022: 75-91.

[31] TYTLER A F. Essay on the Principles of Translation (3rd rev. ed., 1813) [M]. Amsterdam: John Benjamins Publishing Company, 1978.